AGGREGATION CREATE LOOSE ESTIMATES

AM COST ESTIMATOR

AM COST ESTIMATOR

Phillip F. Ostwald

Fourth Edition

Penton Education Division, 1100 Superior Avenue, Cleveland, Ohio 44114

LIBRARY OF CONGRESS

Ostwald, Phillip F., date
   AM Cost Estimator / Phillip F. Ostwald -- 4th Edition
      v.:ill. ; 28 cm.
Current Printing(last digit):
10 9 8 7 6 5 4 3 2

Library of Congress   [8401]

Copyright© 1981, 1983, 1985, 1988, owned by Penton Publishing, Inc.
All rights reserved. Printed in the United States of America. Except as
permitted under the United States Copyright Act of 1976, no part of this
publication may be reproduced or distributed in any form or by any means,
or stored in a data base or retrieval system without the prior written
permission of the publisher.

The publisher and author make no statement regarding the accuracy of these
data. Derived from many sources, tests, and manufacturer's
recommendations, the data may be inappropriate under various
circumstances. Users of the information are thus cautioned that the
information contained herein should be used only as an estimate where the
user judges it to be appropriate.

ISBN 0-932905-06-4

Previously 0-07-606966-4 (McGraw-Hill)

Contents

	Preface	**xi**
	A Letter from the Publisher	**xii**
SECTION I.	<u>Instructions</u>	**1**
SECTION II.	Productive Hour Costs	**21**
SECTION III.	Material Costs	**45**
SECTION IV.	Operation Estimating Data	**53**

1. Sawing and cutting **55**
    - 1.1 Power bandsaw cutoff, countour band sawing, and hacksaw machines **55**
    - 1.2 Abrasive saw cutoff machines **60**
    - 1.3 Oxygen cutting processes **63**
    - 1.4 Computer aided torch cutting processes **66**

2. Molding machines **69**
    - 2.1 Plastics molding bench **69**
    - 2.2 Plastic preform molding machines **71**
    - 2.3 Thermoplastic injection molding machines **72**
    - 2.4 Thermosetting plastic molding machines **76**
    - 2.5 Thermoforming machines **79**
    - 2.6 Glass-cloth layup table **81**
    - 2.7 Extrusion molding machines **82**
    - 2.8 Blow molding machines **85**
    - 2.9 Hot- and cold-chamber die casting machines **86**
    - 2.10 Isostatic molding press **91**

3. **Presswork** **93**
   3.1 Power shear machines **93**
   3.2 Punch press machines (first operations) **98**
   3.3 Punch press machines (secondary operations) **101**
   3.4 Turret punch press machines **105**
   3.5 Single-station punching machines **107**
   3.6 Power press brake machines **110**
   3.7 Jump shear, kick press, and foot brake machines **112**
   3.8 Hand-operated brake, bender, punch press, multiform, coil winder, shear, straightener, and roller machines **115**
   3.9 Nibbling machines **118**
   3.10 Tube Bending machines **120**
   3.11 Ironworker machines **123**

4. **Marking** **127**
   4.1 Marking Bench & Machines **127**
   4.2 Screen printing bench and machines **129**
   4.3 Laser marking machines **131**
   4.4 Pad printing machines **134**

5. **Hotworking** **137**
   5.1 Forging machines **137**
   5.2 Explosive forging machines **144**

6. **Turning** **147**
   6.1 Engine lathes **147**
   6.2 Turret lathes **151**
   6.3 Vertical turret lathes **158**
   6.4 Numerically controlled turning lathes **161**
   6.5 Numerically controlled chucking lathes **168**
   6.6 Single-spindle automatic screw machines **173**
   6.7 Multispindle automatic screw machines **182**

7. **Milling machines** **187**
   7.1 Milling machines setup **187**
   7.2 Knee and column millling machines **189**
   7.3 Bed milling machines **195**
   7.4 Vertical-spindle ram-type milling machines **198**
   7.5 Router milling machines **202**
   7.6 Special milling machines **205**
   7.7 Hand milling machines **209**

8. **Machining centers** **213**
   8.1 Machining centers **213**
   8.2 Rapid travel and automatic tool changer elements **217**

9. **Drilling** **221**
   9.1 Drilling machines setup and layout **221**
   9.2 Sensitive drill press machines **223**
   9.3 Upright drilling machines **227**
   9.4 Turret drilling machines **231**
   9.5 Cluster drilling machines **237**
   9.6 Radial drilling machines **241**

10. Boring machines **247**
   10.1 Horizontal milling, drilling, and boring machines **247**
   10.2 Boring and facing machines **253**

11. Machining and tool replacement **255**
   11.1 Turn, bore, form, cutoff, thread, start drill, and break edges **255**
   11.2 Drill, ream, counterbore, countersink, tap, and microdrill **261**
   11.3 Face, side, slot, form, straddle, end, saw and engrave milling **268**
   11.4 Tool life and replacement **275**

12. Broaching machines **283**
   12.1 Broaching machines **283**

13. Grinding machines **287**
   13.1 Cylindrical grinding machines **287**
   13.2 Centerless grinding machines **293**
   13.3 Honing machines **296**
   13.4 Surface grinding machines **298**
   13.5 Internal grinding machines **306**
   13.6 Free abrasive grinding machines **309**
   13.7 Disk grinding machines **312**
   13.8 Vertical internal grinding machines **316**
   13.9 Numerically controlled internal grinding machines **319**

14. Gear Cutting **323**
   14.1 Gear shaper machines **323**
   14.2 Hobbing machines **327**

15. Thread cutting and form rolling **333**
   15.1 Thread cutting and form rolling machines **333**

16. Welding and joining **337**
   16.1 Shielded metal-arc, flux-cored arc, and submerged arc welding processes **337**
   16.2 Gas metal-arc and gas tungsten-arc welding processes **346**
   16.3 Resistance spot welding machines **350**
   16.4 Torch, dip, or furnace brazing processes **354**
   16.5 Ultrasonic plastic welding machines **357**

17. Furnaces **361**
   17.1 Heat treat furnaces **361**

18. Deburring **365**
   18.1 Drill press machine deburring **365**
   18.2 Abrasive belt machine deburring **368**
   18.3 Pedestal-machine deburring and finishing **372**
   18.4 Handheld portable-tool deburring **375**
   18.5 Hand deburring **379**
   18.6 Plastic material deburring **381**
   18.7 Loose-abrasive deburring processes **383**
   18.8 Abrasive-media flow deburring machines **386**

19. Chemical machining, printed circuit board, production, and non-traditional machining **389**
    19.1 Chemical machining and printed circuit board fabrication **389**
    19.2 Automatic lamination machines (for printed circuit boards) **397**
    19.3 Ultraviolet printing machines (for printed circuit boards) **398**
    19.4 Conveyorized developing machines (for printed circuit boards) **399**
    19.5 In-line plating processes (for printed circuit boards) **401**
    19.6 Electrical discharge machining **405**
    19.7 Traveling wire electrical discharge machining **408**

20. Finishing **413**
    20.1 Mask and unmask bench **413**
    20.2 Booth, conveyor and dip painting, and conversion processes **417**
    20.3 Metal chemical cleaning processes **425**
    20.4 Blast-cleaning machines **428**
    20.5 Metal electroplating and oxide coating processes **432**

21. Assembly **439**
    21.1 Bench assembly **439**
    21.2 Riveting and assembly machines **442**
    21.3 Robot machines **445**

22. Inspection **449**
    22.1 Machines, process, and bench inspection **449**
    22.2 Inspection table and machines **452**

23. Electronic fabrication **461**
    23.1 Component sequencing machines **461**
    23.2 Component insertion machines **463**
    23.3 Axial-lead component insertion machines **466**
    23.4 Light-directed component-selection insertion machines **469**
    23.5 Printed circuit board stuffing **473**
    23.6 Wave soldering machines **476**
    23.7 Vertical solder coating machines **481**
    23.8 Wire harness bench assembly **484**
    23.9 Flat cable connector assembly **490**
    23.10 D-subminiature connector assembly **492**
    23.11 Single wire termination machines **495**
    23.12 Component lead forming machines **498**
    23.13 Printed circuit board drilling machines **499**
    23.14 Shearing, punching, and forming machines **502**
    23.15 Resistance micro-spot welding **505**
    23.16 Turret welding machines **507**
    23.17 Wafer dicing machines **509**
    23.18 Coil winding machines **512**

24. Packaging **517**
    24.1 Packaging bench and machines **517**
    24.2 Corrugated-cardboard packaging conveyor-assembly **522**
    24.3 Case packing conveyor **526**
    24.4 Foam-injected box packaging bench **527**

25. Tooling **531**
    25.1 Dies, jigs and fixtures **531**

Appendix **549**
Conversion **549**
Index **557**

Preface

This fourth edition of the *American Machinist Cost Estimator* has some dramatic changes. When the book was first published several years ago, little did we realize the effect of the personal computer upon the nature of cost estimating. Recently, *American Machinist*, with my cooperation, developed software sold separately entitled *AM Cost Estimator Software*. The software development encouraged rethinking the spreadsheet that we have been using. Those who have used this databook regularly will notice changes in column headings in the examples that are provided in the labor estimating data section. Additional headings including adjustment factor, cycle minutes, and setup hours have been added. This new spreadsheet style is easier to use and simplifies understanding. The software and manual methods are identical in the use of these tables headings.

The operations cost section has been dropped. Only a few estimators consistently used the equations that were provided. In addition, the equations ignored the experience and understanding of the estimator. The operations cost section went the way of the blacksmith.

The data used in this book and in the software are identical. Information regarding the software may be obtained by writing to Penton Education Division, 1100 Superior Avenue, Cleveland, Ohio 44114. You may also call 216-696-7000 directly for information. The software is suitable for an IBM-compatible micro computer. This software package is independent of this databook or it can be used with it. The price of this book does not include the floppy disks and the accompanying specially developed documentation.

This databook should be used with pad, pencil, and calculator. The estimator will find that the new spreadsheet approaches add clarity. Some estimators are using the book in conjunction with Lotus™ or other software spreadsheets instead of a company form. The *AM Cost Estimator Software* also is compatible with Lotus™. Whether the method is manual or electronic, the intent is the same for the data base. The objectives are accuracy, consistency, ease of understanding, and a national norm for comparison.

Commercial and nondefense industries were the original target market for the *AM Cost Estimator*. But in recent years the Department of Defense has encouraged the adoption of engineered estimates and has promulgated Military Standard 1567A. This data base offers the user the potential for compliance with that standard. Many military contractors and DOD agencies, auditors, and depots are now using the *AM Cost Estimator*, which we find gratifying. Their applications are giving new importance to the data base.

Most estimators use Section IV, "Labor Estimating Data," which is the largest part of the databook. This section has been increased by 13 new technologies. Many of these technologies are recent while others are divisions of previous technologies. This highlights two of my objectives: to expand into the leading-edge technologies and to provide better coverage and attention to existing and well-known machining, processing, or bench work. A new chapter dealing with tooling has been added that is suitable for job shops and product manufacturers. This follows from our desire to provide information as we develop it. Manufacturing is diversified, especially when one considers the variety of ways to produce a product.

For this fourth edition we have added new information for time estimating in Section IV. This time section has over 15,000 time values specified for 126 machine, process, and bench identifications. I plan to continue it as the most comprehensive and authoritative work of its kind. Despite these numbers of items, I recognize the unending requirement for more coverage. Many have asked me to include technologies that are considered "rare" by others. Such inclusion remains a long-term objective. Despite the shortcomings of incomplete coverage, an experienced estimator understands that estimating data are always incomplete. An accurate, complete, and broad data base remains the constant challenge.

The productive hour costs have been expanded. These costs consider distinctions for area and increases in wage rate. Wage rate differences in an economic catch-basin can wander as much as $4 for each of the values listed. The productive hour cost is targeted for the in-between winter period of the 1987 and 1988 years. The estimator is able to use this information for estimates; however, it is also possible to "backcast" or "forecast" with this information. Simple methods are suggested that allow the estimator to choose this information or make modifications. The methods are presented in the Instructions.

The *AM Cost Estimator* has been used for cost estimating of direct labor cost and time, material cost, break-even, make-vs-buy, and return-on-investment types of analysis for equipment purchases, operations, parts, and manufacturing technologies. It aids in the pricing decision, and reports from estimators indicate that it was indispensable for this work.

Both large and small companies, government, and colleges are using the data. Some firms that have extensive information already developed are finding that the *Estimator* helps to "complete" their data-base requirements. Sometimes it is used as the only opinion. Small- and medium-sized companies, which do not have sufficient or trained staff available to obtain these data, are relying upon the *Estimator* for their estimates. Irrespective of company size, after a period of self-study the estimator will find that it is easy to use for estimating manufacturing time and costs.

The research and technical effort to expand the *Estimator* is considerable. The many contributions by practitioners, machine tool builders, associations, and others interested in the art and science of manufacturing estimating are appreciated. As with the former editions, I continue to indicate my respect to those professionals who do cost estimating.

Phillip F. Ostwald
Boulder, Colorado

A Letter from the Publisher

American Machinist is pleased to introduce the author of our fourth *AM Cost Estimator*, Dr. Phillip F. Ostwald. He is Professor of Mechanical and Industrial Engineering in the Department of Mechanical Engineering at the University of Colorado, Boulder. A respected authority, he is the originator of the popular "Manufacturing Cost Estimating Seminar." Dr. Ostwald is the author of many papers, technical reports, and several books including *Cost Estimating*, Second Edition, published by Prentice-Hall, which provided the first unified textual treatment of the topic. He edited *Manufacturing Cost Estimating* for the Society of Manufacturing Engineers and is the co-author of *Manufacturing Processes, Eighth Edition*, published by John Wiley & Sons.

Dr. Ostwald is a member of many national and professional panels and committees. The Work Measurement and Methods Division of the Institute of Industrial Engineers citing the *American Machinist Cost Estimating Guide*, selected him for the 1983 Phil E. Carroll Award. The Society of Manufacturing Engineers selected him for the 1982 Sargent Americanism Award, which honored him for his writing and for citing the importance of profit and free enterprise -- two democratic ideals. He is a consultant and President of Costcom Inc.

Joseph DiFranco, Publisher
American Machinist
A Penton Publication

Instructions

INSTRUCTIONS

Using the *AM Cost Estimator*

Introduction

The *AM Cost Estimator* provides estimating information for direct labor and direct material. Estimators, engineers, accountants, financial analysts, foremen, superintendents, office managers, controllers, and other professionals who do this work will find that estimating with the *AM Cost Estimator* is a simple process. The *American Machinist Cost Estimator,* fourth edition, continues on the course of providing estimating information for the many professionals who work in this important occupation. While these instructions are directed to an "estimator," it is our observation that many titles and types of backgrounds do this work. For the sake of simplicity, we define an estimator as one who needs to appraise the time and/or cost of manufactured products. The reasons for this business appraisal are also equally diverse.

The *AM Cost Estimator* has been used for cost estimating of direct labor cost and time, material cost, pricing, break-even, make-vs-buy, and return-on-investment types of analysis for equipment purchases, operations, parts, and manufacturing technologies. Both large and small companies, government, and colleges are using the data. Some firms which have extensive information already developed are finding that the *Cost Estimator* helps to "complete" their database requirements. Sometimes the *AM Cost Estimator* is used as the only information. Small- and medium-sized companies, which do not have sufficient or trained staff available to obtain these data, are relying upon it for their estimates. Irrespective of company size, after a period of self-study, the estimator will find that the *AM Cost Estimator* is easy to use for estimating manufacturing time and costs.

The book is composed of four sections:

I. Instructions
II. Productive Hour Costs
III. Material Costs
IV. Labor Estimating Data

Instructions (Section I)

This section, Section I, is a "how-to-do-it" discussion which explains the remainder of the *AM Cost Estimator.* By using the information in this section, you can determine the current or future cost or time for operations, parts, products, or fabrication services. Cost and time are vital to thousands of firms in the United States. Companies depend upon estimating information for pricing, vendor-cost studies, procurement cost estimating, operations sheet preparation, planning, value engineering, evaluation studies, and performance checking. With intelligent application, you can use this information for wide-ranging cost determination.

Productive Hour Costs (PHC) are presented in Section II. This information is time-based to winter 1987–1988. Adjustments to these Productive Hour Costs are possible. Multiplying by indexes will change these values to other time periods or locations. Section II consists of 126 machines, processes, and bench work identifications for 23 cities or cities and regions. There are a total of 2898 entries.

Material costs are listed in Section III and also are time-based to winter 1987–1988. Some 55 different materials, prices, or estimates are provided. Each material has a quantity and price value. These weights and prices range from the cost for a single bar, for example, to a mill quantity.

Labor estimating information is provided in Section IV. These data have the dimension of time. Setup work is listed in hours while minutes for cycle work are adopted. This new fourth edition has increased coverage to 126 machines, processes, and bench work. Over 15,000 individual time values are available for selection.

The data in Section IV can be used with pad and pencil or with the *AM Cost Estimator Software*. This manual is used with both methods. The data are the same and the instructions provided here are compatible with the software approach; however, the entries and the calculations are with pad or company form, pencil, and calculator. The computer is a faster method, and the instructions for the computer method are provided in the documentation that accompanies the software disks. This book is sold separately of the software, unless the software disks (there are two disks) are purchased, in which case the *AM Cost Estimator* book is included with the disks and the documentation. Information about the software is available from *Penton Education Division* by calling 216-696-7000, or writing to 1100 Superior Avenue, Cleveland, Ohio 44114.

The *AM Cost Estimator* leads to estimates for direct labor and direct material. We define direct labor as "touch labor." This means that the machine operator is working on parts that are sold to customers. An example of direct labor is an NC machine operator. The Material Cost Section (Section III) also provides information useful for direct materials. An example is the part that is being sold to a customer and direct material is the raw material required for the part.

The price that a company bids for work includes direct labor, direct material, overhead, and profit. This *AM Cost Estimator* does not provide information for overhead, as that is special to each company. There are various kinds of overhead calculations, and the *AM Cost Estimator* will work with any style of overhead determination.

Figure 1 shows the modern production systems. This database is compatible with each of these systems. In effect, direct labor and direct material are important considerations for these production systems. If the operation is robotic controlled, the cycle time becomes important. For operations where direct labor is the most significant component of cost, these data are vital. The database is compatible with the types of industry such as job lot, moderate, and mass production. Too often the estimator believes that his or her industry or plant is "different," but we suggest that estimat-

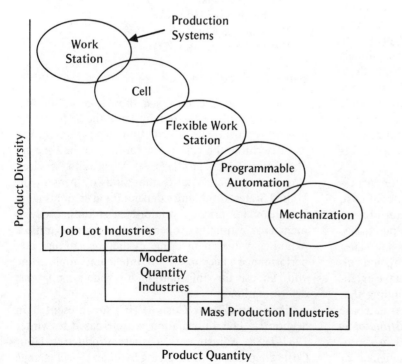

FIGURE 1 Interaction of product quantity, diversity, type of industry, and production system.

ing information is universal and can be used in a variety of plant, production system, quantity, and product circumstances.

Figure 2 gives the composition of a total product cost estimate. Direct labor and direct material are important because they form the foundation of much of the bid and price analysis. Overhead is frequently related to direct labor. Because other cost elements are internal to a company, this book is only concerned with information for labor and material.

Productive Hour Costs (PHC) Estimating Instructions (Section II)

Productive Hour Costs (PHC) are similar to direct wages. The information listed in Section II is identified by a machine, process, or bench description. Each row of PHC has a number, such as "3.4 Turret punch press." A total of 23 locations are shown across the top of the pages. For example, the PHC for the turret punch press, 3.4, on page 23, is $10.60 for Atlanta and $11.54 for Baltimore.

The information is arranged by machine, process, or bench number. These numbers correspond to data in Section IV. The turret punch press PHC is the labor necessary to operate a similarly titled machine given in Section IV. Thus, it is simple to multiply total time of the operation by the PHC and find the cost for a productive operation. Suppose that the lot time for a turret punch operation is 7.62 hours and the job is in Baltimore. The productive cost is 7.62 × $11.54 = $87.93. Other areas could be used as well.

A city or city and state location is listed across the top of the PHC tables. These locations correspond to the Bureau of Labor Statistics' Standard Metropolitan Statistical Areas. The area includes contiguous counties

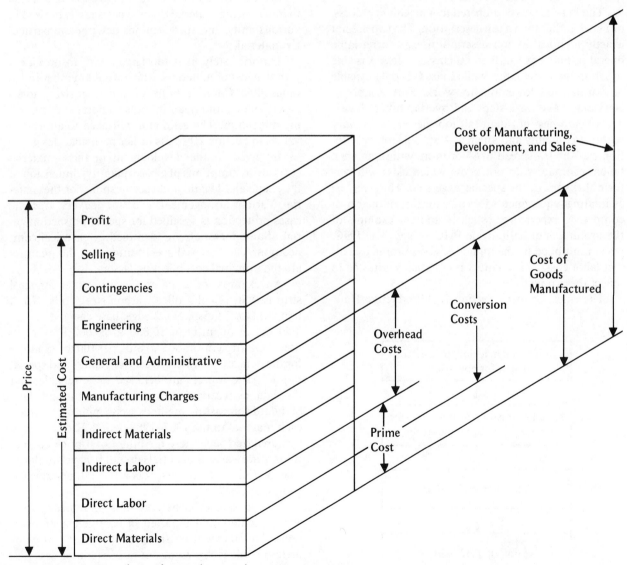

FIGURE 2 Composition of cost element for a product cost estimate.

if there is an economic catch-basin relationship, and it may cross state boundaries. Each city's data may also be valid for other cities within a limited radius.

The source of data is the Bureau of Labor Statistics (BLS), U.S. Department of Labor. The BLS data are derived from industrial surveys of companies with more than 100 employees. The BLS data were already out of date when they were published. They are inappropriate for estimating future costs. Therefore, these data are analyzed and projected to the 1987–1988 period. Time-adjusted values are given in Section II.

The PHC data do not include shift differential, overtime, paid holidays and vacations, health insurance, and retirement plans. These costs are part of the overhead of direct labor. Bonuses and end-of-year profit sharing are also not considered. However, cost-of-living adjustments, as they are reported in earnings, are included. Time and incentive methods of payments are averaged in the analysis and are not separated for the Productive Hour Cost (PHC).

The PHC is cross numbered to a machine, process, or bench rather than a job description. The two are not exactly identical. A job description, e.g., turret lathe operator, may have A, B and C classes. Class A is able to set up the turret lathe, while Class C is only capable of routine and lower quality work. Furthermore, a class may have wage steps and overlap other classes. Thus, earnings of identical job descriptions vary within a company and an area. A specific job description can show a spread of $4 or more within an area. Often, estimators do not know which class will perform the work or the specific wage step. They may be estimating requirements for cities outside of their own company's experience, which is another example of the usefulness of nationwide PHC values. The PHC, when multiplied by the lot hours, gives productive direct-labor cost. This quantity is eventually adjusted to reflect the productivity.

The rules for using Section II, "Productive Hour Costs," are given in Figure 3. From the operation process sheet, the machine or process, or bench is selected to do the routing of the part. This machine has a number, and that number, 3.11 Ironworker, for example, corresponds to a PHC. The estimator may use his or her own labor wages, but in the absence of that information the PHC is selected for the nearest like city.

Consider this example: A resistance spot welding operation is estimated to require 2.84 minutes for the cycle with a setup of .35 hour. A lot of 8 units will result in a total of .73 hour. Our estimator knows that the work will be done in Pittsburgh, and the PHC number for resistance spot welding is 16.3. The PHC as copied from the tables on page 35 is $11.56. Then, $8.44 (= 0.73 × $11.56) is direct-labor cost for this operation and location.

Material Estimating Instructions (Section III)

Section III lists winter 1987–1988 costs for selected manufacturing materials. These costs are expressed in various units, and specifications describe the particular material.

Unfortunately, it is too large a task to provide a complete list of materials, which would number in the thousands. Complications of differing sizes, tolerances, grades, and quantity cause a number of reporting difficulties. The estimator will need to adjust the values in Section III to his or her particular design.

Involved in the development of these material costs are assumptions of a base quantity, dimension of the shape and length, and specifications for the material. Various arrangements exist for delivery. In some cases protection is specified for special-finished materials. Some materials are identified by chemical composition. The material specification and the quantity are the principal cost-affecting factors.

An example of a material specification for steel strip is: "Strip, cold-rolled, carbon steel; coils, No. 4 temper, No. 2 finish, No. 3 edge, base chemistry, 6 × .050 in., in quantity of 10,000 to 20,000 lb, mill to user, FOB (freight-on-board) mill, per 100 lb." Sources for the data are commercial service centers and special inquiry. The sources are believed to be reliable, and the data, as received, were current or of recent date. Standard forecasting methods were applied to update each material to the 1987–1988 period. Each material was analyzed separately for future cost/price competition since various materials do not behave similarly. The *AM Cost Estimator* reports the time-adjusted values.

The unadjusted costs were those of representative service centers and reported in commercial transactions. To the extent possible, the *AM Cost Estimator* material values represent out-of-pocket cost, less discounts, allowances, and other deals, because the originating information was requested in this form. The

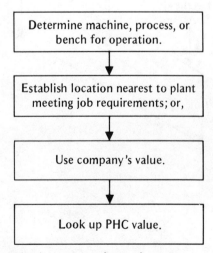

FIGURE 3 Rules for using productive hour costs.

author recommends the out-of-pocket cost policy for estimating future cost. Other cost estimating policies may be inconsistent with this practice.

To avoid the effect of transportation costs, costs are generally freight-on-board (FOB) service center or warehouse or central marketing point. However, delivery prices are presented when the customary practice of the industry is to quote on this basis.

Estimating the materials is done for direct materials, i.e., materials appearing in the product. Customarily, the weight, volume, length, or surface area is determined from drawings. This theoretically computed quantity is then increased by losses such as waste, scrap, and shrinkage. A general formula for material estimating is

$$C_{dm} = W(1 + L_1 + L_2 + L_3)C_m$$

where C_{dm} = cost of direct material for a part
W = theoretical finished weight in compatible dimensions.
L_1 = percentage loss due to scrap which is a consequence of errors in manufacturing or engineering.
L_2 = percentage loss due to waste which is caused by manufacturing process such as chips, cutoff, overburden, skeleton, dross, etc.
L_3 = percentage loss due to shrinkage, theft, or physical deterioration.

These three losses (L_1, L_2, L_3) are determined by measurement or historical information. Once these losses are computed and the decimal is added to 1, a material estimate is possible using the formula.

The value for C_m would be taken from the tables given in Section III, "Material Costs."

Material estimating practices are summarized in Figure 4. The materials as reported in the *AM Cost Estimator* are classified as ferrous and non ferrous, the principal classifications used for manufacturing. They are organized as follows:

Ferrous Products
    Structural shapes
    Plate Bar
    Sheet, strip, and bands
    Pipe, tubing and wire

Nonferrous products

These materials are arranged in order of increasing cost within the above categories.

The small requirement's cost is a feature of the *AM Cost Estimator*. The estimate may only require one bar, sheet, tube, etc., instead of a large mill run. In many cases, a range of weight starting with a minimum order extending to a base quantity is given. In

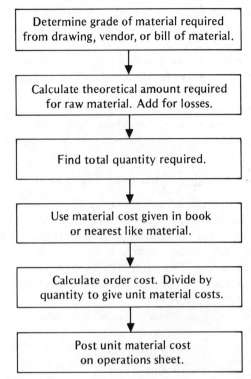

FIGURE 4 Rules for using material costs.

the case of ASTM A-36 plate, for example, the reduction between one plate and a base quantity of 10,000 lb is 4.8%, while for hot-rolled bar, 1½-in. OD, AISI 1020, the reduction is 225% from one bar to base quantity. Usually the reduction is significant.

The amount of material is important to the estimate, since the estimator must know whether one lot run, one order, one year's requirements, or a model life is selected as the quantity upon which to base an estimate. It is the author's recommendation to use the anticipated out-of-pocket material quantity purchase for the material as the more significant factor in choosing a quantity.

In estimating material unit cost, cutting, or shearing charges, delivery of material in a single or consolidated shipment and terms of the purchase agreement influence the cost. These are the values expressed in $/100 lb, or $/1 lb and reported in the *Estimator*.

Practices vary among suppliers of these materials, which are too numerous or minor to mention. However, shipping costs are an important part of the cost and, in some cases, a service center may allow the combining of all orders, while another may limit the consolidation of different materials for purposes of weight rate determination of shipment costs. In almost all cases, this book indicates the freight cost stipulations. FOB mill, FOB service center, or FOB shipment point is exclusive of costs to ship, while FOB destination implies that freight costs have been included in the estimated values. In this latter situation, obviously the distance from the supplier to the destination is a factor in the values presented. It is the practice of the nonferrous industry to provide information in terms of FOB destination.

The base quantity is that weight at which price is no longer sensitive to additional amounts of purchase order. Usually, a single base quantity is given. Any exceptions noted are due to base amounts which may vary among suppliers or where base quantities are indeterminate.

Labor Estimating Instructions (Section IV)

The information for estimating direct-labor time is given in Section IV, "Labor Estimating Data." These data have been constructed with ease of estimating in mind, while being sensitive to minor differences in manufacturing. This sensitivity is made obvious by various selections of the element for a specific machine, process, or bench.

Figure 5 lists the rules for using the labor estimating data.

Sources of information

The methods of measurement used to find the information for estimating include time study, predetermined motion time data systems, laboratory investigation, manufacturer's recommendations, and judgment. One or more methods may have been adopted for a set of element estimating data. Consensus analysis is used for consolidation when an abundance of data were available.

Dimensions of the information

The information provided in Section IV, Labor Estimating Data, uses the units of time: hours for setup, and minutes for the elements of the operation. This type of presentation is unaffected by inflation or PHC changes. Hours (hr) are appropriate for setup because this is the customary expression for setup operations in the United States. Moreover, decimal minutes (min), rather than seconds (sec) or hours, are adopted for the elements of an operation because they are more easily understood. Estimators relate to the "minute" more easily than to other time dimensions.

Universal information

The approach used in this Estimator assumes that plant differences for trained operators of a specific machine, process, or bench are minor for the purposes of estimating. However, differences, say for handling, are significant for different classes of machines. For instance, the time it takes to move and load a part on the punch press is significantly different for an open back inclinable punch press than it is for a turret

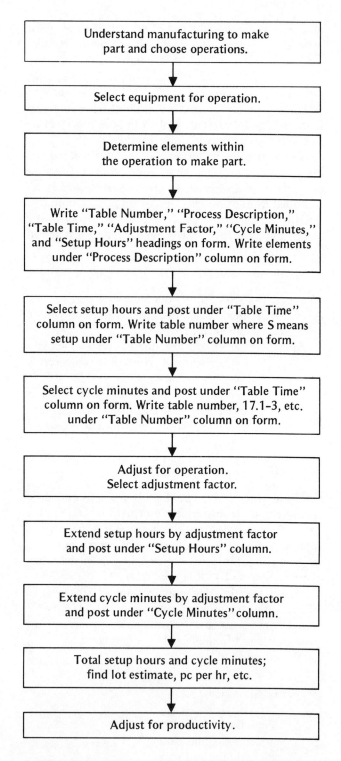

FIGURE 5 Rules for using element estimating data.

punch press. However, the operator time needed to load a turret punch press is similar for different plants or regions. Also, the elemental estimating time for handling a part for a bed milling machine is about the same for plant A in city Z as it is for plant B in city Y.

Operations sheet

The operations sheet is fundamental to manufacturing planning. It is also called a "route sheet," "traveler," or "planner." There are many styles, and each plant has its own form. The purpose of the operations sheet, however, is the same:

- To select the machine, process, or bench which is necessary for converting the material into other forms
- To provide a description of the operations and tools
- To indicate the time for the operation.

The order of the operations is special too, and this sequence indicates the various steps in the manufacturing conversion.

Each operations sheet has a title block indicating the material kind, part number, date, quantity, estimator, and other information that may be essential to the company. Following this part of the form, the instructions to the plant are provided. For cost estimating purposes, consider the following:

Suppose that you want to machine an aluminum casting which is on consignment to your company. The company is called SOHO, and the part number is unknown. This casting is a consignment material, and it is a part of a larger product. It will have a bored hole, enlarged and deburred, and it will be packaged in a carton. Of course, parts and assemblies are more complicated than this, this is only to demonstrate the process of estimating. A typical, simple operations work sheet is shown in Figure 6.

Completing the operations sheet

The operations sheet ① begins in the upper left-hand corner of Figure 6. The final product name ② is oftentimes given along with the assembled product ③. But the operatons sheet is for a specific part name ④, which in this case is called SOHO. A part number ⑤ can be identified and listed if available. The part number and name are removed from the design and repeated on the operations sheet title block. The estimator should indicate the plant location ⑩, lot quantity ⑦, and material specification ⑧. Knowing the final amount of material required by the design, the estimator adds material to cover losses for scrap, waste and shrinkage. Then, the estimator will multiply by the cost per lb rate of the material, as given by the material costs shown in Section III of the *Cost Estimator*. A unit material cost is required for entry ⑨.

The sequence of the operation number (op. no.), and selection of the Machine, Process, or Bench are made to manufacture the part. These are required at circle ⑪ and are shown specifically at ⑱ and ⑲. A complete operations sheet will show this column. Notice that the column titled "Process Op. No." is not repeated in the many examples provided later to conserve space. Even though numbers are vital in operations planning, they are of less importance in detailing estimating once that operation has been selected.

The column titled "table number" ⑫ corresponds to the table number indicated by the *AM Cost Estimator*. For example, Table 7.4 refers to the Ram Milling Machine class. The operations sheet column titled "description" ⑬ lists the instructions that the shop will follow in making the part or subassembly. For Op. No. 10, the instructions to the shop are listed. The description gives a listing of the elements that are pertinent to estimating the operation. Many examples estimate operations throughout the *Estimator*.

The "table time" column ⑭ is a listing of the values removed from the tables given throughout the *Estimator*. These time values are posted in this column. The adjustment factor ⑮ column operates upon the time column. Once adjusted, the time is either entered into the cycle minutes ⑯ or the setup hours ⑰ column. (The adjustment factor column is discussed more later.)

The columns titled "cycle minutes" and "setup hours" are very important. The instructions that follow describe the methods and selection of the elements and time that are necessary to manufacture the part for that operation. All of the tables in Section IV pertain to the determination of the values that are listed in these two columns.

The sequence number ⑱ of the operation is given in the left-hand column along with the equipment ⑲ necessary for the operation. All table number selections are made for an operation and are shown in circle ⑳.

The total of the cycle minutes ㉓ and the setup hours ㉓ are summed. The lot estimate is calculated and presented ㉒, and the dimensions are in hours. The lot estimate for Operation 10 is 3.32 hours.

The total of "lot estimate" is a computation that is shown on the operations sheet. It is explained in the discussion dealing with the appendix. The calculation, however, is made using the setup, unit estimate, and the lot quantity.

The information for estimating direct-labor time is given in Section IV, "Labor Estimating Data." These data have been constructed with ease of estimating in mind, yet are sensitive to minor differences in manufacturing. This sensitivity is made obvious by various selections of the elements for a specific machine, process, or bench.

This operations sheet can be altered to consider simple assemblies or complicated products, but the approach remains the same. The purpose of the *Cost Estimator* is to provide time or cost for the direct labor or material component of the product. Preparation of the operations sheet is important to find part operational costs. Notice that the part cost is the sum of the

① OPERATIONS SHEET

② Product Name: __X-152__
③ Product Part No.: __2224534-03__
④ Part Name: __SOHO__
⑤ Part No.: __(Unknown)__
⑥ Part Material Cost: __$0.000 (Consignment)__
⑦ Lot Quantity: __87__
⑧ Material: __Aluminum casting__
⑨ Unit Material Cost: __0.00__
⑩ Plant Location: __Chicago__

⑪ Process Op. No.	⑫ Table Number ⑳	⑬ Process Description	⑭ Table Time	⑮ Adjustment Factor	⑯ Cycle Minutes	⑰ Setup Hours
⑱ 10 Vertical Milling Machine ⑲	7.4	Ram Milling				
	7.1-S1	Table setup, part length >12 in.	1.45			1.45
	7.1-S2	Make piece	0.02			0.04
	7.1-S3	Tolerance addition	0.13			0.13
	7.4-1	Pick up, move	0.20		0.20	
	7.4-2	Clamp, unclamp, star wheel clamping				
	7.4-1	Pry part out	0.09	2	0.18	
	7.4-4	Start and stop mach.	0.06		0.06	
	8.2-1	Traverse element	0.06		0.06	
	7.4-5	Clean, lubricate	0.08	2	0.16	
	11.3-4	End milling, 2-in length of cut. Depth of cut = 1/32 Weight = 6.28 lb	0.05		0.05	
	7.4-5	Clean area	0.01		0.02	
	11.4-2A	Mill tool life	0.15	2	0.15	
	22.1-3	Inspection ㉑	0.00		0.00	
		Total	0.29		0.29	
		Lot Estimate ㉒	3.32 hr		㉓ 1.17	1.62 ㉓
20 Deburr Bench	18.5	Hand Deburring				
	18.5-S	Setup				0.05
	18.5-1	Handling, repos Box l+w+H=22.3	0.05			
	18.5-2	Tool handling	0.15		0.15	
	18.5-3B	Handle tools twice	0.04	2	0.08	
	18.5-1	Holes over 1/2 in.	0.08		0.08	
	18.5-4A	Handling, repos	0.05		0.05	
		Break edges	0.11		0.11	
		Total			0.47	
		Lot Estimate	0.73 hr			0.05
30 Package Bench	24.1	Pack				
	24.1-S	Setup				0.15
	24.1-1	Order paperwork	0.15			
	24.1-2	Get and position	2.00			
	24.1-10	Paper carton	0.09	4	0.50	
	24.1-10	Paper carton	1.51	4	0.38	
	24.1-16	Miscellaneous	0.22	4	0.05	
			0.26		0.06	
		Total			1.09	
		Lot Estimate	1.73 hr			0.15

FIGURE 6 A typical operations (or route) sheet for estimating purposes.

operational costs, which allows us to concentrate on the important steps that are necessary for estimating operations. Henceforth, we will not show the entire operations sheet. Instead, we will provide only that information necessary for estimating.

This example is extremely simple, but it points out that operations planning precedes cost estimating. This book does not provide the logic that goes into the planning process. After all, cost estimating is difficult enough. Once the operational sequence, the selection of the machine, process, or bench, and a basic description of the work has been roughed out, cost estimating begins.

Breaking Down an Operation into Elements

Detail estimating requires that the operation be "detailed" or reduced to its elements. For example, the vertical milling operation for one hole could be visualized into details, or as we choose to call them "elements."

The selection of these elements vary, and their choice depends on the particular machine, process, or bench being used. Element estimating, as advocated here, balances the number of elements that are eligible for selection. Too many choices increase the estimating cost. Although there is no proof that excessive amounts of detail, beyond those provided in this *Estimator,* reduce estimating errors, it is certain that extensive detail increases estimating costs.

Before elemental estimating can begin, one must have:

- Engineering drawings
- Quantity
- Material specifications
- Listing and specifications of the company machines, processes, and benches
- Operations sheet selecting the operations.

The preparation of the operations may occur at the same time as direct-labor estimating. Once the opertations sheet lists the operations, the estimator can refer to the machine, process, or bench in this *Estimator* which coincides with the one designated by the operations sheet to do the work. The operation is broken down or "detailed" into elements which are described by the estimating table. These elements may be listed on a standardized company form. Marginal jottings on the operations sheet may suffice or even scratch-pad calculations may be followed. Additionally, *American Machinist* produces specific software that "details" the elements. The purpose of the formal or informal elemental breakdown is identical: a listing of elements that will do the work is visualized, and this listing is coordinated with the estimating tables in Section IV.

Setup and Operation Elements

The various machine, process, and bench tables provided in Section IV list setup and cycle elemental times.

Setup includes work to prepare the machine, process, or bench for product parts or the cycle. Starting with the machine, process, or bench in a neutral condition, setup includes punch in/out, paperwork, obtaining tools, positioning unprocessed materials nearby, adjusting, and inspecting. It also includes return tooling; cleanup; and teardown of the machine, process or bench to a neutral condition ready for the next job. Unless otherwise specified, the setup does not include the time to make parts or perform the repetitive cycle. If scrap is anticipated as a consequence of setup, the estimator may optionally increase the time allotment for unproductive material.

Setup estimating is necessary for job shops and companies whose parts or products have small to moderate quantity production. As production quantity increases, the effect of the setup value lessens its prorated unit importance, although its absolute value remains unchanged. Setup values may not be estimated for some very large quantity estimating. In these instances, setup is handled through overhead practices. Our recommendation is to estimate setup and to allocate it to the operation because it is a more accurate practice than costing by overhead methods. This recommendation applies equally to companies manufacturing their own parts or products and to vendors bidding for contract work.

Some operations may not require setup. Flexible manufacturing systems, continuous production, or combined operations may not need setup time. Nonetheless, even a modest quantity may be appropriated in these circumstances. Discussion of the details regarding setup is given for each machine, process or bench.

Cycle time or run time is the work needed to complete one unit after the setup work is concluded. It does not include any element involved in setup. Each example in Section IV provides a unit estimate. Besides finding a value for the operational setup, the estimator finds a unit estimate for the work from the listed elements, which is called estimating minutes. The term "estimating minutes" implies a national norm for trained workers. It includes allowances that take into account personal requirements such as fatigue, where work effort may be excessive due to job conditions and environment, and legitimate delays for operation related interruptions. Since the allowances are included in the time for the described elements, and therefore part of the allowed time for several elements and hence several or many operations, then the allowed time is fair. The concept of fairness implies that a worker can generally perform the work throughout the day.

Organization of the Information

Section IV includes 126 sets of estimating information which are presented in a similar way. Each set provides a description, estimating data discussion, examples and a table.

The description discusses the machine, process, or bench to be estimated. Sometimes a photograph has been included as a courtesy by the manufacturer for instructional purposes. The description is intentionally brief since the estimator is usually familiar with the equipment for the work being estimated.

The estimating data discussion describes most of the elements listed in the table in language that is simple and easily understood. Though misinterpretation is possible, awareness of manufacturing language is a requisite to estimating. The estimator must know the machines, processes, benches, and their associated elemental language.

In some cases, discussion is expanded in another estimating set. For example, the estimating data discussion provided for knee and column milling machines, page 189, is also appropriate for other milling machines. Consistency of terms and practices in describing the elements is followed throughout the *AM Cost Estimator*.

Each set gives examples of how the data are selected and how the estimate is made. A typical problem describing material, special tooling, quantity, and circumstances affecting the work may ask to find and estimate the unit estimate, lot time, prorated unit time, hr/1 unit, hr/100 units, pc/hr, etc. Following the problem statement there are six columns labeled "Table No.," "Process Description," "Table Time," "Adjustment Factor," "Cycle Minutes," and "Setup Hours."

The table is the listing of the estimating data. Various setup possibilities are given. Estimating minutes are provided for a variety of work elements, and the estimator using the elemental breakdown selects elemental time from the many possibilities. For example, see Table 9.3 on page 229. This table number is what is posted on the estimating summary.

Let us return to the examples. Remember that each example provides columns headed by "Table No.," "Process Description," "Table Time," "Adjustment Factor," "Cycle Minutes," and "Setup Hours." The "Table No." column identifies the estimating table and element number. For example, if a sheet metal operation of braking were necessary, the table number "3.6" would be first posted. Similarly, for a drill press operation, the number "9.3" would be written on the row corresponding to the machine selection. Notice for any estimating table that clusters of elements have a number too, starting with "1," "2," "3," etc. These clusters are generally related. The element "handle" may have many possibilities and may be listed as element no. 1. Following the machine no. we list the element no. and it is preceded by a dash. For example, 3.6-1 is a power press brake element called "brake." Also, 9.3-2 is a cluster of elements for "clamp and unclamp" for the upright drilling machine.

The "Process Description" column lists the elements of the operation, which correspond to the element descriptions listed in the estimating tables. This is an elaboration of the description given in the operations planning sheet described above. There is no basic difference; however, the number of lines or elements are greater for estimating than for planning. The "Process Description" may also indicate additional information such as length of cut, tooling used, type of NC manuscript order, etc.

The "Table Time" column is a listing of the number selected from the table corresponding to the value listed for the element. The time units for setup is hours (hr), while the time units for cycle elements is minutes (min). These values are listed under the heading "time." There is no distinction made for the time dimension in this column.

Understanding the "Adjustment Factor"

The column headed by "Cycle Minutes" is a product of the adjustment factor and the time column for the cycle elements. For instance, an end milling operation may have a machining rate of .01 min/in. The adjustment factor will provide the length of cut, say 2 in. The product of .02 (= .01 × 2) is entered in the cycle minutes column.

The column headed by "Setup Hours" is entered after the setup hours and any adjustment factor corrections have been made. Notice in Figure 6 where the setup element "make piece" of .02 is adjusted for a setup hours column entry of .04.

You have read the description of the algebraic steps handled by the adjustment factor. It is an important part of the method which increases your accuracy of application. This step is an essential consideration for every estimate.

Whenever element times are to be adjusted, the factor column is used. When the data tables are used, often dimensions are given with the data. In machining, for example, the data might be expressed as "min/in.". In milling, the length of the safety stock, approach, milled length for the part, and overtravel, might be 17.5 in., inclusive; that 17.5 must be entered as the adjustment factor and multiplication after the value is posted to the cycle minutes column.

In drilling, the dimension given in the data table is

for min/in. It is necessary to multiply this value by the length of the hole plus the drill-point approach. This length value is entered in the adjustment factor column. The multiplied entry as shown in the cycle column is the time for one hole. Multiplying by the number of holes in the adjustment factor column is necessary.

The adjustment factor can be used to adjust for different speeds and feeds as well. The *AM Cost Estimator* gives the speeds and feeds that are related to the time data for many materials. Explanation of the data is often given in the discussion of estimating data found in Section IV. In rough turning and facing of low-carbon steel (free machining), the peripheral velocity is 500 surface feet per min (sfpm) for carbide. These data are shown in Table 11.1-2 for that material. Now, suppose that your speed is 450 sfpm. The adjustment factor is 0.9 or 450/500. A similar adjustment might be made for feeds. The data used for many of the calculations are discussed and given in the estimating data discussion part of this book. You will discover, however, that adjustments for most speeds and feeds are unnecessary.

Other conditions, too, are satisfied by the adjustment factor. For example, consider a drilling operation where several flat parts are stacked. Many of the time elements are described for the entire stack, as if the stack is one part. Use the adjustment factor to divide by the number of flat parts in the stack. Manufacturing operation elements are readily divided by the adjustment factor capability. There are numerous opportunities where one major workpiece is handled, and several parts may result. Handling of one sheet in shearing, or handling strip stock in punch press work is typical. One sheet may produce two or more strips. Sheet handling time is divided by the number of parts on the sheet for the time per part.

The reverse situation is also an opportunity for the adjustment factor. If two or more parts will be welded to result in one final unit, for example, the adjustment factor can compensate for this combination.

Perhaps you will be working with materials that require special care or effort. In those cases, increase the element time by an adjustment factor appropriate to your experience.

The best advice for using the adjustment factor is to notice the dimension scale of the table elements. This might be min/in., each occurrence, etc. Then remember that the cycle estimate is to result in "one" unit for final output. If your product is processed as a "multiple" up to the last operation where the pieces are parted, use the adjustment factor to divide for that style of processing. Similarly, if multiple parts are assembled into one final unit, the adjustment factor should be used to generate the correct value.

Remember, too, the adjustment factor default is a hidden "one," which leaves the time "unadjusted." So no entry in the adjustment factor column allows you to post values without change from the time column to either the cycle minutes or setup hours columns.

The examples conclude by always summing the elemental time and giving the "unit estimate," expressed in minutes, for the cycle time of one part for one operation and the total of the setup hours. "Cycle" and "unit" mean the same thing. One cycle of run time is equal to one unit. The lot time includes a setup time, and is calculated using $SU + N \times RT/1$, where N is the lot quantity and SU (setup) and RT (cycle time) are in equivalent dimensions of hours.

Let's return to the aluminum casting example described above for a simple operations sheet. Now we wish to add those additional and important estimating features we have now discussed.

Notice the variety in the table numbers in the column titled "Table Number." We used several tables in constructing this estimate, such as 7.1, 7.3, 8.2, and 11.3. In many cases the estimating data are used in a variety of circumstances. Machine time is necessary for several machine tools, and this information is collected in a centralized point, i.e., Tables 11.1, 11.2, and 11.3. In many cases the data are universal and can be used for a variety of circumstances.

The examples that are provided for each of the element estimating sets are composed of the six columns and are useful guides for constructing other estimates. You will need some self study here.

Cost Summary

Once the Operations Sheet is completed the next step is to transfer some of the same information to the Cost Summary sheet. A typical Cost Summary sheet is given by Figure 7. This sheet is representative of the printout generated by the *AM Cost Estimating Software*. Additional information is available from *Penton Education Division* by calling 216-696-7000.

The cost summary ㉔ is for the part SOHO and the header information is repeated. It is an important principle of this author that estimating for manufacturing requires that each operation be estimated. Each operation ⑱ is identified and these correspond to the basic Operations Sheet. The *AM Cost Estimator* Table No. ⑫ identifies the basic data set for that operation. Circle ㉕ specifies the description of the machine, process, or bench necessary to perform the operation. Lot hours ㉘ are transferred from the Operations Sheet. These lot hours differentiate between quantities. For low quantity the setup becomes more important while the cycle minute influences the lot hours if the quantity is large. Whether the part is for small or large quantity, the method is acceptable.

② COST SUMMARY

② Product Name: __X-152__
③ Product Part No. __224534-03__
④ Part Name: __SOHO__

⑤ Part No. ____(Unknown)____
⑥ Part Material Cost: __$0.00 (Consignment)__
⑦ Lot Quantity: __87__

⑧ Material: __Aluminum casting__
⑨ Unit Material Cost: __0.00__

⑱ Operation Number	⑫ Cost Estimator Table No.	㉕ Machine, Process, or Bench Description	㉑ Lot Hours	㉖ Productive Hour Cost ($)	㉗ Total Operation Cost ($)
10	7.4	Vertical-Spindle	3.32	43.50	144.27
20	18.5	Hand Deburring	0.73	21.35	15.62
30	24.1	Bench, Machines	1.73	27.75	47.92
		㉘ Total Lot Hours:	5.77		
		Total Operational Productive Hour Cost ($):			㉙ 207.81
		Unit Operational Productive Hour Cost ($):			㉚ 2.39
		㉛ Unit Material Cost ($):			0.00
		㉜ Total Direct Cost Per Unit ($):			2.39
		㉝ Total Job Cost ($):			207.81

FIGURE 7 A typical cost summary for estimating.

Even with a very large quantity, such as mechanization would require, the system is acceptable.

Productive Hour Cost (PHC) is entered as circle ㉖. These values could be from the PHC section, or they may be company values. The values shown for SOHO are company values, and they are the cost for the manpower and the machines. Overhead is included for this case. These company values are calculated by accounting.

The lot hours are multiplied by the PHC and give the total operational cost shown as column ㉗. For example, $3.32 \times 43.50 = \$144.27$. The sum of the operation cost is given by the total operational productive hour cost identified as circle ㉙. The value of $207.81 when divided by the lot quantity of 87 ⑦ gives the unit operational productive hour cost of $2.39. In this case, the value is for labor and the machine process cost. Because the material is a consignment between the buyer and the manufacturer, no cost is assessed for unit material cost identified by ㉛. If a unit cost material exists, it is entered here. The sum of the material and the unit operational productive hour cost gives the total direct cost per unit ㉜. This value is multiplied by the lot quantity and the total job cost is entered as ㉝.

This cost summary is used to provide information to the bill of material cost summary, which is the means of collecting all costs to obtain the full cost. Except in the case of a single part, the bill of material is a vital and important document. For those manufacturers who only produce a single part, the cost summary is adequate since it provides the total job cost.

Rules for Using the Elemental Tables

Published estimating data as presented in the tables of Section IV are of two types: (1) a listing and (2) row and column tabulations. There are suggested rules for using these tables. The listing identifies the element. Jotting down the exact number without alteration is the recommended practice. If the value of the time driver falls between two neighboring numbers for a row-and-column table, adopt the higher value rather than interpolating or selecting the lower value even though it may be closer. Interpolation is time consuming, and using the higher value usually does not introduce any significant error to the application total. Of course, these rules may be adjusted in unusual cases.

If the value of the time driver is smaller than any of the row-and-column numbers available, use the initial value, as it may represent the threshold time for the element. If the value of the time driver is above the range of the table, a tendency exists to extrapolate. An "add'l 1" (meaning each additional 1) may be provided, and the extrapolation slope is provided. The "add'l 1" provides the slope of the data or equation at the maximum table value. The extrapolated quantity is added to the maximum listed value of the table. In some cases, an "add'l 1" is not listed, as there may be a natural manufacturing barrier, or our data sources may not have extended into those regions. Extensions in these cases may require slope extension or judgment by the estimator.

Consider this example. A part will be handled and there is occasional repositioning during the handling. The data for the element is given as:

$L + W + H$	15.0	15.5	16.1	17.4	18.2	Add'l 1
Min	.07	.08	.09	.10	.11	.013

The $L + W + H$ (refers to the box dimensions of the length, width, and height) of the part are added. For instance, suppose that a part had a $L + W + H$ girth dimension of 13.7 in., then the time posted in the "Table Time" column is .07 min. We would use the minimum initial value. Suppose that the part girth is 15.6 in., then the practice is to go to the next higher time value and adopt the time of .09 min for the estimate. Suppose that the value of the girth is 19.1 in., the trick is to take .11 and add that to $(19.1 - 18.2)$ .013 or for an answer of .122. Follow this process when using the tables.

These rules encourage speed in application and consistency. It is important to be able to reconstruct an estimate with a minimum of paperwork. Consistency becomes more possible if two or more estimators derive an equal or similar estimate for the same operation. This can be aided by crosstalk within the estimation team. Rules for using the tables in Section IV can substantially improve the application success.

Determining the Time Drivers

When the estimator determines that a listed element is necessary, such as clamp and unclamp, the time for that element is posted on the operations sheet. There is often only one choice for these listings. But in using the tables a different approach is necessary. Usually the tables have increasing time values, and the selection is driven by a variable that corresponds to the product, process, or the manufacturing itself. Many, many variables are used throughout this manual, and some are lb, in.^{2}, in., ft, $L + W + H$, $L + W$, etc. These variables are in the context of the machine, process, or bench and thus are special to its context. Accuracy is increased if there are several or more time drivers that are necessary for a part.

We now discuss lookup rules for applying these tables. The use of these tables requires a "time driver." But once the time driver is selected, the rules state that if the time driver required for the operation estimate is less than the minimum value of the time driver for the table, then the minimum time value is posted. This minimum time value serves as the fixed time for the element. If the time driver for the estimate falls in between neighboring driver values, then the higher time is posted. This discourages the negligible improvement that might accrue from interpolation, which is difficult to do without mistake and is too slow for the increase in accuracy. Also the tables have been arranged for the habit of jumping to the next higher value. If the time driver is above any of the table values, the estimator will be required to extrapolate by the methods discussed. Many examples illustrate these simple table lookup rules.

The determination of the driver values can only be made by the estimator from knowledge of the drawing or engineering requirement and the manufacturing by the machine, process, or bench. One driver that has been discussed is the length of the machining cut. The length of cut is the sum of the safety stock, approach, drawing length, and overtravel. Details are given later about this driver. But many of the drivers deal with size or weight of the object being processed for that operation. Size is expressed by many variables, such as weight, area, volume, girth, length, etc. Sometimes increased accuracy will not materialize from excessively detailed calculations of these drivers, especially in the context of the table lookup rules. The purpose of the table lookup rules is to encourage speed, accuracy, and consistency. Thus the design variables need only be determined to the extent that they satisfy the table lookup rules.

Weight is a frequently used driver especially for operator handling. Essentially this involves finding the volume of the object and multiplying by the density for that material. After this calculation has been made repeatedly, the estimator will have a good feel for the weight, and perhaps the weight can be judged by the experience of previous calculations. Girth is variable that is used for sheet metal handling. Girth, or $L + W + H$, is the outside dimensions of a box that will enclose the object. The girth of a sphere is 3 times the diameter (3 dia). Girth, as a three-dimensional value, can be found by adding the maximum dimensions in the XYZ plane. The term "box volume" is sometimes used. To find the displacement volume of an irregularly shaped object is time consuming, but box volume is the maximum volume of a box that will envelope the part. Box volume is the XYZ product from their maximum dimension. It is a fast calculation.

The purpose of having time drivers such as box volume, girth, etc., is to encourage speed in estimating. These kinds of drivers are conveniently correlated to the tables. Their quick calculation encourages speed in estimating without reducing accuracy or consistency.

Incomplete Information

Unfortunately, estimating data are seldom complete. If the element description of the operation uncovers requirements not included in the estimating data, the estimator may choose elements from other data that meet the requirement. If information is judged deficient by the estimator, the opportunity to adjust the value either up or down is available by use of the adjustment factor. As a final resort, and oftentimes a necessary action, the estimator must use individual judgment to estimate missing elemental times.

Adjusting Elemental Estimates to Actual Times

Variations in plant productivity have a effect on the times. Productivity must either be measured against the estimated values of these data, or estimated for each plant. Consistency in estimating time is realized by using simple productivity factors. The overall adjustment of a part(s), lot(s), or product(s) time by the productivity factor gives the realized experience. Bulk adjustments, meaning an overall productivity factor adjustment, are superior to individualized, adjusted estimates.

An estimate represents the time which is determined to be necessary for trained operators to perform given operations. The author believes that the optimum procedure will estimate operations, parts, lots, or products for normally efficient plant performance using the national norm data presented here. For the purpose of direct-labor estimating out-of-pocket or actual time, the estimated total for operations, parts, lots, or products is adjusted by a productivity factor. The productivity factor is found by using the relationship,

$$PF = AT/ET$$

where PF = productivity factor, dimensionless number
AT = sum of actual time total for operations, parts, lots, or products
ET = sum of estimated times from *AM Cost Estimator* operations, parts, lots, or products.

The AT can be determined from time-keeping records, foreman reports, time and production studies, dropoff counts of parts from production equipment, or qualified opinion. Of course, the AT is matched timewise against the ET. One may determine the productivity factor from a single sample or it may be periodically evaluated. This author doubts whether a steady-state productivity factor is ever achievable because of plantwide interruptions, scheduling fluctuations, and changes in supervision and manpower. Nonetheless, spotchecks can be useful to monitor plant performance. Remember that local conditions determine if the future productivity factor is identical to past evaluations. In some plants and companies, direct-labor totals cannot be accurately determined and calculation of the productivity factor (*PF*) is not possible.

This productivity factor is sometimes referred to as the Realization Factor. Some companies may have a standard cost scheme. Instead of measuring productivity factors, an accounting labor variance, either favorable or unfavorable, is determined. The *AM Cost Estimator* is consistent with standard cost plans, and may be used keeping in mind that labor variance, not a productivity factor, is measured periodically.

Bill of Material (BOM)

The estimating of labor and material cost and its extension by overhead calculations will lead to the quantity known as full cost. This, in turn, will be increased for profit to give price. Before the routine is executed it is necessary to find the total bill of material cost for several or many parts, subassemblies, and major assemblies. The bill of material explosion is unnecessary if the manufacturer only sells single-item parts as the cost estimate serves as the principal summary document for price setting. But in the case of several or many parts and assemblies it is necessary to organize the cost estimates effectively. A scheme of doing this organization is handled by a bill of material.

An instructional bill of material is described by a graphical "tree" shown by Figure 8. Larger bills of material will follow these same practices. A computer printout, indented for several levels, is the typical presentation instead of a tree.

A tree is described by levels, three in the case of Figure 8. The top level, for this example, is the final product and is the sum of the costs of the components and assemblies below this top level. Each box of a tree contains a part number and part name, although this is not mandatory. The convention may also give the number of units that are required for the part in the next higher assembly. The number of units required for the next higher assembly is a number given to the left of each box. For example, one unit of final product (Model X 152) requires one unit of the airframe product, one unit of SOHO, and eight units of a hatch assembly. The assembly of the second level boxes gives the level one box or the final product. Notice that SOHO does not have any lower parts. SOHO is the part name of an example previously estimated by the operations sheet.

Beginning with the second and the third level, labor cost and material cost are indicated below each box. In some cases material cost (*M*) and labor cost (*L*) equals 0. A zero value for material could indicate a consignment while zero cost for labor might indicate that the part was a purchased item.

Costs roll up from the bottom. For instance, the airframe assembly 50530, which has an estimated ma-

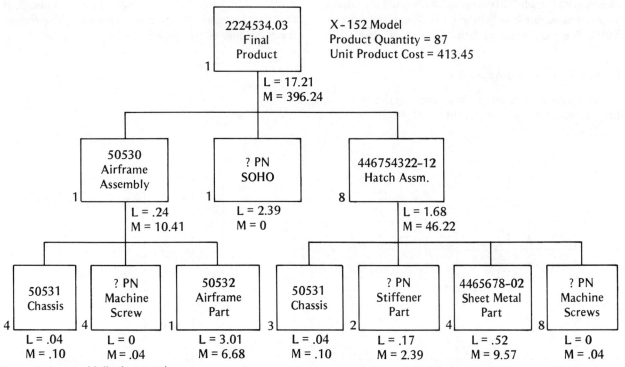

FIGURE 8 A costed bill-of-material tree.

terial cost of $10.41, is the sum of the labor and material costs adjusted for quantity of the lower levels. A labor cost of $0.24 is estimated to assemble the lower levels, which become the 50530 assembly. Level 2 labor and material estimated costs roll up to level one material cost after they have been adjusted for quantity.

The bill-of-material tree may show common materials. A purchased machine screw is used in two different locations in the third level. While the tree is a graphical illustration, more often the bill of material is provided as a printout. The corresponding printout for Figure 8 is shown in Figure 9. Level 1 is the final product. Level 2 column indicates the next lower subassembly, while level 3 is the raw material or purchased parts for level 2. The headings for the costed bill of material indicate their purpose. The UM BOM provides the unit of measure for the bill of material which can have a variety of dimensions. Next assembly quantity and lot quantity are important calculations. The machine screw which appears in two places has different next assembly requirements, but has quantity summed for the lot. It is this summed quantity that is used for the estimated quantity when calculating the lot for production. The total lot hours and the unit material cost are values that the estimate will provide.

The bill of material is an important list since other working documents use it as a central source for information. A material list can be found from the BOM. This is used to fill purchase orders if the item is purchased, or start production of the part if the part is made internally. Information for material yield such as scrap, waste, and shrinkage are additional estimated values that contribute to the cost of material for the BOM. A typical material list is shown by Figure 10.

Instructions for the Appendix

The Appendix is useful for converting the cycle estimate expressed in minutes into hr/1, hr/10, hr/100, hr/1000, or hr/10,000 units. The estimated cycle minute is considered the "precision point" and the conversions are rounded. The author recommends estimating cycle time in minutes. The cycle time estimate is calculated in minutes, but converted to hr/1, hr/10, hr/100, hr/1000, or hr/10,000 units. The conversion is handled very simply in the Appendix. A number of choices such as hr/1, hr/10, hr/100, hr/1000, or hr/10,000 are available.

The company will adopt a denominator (1, 10, 100, 1000, or 10,000) and should use it consistently for every estimate. This uniform policy prevents careless calculations. The number of decimals for the hour estimate depends upon the denominator. A 1.87-min estimate converted to hr/100 units shows 3.117, while for hr/1000 units it is 31.1667. The increasing number of decimals is intentional. This precision, though exaggerated with respect to the original measurements, is required for time and cost sensitivity. For example, if the unit estimate is .0009 min, after entry on the left margin in the Appendix, we would read .002 hr/100, 0.150 hr/1000, or .15000 hr/10,000 units. If the practice is to select the 1000 units, the estimator would record .0150 hr/1000 units.

If the total cycle estimate is .0019, then we have

	Cycle Minutes	Hr/1000
	.001	.0167
	.0009	.0150
Total	.0019	.0317

Continuing, assume a total cycle estimate of 2.01 min. In this unit estimate range, only 2.00 and 2.04 min are shown as entries in the Appendix. The following procedure is used to find the hr/100.

	Cycle Minutes	Hr/100
	2.00	3.333
	.01	.017
Total	2.01	3.350

Level 12345	Part Number-RV	Part No. Description	UM BOM	Next Assm. Quantity	Lot Qty	Total Lot Hours	O P	Unit Material Cost	Unit Labor Cost	Total Unit Cost	Next Assm. Material Cost	Optional First Overhead Rate	Optional Second Overhead Rate	Optional Extended Cost
1	2224534-03	X-152 Model	Ea	0	87		M	396.24	17.21	413.45	—		—	
2	SOHO	Unknown	Ea	1.000	87	5.77	M	0	2.39	2.39	2.39			
2	50530	Assembly	Ea	1.000	87	.84	M	10.41	.24	10.65	10.65			
3	50531	Chassis	Ea	4.000	609	.89	M	.10	.04	.14	.56			
3	Screw	Mach. Screw	Ea	4.000	1044	0	B	.04	0	.04	.16			
3	50532	Airframe Pt.	Ea	1.000	87	8.73	M	6.68	3.01	9.69	9.69			
2	446754322-12	Hatch Assm.	Ea	8.000	696	35.3	M	46.22	1.68	47.90	383.20			
3	4465678	Sheet Metal	Ea	4.000	348	6.57	M	9.57	.52	10.09	40.36			
3	Stiffener	Stiffener	Ea	2.000	174	1.08	M	2.39	.17	2.56	5.12			
3	50531	Sheet Metal	Ea	3.000	609	.89	M	.10	.04	.14	.42			
3	Screw	Mach. Screw	Ea	8.000	1044	0	B	.04	0	.04	.32			

FIGURE 9 A costed bill of material.

Level 12345	Part Number-RV	Part No. Description	UM ML	MS	Lot Qty	Material Specification	Purchase Unit Material Cost	Net Unit Material Requirement	Yield Scrap	Yield Waste	Yield Shrinkage	Final Unit Material Requirement	Material Unit Cost	Material Lot Cost
2	Unknown	SOHO	Ea.	4	87	Consignment	0	0	0	0	0	87	0	0
3	50531	Chassis	Lb.	B	609	5052 Alum.	.7500	.1294	.01	.02	0	.1333	.10	456.75
3	Screw	Mach. Screw	Ea.	B	1044	10-32, 1 in.	.04	1044	0	0	0	1044	.04	41.76
3	50532	Air Frame	Lb.	B	87	8126 Steel	1.11	5.6774	.01	.02	.03	6.01	6.68	581.16
3	4465678-02	Sheet Metal	Lb.	B	348	Titanium	45.78	.1560	.07	.10	.17	.2090	9.57	3330.36
3	Unknown	Stiffener	Lb.	B	178	304 Stainless	4.57	.5097	.005	.014	.007	.5230	2.39	415.86

FIGURE 10 A material list.

Another example of 9.25 total cycle minutes will give for the conversion to hr/10 as

	Cycle Minutes	Hr/10
	9.00	1.50
	.25	.04
Total	9.25	1.54

Suppose that an estimate of 36.17 min was determined for an operation. Although it may not be a difficult matter to convert this by means of a calculator, it is as simple to use tables, such as those given by the Appendix. This would be found as:

	Cycle Minutes	Hr/1	Hr/10	Hr 100
	30.	.5	5.00	50.000
	6.	.1	1.00	10.000
	.17	0	.03	.283
Total	36.17	.6	6.03	60.283

Certainly the estimator would not have a need for the three values, but after selection of a policy, there should be a routine for finding the appropriate hour estimate. Notice the increasing number of decimals. In fact, if it is company policy to estimate in units of one (second column from left), then it is unnecessary to evaluate estimates or minor variations expressed in hundredths of a min., i.e., .17 min.

Pieces per hour (pc/hr) should not be used for calculating estimates. We sometimes refer to pc/hr as a "shop estimate" since it is a crude approximation of output. It is provided in the extreme righthand column of the Appendix and is related to the cycle estimate as expressed in minutes. Sometimes, shop people prefer an expression of "pieces per hour" because of their familiarity. However, the "pc/hr" measure for official cost-estimating documents is not as preferred as an hour estimate.

Many of the examples ask for lot estimates. For example, a job is estimated to have 6.83 min of cycle, and 6.15 hr for the setup. What is the lot time for the job? We use the relationship:

$$\text{Lot time} = SU + N(RT/1)$$

where lot time = total hr for the operation
N = lot quantity
RT = hr/1

Continuing the problem with the 36.17 min operation estimate. Assume that the operation requires 1.75 hour for setup. What is the lot time for 72 units?

	Hr/1	Hr/10	Hr/100
RT/1	.6	.603	.60283
N × RT/1	43.20	43.42	43.40
SU	1.75	1.75	1.75
Lot hr	44.95	45.17	45.15

These lot hr are then multiplied by the PHC as given by the *AM Cost Estimator* or the estimator may use the plant machine hr or wage rate.

$$LOT\ HRS. = SU + N\left(\frac{CM}{60}\right)$$

Productive Hour Costs

II

MACHINE, PROCESS, OR BENCH		ATLANTA	BALTIMORE	BOSTON	BUFFALO	CHICAGO	CLEVELAND
1.1	Power bandsaw cutoff, contour bandsaw, and hacksaw	8.83	10.16	8.47	10.14	9.31	9.89
1.2	Abrasive saw cutoff	8.83	10.16	8.47	10.14	9.31	9.89
1.3	Oxygen cutting	9.22	9.76	9.78	11.81	10.90	10.81
1.4	Computer aided torch cutting	9.22	9.76	9.78	11.81	10.90	10.81
2.1	Plastics molding	9.80	9.76	11.57	11.46	10.77	11.38
2.2	Plastic preform molding	9.63	10.68	11.33	11.23	10.63	11.15
2.3	Thermoplastic injection molding	9.80	11.02	11.68	11.57	10.89	11.51
2.4	Thermosetting plastic molding	9.80	11.02	11.68	11.57	10.89	11.51
2.5	Thermolding	9.80	11.02	10.49	11.22	10.55	11.15
2.6	Glass cloth layup	9.80	10.71	11.33	11.22	10.55	11.15
2.7	Extrusion	9.80	11.02	10.49	11.22	10.55	11.15
2.8	Blow molding	9.80	11.02	10.49	11.22	10.55	11.15
2.9	Hot and cold chamber die casting	10.45	11.02	12.53	12.41	11.67	12.34
2.10	Isostatic molding press	9.80	11.02	10.49	11.22	10.55	11.15
3.1	Power shear	10.60	10.77	10.74	11.93	11.16	11.55
3.2	Punch press (first operation)	10.45	10.77	10.85	11.82	11.04	11.43
3.3	Punch press (secondary operations)	10.45	10.77	10.85	11.82	11.04	11.43
3.4	Turret punch press	10.60	11.54	10.85	12.05	11.27	11.70
3.5	Single-station punching	10.45	10.77	10.71	11.82	11.07	11.43
3.6	Power press brake	10.60	10.77	10.59	11.71	10.93	11.36
3.7	Foot brake, kick press, and jump shear	8.44	10.42	9.69	10.14	9.23	9.23
3.8	Hand-operated brake, bender, punch press, multiform, coil winder, shear, straightener, and roller	8.44	10.36	9.69	10.14	9.23	9.23
3.9	Nibbling	8.44	10.36	9.69	10.14	9.23	9.23
3.10	Tube bending	10.60	10.77	10.59	11.71	10.93	11.36

MACHINE, PROCESS, OR BENCH		ATLANTA	BALTIMORE	BOSTON	BUFFALO	CHICAGO	CLEVELAND
3.11	Ironworker	10.60	10.77	10.59	11.71	10.93	11.36
4.1	Marking	8.44	10.36	9.69	10.14	9.23	9.23
4.2	Screen printing	9.89	10.59	10.59	11.71	10.89	11.51
4.3	Laser marking	8.44	10.36	9.69	10.14	9.23	9.23
4.4	Pad printing	9.89	10.59	10.59	11.71	10.89	11.51
5.1	Forging	13.35	13.80	11.46	14.24	15.44	14.35
5.2	Explosive forging	12.05	13.37	11.66	12.33	13.02	13.81
6.1	Engine lathe	13.80	13.67	12.49	13.75	13.73	13.33
6.2	Turret lathe	12.05	13.37	11.66	12.33	13.02	13.81
6.3	Vertical turret lathe	12.05	13.37	11.66	12.33	13.02	13.81
6.4	Numerical controlled turning lathe	11.89	11.18	11.11	13.69	13.34	13.63
6.5	Numerical controlled chucking lathe	11.89	11.18	11.11	13.69	13.34	13.63
6.6	Single-spindle automatic screw machine	13.33	14.65	11.46	14.24	15.44	14.35
6.7	Multispindle automatic screw machine	13.33	14.65	11.46	14.24	15.44	14.35
7.1	Milling machine setup	12.53	13.46	12.86	13.30	13.53	14.45
7.2	Knee and column milling	12.53	13.46	12.86	13.30	12.78	14.45
7.3	Bed milling	12.05	13.46	9.05	11.93	10.82	12.09
7.4	Vertical-spindle ram-type milling	12.53	11.57	12.85	13.39	12.78	14.45
7.5	Router milling	11.89	11.57	7.97	10.14	11.47	9.65
7.6	Special milling	12.05	11.57	9.05	10.14	10.82	12.09
7.7	Hand miller	11.89	10.15	9.05	10.14	11.47	9.65
8.1	Machining centers	12.53	13.46	11.66	13.69	11.86	13.17
8.2	Rapid travel and automatic tool changer elements	12.53	13.46	11.11	13.69	11.86	13.17
9.1	Drilling machine setup and layout	10.84	12.53	10.23	11.49	13.53	13.83
9.2	Sensitive drill press	10.84	10.15	9.28	11.49	10.24	7.73
9.3	Upright drilling	12.36	11.35	10.23	11.49	12.83	12.62

MACHINE, PROCESS, OR BENCH	ATLANTA	BALTIMORE	BOSTON	BUFFALO	CHICAGO	CLEVELAND
9.4 Turret drilling	12.36	12.53	10.76	11.49	13.34	15.28
9.5 Cluster drilling	12.36	12.53	10.76	11.49	13.34	15.28
9.6 Radial drilling	12.69	12.53	10.22	12.54	13.41	14.77
10.1 Horizontal milling, drilling and boring	13.16	14.65	16.09	13.30	13.50	14.45
10.2 Boring and facing	13.16	13.37	12.51	13.30	13.02	13.90
11.1 Turn, bore, form, cutoff, thread, start drill, and break edges	12.05	13.52	11.17	13.75	12.04	13.17
11.2 Drill, counterbore, ream, countersink, and tap	11.64	12.03	9.76	12.54	12.26	11.39
11.3 Face, side, slot, form, straddle, end, saw, and engrave milling	12.20	12.49	9.96	13.31	11.69	12.07
11.4 Tool life and replacement	11.25	12.68	10.53	13.19	12.00	12.22
12.1 Broaching	12.86	13.05	12.40	11.94	13.47	13.33
13.1 Cylindrical grinding	12.86	14.16	13.36	13.17	14.42	13.51
13.2 Centerless grinding	11.25	13.41	12.05	11.94	12.26	12.07
13.3 Honing	9.63	13.25	8.89	10.54	10.40	10.14
13.4 Surface grinding	11.25	14.16	13.36	13.17	14.42	13.51
13.5 Internal grinding	11.25	14.16	13.36	13.17	14.42	13.51
13.6 Free abrasive grinding	9.63	13.25	8.89	10.54	10.40	10.14
13.7 Disk grinding	9.63	13.25	8.89	10.54	10.40	10.14
13.8 Vertical internal grinding	11.25	14.16	13.36	13.17	14.42	13.51
13.9 Numerical controlled internal grinding	11.25	14.16	13.36	13.17	14.42	13.51
14.1 Gear shaper	12.05	14.16	12.40	13.31	13.47	13.48
14.2 Hobbing	12.05	13.47	12.40	13.31	13.47	13.75
15.1 Thread cutting and form rolling	10.45	11.21	12.05	11.94	11.22	11.87
16.1 Shielded metal arc, flux-cored arc, and submerged welding	10.45	12.21	12.47	12.91	11.35	14.45
16.2 Gas metal-arc welding and gas tungsten-arc welding	10.45	12.21	12.47	12.91	11.35	14.45

Productive Hour Costs

MACHINE, PROCESS, OR BENCH	ATLANTA	BALTIMORE	BOSTON	BUFFALO	CHICAGO	CLEVELAND
16.3 Resistance spot welding	10.04	12.12	8.70	13.84	10.90	10.81
16.4 Torch, dip, and furnace brazing	10.04	9.76	8.70	11.81	10.90	10.81
16.5 Ultrasonic plastic-welding	10.04	9.76	8.70	11.81	10.90	10.81
17.1 Heat treat furnaces	10.27	9.76	9.49	11.45	10.58	10.49
18.1 Drill press deburring	7.62	8.78	8.53	10.80	7.83	9.79
18.2 Abrasive belt deburring	8.35	10.16	10.02	10.89	11.40	13.40
18.3 Pedestal deburring and finishing	8.27	9.28	10.14	10.89	11.64	14.56
18.4 Handheld portable tool deburring	7.62	9.24	8.53	10.80	7.83	9.79
18.5 Hand deburring	7.62	9.24	8.53	10.80	7.83	9.79
18.6 Plastic material deburring	8.27	9.24	8.53	10.80	7.83	9.79
18.7 Loose abrasive deburring	8.35	11.59	8.47	10.80	9.31	9.89
18.8 Abrasive-media flow deburring	8.35	11.59	8.47	10.80	9.31	9.89
19.1 Chemical machining and printed circuit boards	9.67	10.16	12.53	12.47	11.73	12.40
19.2 Automatic lamination for printed circuits	9.67	10.16	12.53	12.47	11.73	12.40
19.3 Ultraviolet printing for printed circuits	9.67	10.16	12.53	12.47	11.73	12.40
19.4 Conveyorized developing for printed circuits	9.67	10.16	12.53	12.47	11.73	12.40
19.5 In-line plating processes for printed circuits	9.67	10.16	12.53	12.47	11.73	12.40
19.6 Electrical discharge machining	12.05	13.47	12.40	13.31	13.47	13.75
19.7 Traveling wire electrical discharge machining	12.05	13.47	12.40	13.31	13.47	13.75
20.1 Mask and unmask bench	9.32	10.16	11.58	11.46	10.77	11.38
20.2 Booth, conveyor, and dip painting and conversion	9.32	9.24	12.60	12.47	11.73	12.40
20.3 Metal chemical cleaning	9.32	9.24	10.14	12.47	11.00	12.40
20.4 Blast cleaning	9.32	11.12	11.80	11.71	11.00	11.63
20.5 Metal electroplating and oxide coating	9.16	11.59	12.29	12.18	11.45	12.09

MACHINE, PROCESS, OR BENCH		ATLANTA	BALTIMORE	BOSTON	BUFFALO	CHICAGO	CLEVELAND
21.1	Bench assembly	8.35	9.24	9.55	11.68	10.42	11.38
21.2	Rivet and assembly machines	8.67	9.24	9.55	11.68	10.42	11.38
21.3	Robot	8.67	9.24	12.05	11.94	11.49	11.87
22.1	Machine, processs, and bench inspection	11.00	11.24	10.97	12.28	11.34	11.83
22.2	Inspection table	13.17	12.87	12.22	13.53	12.98	14.49
23.1	Component sequencing	10.45	11.21	12.05	11.94	11.22	11.87
23.2	Component insertion	10.45	11.21	12.05	11.94	11.22	11.87
23.3	Axial-lead component insertion	10.45	11.21	12.05	11.94	11.22	11.87
23.4	Light-directed component insertion	10.45	11.21	12.05	11.94	11.22	11.87
23.5	Printed circuit board stuffing	8.35	9.24	9.55	11.68	10.42	11.38
23.6	Wave soldering	10.04	9.76	8.70	11.81	10.90	10.81
23.7	Vertical solder coating	10.04	9.76	8.70	11.81	10.90	10.81
23.8	Wire harness bench assembly	8.35	9.24	9.55	11.68	10.42	11.38
23.9	Flat cable connector assembly	8.35	9.24	9.55	11.68	10.42	11.38
23.10	D-subminiature connector	8.35	9.24	9.55	11.68	10.42	11.38
23.11	Single-wire termination	8.35	9.24	9.55	11.68	10.42	11.38
23.12	Component lead forming	8.35	9.24	9.55	11.68	10.42	11.38
23.13	Printed circuit board drilling	12.36	12.53	10.76	11.49	13.34	15.28
23.14	Shearing, punching, forming	10.45	10.77	10.71	11.82	11.07	11.43
23.15	Resistance micro-spot welding	10.04	9.76	8.70	11.81	10.90	10.81
23.16	Turret welding	10.04	9.76	8.70	11.81	10.90	10.81
23.17	Wafer dicing	8.83	10.16	8.47	10.14	9.31	9.89
23.18	Coil winding	8.83	10.16	8.47	10.14	9.31	9.89
24.1	Packaging bench and machines	7.87	8.33	8.53	10.80	7.83	9.79
24.2	Corrugated-cardboard packaging	7.87	8.33	8.53	10.80	7.83	9.79
24.3	Case packing conveyor	7.87	8.33	8.53	10.80	7.83	9.79
24.4	Foam-injected box packaging	7.87	8.33	8.53	10.80	7.83	9.79
25.1	Tools, dies, jigs, and fixtures	13.65	15.18	12.60	15.66	16.97	15.78

MACHINE, PROCESS, OR BENCH	DALLAS/ FT. WORTH	DENVER	DETROIT	HARTFORD	HOUSTON	LOS ANGELES/ LONG BEACH
1.1 Power bandsaw cutoff, contour bandsaw, and hacksaw	8.45	6.46	10.12	9.62	8.84	7.66
1.2 Abrasive saw cutoff	8.45	6.46	10.12	9.62	8.84	7.66
1.3 Oxygen cutting	9.56	9.86	14.65	9.50	14.35	9.82
1.4 Computer aided torch cutting	9.56	9.86	14.65	9.50	14.35	9.82
2.1 Plastics molding	9.66	9.53	14.82	11.58	12.75	11.25
2.2 Plastic preform molding	9.46	9.34	14.52	11.34	12.56	10.97
2.3 Thermoplastic injection molding	9.75	9.63	14.98	11.71	12.96	11.26
2.4 Thermosetting plastic molding	9.75	9.63	14.98	11.71	12.96	11.26
2.5 Thermomolding	9.46	9.34	14.52	11.34	12.56	10.97
2.6 Glass cloth layup	9.46	9.34	14.52	11.34	12.56	10.97
2.7 Extrusion	9.46	9.34	14.52	11.34	12.56	10.97
2.8 Blow molding	9.46	9.34	14.52	11.34	12.56	10.97
2.9 Hot and cold chamber die casting	10.46	10.33	16.05	12.55	13.89	12.20
2.10 Isostatic molding press	9.46	9.34	14.52	11.34	12.56	10.97
3.1 Power shear	10.24	9.96	14.82	9.82	13.14	11.89
3.2 Punch press (first operation)	10.14	9.90	14.65	10.41	12.99	11.79
3.3 Punch press (secondary operations)	10.14	9.90	14.65	10.41	12.99	11.79
3.4 Turret punch press	10.35	10.10	14.82	10.41	13.27	12.03
3.5 Single-station punching	10.14	9.90	14.65	10.41	12.99	11.79
3.6 Power press brake	10.02	9.82	14.26	10.69	12.86	11.66
3.7 Foot brake, kick press, and jump shear	7.88	7.17	10.60	9.62	9.99	8.28
3.8 Hand-operated brake, bender, punch press, multiform, coil winder, shear, straightener, and roller	7.88	7.17	10.60	9.62	9.99	8.28
3.9 Nibbling	7.88	7.17	10.60	9.62	9.99	8.28
3.10 Tube bending	10.02	9.82	14.26	10.69	12.86	11.66

MACHINE, PROCESS, OR BENCH		DALLAS/ FT. WORTH	DENVER	DETROIT	HARTFORD	HOUSTON	LOS ANGELES/ LONG BEACH
3.11	Ironworker	10.02	9.82	14.26	10.69	12.86	11.66
4.1	Marking	7.88	7.17	10.60	9.62	9.99	8.28
4.2	Screen printing	9.75	9.63	14.98	11.71	12.96	11.26
4.3	Laser marking	7.88	7.17	10.60	9.62	12.56	8.28
4.4	Pad printing	9.75	9.63	14.98	11.71	12.96	11.26
5.1	Forging	14.29	13.28	15.26	11.34	14.57	15.34
5.2	Explosive forging	13.75	12.27	17.40	11.83	14.57	14.43
6.1	Engine lathe	13.28	12.71	19.23	11.83	14.69	15.34
6.2	Turret lathe	13.75	12.27	17.40	11.83	14.57	14.43
6.3	Vertical turret lathe	13.75	12.27	17.40	11.83	14.57	14.36
6.4	Numerical controlled turning lathe	13.19	11.38	14.05	10.13	15.21	17.60
6.5	Numerical controlled chucking lathe	13.19	11.38	14.05	10.13	15.21	17.60
6.6	Single spindle automatic screw machine	14.29	14.69	20.82	11.34	14.38	15.76
6.7	Multispindle automatic screw machine	14.29	14.69	20.82	11.34	14.38	15.76
7.1	Milling machine setup	12.10	13.17	18.24	12.53	14.86	12.08
7.2	Knee and column milling	13.83	13.17	18.24	11.20	14.86	14.65
7.3	Bed milling	12.37	10.76	14.60	9.59	13.83	10.19
7.4	Vertical-spindle ram-type milling	13.83	10.76	18.24	11.20	14.86	14.65
7.5	Router milling	12.37	8.38	13.40	9.62	10.50	7.67
7.6	Special milling	12.37	11.69	14.60	9.59	13.83	11.49
7.7	Hand miller	8.45	8.94	13.40	9.62	10.50	7.67
8.1	Machining centers	13.74	11.88	14.27	10.34	14.89	13.40
8.2	Rapid travel and automatic tool changer elements	13.74	9.23	14.27	10.34	14.89	13.40
9.1	Drilling machine setup and layout	12.09	9.23	16.22	12.53	12.71	12.08
9.2	Sensitive drill press	8.65	10.93	15.44	10.28	10.19	10.65
9.3	Upright drilling	12.21	10.86	16.22	10.28	13.35	14.76

Productive Hour Costs

MACHINE, PROCESS, OR BENCH	DALLAS/FT. WORTH	DENVER	DETROIT	HARTFORD	HOUSTON	LOS ANGELES/LONG BEACH
9.4 Turret drilling	13.74	10.86	14.21	10.28	14.20	14.76
9.5 Cluster drilling	12.21	10.86	15.44	10.28	13.35	10.65
9.6 Radial drilling	12.44	12.37	17.44	11.23	13.80	13.92
10.1 Horizontal milling, drilling and boring	13.83	12.78	18.24	11.20	14.86	14.65
10.2 Boring and facing	13.75	12.78	18.24	11.20	14.86	14.65
11.1 Turn, bore, form, cutoff, thread, start drill, and break edges	11.61	11.37	18.32	11.68	12.96	13.33
11.2 Drill, counterbore, ream, countersink, and tap	9.97	9.23	16.22	10.28	12.45	11.16
11.3 Face, side, slot, form, straddle, end, saw, and engrave milling	13.11	11.97	15.41	10.40	13.05	10.83
11.4 Tool life and replacement	11.54	10.86	16.65	10.79	12.82	11.78
12.1 Broaching	13.06	12.39	18.33	11.22	14.41	14.43
13.1 Cylindrical grinding	13.42	13.46	18.65	11.61	14.82	15.39
13.2 Centerless grinding	10.80	10.40	14.12	12.21	14.83	11.49
13.3 Honing	7.30	7.00	10.05	9.06	9.37	7.82
13.4 Surface grinding	13.42	13.46	18.65	11.61	14.82	15.39
13.5 Internal grinding	13.42	13.41	18.65	11.61	14.82	15.39
13.6 Free abrasive grinding	7.30	7.00	10.05	9.06	9.36	7.82
13.7 Disk grinding	7.30	7.00	10.05	9.06	9.36	7.82
13.8 Vertical internal grinding	13.42	13.41	18.65	11.61	14.82	15.39
13.9 Numerical controlled internal grinding	13.42	13.41	18.65	11.61	14.82	15.39
14.1 Gear shaper	13.20	12.51	18.54	11.34	14.49	14.57
14.2 Hobbing	13.20	12.51	18.54	11.34	14.49	14.57
15.1 Thread cutting and form rolling	10.06	9.38	15.44	12.07	13.35	11.72
16.1 Shielded metal arc, flux-cored arc, and submerged welding	13.20	11.67	14.78	11.61	14.93	13.79
16.2 Gas metal-arc welding and gas tungsten-arc welding	13.20	11.67	14.78	11.61	14.93	13.79

MACHINE, PROCESS, OR BENCH	DALLAS/ FT. WORTH	DENVER	DETROIT	HARTFORD	HOUSTON	LOS ANGELES/ LONG BEACH
16.3 Resistance spot welding	9.11	9.86	15.62	9.33	14.35	14.76
16.4 Torch, dip, and furnace brazing	9.56	9.86	14.65	9.50	11.57	9.82
16.5 Ultrasonic plastic-welding	9.56	9.86	14.65	9.50	11.57	9.82
17.1 Heat treat furnaces	9.27	9.55	14.21	9.21	11.24	10.95
18.1 Drill press deburring	7.65	7.42	11.82	9.89	9.99	8.04
18.2 Abrasive belt deburring	8.72	7.79	13.25	9.74	10.21	9.87
18.3 Pedestal deburring and finishing	9.23	7.79	13.67	9.74	10.21	9.87
18.4 Handheld portable tool deburring	7.65	7.42	11.82	9.89	9.99	8.04
18.5 Hand deburring	7.65	7.42	11.82	9.89	9.99	8.04
18.6 Plastic material deburring	7.65	7.42	11.82	9.89	9.99	8.04
18.7 Loose abrasive deburring	8.45	7.79	10.12	9.62	8.84	11.72
18.8 Abrasive-media flow deburring	8.45	7.79	10.12	9.62	8.84	11.72
19.1 Chemical machining and printed circuit boards	10.46	10.33	16.05	12.61	12.82	12.20
19.2 Automatic lamination for printed circuits	10.46	10.33	16.05	12.61	12.82	12.20
19.3 Ultraviolet printing for printed circuits	10.46	10.33	16.05	12.61	12.82	12.20
19.4 Conveyorized developing for printed circuits	10.46	10.33	16.05	12.61	12.82	12.20
19.5 In-line plating processes for printed circuits	10.46	10.33	16.05	12.61	12.82	12.20
19.6 Electrical discharge machining	13.20	12.51	18.54	11.34	14.49	14.57
19.7 Traveling wire electrical discharge machining	13.20	12.51	18.54	11.34	14.49	14.57
20.1 Mask and unmask bench	9.66	9.53	14.82	11.58	12.82	11.25
20.2 Booth, conveyor, and dip painting and conversion	10.46	9.53	16.05	12.55	12.82	11.25
20.3 Metal chemical cleaning	10.25	10.13	15.75	12.31	12.82	11.25
20.4 Blast cleaning	9.86	9.74	15.13	12.31	12.82	11.25
20.5 Metal electroplating and oxide coating	10.25	10.13	15.75	12.31	12.82	11.96

MACHINE, PROCESS, OR BENCH		DALLAS/FT. WORTH	DENVER	DETROIT	HARTFORD	HOUSTON	LOS ANGELES/LONG BEACH
21.1	Bench assembly	9.19	10.70	15.62	9.95	13.50	9.66
21.2	Rivet and assembly machines	9.19	10.70	15.62	10.21	13.50	9.66
21.3	Robot	10.06	10.70	15.44	10.21	13.35	11.72
22.1	Machine, process, and bench inspection	10.68	10.71	16.19	11.08	12.33	11.42
22.2	Inspection table	13.19	11.75	17.94	10.66	12.31	15.49
23.1	Component sequencing	10.06	9.38	15.44	12.07	13.35	11.72
23.2	Component insertion	10.06	9.38	15.44	12.07	13.35	11.72
23.3	Axial-lead component insertion	10.06	9.38	15.44	12.07	13.35	11.72
23.4	Light-directed component	10.06	9.38	15.44	12.07	13.35	11.72
23.5	Printed circuit board stuffing	9.19	10.70	15.62	9.95	13.50	9.66
23.6	Wave soldering	9.56	9.86	14.65	9.50	11.57	9.82
23.7	Vertical solder coating	9.56	9.86	14.65	9.50	11.57	9.82
23.8	Wire harness bench assembly	9.19	10.70	15.62	9.95	13.50	9.66
23.9	Flat cable connector assembly	9.19	10.70	15.62	9.95	13.50	9.66
23.10	D-subminiature connector	9.19	10.70	15.62	9.95	13.50	9.66
23.11	Single-wire termination	9.19	10.70	15.62	9.95	13.50	9.66
23.12	Component lead forming	9.19	10.70	15.62	9.95	13.50	9.66
23.13	Printed circuit board drilling	13.74	10.86	14.21	10.28	14.20	14.76
23.14	Shearing, punching, forming	10.14	9.90	14.65	10.41	12.99	11.79
23.15	Resistance micro-spot welding	9.56	9.86	14.65	9.50	11.57	9.82
23.16	Turret welding	9.56	9.86	14.65	9.50	11.57	9.82
23.17	Wafer dicing	8.45	6.46	10.12	9.62	8.84	7.66
23.18	Coil winding	8.45	6.46	10.12	9.62	8.84	7.66
24.1	Packaging bench and machines	7.65	7.42	11.82	9.89	9.99	8.04
24.2	Corrugated-cardboard packaging	7.65	7.42	11.82	9.89	9.99	8.04
24.3	Case packing conveyor	7.65	7.42	11.82	9.89	9.99	8.04
24.4	Foam-injected box packaging	7.65	7.42	11.82	9.89	9.99	8.04
25.1	Tools, dies, jigs, and fixtures	15.71	14.60	16.78	12.47	16.02	18.90

MACHINE, PROCESS, OR BENCH		MILWAUKEE	MINNEAPOLIS/ST. PAUL	NEWARK	NEW YORK CITY & NEW JERSEY	PHILADELPHIA & NEW JERSEY	PITTSBURGH
1.1	Power bandsaw cutoff, contour bandsaw, and hacksaw	11.90	10.50	9.41	7.68	8.20	12.08
1.2	Abrasive saw cutoff	11.90	10.50	9.41	7.68	8.20	12.08
1.3	Oxygen cutting	10.70	9.67	11.06	9.73	10.81	11.56
1.4	Computer aided torch cutting	10.70	9.67	11.06	9.73	10.81	11.56
2.1	Plastics molding	13.39	12.14	12.85	9.36	11.66	12.55
2.2	Plastic preform molding	13.13	11.99	12.66	9.16	11.38	12.28
2.3	Thermoplastic injection molding	13.53	12.38	12.99	9.45	11.75	12.67
2.4	Thermosetting plastic molding	13.53	12.38	12.99	9.45	11.75	12.67
2.5	Thermomolding	13.13	11.99	12.66	9.16	11.38	12.28
2.6	Glass cloth layup	13.13	11.99	12.66	9.16	11.38	12.28
2.7	Extrusion	13.13	11.99	12.66	9.16	11.38	12.28
2.8	Blow molding	13.13	11.99	12.66	9.16	11.38	12.28
2.9	Hot and cold chamber die casting	14.52	13.30	13.93	10.04	12.59	13.58
2.10	Isostatic molding press	13.13	11.99	12.66	9.16	11.38	12.28
3.1	Power shear	14.21	12.49	9.49	10.78	9.67	14.69
3.2	Punch press (first operation)	14.07	12.37	9.41	10.68	9.56	14.55
3.3	Punch press (secondary operations)	14.07	12.37	9.41	10.68	9.56	14.55
3.4	Turret punch press	14.35	12.63	9.58	10.90	9.77	14.84
3.5	Single-station punching	14.07	12.37	9.41	10.68	9.56	14.55
3.6	Power press brake	13.93	12.26	9.30	10.57	9.47	14.40
3.7	Foot brake, kick press, and jump shear	12.91	10.45	8.77	7.45	8.17	11.13
3.8	Hand-operated brake, bender, punch press, multiform, coil winder, shear, straightener, and roller	12.91	10.45	8.77	7.45	8.17	11.13
3.9	Nibbling	12.91	10.45	8.77	7.45	8.17	11.13
3.10	Tube bending	13.93	12.26	9.30	10.57	9.47	14.40

MACHINE, PROCESS, OR BENCH		MILWAUKEE	MINNEAPOLIS/ ST. PAUL	NEWARK	NEW YORK CITY & NEW JERSEY	PHILADELPHIA & NEW JERSEY	PITTSBURGH
3.11	Ironworker	13.93	12.26	9.30	10.57	9.47	14.40
4.1	Marking	12.91	10.45	8.77	7.45	8.17	11.13
4.2	Screen printing	13.53	12.38	13.18	9.45	11.75	11.13
4.3	Laser marking	12.91	10.45	12.66	7.45	8.17	11.13
4.4	Pad printing	13.53	12.38	13.18	9.45	11.75	11.13
5.1	Forging	14.94	14.37	13.98	12.53	13.03	14.01
5.2	Explosive forging	14.48	13.97	12.90	12.56	13.61	13.57
6.1	Engine lathe	14.04	13.37	13.90	12.09	14.00	13.57
6.2	Turret lathe	14.48	13.97	12.90	12.56	13.61	13.57
6.3	Vertical turret lathe	14.48	13.97	12.90	12.56	13.61	13.57
6.4	Numerical controlled turning lathe	15.67	12.16	12.95	12.44	13.54	14.45
6.5	Numerical controlled chucking lathe	15.67	12.16	12.95	12.44	13.54	14.45
6.6	Single spindle automatic screw machine	14.94	14.37	12.90	11.21	13.35	13.57
6.7	Multispindle automatic screw machine	14.94	14.37	12.90	11.21	13.35	13.57
7.1	Milling machine setup	15.99	12.10	12.39	12.64	12.27	13.57
7.2	Knee and column milling	14.26	12.80	12.39	12.64	13.44	13.57
7.3	Bed milling	14.14	12.77	10.85	10.19	12.35	13.06
7.4	Vertical-spindle ram-type milling	14.26	13.41	12.39	12.64	13.44	13.57
7.5	Router milling	11.90	10.50	9.41	8.62	8.20	13.57
7.6	Special milling	14.14	12.77	12.39	10.19	13.44	13.57
7.7	Hand miller	11.90	10.50	9.41	8.62	8.20	12.08
8.1	Machining centers	14.59	13.41	12.39	12.64	12.27	14.01
8.2	Rapid travel and automatic tool changer	14.59	13.41	12.39	12.65	12.27	14.01
9.1	Drilling machine setup and layout	15.99	12.10	8.90	11.39	12.25	13.34
9.2	Sensitive drill press	12.93	10.50	9.18	7.72	8.20	12.08
9.3	Upright drilling	13.58	12.77	9.18	11.39	12.25	13.34

MACHINE, PROCESS, OR BENCH	MILWAUKEE	MINNEAPOLIS/ST. PAUL	NEWARK	NEW YORK CITY & NEW JERSEY	PHILADELPHIA & NEW JERSEY	PITTSBURGH
9.4 Turret drilling	15.42	12.32	8.90	7.72	12.27	13.06
9.5 Cluster drilling	12.93	12.15	9.18	11.39	12.81	13.06
9.6 Radial drilling	13.06	13.27	9.18	11.39	12.69	13.34
10.1 Horizontal milling, drilling, and boring	14.26	13.41	12.39	12.64	13.44	14.01
10.2 Boring and facing	14.26	13.41	12.39	12.64	13.44	14.01
11.1 Turn, bore, form, cutoff, thread, start drill, and break edges	13.73	13.67	13.87	11.05	13.60	13.25
11.2 Drill, counterbore, ream, countersink, tap	13.48	13.27	8.90	9.41	12.77	13.06
11.3 Face, side, slot, form, straddle, end, saw, and engrave milling	14.19	13.41	13.11	10.48	13.28	13.57
11.4 Tool life and replacement	13.85	13.46	11.95	10.33	13.22	13.30
12.1 Broaching	14.31	13.67	13.57	11.94	13.21	13.60
13.1 Cylindrical grinding	15.15	14.39	13.98	12.10	13.01	14.85
13.2 Centerless grinding	12.63	12.77	13.39	9.75	12.03	14.31
13.3 Honing	11.90	10.50	9.41	7.68	8.20	12.08
13.4 Surface grinding	15.15	13.37	13.98	12.10	13.01	14.85
13.5 Internal grinding	15.15	13.37	13.98	12.10	13.01	14.85
13.6 Free abrasive grinding	11.90	10.50	9.41	7.68	8.20	12.08
13.7 Disk grinding	11.90	10.50	9.41	7.68	8.20	12.08
13.8 Vertical internal grinding	15.15	13.37	13.98	12.10	13.01	14.85
13.9 Numerical controlled internal grinding	15.15	13.37	13.98	12.10	13.01	14.85
14.1 Gear shaper	14.45	13.81	13.70	12.08	13.33	13.73
14.2 Hobbing	14.45	13.81	13.70	12.08	13.33	13.73
15.1 Thread cutting and form rolling	13.95	12.77	13.39	9.75	12.11	13.06
16.1 Shielded metal arc, flux-cored arc, and submerged welding	14.62	13.47	13.30	10.87	13.82	13.98
16.2 Gas metal-arc welding and gas tungsten-arc welding	14.62	13.47	13.30	10.87	13.82	13.98

MACHINE, PROCESS, OR BENCH	MILWAUKEE	MINNEAPOLIS/ST. PAUL	NEWARK	NEW YORK CITY & NEW JERSEY	PHILADELPHIA & NEW JERSEY	PITTSBURGH
16.3 Resistance spot welding	10.70	13.20	12.27	9.73	9.12	11.56
16.4 Torch, dip, and furnace brazing	14.59	12.45	11.06	11.14	10.81	12.89
16.5 Ultrasonic plastic-welding	14.59	12.45	11.06	11.14	10.81	12.89
17.1 Heat treat furnaces	14.16	12.09	10.74	10.82	10.49	12.51
18.1 Drill press deburring	12.51	9.11	9.12	7.71	8.89	10.36
18.2 Abrasive belt deburring	13.03	9.29	11.40	9.45	11.15	10.76
18.3 Pedestal deburring and finishing	13.03	9.29	11.40	9.54	11.27	11.13
18.4 Handheld portable tool deburring	12.51	9.11	9.12	7.71	8.89	10.36
18.5 Hand deburring	12.51	9.11	9.12	7.71	8.89	10.36
18.6 Plastic material deburring	12.51	9.11	9.12	7.71	8.89	10.36
18.7 Loose abrasive deburring	11.90	9.46	9.41	7.68	8.20	12.08
18.8 Abrasive-media flow deburring	11.90	9.46	9.41	7.68	8.20	12.08
19.1 Chemical machining and printed circuit boards	14.52	13.29	13.93	9.30	12.65	13.06
19.2 Automatic lamination for printed circuits	14.52	13.29	13.93	9.30	12.65	13.06
19.3 Ultraviolet printing for printed circuits	14.52	13.29	13.93	9.30	12.65	13.06
19.4 Conveyorized developing for printed circuits	14.52	13.29	13.93	9.30	12.65	13.06
19.5 In-line plating processes for printed circuits	14.52	13.29	13.93	9.30	12.65	13.06
19.6 Electrical discharge machining	14.45	13.81	13.70	12.08	13.33	13.73
19.7 Traveling wire electrical discharge machining	14.45	13.81	13.70	12.08	13.33	13.73
20.1 Mask and unmask bench	13.39	12.26	12.85	9.36	11.61	12.55
20.2 Booth, conveyor, and dip painting and conversion	14.52	13.29	14.00	10.17	12.65	12.55
20.3 Metal chemical cleaning	14.23	12.26	13.65	9.94	12.34	12.79
20.4 Blast cleaning	13.67	12.26	13.12	9.55	11.86	12.79
20.5 Metal electroplating and oxide coating	14.23	13.29	13.65	9.94	11.86	12.79

MACHINE, PROCESS, OR BENCH		MILWAUKEE	MINNEAPOLIS/ ST. PAUL	NEWARK	NEW YORK CITY & NEW JERSEY	PHILADELPHIA & NEW JERSEY	PITTSBURGH
21.1	Bench assembly	12.80	10.69	8.73	9.01	9.72	11.55
21.2	Rivet and assembly machines	12.80	12.06	8.73	9.01	9.72	11.55
21.3	Robot	13.95	12.77	13.39	9.75	12.11	13.06
22.1	Machine, process, and bench inspection	13.53	12.45	12.26	9.90	11.30	13.04
22.2	Inspection table	14.07	12.63	11.60	12.27	12.64	14.01
23.1	Component sequencing	13.95	12.77	13.39	9.75	12.11	13.06
23.2	Component insertion	13.95	12.77	13.39	9.75	12.11	13.06
23.3	Axial-lead component insertion	13.95	12.77	13.39	9.75	12.11	13.06
23.4	Light-directed component insertion	13.95	12.77	13.39	9.75	12.11	13.06
23.5	Printed circuit board stuffing	12.80	10.69	8.73	9.01	9.72	11.55
23.6	Wave soldering	14.59	12.45	11.06	11.14	10.81	12.89
23.7	Vertical solder coating	14.59	12.45	11.06	11.14	10.81	12.89
23.8	Wire harness bench assembly	12.80	10.69	8.73	9.01	9.72	11.55
23.9	Flat cable connector assembly	12.80	10.69	8.73	9.01	9.72	11.55
23.10	D-subminiature connector	12.80	10.69	8.73	9.01	9.72	11.55
23.11	Single-wire termination	12.80	10.69	8.73	9.01	9.72	11.55
23.12	Component lead forming	12.80	10.69	8.73	9.01	9.72	11.55
23.13	Printed circuit board drilling	15.42	12.32	8.90	7.72	12.27	13.06
23.14	Shearing, punching, forming	14.07	12.37	9.41	10.68	9.56	14.55
23.15	Resistance micro-spot welding	14.59	12.45	11.06	11.14	10.81	12.89
23.16	Turret welding	14.59	12.45	11.06	11.14	10.81	12.89
23.17	Wafer dicing	11.90	10.50	9.41	7.68	8.20	12.08
23.18	Coil winding	11.90	10.50	9.41	7.68	8.20	12.08
24.1	Packaging bench and machines	12.51	9.11	9.12	7.71	8.89	10.36
24.2	Corrugated-cardboard packaging	12.51	9.11	9.12	7.71	8.89	10.36
24.3	Case packing conveyor	12.51	9.11	9.12	7.71	8.89	10.36
24.4	Foam-injected box packaging	12.51	9.11	9.12	7.71	8.89	10.36
25.1	Tools, dies, jigs, and fixtures	16.44	15.81	15.39	13.79	14.34	15.42

MACHINE, PROCESS, OR BENCH	PORTLAND & WASHINGTON	ST. LOUIS & ILLINOIS	SAN FRANCISCO & OAKLAND	TULSA	WORCESTER
1.1 Power bandsaw cutoff, contour bandsaw, and hacksaw	10.73	12.21	11.73	8.73	10.19
1.2 Abrasive saw cutoff	10.73	12.21	11.73	8.73	10.19
1.3 Oxygen cutting	11.23	10.60	11.95	11.25	11.68
1.4 Computer aided torch cutting	11.23	10.60	11.95	11.26	11.68
2.1 Plastics molding	11.63	11.81	13.39	10.65	10.80
2.2 Plastic preform molding	11.38	11.55	13.11	11.73	8.20
2.3 Thermoplastic injection molding	11.74	11.93	13.53	15.14	10.80
2.4 Thermosetting plastic molding	11.74	11.93	13.53	15.14	10.80
2.5 Thermomolding	11.38	11.55	13.11	9.70	8.20
2.6 Glass cloth layup	11.38	11.55	13.11	9.70	10.80
2.7 Extrusion	11.38	11.55	13.11	9.70	8.20
2.8 Blow molding	11.38	11.55	13.11	9.70	8.20
2.9 Hot and cold chamber die casting	12.58	12.78	14.51	16.14	11.69
2.10 Isostatic molding press	11.38	11.55	13.11	9.70	8.20
3.1 Power shear	11.82	12.28	12.45	10.43	11.84
3.2 Punch press (first operation)	11.70	12.16	12.45	10.68	11.92
3.3 Punch press (secondary operations)	11.70	12.15	12.45	10.68	11.92
3.4 Turret punch press	11.93	12.39	12.45	10.58	11.84
3.5 Single-station punching	11.70	12.28	12.13	10.46	11.84
3.6 Power press brake	11.58	12.03	12.13	10.46	11.84
3.7 Foot brake, kick press, and jump shear	9.04	10.17	11.79	9.70	10.70
3.8 Hand-operated brake, bender, punch press, multiform, coil winder, shear, straightener, and roller	9.04	10.17	11.79	9.70	10.70
3.9 Nibbling	9.04	10.17	11.79	9.70	10.70

MACHINE, PROCESS, OR BENCH		PORTLAND & WASHINGTON	ST. LOUIS & ILLINOIS	SAN FRANCISCO & OAKLAND	TULSA	WORCESTER
3.10	Tube bending	11.58	12.03	12.13	10.46	11.84
3.11	Ironworker	11.58	12.03	12.13	10.46	11.84
4.1	Marking	9.04	10.17	11.79	9.70	10.70
4.2	Screen printing	11.74	10.17	12.45	10.75	10.91
4.3	Laser marking	9.04	10.17	11.79	9.70	10.70
4.4	Pad printing	11.74	10.17	12.45	10.75	10.91
5.1	Forging	12.76	13.89	15.80	12.75	13.53
5.2	Explosive forging	12.99	13.89	15.80	12.92	13.68
6.1	Engine lathe	12.99	13.89	15.80	13.41	13.65
6.2	Turret lathe	12.99	13.89	15.80	12.92	13.68
6.3	Vertical turret lathe	12.99	13.89	15.80	12.92	13.68
6.4	Numerical controlled turning lathe	13.06	13.62	15.80	12.76	13.14
6.5	Numerical controlled chucking lathe	13.06	13.62	15.80	12.76	13.14
6.6	Single spindle automatic screw machine	12.76	13.89	15.80	12.75	13.53
6.7	Multispindle automatic screw machine	12.76	13.89	15.80	12.75	13.53
7.1	Milling machine setup	13.06	14.19	15.80	12.94	13.49
7.2	Knee and column milling	13.06	14.19	15.80	12.94	13.49
7.3	Bed milling	12.11	13.08	13.95	11.09	10.78
7.4	Vertical-spindle ram-type milling	13.06	14.19	15.80	12.94	13.49
7.5	Router milling	8.77	11.08	11.73	8.73	10.19
7.6	Special milling	8.77	13.08	11.64	11.09	13.49
7.7	Hand miller	8.77	10.51	11.73	8.73	10.19
8.1	Machining centers	12.62	13.62	15.80	12.76	13.30
8.2	Rapid travel and automatic tool changer elements	12.62	13.62	15.80	12.76	13.30
9.1	Drilling machine setup and layout	12.60	13.03	15.80	11.97	12.01

Productive Hour Costs

MACHINE, PROCESS, OR BENCH	PORTLAND & WASHINGTON	ST. LOUIS & ILLINOIS	SAN FRANCISCO & OAKLAND	TULSA	WORCESTER
9.2 Sensitive drill press	12.11	12.29	11.73	10.04	11.50
9.3 Upright drilling	12.60	11.63	15.80	11.74	12.01
9.4 Turret drilling	12.60	11.63	15.80	11.74	11.50
9.5 Cluster drilling	12.60	11.63	15.80	11.74	11.50
9.6 Radial drilling	12.77	11.63	15.80	11.97	13.53
10.1 Horizontal milling, drilling, and boring	13.06	14.19	15.80	12.94	13.49
10.2 Boring and facing	13.06	14.19	15.80	12.94	13.49
11.1 Turn, bore, form, cutoff, thread, start drill, and break edges	12.99	13.89	15.80	12.44	12.94
11.2 Drill, counterbore, ream, countersink, and tap	12.60	13.89	15.80	11.00	11.80
11.3 Face, side, slot, form, straddle, end, saw, and engrave milling	13.06	14.19	15.80	12.94	12.13
11.4 Tool life and replacement	12.88	13.96	15.80	12.14	12.28
12.1 Broaching	12.63	13.48	15.03	12.40	13.20
13.1 Cylindrical grinding	13.15	13.89	15.80	12.51	13.53
13.2 Centerless grinding	12.11	12.29	13.95	11.09	12.05
13.3 Honing	13.15	12.21	11.73	8.73	9.68
13.4 Surface grinding	13.15	12.21	15.80	12.51	13.53
13.5 Internal grinding	13.15	12.29	15.80	12.51	13.53
13.6 Free abrasive grinding	13.15	12.21	11.73	8.73	9.68
13.7 Disk grinding	13.15	12.21	11.73	8.73	9.68
13.8 Vertical internal grinding	13.15	12.29	15.80	12.51	13.53
13.9 Numerical controlled internal grinding	13.15	12.29	15.80	12.51	13.53
14.1 Gear shaper	12.76	13.89	15.80	12.53	13.53
14.2 Hobbing	12.76	13.89	15.18	12.53	13.53
15.1 Thread cutting and form rolling	12.11	12.21	13.25	11.09	11.25

MACHINE, PROCESS, OR BENCH	PORTLAND & WASHINGTON	ST. LOUIS & ILLINOIS	SAN FRANCISCO & OAKLAND	TULSA	WORCESTER
16.1 Shielded metal arc, flux-cored arc, and submerged welding	13.15	13.47	11.53	12.82	12.63
16.2 Gas metal-arc welding and gas tungsten-arc welding	13.15	13.47	11.53	12.82	12.63
16.3 Resistance spot welding	12.20	10.72	12.45	11.26	12.50
16.4 Torch, dip, and furnace brazing	11.60	10.60	11.15	10.92	12.34
16.5 Ultrasonic plastic-welding	11.60	10.60	11.15	10.92	12.34
17.1 Heat treat furnaces	12.20	10.60	11.53	10.59	12.34
18.1 Drill press deburring	10.40	10.28	11.15	7.95	9.81
18.2 Abrasive belt deburring	10.73	12.24	11.15	8.73	10.28
18.3 Pedestal deburring and finishing	10.85	12.24	11.15	8.73	10.28
18.4 Handheld portable tool deburring	10.40	10.28	11.15	7.95	9.81
18.5 Hand deburring	10.40	10.28	11.15	7.95	9.81
18.6 Plastic material deburring	10.40	10.28	11.15	7.95	9.81
18.7 Loose abrasive deburring	12.11	12.21	11.28	8.73	11.25
18.8 Abrasive-media flow deburring	12.11	12.21	11.28	8.73	11.25
19.1 Chemical machining and printed circuit boards	11.48	12.29	11.66	11.09	10.19
19.2 Automatic lamination for printed circuits	11.48	12.29	11.66	11.09	10.19
19.3 Ultraviolet printing for printed circuits	11.48	12.29	11.66	11.09	10.19
19.4 Conveyorized developing for printed circuits	11.48	12.29	11.66	11.09	10.19
19.5 In-line plating processes for printed circuits	11.48	12.29	11.66	11.09	10.19
19.6 Electrical discharge machining	12.76	13.89	15.18	12.53	13.53
19.7 Traveling wire electrical discharge machining	12.76	13.89	15.18	12.53	13.53
20.1 Mask and unmask bench	11.48	12.29	11.28	8.73	10.19

	MACHINE, PROCESS, OR BENCH	PORTLAND & WASHINGTON	ST. LOUIS & ILLINOIS	SAN FRANCISCO & OAKLAND	TULSA	WORCESTER
20.2	Booth, conveyor, and dip painting and conversion	11.48	12.29	11.93	10.68	11.25
20.3	Metal chemical cleaning	11.48	12.29	11.93	10.68	11.25
20.4	Blast cleaning	11.48	12.21	12.61	9.70	11.25
20.5	Metal electroplating and oxide coating	11.54	12.29	12.61	10.68	11.25
21.1	Bench assembly	11.56	11.13	10.75	10.60	10.84
21.2	Rivet and assembly machines	12.55	11.13	11.15	10.60	10.82
21.3	Robot	12.55	12.29	11.53	10.60	11.25
22.1	Machine, process, and bench inspection	12.57	12.79	13.95	10.87	11.68
22.2	Inspection table	13.37	13.28	14.55	12.92	13.80
23.1	Component sequencing	12.11	12.21	13.25	11.09	11.25
23.2	Component insertion	12.11	12.21	13.25	11.09	11.25
23.3	Axial-lead component insertion	12.11	12.21	13.25	11.09	11.25
23.4	Light-directed component insertion	12.11	12.21	13.25	11.09	11.25
23.5	Printed circuit board stuffing	11.56	11.13	10.75	10.60	10.84
23.6	Wave soldering	11.60	10.60	11.15	10.95	12.34
23.7	Vertical solder coating	11.60	10.60	11.15	10.92	12.34
23.8	Wire harness bench assembly	11.56	11.13	10.75	10.60	10.84
23.9	Flat cable connector assembly	11.56	11.13	10.75	10.60	10.84
23.10	D-subminiature connector	11.56	11.13	10.75	10.60	10.84
23.11	Single-wire termination	11.56	11.13	10.75	10.60	10.84
23.12	Component lead forming	11.56	11.13	10.75	10.60	10.84
23.13	Printed circuit board drilling	12.60	11.63	15.80	11.74	11.50
23.14	Shearing, punching, forming	11.70	12.28	12.13	10.46	11.84
23.15	Resistance micro-spot welding	11.60	10.60	11.15	10.92	12.34

MACHINE, PROCESS, OR BENCH	PORTLAND & WASHINGTON	ST. LOUIS & ILLINOIS	SAN FRANCISCO & OAKLAND	TULSA	WORCESTER
23.16 Turret welding	11.60	10.60	11.15	10.92	12.34
23.17 Wafer dicing	10.73	12.21	11.73	8.73	10.19
23.18 Coil winding	10.73	12.21	11.73	8.73	10.19
24.1 Packaging bench and machines	10.40	10.28	10.75	7.95	9.81
24.2 Corrugated-cardboard packaging	10.40	10.28	10.75	7.95	9.81
24.3 Case packing conveyor	10.40	10.28	10.75	7.95	9.81
24.4 Foam-injected box packaging	10.40	10.28	10.75	7.95	9.81
25.1 Tools, dies, jigs, and fixtures	14.03	15.30	17.39	14.03	14.89

Material Costs

Description	Quantity	1987–1988 Cost
FERROUS PRODUCTS **Bar** Bar, hot-rolled, carbon steel; 1½-in. round OD, 16- or 20-ft lengths; AISI specification 1020, special quality. FOB service center for less than base quantity; FOB mill for base quantity.	*Weight* 120.3 lb 2,000 lb 6,000 lb Base quantity: 10,000 lb	*Costs: $ per 100 lb* 67.42 23.83 21.92 20.75
Bar, hot-rolled, carbon steel; ¾-in. square; AISI specification 1020, 20-ft lengths. FOB service center for quantities listed.	*Weight* 38.2 lb 2,000 lb 6,000 lb Base quantity: 10,000 lb	*Cost: $ per 100 lb* 88.28 26.18 23.31 21.40
Bar, hot-rolled, carbon steel; AISI specification M1020, merchant quality, 1-in. round. 20-ft random lengths. Standard packaging FOB service center for less than base quantity; FOB mill for base quantity.	*Weight* 53.4 lb 500 lb 2,000 lb 4,000 lb 6,000 lb Base quantity: 10,000 lb	*Cost: $ per 100 lb* 72.02 38.40 24.80 24.21 31.33 20.75
Bar, hot-rolled carbon steel, flat 1½ × 5 in. × 16- or 20-ft lengths; ASTM specification A-36. FOB warehouse for less than base quantity; FOB mill for base quantity.	*Weight* 504 lb 2,000 lb 6,000 lb Base quantity: 10,000 lb	*Cost: $ per 1 lb* .61 .43 .41 .39
Bar, hot-rolled steel, 2 × 2 in. × 16- or 20-ft lengths; AISI specification M1020, merchant quality. FOB warehouse for less than base quantity; FOB mill for base quantity.	*Weight* 272 lb 2,000 lb 6,000 lb Base quantity: 10,000 lb	*Cost: $ per 100 lb* 82.35 41.94 39.16 37.20
Bar, cold-finished, carbon steel, ⅝-in. round OD × 10- or 12-ft. lengths; AISI specification C1215, standard quality. FOB service center for less than base quantity; FOB mill for base quantity.	*Weight* 12.5 lb Base quantity: 10,000 lb	*Cost: $ per 100 lb* 129.96 51.10
Bar, cold-finished carbon steel; 1-in. OD × 10- or 12-ft lengths, specification 12L14. FOB warehouse for less than base quantity; FOB mill for base quantity.	*Weight* 32.04 lb 2,000 lb 6,000 lb Base quantity: 10,000 lb	*Cost: $ per 100 lb* 90.12 42.63 46.77 40.05

Description	Quantity	1987–1988 Cost
Bar, cold-finished. 2-in. round × 12-ft long; AISI specification 1215. FOB service center.	*Weight* 128.2 lb 2,000 lb 6,000 lb Base quantity: 10,000 lb	*Cost: $ per 1 lb* 1.03 .40 .38 .38
Bar, cold-finished, carbon steel; ⅝-in. round OD × 10- or 12-ft lengths; AISI specification C1117, standard quality. FOB service center for less than base quantity; FOB mill for base quantity.	*Weight* 12.5 lb 2,000 lb Base quantity: 6,000 lb	*Cost: $ per 1 lb* 1.35 .42 .40
Bar, hot-rolled, alloy steel; 1½-in. round OD × 20-ft long; AISI specification 4140 oil hardening, annealed, machine straightened. FOB service center for order less than base quantity; FOB mill for base quantity.	*Weight* 120 lb 2,000 lb 6,000 lb Base quantity: 10,000 lb	*Cost: $ per 1 lb* 2.10 .45 .40 .40
Bar, cold-finished, carbon; 1½-in OD × 20-ft length; AISI specification 1045, turned, ground and polished. FOB service center.	*Weight* 120.2 lb 2,000 lb 6,000 lb Base quantity: 10,000 lb	*Cost: $ per 1 lb* 1.46 .50 .56 .52
Bar, cold-finished, alloy steel; ¾-in. round OD × 10- or 12-ft random lengths; AISI specification 8620, annealed. FOB service center for less than base quantity; FOB mill for base quantity.	*Weight* 18.04 lb 2,000 lb 6,000 lb Base quantity: 10,000 lb	*Cost: $ per 1 lb* 2.66 .51 .46 .46
Bar, hot-rolled, annealed, alloy steel; 3-in. round OD × 20-ft random lengths; ASTM specification 4340, commercial quality. FOB warehouse for small orders; FOB mill for base quantity.	*Weight* 481 lb 1,000 lb 5,000 lb Base quantity: 10,000 lb	*Cost: $ per 1 lb* 1.67 .80 .57 .52
Bar, cold-drawn, annealed, alloy steel; 4-in round OD × 12-ft stock lengths (approx.); AISI specification 41L40, commercial quality. FOB warehouse for less than mill quantity; FOB mill for base quantity.	*Weight* 513.2 lb 2,000 lb Base quantity: 6,000 lb	*Cost: $ per 1 lb* 1.04 .63 .58
Bar, alloy, aircraft quality, annealed cold-finished, 3-in. OD × 10- or 12-ft lengths; E 4340, FOB service center for less than base quantity; FOB mill for base quantity.	*Weight* 240 lb 2,000 lb 6,000 lb Base quantity: 10,000 lb	*Cost: $ per 1 lb* 1.66 1.23 1.19 1.16

Description	Quantity	1987–1988 Cost
	Weight	*Cost: $ per 1 lb*
Bar, hot-rolled, stainless steel; type 304, 2-in. round OD × 10- or 12-ft lengths; forging quality, unannealed, base packaging. FOB service center for less than base quantity; FOB mill for base quantity.	128 lb 2,000 lb 6,000 lb Base quantity: 10,000 lb	3.21 1.70 1.66 1.62
	Weight	*Cost: $ per 1 lb*
Bar, cold-finished, annealed centerless ground, stainless type 303, 1-in. round OD × 10- to 22-ft mill lengths, boxed. FOB service center for less than base quantity; FOB mill for base quantity.	58.96 lb 2,000 lb 6,000 lb Base quantity: 10,000 lb	3.78 1.79 1.77 1.73
	Weight	*Cost: $ per 1 lb*
Bar, tool steel, cold-finished; ground 1-in. round OD, 100/144-in. mill lengths, annealed; AISI specification 0 1 grade (C .9%, Cr .5%, W .50%). FOB shipping point.	32 lb 250 lb 1,000 lb Base quantity: 1,000 lb	4.82 4.56 4.35 4.14
	Weight	*Cost: $ per 1 lb*
Bar, tool steel, alloy, oil-hardening die steel; 2-in. round OD, 100/144-in. mill lengths, annealed; AISI 0 1 grade (C .90%, Mn 1.00%, Cr .50%, W .50%). FOB service center.	288.7 lb 500 lb Base quantity: 1,000 lb	4.82 4.56 4.14
	Weight	*Cost: $ per 1 lb*
Bar, tool steel, cold finished; ground 1 × 2-in. flats, 100/144-in. mill lengths; AISI specification A2 grade (C 1.00%, Cr 5.00%, MO 1.00%). FOB shipping point.	81.7 lb 250 lb Base quantity: 1,000 lb	9.30 8.84 8.39
	Weight	*Cost: $ per 1 lb*
Bar, tool steel, cold finished; 1-in. round OD, 100/144-in. mill lengths, annealed. AISI specification M2 grade (C .85%, Cr 4.00%, W 6.00%, Mo 5.00%, V 2.00%). FOB shipping point.	34 lb 250 lb Base quantity: 1,000 lb	5.92 5.50 5.27
Rod	*Weight*	*Cost: $ per 1 lb*
Rod, 7/32-in. diameter, 12- to 20-ft lengths, cold-rolled carbon steel; C1018 industrial or standard quality. FOB service center; FOB mill for base quantity.	2.6 lb 500 lb 1,000 lb 2,000 lb 4,000 lb Base quantity: 6,000 lb	1.47 .75 .61 .56 .54 .52
Sheet	*Weight*	*Cost: $ per 1 lb*
Sheet, hot-rolled, low carbon steel, commercial quality; cut length, .1196-in. minimum and theoretical minimum weight, 48 × 120 in., cut edge, not pickled, base chemistry. FOB service center for less than base quantity; FOB mill for base quantity.	200 lb 2,000 lb Base quantity: 5,0000 lb	1.06 .48 .41

Description	Quantity	1987–1988 Cost
	Weight	*Cost: $ per 1 lb*
Bands (sheets), hot-rolled, carbon steel; 14 ga or heavier, 24- to 72-in. width, base chemistry, base quantity. Standard tolerances, not edge trimmed, end chopped, temper rolled, or further processed in any manner. FOB service center.	500 lb 2,000 lb 6,000 lb Base quantity: 10,000 lb	.49 .35 .31 .31
	Weight	*Cost: $ per 1 lb*
Sheet, cold-rolled, carbon steel; .0344-in. minimum, theoretical minimum weight, 42-in. wide coil (200 lb per in. of width or over), base chemistry, commercial quality, controlled surface texture, surface condition, flatness, limitations and tempers, bare (unwrapped) wire or banded without skids or platforms. FOB service center or mill.	Base quantity: 10,000 lb	.37
	Weight	*Cost: $ per 1 lb*
Sheet, cold-rolled, carbon steel, commercial quality, No. 10 (.1345 in.) 36 × 96 in. FOB service center for less than base quantity; FOB mill for base quantity.	135 lb 2,000 lb 6,000 lb Base quantity: 10,000 lb	.98 .50 .43 .43
	Weight	*Cost: $ per 1 lb*
Sheet, hot-rolled carbon steel; 7 ga × 36 × 96 in., commercial quality to maximum carbon of 0.15%. FOB warehouse for small quantity orders; FOB mill for base quantity.	500 lb 2,000 lb 6,000 lb Base quantity: 10,000 lb	.66 .48 .41 .41
	Weight	*Cost: $ per 1 lb*
Sheet, hot-rolled, low carbon steel; 48 × 120 in., (.0747-in.) 14 ga, pickled and oiled, sheet or coil. FOB service center for less than base quantity. FOB mill for base quantity.	196.9 lb 2,000 lb 6,000 lb Base quantity: 10,000 lb	1.46 .47 .47 .37
	Weight	*Cost: $ per 100 lb*
Sheet, high strength; 72 × 240 in. ASTM specification A-607 type 1 grade 50, (3/16 in.). FOB service center for less than base quantity; FOB mill for base quantity.	300 lb 2,000 lb Base quantity: 6,000 lb	43.76 41.76 40.76
	Weight	*Cost: $ per 100 lb*
Sheet, galvanized, flat, carbon steel; .0262-in–.0329-in. minimum, theoretical minimum weight, 36 × 96 in., commercial coating (standard 1¼ oz), commercial quality, bare (unwrapped), wire or banded. FOB service center for less than base quantity; FOB mill for base quantity.	2,000 lb 6,000 lb Base quantity: 20,000 lb	36.45 33.45 31.45

Description	Quantity	1987–1988 Cost
	Weight	*Cost: $ per 1 lb*
Sheet, cold-rolled, stainless steel; type 304, 2B finish, 24 ga × 48 × 120 in. FOB service center for less than base quantity; FOB mill for base quantity. Provide protection on skid platforms for larger orders.	24.7 lb Base quantity: 20,000 lb	2.96 1.21
Plate	*Weight*	*Cost: $ per 100 lb*
Plate, carbon steel; 84 × ½ × 240 in; ASTM specification A-36. FOB service center for less than base quantity; FOB mill for base quantity.	2,858 lb Base quantity: 6,000 lb	21.65 20.65
	Weight	*Cost: $ per 1 lb*
Plate, carbon steel, hot-rolled; .40 to .50%, ½ × 60 in. × 20 ft. FOB service center for less than base quantity; FOB mill for base quantity.	4,084 lb 6,000 lb Base quantity: 10,000 lb	.51 .45 .41
	Weight	*Cost: $ per 1 lb*
Plate, steel, high-strength, low-alloy; ASTM specification A-572, grade 50, ½ × 96 × 240 in. FOB service center for less than base quantity; FOB mill for base quantity.	2,858 lb Base quantity: 5,000 lb	.39 .38
	Weight	*Cost: $ per 1 lb*
Plate, carbon steel; ASTM specification A-285, pressure vessel quality, grade C, 3/16 × 84 × 240-in. length. FOB service center for less than base quantity; FOB mill for base quantity.	1,072.4 lb 2,000 lb 6,000 lb Base quantity: 10,000 lb	.36 .35 .33 .31
	Weight	*Cost: $ per 1 lb*
Plate, hot-rolled, stainless steel; type 304, ¼ × 72 × 240 in., annealed and pickled. FOB service center for order less base quantity; FOB mill for base quantity.	1,339 lb 6,700 lb Base quantity: 10,000 lb	1.39 1.36 1.31
Structural shapes	*Weight*	*Cost: $ per 1 lb*
Structural steel shape, carbon steel; angle 3 × 3 × ¼ in., 4.90 lb per ft, 40-ft lengths; ASTM specification A-36. FOB mill for base quantity.	196 lb 2,000 lb 6,000 lb Base quantity: 10,000 lb	.30 .23 .21 .19
	Weight	*Cost: $ per 1 lb*
Structural steel shape, carbon steel; C-channel, C 3 × 4.1, 4.1 lb per ft, 40-ft lengths; ASTM specification A-36. FOB service center for less than base quantity; FOB mill for base quantity.	164 lb 2,000 lb 6,000 lb Base quantity: 10,000 lb	.30 .23 .21 .19

Description	Quantity	1987–1988 Cost
Structural shapes, carbon steel; 6 × 4 × ½-in., angles, 350/400-in. long; ASTM specification A-7. FOB service center for less than base quantity; FOB mill for base quantity.	*Weight* 500 lb 2,000 lb 4,000 lb 6,000 lb Base quantity: 10,000 lb	*Cost: $ per 100 lb* .31 .25 .24 .23 .21
Structural steel shape, carbon steel, 8-in. wide flange, 24 lb per ft wide flange section × 20 ft; ASTM specification A-36. FOB service center for less than base quantity; FOB mill for base quantity.	*Weight* 480 lb 2,000 lb Base quantity: 10,000 lb	*Cost: $ per 100 lb* .31 .30 .26
Pipe, tube Tubing, mechanical, carbon steel, electric weld; 1½-in. OD × 14 ga, 1.256 lb per ft, random mill lengths. FOB service center for less than mill quantity; FOB mill for base quantity.	*Weight* 2,000 lb 6,000 lb Base quantity: 10,000 lb	*Cost: $ per 100 lb* 1.04 .96 .92
NONFERROUS PRODUCTS Bar, aluminum; 1 × 2 in. × standard stock lengths of 12 ft; specification 6061-T6511. FOB destination.	*Weight* 28.2 lb 500 lb 2,000 lb Base quantity: 6,000 lb	*Cost: $ per 1 lb* 3.07 1.71 1.38 1.30
Bar, aluminum; 2-in OD × 12 ft (approx.); specification 2024 T4 rounds. FOB destination.	*Weight* 45.8 lb 500 lb 2,000 lb Base quantity 5,000	*Cost: $ per 1 lb* 3.97 2.61 2.28 2.20
Sheet, aluminum, heat treatable, mill finish; 0.090 × 48 × 144 in.; specification 6061-T6, bare. FOB destination.	*Weight* 61 lb 500 lb 2,000 lb Base quantity: 5,000 lb	*Cost: $ per 1 lb* 3.11 1.75 1.42 1.34
Sheet, flat, aluminum; .125 × 48 × 144 in., mill finish; specification 5052-H32. FOB destination.	*Weight* 84 lb 500 lb Base quantity: 2,000 lb	*Cost: $ per 1 lb* 2.78 1.16 1.11
Sheet, flat aluminum; .313 × 48 × 144 in.; specification 2024-T3. FOB destination.	*Weight* 219 lb 500 lb 2,000 lb Base quantity: 10,000 lb	*Cost: $ per 1 lb* 3.74 3.00 2.73 2.64

Description	Quantity	1987–1988 Cost
	Weight	*Cost: $ per 1 lb*
Sheet, aluminum, non-heat treatable; .125 × 48 × 144 in., mill finish; specification 3033-H14. FOB destination.	84.5 lb 500 lb Base quantity: 2,000 lb	2.83 1.11 1.06
	Weight	*Cost: $ per 1 lb*
Rod, aluminum 2011-T3 alloy; 1-in OD screw machine stock, 12-ft lengths (approx.), (5.5% copper, .5% bismuth). FOB destination.	116 lb 500 lb 2,000 lb Base quantity: 10,000 lb	2.83 1.59 1.31 1.22
	Weight	*Cost: $ per 1 lb*
Tubing, aluminum hard-drawn; 6063-T832 alloy; 1-in OD × .058-in wall thickness × 12-ft length. FOB destination.	Base quantity: 5,000 lb	3.34
	Weight	*Cost: $ per 1 lb*
Plate, heat treatable aluminum, 1 × 48 × 144 in., aircraft quality; specification 7075-T651, bare. FOB destination.	350 lb 500 lb 2,000 lb Base quantity: 5,000 lb	3.26 2.93 2.59 2.50
	Weight	*Cost: $ per 1 lb*
Plate, aluminum, heat treatable, mill finish; 0.500 × 48 × 144 in.; specification 6061-T651. FOB destination.	338.9 lb 500 lb 2,000 lb Base quantity 5,000 lb	2.21 1.86 1.54 1.45
	Weight	*Cost: $ per 1 lb*
Strip, cartridge brass; 8 in. (wide) × .0160 in. (thick) × coil (26 B & S ga); CDA alloy No. 260 (70% copper, 30% zinc). FOB mill with freight allowed or prepaid.	500 lb Base quantity: 2,000 lb	1.99 1.84
	Weight	*Cost: $ per l lb*
Rod, yellow brass, free cutting; ⅜- to ½-in OD, random lengths, CDA alloy No. 360 (62% copper, 35% zinc, 3% lead). FOB distributor warehouse for less than base quantity; FOB mill for base quantity.	500 lb Base quantity: 2,000 lb	.88 .76
	Weight	*Cost: $ per 1 lb*
Sheet, copper; 16-oz thickness ga, 24 × 96 in or copper strip, .058-in thick × 6-in wide × coil. FOB distributor's warehouse for orders less than base quantity; FOB mill for base quantity.	500 lb Base quantity: 2,000 lb	1.50 1.24
	Weight	*Cost: $ per 1 lb*
Monel metal, No. 400 alloy, cold-rolled sheet; .078 × 48 × 120 in.; FOB warehouse for less than base quantity; FOB mill for base quantity.	500 lb 2,000 lb Base quantity: 5,000 lb	6.90 5.69 5.51

Operation Estimating Data

IV

SAWING and CUTTING

1.1 Power Bandsaw Cutoff, Contour Band Sawing and Hacksaw Machines

DESCRIPTION

An important first operation in any shop is the sawing of materials and bar stock for subsequent machining operations. Although other machine tools can do cutting-off operations, special machines are necessary for production and miscellaneous work. This section discusses straight and contour sawing in metal.

A power bandsaw cutoff machine has a long continuous band with many cutting edges which result in a low tooth load, and it is intended for repetitive cutoff work. The work can be solid, tubing, nested material, angles, structurals, and special geometry. Machines vary but head and stockfeed can be manually or hydraulically servo-controlled. Flatness of ± 0.002 in. s/in. of cut and stock repeatability of ± 0.005 in. are claimed. Sawguides and coolant systems apply cutting fluids directly to the point of cut. Band tension is either adjustable or preset. Bandsaw blades with bimetal construction and proper selection of band velocities affect performance. A power cutoff bandsaw is shown in Figure 1.1A.

Machines similar to Figure 1.1A handle up to 12 × 16-in. rectangular or 12-in. round stock. Indexing up to 24 in. under servo-control aids repeatability of length. Band drive of 85 to 450 fpm (feet per minute) is possible and feed control maintains a constant cutting rate regardless of workpiece cross-section. Vises are quick positioning with hydraulic clamping force.

The vertical band machine (contour machine), although sometimes used for cutoff, has greater application in shaping by sawing off unwanted material, both externally and internally. The band is vertical; the work is carried on a horizontal table and is fed into the band. Machines are available in a wide range of sizes, and throat depth varies from 16 to 60 in. Irregular contours such as slots, miter, and notch are possible. Machining time is usually slower because the machine may be cutting to a layout line. A vertical universal bandsaw is shown in Figure 1.1B. These machines allow the upright column to move parallel with the long dimension of the table or bed. Additionally,

FIGURE 1.1A Horizontal sawing operation using power cutoff bandsaw. *(DoAll Company)*

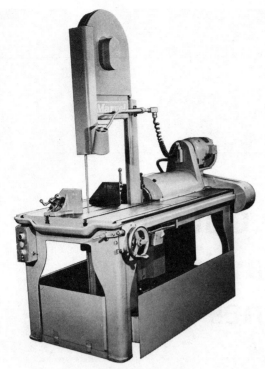

FIGURE 1.1B Universal bandsaw with tilting and traveling feature of upright column. *(Armstrong-Blum Manufacturing Company)*

the upright column can tilt to a maximum of 45° for miter cutting without turning or moving the work.

The reciprocating power hacksaw, which may vary in design from a light-duty, crank-driven saw to a large heavy-duty machine hydraulically driven, has long been a favorite because of its simplicity. Machines are manual, semi-, or fully automatic. Methods of saw feeding vary from gravity to positive or uniform pressure feeds. Quick return strokes are common. Cutting speeds are nominally fixed and depend on the material being cut, usually ranging from 80 to 160 spm (strokes per minute).

Band blades include hardback carbon steel, intermediate-alloy steel, and high-speed steel with welding of a high-speed cutting edge possible. The backing material provides flex life and weldability. Blade width can vary from 1/16 to 2 in., either raker or wave set.

ESTIMATING DATA DISCUSSION

The bandsaw cutoff, contour band, and hacksaw machines are usually ready to use and little time is required for setup. Adjustments may be required for stock stops, band speed, nesting fixtures, vise, feed adjustment, tilt or miter, and dolly discharge. Raw stock, quantity required, its size, convenience, and geometry are the biggest factors. Cranes, power or chain hoists, jib or monorail, fork truck, etc., may bring the material nearby. Material handling cranes can affect the setup estimate.

A major factor is the number of machines an operator will handle. In Element 1 of Table 1.1, load part has provisions for manual, hoist, or chain crane. Cranes are required for the heavier loads. Handling could be done during prior sawing, and thus would not be allowed. If positioning is hydraulically controlled, time is required. Automatic provisions for discharge, such as tipoff, or manual gathering of parts from a chute depends on the machine.

In Elements 10 and 11, Sawing, distinctions are made for three machines. Power bandsawing, similar to Figure 1.1A, uses the first table, and sawing time is given in min/sq in. Contour or universal bandsawing uses a slower rate and sometimes the saw follows a scribe line. The data are for a 1 in. × .035 in. × 144 in. blade. Optimum cutting rates are obtained by sawing solid materials because many teeth are uniformly loaded, but in pipe, tubing, and structural sections, about one-fifth of the total cross-sectional area is metal. A limited number of teeth (minimum of 3) are engaged while sawing sections, the loading per tooth increases and a reduction in cutting rate must be made. The adjustment factor increases the cutting time, and the data, given for 5-in. diameter solids, is necessary. The contour bandsawing time is for straight cuts using a 1 × .035 in. blade with contour sawing attachment. For contours, multiply time by adjustment factor according to blade width. Note the following table of adjustments for power bandsaw and contour operations:

Wall Thk	Factor Cutting Time
up to 3/16	2.5
3/16–3/8	2.0
3/8–5/8	1.7
5/8–1	1.4
over 1	1.4
Width	Factor Cutting Time
1	1.2
3/4	1.3
1/2	2.0
3/8	2.9
1/4	4.0

Blade life, Element 9, depends on the machine blade length and is given as area in sq in. of sawing. The time to change the blade is divided by the number of pieces. Weld time is separate from installing blade.

Power hacksawing data are given for stock size and can be used directly. The sawing data must be increased by 50% for medium machines and 100% for light gravity machines. If irregular shapes are cut, the times must be multiplied by the factors such as brass, 1.13; steel, 1.2; and die steel, 1.5.

EXAMPLES

A. Estimate the time to cutoff SAE 1035 8-in. solid OD bar stock 5.75-in. long. Raw stock is purchased in 10-ft lengths, and the traveler calls for 26 pc. Tolerances as customarily found with power bandsaw cutoff machines are adequate. A 1-in. band is used and a kerf of 0.035 in. requires 12.53 ft of length, or two 10-ft bars. One full time operator is assumed.

Table Number	Process Description	Table Time	Adjustment Factor	Cycle Minutes	Setup Hours
1.1-S	Setup	.17			.17
1.1-1B	Handle 10-ft. bar with double hitch sling. Two bars required	3.38	$2/26$	.26	
1-1-2B	Automatic stock feed	.06		.06	
1.1-3	Open and load vise	.01		.01	
1.1-4	Raise head	.04		.04	
1.1-6	Start and stop blade	.04		.04	
1.1-10	Sawing, 1 pc., 06 × 50.27 sq in. bandsaw cutoff	.06	50.27	3.02	
1.1-9	144-in. blade life, 26 × 50.27 sq in. = 1307 sq in. of work material, about ¼ of blade life	1.20	$1/26 \times 1/4$	.01	
1.1-7B	Unload part, 86.71b, tipoff	.15		.15	
	Total			3.59	.17
	Total estimate for lot	1.73hr			

B. A die block, ½ in.-thick, has an internal cut of straight perimeter of 18 in. and 4 corner radii of 1 in. each. A hole is predrilled to allow passage of the blade. A contour bandsaw is used, and quantity of one is required. The bandsaw width is ⅜ in.

Table Number	Process Description	Table Time	Adjustment Factor	Cycle Minutes	Setup Hours
1.1-S	Setup	.17			.17
1.1-1A	Handle, 6in. × 6 in. block, 5.2lb	.08		.08	
1.1-9	Insert and weld blade, 2 assumed	6.36		6.36	
1.1-10	Saw 18 in. straight	.38	$18 \times 1/2 \times 4$	13.68	
1.1-10	Saw four ½-in. radii, or a total circumference distance of 3.14	.38	$3.14 \times 1/2 \times 4$	2.39	
1.1-6	Start and stops	.04	2	.08	
1.1-7A	Unload	.08		.08	
1.1-8	Air-clean table, wipe excess oil	.62		.62	
	Total			23.29	.17
	Total estimate	.56 hr			

C. Four 3 × 3 × ¼-in. cold rolled steel angles are to be nested and sawed simultaneously by a large power hacksaw. The shop order requires 1600 parts. Stock size is 8 ft and the length of each part is 12 in. Each angle has a cross-sectional area of 1.44 sq in. and weighs 4.9 lb/ft.

Table Number	Process Description	Table Time	Adjustment Factor	Cycle Minutes	Setup Hours
1.1-S	Setup	.17			.17
1.1-1A	Each 8-ft angle weighs 39.2 lb, and manual nesting time is .40 for the first 35 lb. and .010 for each add'l lb, or .44 min divided by 8 or	.44	$1/8$	.06	
1.1-2	Position against stop	.04		.04	
1.1-3	Open and close vise	.14		.03	
1.1-4	Raise and lower blade	.04	$1/2$	.02	
1.1-6	Start and stop blade	.01		.01	
1.1-11	Saw using an irregular shape factor	1.54	$1.44 \times .54$	.89	
1.1-9	Blade life for 16-in. hacksaw 1200/1.44 = 833 parts before blade change, assume 2 blades	1.20	$1/800$	.01	
1.1-8	Air-clean part	.13		.13	
1.1-7A	Unload part, gather	.07		.07	
	Total			1.26	.17
	Total estimate for lot	33.77 hr			

TABLE 1.1 POWER BANDSAW CUTOFF, CONTOUR BAND SAWING AND HACKSAW MACHINES

Setup   .17 hr

Operation elements in estimating minutes

1A. Load part or stock on table or rack

Weight	3	8	12	25	35	Add'l
Manual-one	.04	.08	.12	.16	.43	.011
Manual-nest	.04	.07	.10	.13	.40	.010

B.

Ft	5	25
Sling, Hoist, Move	3.38	3.75

C.

Ft	5	15
Jib Crane, Chain	.64	.70

2A. Position against stop

Weight	3	8	12	25	35	Add'l
Manual-one	.05	.06	.07	.11	.13	.003
Manual-nest	.03	.04	.05	.09	.11	.003

B.

Part Length, in.	3	6	12	24
Automatic Stock Feed	.03	.06	.12	.23

3. Open and close vise
   - Vise handle 4 turns   .14
   - Quick release   .06
   - Automatic   .01

4. Raise or lower blade or head
   - Set to work   .04

5. Position table or miter for cut   .06

6. Start and stop blade for sawing   .04

7A. Unload part

Weight	3	8	12	25	35	Add'l
Manual-stock	.04	.08	.12	.16	.43	.011
Manual Gather	.03	.07				

B. Dolly   .21
     Tipoff   .15

Ft	10	15
Jib Crane	1.00	1.08

8. Air-clean part, vise, or table .13
    Cloth-clean table or vise .17
    Wipe part with hand .05
    Wipe excess oil .49

9. Saw blade installation and life

Blade Length	16	60	144	180
Life, Area, Sq In.	1200	2250	5400	6750
Time to Change Blade			1.20	
Weld Blade Together			1.98	

10. Power bandsaw cutoff

Material	In. Thk	Cutting Rate Min/Sq In.
1008–1013	less than 1	.10
150–175	1–3	.07
Bhn	3–6	.06
	6+	.07
1015–1035	less than 1	.09
160–175	1–3	.07
Bhn	3–6	.06
	6+	.06
1040–1092	less than 1	.16
160–205	1–3	.14
Bhn	3–6	.10
	6+	.11
1115–1132	less than 1	.09
140–165	1–3	.07
Bhn	3–6	.06
	6+	.06
3115–3130	less than ½	.23
180–220	½–1	.23
Bhn	1–3	.16
	3–6	.16
	6+	.19
Die steels	less than ½	.38
217–241	½–1	.38
Bhn	1–3	.33
HSS tool	3–6	.33
Steels	6+	.46
Ni Mo Steel	less than ½	.20
4608–4621	½–1	.23
190–210	1–3	.21
Bhn	3–6	.21
	6+	.26

11. Power hacksawing, min.

Stock Size	Brass		1015–1015 1115–1132		1320 3130		4608 Die steels	
	Rd.	Sq.	Rd.	Sq.	Rd.	Sq.	Sq.	Rd.
½	.20	.20	.24	.24	.55	.55	.60	.60
1	.36	.40	.40	.48	.82	1.07	1.01	1.30
2	.84	1.16	1.19	1.66	1.67	2.31	4.21	5.38
3	2.30	3.01	3.35	4.41	4.61	6.06	9.56	12.18
4	4.33	5.60	6.37	8.26	8.74	11.31	17.04	21.70
5	6.95	8.93	10.26	13.21	14.04	18.06	26.65	33.94
6	10.14	13.00	15.01	19.26	20.51	26.31	38.39	48.90
per sq in.	.36		.54		.73		1.36	

1.2 Abrasive Saw Cutoff Machines

DESCRIPTION

The abrasive saw cutoff machine can be hand-operated or power-driven for downstroke cutting. Figure 1.2, a picture of a manual machine, allows for either wet or dry cutting. The machine is designed for cutting barstock or conventional or nonconventional geometry to length. The cutting tool is an abrasive wheel and wheel speeds and horsepower are important to productivity. Wheel speeds up to 16,000 fpm are possible with certain wheel materials. Abrasive wheels can be reinforced, regular or have special bonds. Options exist for manual or air-operated vises.

Machines, both smaller and larger than Figure 1.2, can be purchased with equivalent performance. The wheels may use a bond of shellac, resinoid, or plastic. Shellac-bonded wheels are soft for cutting hard materials, the resinoid-bonded wheels are for dry cutting and high-speed production, and the rubber-bonded wheels are best for wet cutting. Materials cut include metals, plastics, etc.

ESTIMATING DATA DISCUSSION

The abrasive saw is usually ready to use and little time is required for setup. Adjustments are required for the stock stops of various lengths. Changing a wheel takes 3 min, and this is usually prorated into the run time by the number of parts. For short runs, the change wheel can be a part of the setup for convenience in estimating.

Stock racks are usually next to the machine or the material is trucked nearby. Loading of bar, plate, angles, tube stock, or special materials is similar to that of the hacksaw. The average floor-to-floor time for handling is given by Element 1 and is based upon weight in lb.

Element 7, if it is a special length or geometry, must be divided by the number of parts. The length is divided by the drawing dimension plus stock for later facing, if required for tolerances greater than commercial tolerances, and the kerf thickness, .020 to ⅜ in.

FIGURE 1.2 Manually operated abrasive cutoff machine with 20-in. cutoff wheel. (A.P. de Sanno & Sons, Inc.)

The abrasive sawing time is influenced by part geometry, physical properties of the material, finish of part (burr-free, heat checking, tolerance), and wheel condition. The time to cut is usually incidental to overall floor-to-floor time and depends on material, machine horsepower, and wet or dry cut. Speeds and feeds, as usually understood in other machining operations, are not as significant with abrasive sawing. The downfeed is considered important. Cost of additional feed time required due to wheel wear is usually negligible. A grinding ratio of volume of metal cut to volume of wheel consumed varies with downfeed.

Abrasive sawing, as given in Table 1.2, is a composite of several machines and materials. For gang sawing of two or more parts, find total area and divide by number of parts in vise. Gang loading can increase handling time above values listed.

EXAMPLES

A. Find the estimated time for 125 pc of 1½ in. OD CRS bar in 6-in. lengths. The stock is 12-ft long and the abrasive disk is ³⁄₃₂ in. thk. Commercial tolerances.

Table Number	Process Description	Table Time	Adjustment Factor	Cycle Minutes	Setup Hours
1.2-S	Setup	.17			.17
1.2-1A	Load and unload piece, (76-lb bar, 3.2-lb pc)	.14		.14	
1.2-7A	Cutoff part	.16		.16	
1.2-6	Wheel wear (20 min of wheel life from .16 × 125 and 20/30 × 3 min divided by 125)	.02		.02	
	No checking allowances	00		0	
	Total			.32	.17
	Total estimated time for lot	.84 hr			

B. Find the estimates for 40 pc of 2½ in. × 7-in. long square CRS material. (Density equals .28) Area = 6.25 in.²; bar weighs 210 lb; bar is 10 ft long.

Table Number	Process Description	Table Time	Adjustment Factor	Cycle Minutes	Setup Hours
1.2-S	Setup total	.17			.17
1.2-1B	Load, unload bar	1.50	¹⁄₁₆	.094	
1.2-2	Vise	.20		.20	
1.2-3	Start, stop blade	.04		.04	
1.2-4	Coolant lines	.04		.04	
12.-7A	Saw, 6.25 in.²	.41		.41	
	Total			.79	.17
	Lot hours	.69 hr			

TABLE 1.2 ABRASIVE SAW CUTOFF MACHINES

Setup .17 hr

Operation elements in estimating minutes

1A. Load stock on side table, position stock against stop, advance and retract disk, part aside (use part weight)

Weight	.1	2.0	4.4	8.1	13.1	18.0	24.2	31.6	40.2
Min	.10	.12	.14	.17	.21	.25	.20	.36	.43

Weight	51.4	63.7	78.5	97.0	119.3
Min	.52	.62	.74	.89	1.07

B. Load bar, angle, pipe on side table
 Light, up to 25 lb   .25
 Medium, 26–100 lb   .50
 Heavy, over 100 lb   1.50

2. Vise, tighten and loosen on stock, position stock against stop, advance grinder disk to work, retract disk, and part aside   .20

3. Start and stop blade   .04

4. Position and clear coolant line   .04

5. Air clean
 Part   .06
 Part and vise   .08

6. Change wheel   3.00
 Life of abrasive disks
  hard steel   30.00
  soft steel   15.00

7A. CRS material abrasive sawing

Saw Dimension	½	1	1½	2	3
Round	.04	.09	.16	.24	.41
Square	.04	.11	.19	.28	.49
Hexagon	.04	.10	.17	.26	.44

B. Hard steel and tungsten material

Saw dimension	½	1	1½	2	3
Round	.16	.26	.36	.45	.61
Square	.17	.29	.39	.49	.67
Hexagon	.16	.27	.37	.46	.63

C. Rectangular steel bar (CRS Material)

W	Thk	Min	W	Thk	Min
1	¼	.04	2	¼	.07
	½	.07		½	.11
	¾	.09		¾	.15
1½	¼	.06		1	.18
	½	.09		1½	.23
	¾	.12	3	½	.15
	1	.15		1	.23
				1½	.31
				2	.37

D. Steel pipe and structural angles

	1	2	3	4	5	6
Nominal Pipe Size	.22	.30	.39	.45	.50	.56
Equal Side Angle by Average Thk	.21	.28	.42	.48		

1.3 Oxygen Cutting Processes

DESCRIPTION

Oxygen cutting is the burning of iron or steel where the temperature of the metal is raised to the point where it oxidizes in an atmosphere of oxygen. It is always necessary to add some fuel to sustain the reaction. This process is sometimes called flame cutting, a term that is misleading since the flame does not do the cutting. Fuel gases of several varieties may be used. Gouging, grooving, piercing, bevel cutting, and scarfing are processes that use the oxyfuel cutting principles.

Many machine varieties are available. The simplest is the small "tractor type," which has a small motor-driven carriage that follows a track with a speed control. The torch can be positioned in several ways. Shape-cutting machines range from manually controlled machines to a magnetized roller that follows a templet. A photocell tracer can be used to follow a line or edge of a silhouette. X–Y servo motors can be adopted and patterns can be recorded on punched tape for numerical control.

Figure 1.3 is a heavy-duty shape-cutting machine that can use oxyfuel gas or plasma arc. Automatic shape-cutting of rectangles, circles, etc., is possible using optical tracing or NC.

ESTIMATING DATA DISCUSSION

Three values are given for setup as these figures depend upon whether torch cutting is manual, or whether there is machine cutting, with or without a templet. The manufacture of the templet is estimated by using Table 11.1, as given for drill-press and sheet metal work, when templet production is classified as direct labor.

The run-time data cover general oxygen fuel-cutting operations performed with stationary, mobile, or manual flame-cutting machines. The required time to complete a flame-cutting operation varies with material type, condition and thickness, length of cut, tip and fuel, and operating pressures.

In the pickup element, the material is from the table for less than 12 lb, and from the floor if more than 12 lb. Large bars, pieces, or sheets are located on a skid and moved with an electric hoist using hooks. Torch ignition and shutoff may be for each torch, or a flame-tube may allow for one ignition/shutoff. The position tracking or tracer roller are for templet or guided-roller cuts. Torch position adjustment, if not automatically controlled, may be combined with a preheat of material. Also, material preheat has been separated.

FIGURE 1.3 Heavy-duty shape cutting oxyfuel machine. (*Union Carbide Corporation*)

Cutting can be straight-line beveling, circle cutting, hand-guided shape cutting, tracer-templet, or optical-templet controlled. Though the time is affected by several variables, a consolidated time, or min/in., is related to the material thickness for machine or manual cutting. Hole-piercing standards are one per occasion. The place part aside depends upon size of the part, size of equipment, and general shop practice. If cutting is usually poor, it may be necessary to chip or knock scrap from the kerf or edge of the finished part or the raw material. Tong, glove, mark piece, and measure elements may be optionally done during the automatic cut time.

If more than one part is cut simultaneously, the elements must be divided by the number of active torches. Stack cutting is also possible.

EXAMPLES

A. Using a shape-cutting machine where tracing is done by a magnetized rotor which rolls along the edge of a steel templet, estimate the following job. The part is an 11 × 15½-in. rectangle with an irregular 10¼-in. radius off the piece, a 15° angle edge cut, and a 2¹⁷⁄₃₂-in. partial circle. Total cut perimeter is 52⅜ in. Material is ⅝-in. thick and AISI 4130, and is supplied in 48 × 144 in.-plates. A total of 15 parts is required. Two warmups per part are required. Each oxyfuel cut supplies one part and a templet is available.

Table Number	Process Description	Table Time	Adjustment Factor	Cycle Minutes	Setup Hours
1.3-S	Setup	.7			.7
1.3-1	Pickup sheet from stack, hoist, 1.04/15	.07		.07	
1.3-1	Roll to position near stops each time	.64		.64	
1.3-3	Position tracer	.15		.15	
1.3-2	Ignite and shut off torch	.05	3	.15	
1.3-7	Position material, 2 warmups	.70	2	1.40	
1.3-8A	Cut 52⅜ in. of ⅝-in. stock	.08	52.375	4.19	
1.3-4	Reposition stock for next piece	.15		.15	
1.3-5	Part aside, hoist	.86		.86	
1.3-5	Return reduce sheet to pile, hoist	.86	¹⁄₁₅	.06	
1.3-5	Clean center and preheat holes	.36	¹⁄₁₅	.02	
	Total			7.69	.7
	Hr/100	12.817			
	Lot estimate	2.62			

TABLE 1.3 OXYGEN CUTTING PROCESSES

Setup

Hand cutting	.3 hr
Machine cutting	.5 hr
Add'l for templet	.2 hr

Operation elements is estimating minutes

1. Pick up

Weight	12	12–60
Size	2 × 2 ft	3 × 6 ft
Time	.08	.18

Hoist	1.04
Roller	.64

2. Ignite and shutoff torch 05/ea

3. Position tracking roller or tracer, engage, disengage 15

4. Reposition stock on frame to cut next piece .15
   Add for C-clamp if used to hold stock .13

5. Clean center hole of tip with wire, prorate .19
   Clean preheat holes, prorate .17

6. Put on and remove glove
   One .08
   Two .16

7. Position material or torch, warmup material .70
   Warmup material only thk: ½ in., .14; 1 in., .15; 2 in., .18; 3 in., .21

8A. Cut or pierce
   Cut thickness, min/in.

Thk Machine	Thk Manual	Min	Thk Machine	Thk Manual	Min
.1		.06	3.3	2.4	.16
.3		.07	3.8	2.9	.17
.5	.2	.07	4.4	3.3	.19
.7	.4	.08	5.1	3.8	.21
1.0	.6	.09	5.8	4.4	.23
1.3	.9	.10	6.5	5.0	.25
1.6	1.1	.11	7.4		.28
2.0	1.4	.12	8.3		.30
2.4	1.7	.13	Add'l		.030
2.8	2.1	.14		Add'l	.037

B. Pierce hole, each

Thk	Min	Thk	Min
1/16	.19	½	.45
⅛	.23	⅝	.53
¼	.31	¾	.60
⅜	.38	1	.75

9. Mark piece with pencil, scribe, or chalk .05
   Measure material with scale .21
   Add'l dimension .05

10. Part aside
    Toss aside .09
    Place on stack .12
    Hoist .86
    If tong grip required, add .03

11. Knock any scrap loose from part with hammer .12
    Aside scrap
    Pierce to tote box or barrel .12/ea
    Through frame to floor .09/ea

1.4 Computer Aided Torch Cutting Processes

DESCRIPTION

Torch cutting is the burning of iron or steel at the temperature where oxidation occurs. The addition of fuel to sustain the reaction is necessary. Several fuel gases may be used in the process. Propane is the fuel used for these data. Cutting is over a water bath to reduce pollutants and thus improve the working conditions.

Many machines are available, ranging from small motor-driven carriages that follow a track, to electronic photosensors that follow the lines of a drawing. The machine type reported upon by these data is computer assisted automated shape-cutting of rectangles, circles, etc. Optical sensing is able to trace drawing lines. These estimating data are suitable for large bulky and heavy materials. Low quantity is normal.

ESTIMATING DATA DISCUSSION

Two values are given for setup as these figures depend on whether a drawing is necessary or the design can be programmed into the machine. The torch-cutting operation time varies with material type, thickness, burn length, tip and fuel, and operating pressures.

In the pickup element the plate must be moved with a hoist. If the hoist requires one attachment, one value is used. If the hoist requires two attachments because of weight or position, another value is available.

Preheating of a surface includes a test run over the surface to be cut and is dependent upon the surface thickness. A consolidated time, or minutes per inch is used. Similarly, cutting time is in min per in. The adjustment factor will be used to enter the length, and muliplication follows. The product is entered in the cycle minutes column.

Movement of the torch machine is necessary for removing scrap, stock, and part with overhead hoist from bath. A small part may be carried, but distance traveled remains constant, thus only one value is given.

Use of hoist to unload the part is subject to, but not restricted to, one attachment. If two attachments were used for loading, hook up for unloading becomes easier because of area. Thus, single values are given for scrap, part, and stock.

If more than one part is cut simultaneously, the necessary elemental times must be divided by the number of active torches which is entered in the adjustment factor column.

EXAMPLES

A. Estimate the cycle, setup, and lot time for a rectangle 11 × 15.5 in. having an irregular 10¼-in. radius. Total burn length is 52⅜ in. Material is ⅝-in. thick. A drawing is available, which is traced by a photocell. Two parts are cut simultaneously by two torches.

Table Number	Process Description	Table Time	Adjustment Factor	Cycle Minutes	Setup Hours
1.4-S	Setup with drawing	.5			.5
1.4-1	Pickup stock	4.36	½	2.18	
1.4-2	Position torch to drawing	18.02	½	9.01	
1.4-4	Cut part	.128	52.375/2	3.35	
1.4-6	Scrap cutoff	1.98	½	.99	
1.4-5	Move machine clearance	1.50	½	.79	
1.4-6	Dispose of scrap	1.31	½	.66	
1.4-7	Store part	1.74		1.74	
1.4-8	Return stock	3.78	½	1.89	
	Total			20.61	.5
	Lot estimate	1.19 hr			

B. Estimate the cycle, setup, and lot time for a 30-in. OD disk. Material is 6-in. thick. The circle is programmed into memory, and a drawing is unnecessary. Lot size is 3 units. A one-torch machine is to be used.

Table Number	Process Description	Table Time	Adjustment Factor	Cycle Minutes	Setup Hours
1.4-S	Setup	.35			.35
1.4-1	Pickup stock, 2 clamps	15.12	⅓	5.04	
1.4-2	Position torch	1.96		1.96	
1.4-3	Preheat	.137	94.25	12.89	
1.4-4	Torch part	.268	94.25	25.26	
1.4-6	Cut off scrap	1.98	⅓	.66	
1.4-5	Clear machine	1.58	⅓	.53	
1.4-6	Dispose of scrap	1.31	⅓	.44	
1.4-7	Store part	1.74		1.74	
1.4-8	Return stock	3.78	⅓	1.26	
	Total			49.78	.35
	Lot estimate	2.84 hr			

TABLE 1.4 COMPUTER AIDED TORCH CUTTING PROCESSES

Setup
    With drawing template   .5 hr
    Without drawing   .35 hr

Operation elements in estimating minutes

1. Pickup stock, hoist or crane

No. of Attachments	1	2
Min	4.36	15.12

2. Position torch
    Manual   1.96
    Drawing   18.02

3. Preheat cutting surface, min/in.

Thk	Min/In.	Thk	Min/In.	Thk	Min/In.	Thk	Min/In.
2.5	.125	4.25	.131	5.5	.135	7.25	.141
3.0	.127	4.5	.132	6.0	.137	7.5	.142
3.5	.128	5.0	.134	6.5	.139	7.75	.143
4.0	.130	5.25	.134	7.0	.140	8.0	.145

4. Torch, min/in.

Thk	Min/In.	Thk	Min/In.	Thk	Min/In.	Thk	Min/In.
⅝	.128	2⅛	.167	3⅝	.206	5⅛	.245
¾	.132	2¼	.171	3¾	.209	5¼	.248
⅞	.135	2⅜	.174	3⅞	.213	5⅜	.252
1.0	.138	2½	.177	4.0	.216	5½	.255
1⅛	.141	2⅝	.180	4⅛	.219	5⅝	.259
1¼	.145	2¾	.184	4¼	.222	5¾	.261
1⅜	.148	2⅞	.187	4⅜	.226	5⅞	.265
1½	.151	3.0	.190	4½	.229	6.0	.268
1⅝	.154	3⅛	.193	4⅝	.232	6⅛	.271
1¾	.158	3¼	.197	4¾	.235	6¼	.274
1⅞	.161	3⅜	.200	4⅞	.239	6⅜	.278
2.0	.164	3½	.203	5.0	.242	6½	.281

Thk	*Min/In.*	*Thk*	*Min/In.*	*Thk*	*Min/In.*
6⅝	.284	7⅛	.297	7⅝	.310
6¾	.287	7¼	.300	7¾	.313
6⅞	.290	7⅜	.304	7⅞	.316
7.0	.294	7½	.307	8.0	.320

5.	Clear machine out of way	1.58
6.	Scrap cutoff	1.98
7.	Store part on rack, hoist	1.74
8.	Return remaining stock to storage, hoist	3.78

MOLDING MACHINES

2.1 Plastics Molding Bench

DESCRIPTION

Plastic compounds differ greatly and lend themselves to a variety of processing methods. This section is concerned with bench processing where the operator is closely involved. These estimating data are in contrast to semiautomatic or fully automatic methods of plastic processing.

A variety of equipment ranging from hand-operated to automatic presses is available. The simplest type is the hand-operated press. Separate molds, which are loaded and unloaded outside the press, are used and the principal function of the press is to supply pressure. Some presses, also manually controlled, have the molds permanently mounted on the platens.

ESTIMATING DATA DISCUSSION

These data consider a variety of elements, and the properties of the plastic material affect the selection of the elements.

The setup time includes the minimum kinds of elements, since all tooling, materials, etc. are located within the bench area.

The operation may or may not include a designed mold with inserts. For instance, a case-encapsulation would use a container. Thus, the container has a bearing on selection. Mold preparation is considered by Elements 1, 3, 4, 5, and 9. The application of the material may vary depending on whether the material form is granular, powder, liquid, or preform. Inserts, both loose and tight tolerance, are considered. Deburring of flash, opening of holes, etc., or secondary processing operations are found in Table 21.6. Some elements involve waiting for air to escape, cooling, or press-holddown, and their inclusion in the estimate depends on whether other work is provided for this potentially available time.

EXAMPLES

A. A part, having a surface area of 63 sq in., is to be cast with a loose-fitting insert using a hand-operated molding press where platen temperature can range from 200–400°F. The mold is limited to a daylight opening of 18 × 18 × 10 in. Estimate the unit time if the operator waits for preheating and air escape.

Table Number	Process Description	Table Time	Adjustment Factor	Cycle Minutes	Setup Hours
2.1-S	Setup		.1		.1
2.1-1	Clean mold	2.47		2.47	

Table Number	Process Description	Table Time	Adjustment Factor	Cycle Minutes	Setup Hours
2.1-2	Obtain insert	.11		.11	
2.1-2	Place insert relative to mold	.06		.06	
2.1-3	Mold release	3.30		3.30	
2.1-4	Assemble mold	.95		.95	
2.1-4	Move mold to machine	.75		.75	
2.1-5	Preheat	1.16		1.16	
2.1-9	Free pour premixed materials	.44		.44	
2.1-9	Wait for escaping air, cooling	3.30		3.30	
2.1-10	Clean off excess	.17		.17	
2.1-14	Remove part from mold and aside	.18		.18	
	Total			12.89	.1

TABLE 2.1 PLASTICS MOLDING BENCH

Setup .10 hr

Operation elements in estimating minutes

1. Clean mold

Surface Area	0–25	26–50	51–75	Add'l
Min	1.29	2.15	2.47	.02

2. Handle .11
    Turn over piece .03
    Place part or insert relative to 2nd part: loose tol .06
                                                             tight tol .09

3. Mold release and wipe clean

Surface Area	0–25	26–50	51–75	Add'l
Min	1.00	2.50	3.30	.05

4. Assemble mold, small .58
    Assemble mold, medium .95
    Move mold to machine, unload .75

5. Preheat mold, small .85
    Preheat mold, medium 1.16

6. Weigh components and mix
        Foaming, casting, potting, bonding 2.39
        Encapsulation (without mold) .28

7. Sand surface, per sq in. .07

8. Assemble, load, disassemble syringe or gun .55

9. Force plastic into container by syringe or gun
        Small surface area .07
        Large surface area .28
        Remove cover, pour, remove excess, close mold 1.82
        Free pour into container .44
        Apply adhesive by knife or brush, per sq in. .07
        Apply adhesive by dipping .09
        Apply adhesive to groove by syringe gun, per linear in. .09
        Rise of foam in mold, wait for escaping air

Surface Area	0–25	26–50	51–75	Add'l
Min	1.00	2.15	3.30	.05

10. Clean off excess from part — .17

11. Assemble and disassemble screw and nut — .28

12. Cure filled mold — 1.16
    Cure tray of parts — .28

13. Disassemble mold — 1.89

14. Remove part from mold and aside, small — .18
    Complicated or large — .44

2.2 Plastic Preform Molding Machines

DESCRIPTION

Plastic materials differ greatly and can be processed by a variety of methods. Each material is best adapted to one of the methods, although many can be fabricated by several. The material used in most processes is in powder, granular, or liquid form. For some materials there is a preliminary operation of preforming. This is an operation that involves conforming several differently mixed or mulled materials into small pellets of proper size and shape. The preforms have the same density, mass, and speedup as subsequent molding operations by rapid-mold loading, although on some machines the making of preforms is integral to the production of the finished parts, and therefore, these estimating data would not be used. Preforms are used for compression and transfer molding processes. Preforming is usually required to save time and labor in preparing charges of molding materials for multiple cavity molds.

ESTIMATING DATA DISCUSSION

The preform machines are usually setup in a go-mode and a limited amount of work is required for setup. The number of cavities is sometimes fixed with a machine.

The operation estimate is one number only, and it depends on the machine strokes and the number of cavities. If the preform machine is rotary, and there is only one die, then the number of cavities is one. The estimate is for one machine. If more than one machine is tended by the operator, then estimated minutes are divided by that number. A manual machine is assumed to operate at half the speed of an automatic machine. To find the production estimate for machine speeds not given use the following formula:

$$\text{min/preform} = \frac{1}{\text{spm} \times \text{no. of cavities in die}}$$

It is necessary to know the number of preforms to produce one part, including runners and gates. The gram mass of the part is divided by the gram mass of the preform.

EXAMPLES

A. Find the preforms estimate for a quantity of 80,000 finished parts produced by compression-type molds. One operator controls two automatic machines. The machines operate at 25 spm, and the cavity number and size are 3 and ¾ in. OD × 2 in. The gram mass for each preform is 12 and the part is 40 g. Thus, the number of parts is approximately 4 preforms per shot.

Table Number	Process Description	Table Time	Adjustment Factor	Cycle Minutes	Setup Hours
2.2-S	Setup	.2			.2
2.2-1	Operate machine, .013 ÷ 2 machines = .0065				
	.0065 × 4 preforms/unit	.013	4/2	.026	
	Total			.026	.2
	Lot estimate	59.0 hr			

TABLE 2.2 PREFORM MOLDING MACHINES

Setup

Rotary	.2 hr
Multiple cavity	1.0 hr

Operation elements in estimating minutes

1. Time per preform for one machine

	Machine Spm				Manual	
No. Cavities	25	40	50	80	25	40
1	.040	.025	.020	.013	.047	.029
2	.020	.013	.010	.006	.023	.015
3	.013	.008	.007	.004	.015	.009
4	.010	.006	.005	.003		
5	.008	.005	.004	.003		

2.3 Thermoplastic Injection Molding Machines

DESCRIPTION

An injection molding machine plasticizes and injects thermoplastic materials into a mold that is held between clamped platens. Clamping is usually one of two types—mechanical (toggle) or hydraulic. The molding of thermoplastic materials is a time-pressure-temperature dependent process.

Automatic injection molding is often done by machines which perform each operation in sequence with the elements controlled by electromechanical devices. In semiautomatic molding, the press operator removes molded pieces from the press following the cycle, and after closing the gate, starts the next cycle. In fully automatic molding, the operations are clock controlled, and a machine tender checks the pieces from time to time. Hoppers are filled automatically with preconditioned compound.

Presses are available with horizontal or vertical movement. Vertical-movement presses are desirable for insert or loose coring types of molds. Capacities of injection presses are rated in oz of material molded per cycle or in cu in. of material displacement.

Some machines having fast cycle times for injection, cooling, open eject, and close are intended for thin wall disposable parts. For instance, Figure 2.3 involves hydraulic clamping action, infinite stroke adjustment, continuous screw recovery, nozzle shutoff, and prepressurized melt chamber.

ESTIMATING DATA DISCUSSION

The data are average for small part weights and include semiautomatic and automatic machine characteristics. The estimate is selected from the described elements. The molding cycle time may be composed of several of the elements, and it is divided by the number of cavities in the mold. Work elements are possible during the transfer and hold time. If the total time for these work elements is less than the automatic portions of the cycle time, then the basic cycle time need not be increased. If the work time is greater than the automatic machine elements, either increase dwell time or decrease work elements during hold time.

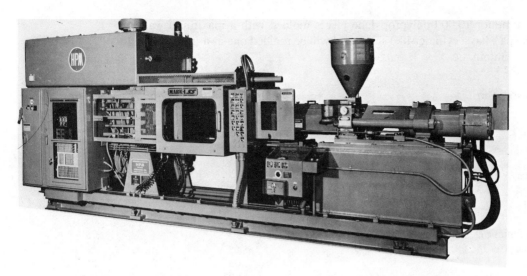

FIGURE 2.3 Injection molding machine, 200-ton clamping force. *(HPM Division, Koehring)*

Timers and automatic cycle equipment are simple to measure in the shop and can supplement these data. Furthermore, the properties of the plastics affect the estimate.

The cycle time of any product in injection molding depends upon product variables such as material, size of part, wall thickness, part geometry and dimension tolerances, and various equipment variables. The materials we consider are polyethylene, polypropylene, polyestryrene (which are called polymaterials), and nylon.

EXAMPLES

A. Estimate the molding time for a polyethylene piece described as flat with a wall thickness of 0.1 in. The mold has 3 cavities. The lot will require 5000 units. Determine if the operator can perform minor finishing work on the part during the machine cycle. A semiautomatic machine is available. The adjustment factor accounts for the three parts per cycle.

Table Number	Process Description	Table Time	Adjustment Factor	Cycle Minutes	Setup Hours
2.3-S	Setup	1.0			1.0
2.3-1	Close Mold	.017	$1/3$	.006	
2.3-2	Injection	.043	$1/3$	.014	
2.3-3	Hold low pressure	0		0	
2.3-4	Screw return	.05	$1/3$	.017	
2.3-5	Holding/cure	.047	$1/3$	.016	
2.3-6	Open mold	.033	$1/3$	.011	
	Subtotal automatic elements	.19		.064	
	Work elements				
2.3-8	Open guard gate, remove parts, close gate	.08	$1/3$	.027	
	Cycle estimate, total			.091	
	Unit estimate				
	Work elements during automatic cycle				
2.3-16	Degate, .07 + 2 × .03	.13		.13	
2.3-18	Stack dispose, .04 + 2 ×.01	.06		.06	
	Subtotal manual elements			.19	

As manual elements of .19 min during the automatic cycle of .19 min. are equal, no additional time is necessary for the cycle time of .27 min.

Unit estimate	.091 min
Hr/1000 units	1.5167
Lot estimate	8.58 hr

Thermoplastic Injection Molding Machines

B. Estimate the time for 10,000 polypropylene pieces molded with a maximum wall thickness of ⅛ in. The part has a cylinder-like shape and will be molded one-at-a-time. Ejection by the machine is automatic.

Table Number	Process Description	Table Time	Adjustment Factor	Cycle Minutes	Setup Hours
2.3-S	Setup	3.0			3.0
2.3-1	Close mold	.017		.017	
2.3-2	Injection	.083		.083	
2.3-3	Boost	.050		.05	
2.3-4	Screw return	.17		.17	
2.3-5	Holding	.33		.33	
2.3-6	Open mold	.05		.05	
2.3-7	Eject	.05		.05	
	Subtotal automatic elements			.75	
	Work elements				
2.3-18	Stack dispose	.05		.05	
2.3-19	Cardboard separation	.003		.003	
2.3-13	Visual inspect	.05		.05	
	Subtotal manual elements			.103	

As manual elements are less than automatic, additional time is unnecessary for the cycle time.

Unit estimate	.75 min
Hr/1000 units	12.5000
Lot estimate	128 hr
Prorated setup and unit estimate	.768 min

C. A polyethylene tray having a wall of 0.095 in., 6 cavities in a mold, and a quantity of 100,000 units is to be estimated. Assume a manually-controlled machine.

Table Number	Process Description	Table Time	Adjustment Factor	Cycle Minutes	Setup Hours
2.3-S	Setup, box-like part	2.0			2.0
2.3-8	Open gate, remove parts, close gate	.08		.08	
2.3-1	Mold close	.017		.017	
2.3-2	Injection	.057		.057	
2.3-3	Hold low pressure/boost	.025		.025	
2.3-4	Screw return	.11		.11	
2.3-5	Hold and cure	.12		.12	
2.3-6	Mold open	.042		.042	
	Subtotal			.451	
	Cycle estimate	.451			
	Work elements during machine cycle				
2.3-16	Degate, first part	.07		.07	
2.3-16	Add'l parts	.05	3	.15	
2.3-18	Stack dispose, first part	.04		.04	
2.3-18	Add'l parts	.01	5	.05	
2.3-19	Cardboard	.001		.001	
	Work elements during automatic time			.311	

As the work elements estimated during the machine cycle are less, no additional time needs to be added to the cycle time of .451 min.

Unit estimate	.075 min
Hr/10,000 units	12.5
Lot estimate	127 hr

D. A thin wall cup, 0.035 in., is made in a 6 cluster die. The material temperature is low to allow faster cooling. Determine a unit estimate since indefinite production requirements eliminate the need for setup. The material is polyethylene. The finished part does not require degating or any further processing. One operator tends 4 machines.

Table Number	Process Description	Table Time	Adjustment Factor	Cycle Minutes	Setup Hours
2.3-1	Mold close	.017	½×4	.0007	
2.3-2	Injection	.050	1⁄24	.0021	
2.3-3	Pressure and boost	.050	1⁄24	.0021	
2.3-4	Screw return	.17	1⁄24	.0071	
2.3-5	Cure polyethylene	.083	1⁄24	.0030	
2.3-6	Mold open	.050	1⁄24	.0021	
2.3-7	Eject	.050	1⁄24	.0021	
	Cycle time			.0196	
	Time per unit, .47/6 (for cluster die)	.078 min			
	Direct-labor unit estimate (for 4 machines)	.020 min			
	Hr/10,000 units	3.33333			

TABLE 2.3 THERMOPLASTIC INJECTION MOLDING MACHINES

Setup

Flat-like	1 hr
Box-like	2 hr
Cylinder-like	3 hr

Operation elements in estimating minutes

		Automatic Machine Elements Maximum Wall Thickness								
		1⁄16			.1			1⁄3		
		Flat	Box	Cyl.	Flat	Box	Cyl	Flat	Box	Cyl.
1.	Mold Close	.017	.017	.017	.017	.017	.017	.017	.017	.017
2.	Injection									
	Poly materials	.033	.042	.050	.043	.057	.070	.050	.067	.083
	Nylon	.017	.025	.033	.017	.030	.043	.017	.033	.050
3.	Hold low pressure/boost									
	Poly materials		.025	.050		.025	.050		.025	.050
4.	Screw return									
	Poly materials	.050	.11	.17	.050	.11	.17	.050	.11	.17
	Nylon	.033	.12	.20	.033	.12	.20	.033	.12	.20
5.	Holding/cure									
	Polyethylene	.017	.050	.083	.047	.12	.18	.067	.16	.25
	Pollypropylene	.050	.11	.17	.090	.18	.27	.12	.23	.33
	Polystyrene	.017	.042	.067	.037	.072	.11	.05	.092	.13
	Nylon	.017	.042	.067	.027	.025	.11	.033	.083	.13
6.	Mold open	.033	.042	.050	.033	.042	.050	.033	.042	.050
7.	Eject		.025	.050		.025	.050		.025	.050

		Manual Elements	
		First Part	Add'l
8.	Open gate, remove parts, close gate	.08	
9.	Lubricate mold (per in., 15 cycles)	.01	
10.	Load powder into hopper (prorated)	.005	
	1 or 2 cavities	.02	
	3 or more cavities	.003	
11.	Cardboard interlayer for dispose	.04	
12.	Pierce holes in part (air fixture)	.05	.03

	Manual Elements	
	First Part	Add'l
13. Visual inspect parts	.04	
14. Twist off part from runner	.06	.03
15. Clip off part from runner	.07	.03
16. Guillotine degate	.11	
17. Saw off gate	.05	.01
18. Stack dispose	.001	.003
19. Cardboard separation		

2.4 Thermosetting Plastic Molding Machines

DESCRIPTION

Thermosetting materials undergo a chemical change on heating and cure to an infusible shape. The two general methods considered for this estimate are compression and transfer molding. In compression molding, the plastic is placed in a heated mold. An upper die moves downward and compresses the material into a shape. Continued heat and pressure produce the chemical reaction to harden the material. In transfer molding, the thermosetting material is loaded into a reservoir, and once a semifluid, the mass is forced through screws into mold cavities.

Figure 2.4 is an automatic compression press with an integral preformer.

ESTIMATING DATA DISCUSSION

Besides the usual chores for setup, the time includes the setup and teardown and heat and cooling of electrically heated molds perhaps to 400°F.

Element 1 deals with operator control of opening and closing of the press. In Element 2, the entry variable is the major transverse section thickness for curing. This element is sensitive to the particular material chosen. If a cooling fixture is used, Element 4 is added. Air cleaning of the mold depends upon the number of cavities. Inserts can be individually hand loaded, or gang-fixture loaded, or loaded singly with a loader. If the molding machine does not produce its own preforms, then Element 7 can be a candidate to include in the estimate.

Element 9 is for hand molding only where compression transfer machines are used. Some ele-

FIGURE 2.4 Automatic compression press with preformer. (Penwalt Corporation)

ments, such as Elements 10 through 12, could be done simultaneously to cure time and is not included if their time is less than the mold time. If the work elements during cure time are greater, increase the cycle time, or reduce the embedded work time during the cure.

EXAMPLES

A. A part is to be compression-molded using premixed powders. The material, a phenolic resin, will have a maximum wall thickness of ⅜ in., and a 2-cavity die is anticipated with typical runners and sprue. As tolerances are considered close, a cooling fixture will be designed for postcuring to prevent distortion. Each part has 2 inserts and lot size is 250 units.

Table Number	*Process Description*	*Table Time*	*Adjustment Factor*	*Cycle Minutes*	*Setup Hours*
2.4-S	Setup	1.5			1.5
2.4-1	Open and close press, per shot	.38	½	.19	
2.4-2	Cure, per shot	1.92	½	.96	
2.4-3	Unload parts, per shot	.05	½	.025	
2.4-4	Load part onto cooling fixture				
	1st	.24	½	.12	
	2nd	.04	½	.02	
2.4-5	Air clean mold, per shot	.15	½	.075	
2.4-6	Load inserts	.05	4/2	.10	
2.4-7	Load powder in cavity, per shot	.20	½	.10	
2.4-9	Hand compression molding	.91	½	.455	
	Molding cycle estimate	4.09 min		2.05	
	Unit estimate	2.05 min			
	Work elements during Element 2				
2.4-10	Load powder in cup and preheat	.10		.10	
2.4-11	Degate part and visual inspect	.03		.03	
2.4-11	Deburr flash and gate	.16		.16	
2.4-4	Unload parts from cooling fixture	.24		.24	
	Unit estimate			.53	

Since total cycle time of work elements during cure time is less, the molding cycle estimate is unadjusted.

	Lot estimate	10.04 hr

B. Transfer molding for a furane resin part is to be estimated. A metal insert, 2 preforms, and a 4-mold cavity die are planned. Wall thickness will be 3/16 in. maximum. Tolerances require a cooling fixture. A quantity of 3200 is planned.

Table Number	*Process Description*	*Table Time*	*Adjustment Factor*	*Cycle Minutes*	*Setup Hours*
2.4-S	Setup	1.5			1.5
2.4-1	Open and close press	.38	¼	.095	
2.4-2	Cure per shot	1.46	¼	.365	
2.4-3	Unload part, per shot	.09	¼	.022	
2.4-4	Load parts on cooling fixture	.32	¼	.08	
2.4-5	Air clean molds	.15	¼	.038	
2.4-6	Load 4 inserts	.05	4/4	.05	
2.4-7	Load 2 preforms, 4 cavities	.05	2	.10	
	Molding cavity estimate	2.5		.75	
	Unit estimate	.625			

Work elements during Element 2

2.4-10	Preheat performs	.09	.09
2.4-10	Preposition preforms	.08	.08
2.4-11	Degate part	.12	.12
2.4-11	Deburr flashing	.28	.28
			.57

Since work elements are less than cure time, no additional time is added to molding estimate.

	Lot estimate	36.17 hr

TABLE 2.4 THERMOSETTING PLASTIC MOLDING MACHINES

Setup   1.5 hr

Operation elements in estimating minutes

1. Open press, close press   .38

2. Cure preheated material

Wall Thickness	Transfer	Compression	Wall Thickness	Transfer	Compression
.06	1.15	1.46	.56	2.46	3.54
.12	1.30	1.71	.62	2.54	3.80
.18	1.46	1.98	.68	2.69	4.06
.25	1.61	2.23	.75	2.85	4.32
.31	1.77	2.50	.81	3.00	4.58
.37	1.92	2.75	.87	3.16	4.84
.43	2.08	3.02	.93	3.35	5.16
.50	2.23	3.28	1.00	3.47	5.36
			Add'l	.30	.55

3. Unload parts
   - Transfer mold, per shot   .09
   - Compression mold, 1 to 3 parts   .05
   - Add'l   .03

4. Load or unload part on cooling fixture
   - 1st part   .20
   - Add'l   .04

5. Air clean mold
   - 1-4 cavities   .15
   - Add'l cavity   .02
   - Air clean knockout pins and occasional raise press   .05

6. Load inserts
   - By hand, per insert   .05
   - Gang loader, per loader   .20
   - Single loader, per loader   .05

7. Load preforms, per preform   .05
   Load powder in cavity
   - Small   .07
   - Large   .20

8. Index mold   .09

	Handmolding Elements	
	Small	Large

9. | | Small | Large |
   |---|---|---|
   | Remove bottom plate | .09 | .17 |
   | Remove force | .09 | .22 |
   | Kick out parts and dispose plate | .11 | .17 |
   | | .07 | .13 |
   | Replace bottom plate | .11 | .22 |
   | Replace force and slide mold into press | | |

   Note: Work elements during cure time

10. Preheat preforms in oven, 1–5 preforms	.07
Add'l	.02
Load powder in cup and preheat, per cup	.10
Reposition preforms convenient	.02
11. Degate part and visual inspect, per part	.03
Deburr flash or gate	
1st part	.10
Add'l	.06
12. Stack inserts	.02
Load inserts into loader, per insert	.05

2.5 Thermoforming Machines

DESCRIPTION

The thermoforming process consists of heating a thermoplastic sheet to its softening temperature and forcing the hot and flexible material against the contours of a mold by means of tools, plug, solid molds, etc., or pneumatic differential air pressure. When held to the mold and allowed to cool, the plastic retains the shape and detail of the mold. This technique is applicable only to thermoplastic materials and not to thermosets, due to the heating and cooling action. In thermoforming, the raw materials do not have sharp melting temperatures and softening is gradual. This is a factor of importance in the forming of the thermoplastic sheet.

Many different techniques and production equipment exist for production thermoforming machines. Methods include straight vacuum, drape, match mold, slip ring, plug assist, and others. Any thermoforming machine must provide a method for heating the sheet, clamping, a device to raise and lower the mold into the plastic sheet or vice versa, vacuum, air pressure, and controls. Machines may be classified according to single-station, in which the sheet is fed and the machine can perform only one operation at a time. A multiple-station can perform two or more operations simultaneously. Figure 2.5 illustrates the process of thermoforming.

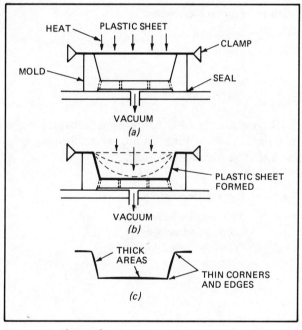

FIGURE 2.5 Thermoforming process.

ESTIMATING DATA DISCUSSION

Setup is provded for one-, two- or three-station machines. A one-stage machine performs loading, heating, forming, and cooling successively. A two-station or double-ended thermoformer shares a common oven. Considering the estimating of direct labor, cycle time is identical to a one-station machine. A three-station machine does have a different cycle for direct labor. The cycle time is the longest elemental time of loading, heating or forming, and cooling. Usually, when wall thickness is less than ½ in., the longest time is for forming and cooling, while for walls thicker than ½ in., the heating takes the longest time.

In a typical operation, the sheet is clamped in a frame, the frame is moved between or under heaters, and to a forming station. Load and unload elements are divided into small and large. A small sheet is less than 12 × 12 in. In Element 2, the forming and cooling station is dependent upon the nature of the mold. "Flat" implies flat-like. "Cavities" are suggestive of a mold requiring deep draw-down. "Medium" is a mold that has flat areas and minor draw-down requirements of a mold.

EXAMPLES

A. Estimate the time to thermoform a tote box where raw material thickness is .2 in., and is 80% polyethylene and 20% filler. A quantity of 1000 units is to be produced in a single-station machine. Also, compare this to a three-station machine. Find lot time for a large sheet. The mold has ridged cavities to strengthen the box.

Table Number	Process Description	Table Time	Adjustment Factor	Cycle Minutes	Setup Hours
2.5-S	One-station setup	2			2.0
2.5-1	Load and unload	.25		.25	
2.5-2	Heat	1.75		1.75	
2.5-2	Form and cool (cavities)	3.66		3.66	
	Total			5.66	2.0
	Hr/100	9.433			
	Lot estimate	96.33 hr			

Three-station machine

Table Number	Process Description	Table Time	Adjustment Factor	Cycle Minutes	Setup Hours
2.5-S	Three-station setup	6			6.0
2.5-2	Forming and cooling	3.66		3.66	
	Unit estimate			3.66	6.00
	Hr/100	6.1			
	Lot estimate	67 hr			

B. A simple box-like part uses polyethylene as the principal material. Thickness of raw material is .3 in. A three-station thermoforming machine is to be used. Find the lot time for 440 units which are flat-like.

Table Number	Process Description	Table Time	Adjustment Factor	Cycle Minutes	Setup Hours
2.5-S	Three-station setup	6			6
2.5-1	Loading and unload	0		0	
2.5-2	Form and cool .3 in.	6.00		6.0	
	Unit estimate			6.0	
	Lot estimate	50 hr			

TABLE 2.5 THERMOFORMING MACHINES

Setup
- 1 Station   2 hr
- 2 Stations   4 hr
- 3 Stations   6 hr

Operation elements in estimating minutes

1. Load and unload station
   - Small sheet   .15
   - Large sheet   .25

2. Heating, forming and cooling

Material	In. Thickness	Heating Station	Forming and Cooling Station		
			Flat	Medium	Cavities
Butyrate	.065	.47	1.00	1.05	1.10
	.100	.95	2.00	2.10	2.20
Polyethylene	.100	1.05	2.00	2.10	2.20
	.135	1.12	2.14	2.25	2.35
	.140	1.31	2.50	2.63	2.75
	.160	1.57	3.00	3.15	3.30
	.175	1.57	3.00	3.15	3.60
	.200	1.75	3.33	3.50	3.66
	.220	1.97	3.75	3.94	4.13
	.260	2.63	5.00	5.25	5.50
	.300	3.15	6.00	6.30	6.60
	.375	5.25	10.00	10.50	11.00

2.6 Glass-Cloth Layup Table

DESCRIPTION

This operation consists of laying successive thicknesses of glass-fiber cloth and resin in place. Care is taken to prevent air entrapment. A paint brush used with a dabbing stroke forces a room-temperature catalyzed resin into the cloth. The resins change from a liquid state to a rigid, predetermined, fixed shape. The methods involved in these data are for hand layup, usually with a mold.

Besides the glass fiber cloth and resin, the equipment usually consists of oven, knife, portable tables, wax, and scissors. Parts requiring wire mesh or other inserts are not considered in these data.

ESTIMATING DATA DISCUSSION

A setup distinction is made between dry- and wet-hand layups. The operation elements usually consist of cut cloth, prepare mold, vacuum layup, cure, remove from mold, clean mold, and identify part. Time for curing of the layup is not included. Part configuration is flat or nearly flat.

Element 1 deals with the dry layup preimpregnated cloth, while Element 2 is a wet layup. Entry variables are area of the layup, sq in., and layers of cloth. If values other than those shown are desired, factors for additional layers of cloth and for 500 sq in. of area are available. The data are essentially for flat work with a minimum of compound surfaces.

EXAMPLES

A. Estimate the time to layup 1750 sq in. of flat glass-fiber for a small quantity of 10. Dry layup will be used for 9 layers of cloth. Estimate lot time.

Table Number	Process Description	Table Time	Adjustment Factor	Cycle Minutes	Setup Hours
2.6-S	Dry layup	.25			.25
2.6-1	Basic 1300 sq in. and 8 layers	103.		103.0	
2.6-1	Adjustment for area, ⅘ × (34.8 + 42.2)/2	30.8		30.8	
2.6-1	Adjustment for layer, (9.3 + 14.2)/2	11.8		11.5	
	Unit estimate	145.6		145.6	.25
	Lot estimate	24.52			

TABLE 2.6 GLASS-CLOTH LAYUP TABLE

Setup
Dry layup .25 hr
Wet layup .35 hr

Operation elements in estimating minutes

1. Dry Layup

	Layers of Cloth							
Area	2	3	4	5	7	8	10	Add'l
20	18.0	18.8	20.3	21.0	22.4	23.2	24.6	.7
100	19.5	21.0	23.0	24.2	26.6	27.8	30.2	1.2
300	23.8	26.6	30.1	32.6	37.7	40.2	45.2	2.5
700	32.2	37.9	44.2	49.4	59.6	64.8	75.1	5.2
1300	45.5	55.4	66.1	75.4	94.0	103	122	9.3
2000	61.3	76.1	92.2	106	134	149	177	14.2
2400	70.9	88.9	108	125	160	177	211	17.2
Add'l 500 sq in.	11.9	15.8	19.8	23.6	31.0	34.8	42.2	

2. Wet layup

	Layers of Cloth							
Area	2	3	4	5	7	8	10	Add'l
20	18.5	19.9	21.7	22.8	24.9	26.0	28.1	1.1
100	20.3	24.1	24.7	26.3	29.6	31.2	34.5	1.6
300	24.9	28.3	32.4	36.3	41.7	44.8	50.9	3.1
700	33.9	40.3	47.5	53.5	65.5	71.4	83.3	6.0
1300	47.9	58.9	70.8	81.2	102	113	133	10.4
2000	64.4	80.9	98.6	114	146	162	193	15.7
2400	74.6	94.4	116	135	173	192	230	19.0
Add'l 500 sq in.	12.5	16.7	20.9	25.0	29.0	33.1	37.1	45.1

2.7 Extrusion Molding Machines

DESCRIPTION

Extrusion converts raw thermoplastics in powdered or granular form into a continuous melt stream which is then formed into a variety of shapes. The process forces hot melt through a die which has an opening similar to the desired cross section. It differs from other molding processes in that it is a continuous process and forms an almost endless product that may be cut, sawed, chopped, rolled, or otherwise altered to a desired length. Commercial extrusion provides ram extrusion, rotating screw, and a single extruder screw operation. Wire covering, sheet and flat film, blown or tubular film, coating, and tube and pipe extrusion are popular profiles.

The screw geometry and material processed are factors affecting speed of production. For the profile pipe and sheet, as estimated here, an average screw design and polystyrene are assumed.

Figure 2.7 is a sketch of sheet extrusion.

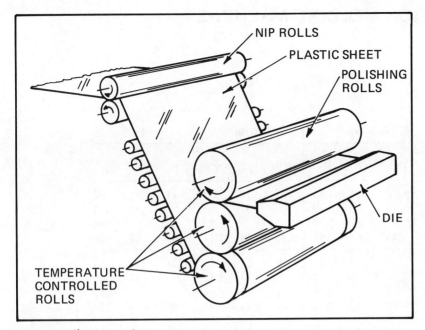

FIGURE 2.7 Sheet extrusion.

ESTIMATING DATA DISCUSSION

Setup of 4 hr is assumed necessary for extruded operations. Materials are polyethylene, polystyrene, butyrate, and vinyl. Two products are considered, i.e., sheets and pipe. The entry variables for Element 1 are thickness, length, and width in in. The values are expressed in min/unit.

Element 2 deals with tube estimating. Extra variables are wall thickness (w.t.), material, and outside diameter (OD).

EXAMPLES

A. Estimate the time to extrude 2000 sheets of polystyrene of ¼-in. wall thickness, 24-in. width, and 24-in. length. Find lot time.

Table Number	Process Description	Table Time	Adjustment Factor	Cycle Minutes	Setup Hours
2.7-S	Setup	4			4.0
2.7-1	Sheet production, ¼ × 24 × 24	.66		.66	
	Lot estimate	26.00 hr		.66	4.0

B. Find the total time to produce 1500 ft of polyethylene tube with a wall thickness of ⅜ in. and dia of 2.5 in.

Table Number	Process Description	Table Time	Adjustment Factor	Cycle Minutes	Setup Hours
2.7-S	Setup	4 hr			4.0
2.7-2A	Feet time	.16	1500	240	
	Estimate for 1500 ft			240	4.0
	Lot estimate	8.00 hr			

Extrusion Molding Machines

TABLE 2.7 EXTRUSION MOLDING MACHINES

Setup **4 hr**

Operation elements in estimating minutes

1. Extrude sheets

Material		*Polyethylene*			*Polystyrene*			*Butyrate*			*Vinyl*		
Thick.	*L* \ *W*	*12*	*24*	*36*	*12*	*24*	*36*	*12*	*24*	*36*	*12*	*24*	*36*
	12	.07	.14	.21	.08	.16	.25	.09	.18	.28	.11	.21	.32
⅛	24		.29	.43		.33	.49		.37	.55		.43	.64
	36			.65			.74			.83			.96
	48	.29	.57	.86	.33	.66	.99	.37	.74	1.11	.43	.85	1.28
	12	.14	.29	.43	.16	.33	.49	.18	.37	.55	.21	.43	.64
¼	24		.57	.86		.66	.99		.74	1.11		.85	1.28
	36			1.29			1.48			1.66			1.92
	48	.57	1.15	1.72	.66	1.32	1.98	.74	1.48	2.21	.85	1.71	2.57
	12	.21	.43	.65	.25	.49	.74	.28	.55	.83	.32	.64	.96
⅜	24		.86	1.29		.99	1.48		1.11	1.66		1.28	1.92
	36			1.93			2.22			2.49			2.89
	48	.86	1.72	2.58	.99	1.98	2.96	1.11	2.21	3.32	1.28	2.57	3.85
	12	.29	.57	.86	.33	.66	.99	.37	.74	1.11	.43	.85	1.28
½	24		1.15	1.72		1.32	1.98		1.48	2.21		1.71	2.57
	36			2.58			2.96			3.32			3.85
	48	1.15	2.29	3.44	1.32	2.63	3.95	1.48	2.95	4.43	1.71	3.42	5.13

2. Extrude tube, min/ft of tube

A.

Material	*Polyethylene*					*Polystyrene*				
w.t \ *OD*	*1.0*	*1.5*	*2.0*	*2.5*	*3.0*	*1.0*	*1.5*	*2.0*	*2.5*	*3.0*
⅛	.022	.034	.047	.059	.072	.025	.039	.054	.068	.083
¼	.037	.063	.088	.113	.138	.043	.072	.101	1.29	.158
⅜	.047	.085	.122	.160	.197	.054	.097	.140	.183	.226
½		.100	.150	.200	.250		.115	.173	.230	.287

B.

w.t.	*Butyrate*					*Vinyl*				
⅛	.028	.044	.060	.076	.092	.033	.051	.070	.089	.107
¼	.048	.081	.113	.145	.177	.056	.093	.131	.168	.205
⅜	.060	.109	.157	.205	.253	.070	.126	.182	.238	.294
½		.129	.193	.258	.322		.149	.224	.299	.374

2.8 Blow Molding Machines

DESCRIPTION

Blow molding is used in manufacturing hollow plastic products. Although differences in processes exist, all have the requirement of a parison, which is a tube-like plastic shape, in common. The parison is inserted into a close mold and air is forced into the parison to expand it onto the surfaces of the mold where it sets up into the finished product.

The parison can be made by extrusion or injection molding. It may be made and immediately used, or it may be premade and later reheated and transferred to the blow mold. However, the basic process remains the same; (1) melt the material; (2) form the molten resin into a parison; (3) seal the end of the parison, except for an area in which blowing air can enter; (4) inflate the parison inside the mold; (5) cool; (6) eject; and (7) trim flash, if necessary. The sequence is automated and machines include controlling instruments which monitor the time for each step, pressure, temperature, and material consumption. Figure 2.8 is a sketch of a three-station injection blow molding process, but other process arrangements are also available.

ESTIMATING DATA DISCUSSION

In blow molding the parison can be produced by extrusion or injection. The process, for estimating purposes, is divided into estimating of parison, production, and blow molding. To find parison production, use Table 2.3 or Table 2.7. For blow molding, the material is high density polyethylene. A minimum wall thickness of 0.010 in. is held. The bottle volume expressed in oz is the variable. Blow molding is a process oriented mainly to high-volume production. In many cases, the equipment is made for production of special bottles, such as the 64-oz beverage bottle.

EXAMPLES

A. Estimate the unit time to produce a round 64-oz juice bottle with a 38-mm neck. The minimum wall thickness is .010 in. Material is high-density polyethylene. Assume that

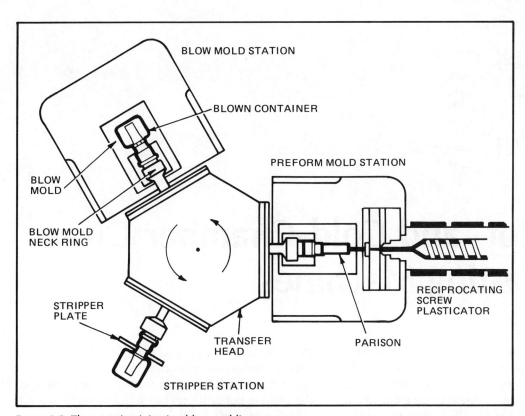

FIGURE 2.8 Three-station injection blow molding process.

the injection process is prior to blow molding of the parison. Find the estimate for 10,000 units, and a lot time for 100,000 containers. Assume consecutive production of the parison and bottle.

Table Number	Process Description	Table Time	Adjustment Factor	Cycle Minutes	Setup Hours
2.3-S	Setup injection of parison	3			3
2.3-1	Mold close, injection	.017		.017	
2.3-2	Injection	.042		.042	
2.3-3	Hold low pressure injection	.025		.025	
2.3-4	Screw return injection	.11		.11	
2.3-5	Holding	.05		.05	
2.3-6	Mold open	.042		.042	
	Subtotal injection			.284	
2.8-S	Setup for blow molding	10			10
2.8-2	Blow molding, subtotal	.248		.248	
	Unit estimate			.532	17
	Hr/10,000 units	88.66666			
	Lot estimate	899.66 hr			

TABLE 2.8 BLOW MOLDING

Setup 10 hr

Operation elements in estimating minutes

1. Injection molding of parison   See Table 2.3

2. Extrusion molding of parison   See Table 2.7

3. Blow molding

Bottle Size, Oz	Min
16	.100
20	.112
24	.124
28	.137
32	.149
48	.199
64	.248

2.9 Hot- and Cold-Chamber Die-Casting Machines

DESCRIPTION

Die casting is a process in which molten metal is forced by pressure into a metal mold known as a die. Because the metal solidifies under a pressure from 80 to 40,000 psi, the casting conforms to the die cavity in both shape and surface finish. Die casting is a widely used permanent-mold process and two methods of this process are hot and cold chamber. The distinction between these two methods is determined by the location of the melting pot. In the hot-chamber method, the melting pot is integral with the machine and the injection cylinder is immersed in the molten metal. The injection cyclinder is actuated by air or hydraulic pressure that forces metal into dies to complete the casting. Machines using the cold-chamber process have a separate melting furnace and metal is introduced into the injection cylinder by manual ladling or through mechanical means. Hydraulic pressure forces the metal into the die. Usual metal pressure ranges from 1500 to 3000 psi for hot-chamber and 2500 to 10,000 psi for cold-chamber machines.

Lot-melting alloys of zinc, tin, and lead are the most widely used materials cast in hot-chamber machines. Die casting of aluminum, magnesium, and brass requires higher pressures, thus melting temperatures are heated in auxiliary furnaces and ladled into the plunger cavity nearby the dies.

Figure 2.9A is a 250-ton cold-chamber machine used to produce small, high-production parts. Shot weights for this size machine range from 2 to 4 lb. Typically, the dies are multicavity, ranging from 6 to 10 cavities per die. Typical wall thicknesses are from .060 to .090 in., and the overall cast surface area is approximately 100 to 125 sq in. for cold chamber. For the hot-chamber method, typical wall thickness are .045 to .090 in. with multiple cavity tooling. The number of cavities ranges from 8 to as high as 24 in zinc alloy.

Figure 2.9B is a 400-ton machine and it is about twice as big as the 250-ton machine. Larger parts are cast on these machines with a typical number of cavities ranging from 2 to 6. With cold-chamber machines the typical wall thickness ranges from .080 to .160 in., while for hot-chamber machines the thickness is .060 to .130 in. Shot weights are 4 to 8 lb, and 6 to 14 lb for the cold- and hot-chamber machines.

Typical parts for the 250-ton cold-chamber machine are small faucet handles, electrical conduit fittings, small gears, hose couplings, etc., and typical shots/hr range from 150 to 250/hr. Hot-chamber machines can manufacture a variety of parts, such as door handles, hardware, clock parts, etc., where good finish and mechanical properties are necessary. Shots/hr would range from 250 to 500.

ESTIMATING DATA DISCUSSION

Setup data include the entry variable of die surface area. It does not include initial tool tryout. The data, as shown by Table 2.9, are for hot- and cold-chamber die-casting machines. Shot and dwell time is average. Automatic, semiautomatic, and manual machines are considered. The times are for a die having one or several units, and the total cycle time must be divided by the number of cavities. Certain elements, air clean die for instance, are given as 1 in 3 average, 0.05 min, the per unit time. The total elemental time is 0.15 min, and if other prorata occurrences are believed more appropriate, the value of 0.15 is divided by the frequency per shot. If for Element 8, load insert, the first insert time is higher, it is possible to have work elements during shot and dwell time, Element 4. If the total time for work elements done during transfer plus hold, as found in Element 9, can be less than or equal to Element 4, the time is done without increasing cost. The unit estimate equals the cycle min divided by the no. of cavities in the casting die.

FIGURE 2.9A Front side of a cold-chamber die casting machine, 250 die lock end force 2.7–5.5 lb aluminum shot weight. *(HPM Division, Koehring)*

FIGURE 2.9B Front side of a 400-ton die casting machine with integral shot end. *(HPM division, Koehring)*

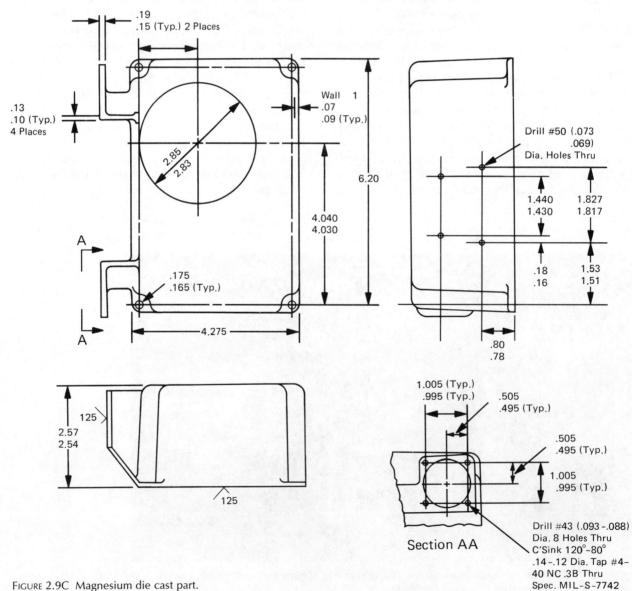

FIGURE 2.9C Magnesium die cast part.

EXAMPLES

A. Estimate a 5-lb aluminum cold-chamber die-cast part where the die has 4 cavities. A 400-ton manual machine and lot of 20,000 units is planned. The die has a large surface area; about 30 sq in. is anticipated.

Table Number	Process Description	Table Time	Adjustment Factor	Cycle Minutes	Setup Hours
2.9-S	Setup	4.50			4.5
2.9-5	Open and close guard door	.10	¼	.025	
2.9-1	Open and close die	.08	¼	.02	
2.9-3	Ladle 5 lb, 20 lb for 4 cavities	.10		.10	
2.9-4	Shot and dwell	.17	¼	.043	
2.9-6	Unload cluster	.09	¼	.023	
2.9-7	Clean and lubricate die	.17	¼	.043	
	Total			.254	4.5
	Hr/100	.423			
	Lot estimate	39.17			

B. A part is to be magnesium die cast, and the part is given as Figure 2.9C. The material is magnesium and information is not provided. Therefore we use the most similar material or aluminum. The quantity for this part is 500. Find the unit cycle and setup hours, and the lot hours.

Table Number	Process Description	Table Time	Adjustment Factor	Cycle Minutes	Setup Hours
2.9-S	Setup for about 102 square in., 1 part per die.	6.00			6.00
2.9-1	Open and close 200-ton press	.04		0.04	
2.9-2	Trip ladle metal	0.16		0.16	
2.9-3	Ladle metal				
	metal weight = .5 lb with runner, sprue	0.06		0.06	
2.9-4	Shot and dwell	0.17		0.17	
2.9-5	Guard door, manual	0.10		0.10	
2.9-6	Eject for part drop	0.04		0.04	
2.9-6	Remove part	0.09		0.09	
2.9-7	Clean, lubricate	0.17		0.17	
	Total			0.83	6.00
	Lot hours	12.92			

TABLE 2.9 HOT- AND COLD-CHAMBER DIE-CASTING MACHINES

Setup

Die surface area:	10 sq in.	1.5 hr
	20 sq in.	3.0 hr
	35 sq in.	5–6 hr

Operation elements in estimating minutes

1. Open and close die

ton	250	400	600	800	1000
min	.04	.08	.12	.14	.16

2. Trip to close, ladle metal, and fire shot .... .16

3. Ladle metal

lb	2	5	10
min	.06	.10	.22

4. Shot and dwell
   Zinc .14
   Aluminum .17

5. Open and close guard door
   Manual .10
   Automatic —

6. Eject for part drop

ton	250	400	600
min	.04	.06	.09

   Machine sense for part drop .02

   Remove part or cluster
   Zinc .05
   Aluminum .09

7. Clean or lubricate
   Air clean die, per unit
   Zinc (1 in 3 ave.) .05
   Aluminum (1 in 1 ave.) .18

   Lubricate die
   Zinc (1 in 5 ave.) .04
   Aluminum (1 in 2 ave.) .17

   Clean and lube die (automatic) .05

   Lubricate shot plunger (1 in 10 ave.) .01

   Skim off slag and clean shot plunger. .005

8. Load insert in die
   1st .14
   Add'l .07

   Load insert in die using holder .30

9. Simultaneous work elements during shot and dwell time
   Break part from runner, Toss .13
   Stack 1st .17
   Add'l .04

   Degate cluster with press .16

   Get and place inserts in holder
   1st .14
   Add'l .07

   Return ladle to melting furnace .08

Note: Visual inspection occasionally done. Also total time for work elements done during shot and dwell time can be equal to or be less than Element 4. If greater than, either increase dwell time or decrease work elements.

2.10 Isostatic Molding Machines

DESCRIPTION

Isostatic molding has a die inside a rubber mold that is subjected to a circumferential and head pressure. A pressure cycle is started with the filling of dry powder into the mold before the pressure is applied. Molded parts range up to a maximum of 4 in. in height and 3 in. in diameter as covered by these data. Pressures up to 9000 psi are used to mold the ceramic body about the die. Once the desired pressure is reached, the mold remains subjected to this pressure for a brief period of time. After pressing is complete, the molded part exits the pressing unit on a carrousel carrying the next mold to be pressed.

ESTIMATING DATA DISCUSSION

Setup time includes obtaining the production order form, selecting the necessary die and mold, installing the dies on the machine, setting the molding pressure, and acquiring the appropriate ceramic body.

Element 1, position mold and remove die, consists of positioning the mold over the die. Following the completion of pressing, it also includes removing the molded part and setting it aside. The time is considered constant.

Element 2 is clean die, wipe and blow with air hose. Time is a variable. The independent variable is girth + number of internal diameter (NDC) changes, $L + W + H + NDC$. Girth is the smallest-sized box which will enclose any shape. In the case of a cylinder, girth is the sum of height plus two diameters. Girth is an easy calculation, and isostatic molded parts are correlated to girth. The number of diameter changes, NDC, is a number which affects the molded part removal and die cleaning time. Element 3 is for positioning the mold over the die plus wait time for press completion. Time includes pressing and rotation of the pressing unit over the carrousel. The time variable is solely dependent on girth. The estimator, in order to expedite application, should use higher table values rather than interpolating. For example, if $L + 2(Dia) + NDC = 7.72$, use 0.17.

Isostatic molding considers ceramic powders, and several types were part of the observations. The data are expressed for a single operator only. No adjustments should be made for two or more operators. No time is allowed for simultaneous molding and filling since these times are embedded in Elements 2 and 3. An occasional spray is used on the die to reduce powder adhesion. Time is included in Element 1.

EXAMPLES

A. A blasting cap, made from AD-85-I ceramic body, is to be molded. The part has a girth of 4.9 in. and has two internal diameter changes. Determine the unit and lot estimates for 8037 parts.

Table Number	Process Description	Table Time	Adjustment Factor	Cycle Minutes	Setup Hours
2.10-S	Setup	.5			.5
2.10-1	Position mold, remove die	.13		.13	
2.10-2	Clean die, $L + 2(Dia) + NDC = 6.9$	.15		.15	
2.10-3	Replace mold over die, $L + 2(Dia) + 4.9$	.09		.09	
	Total			.37	.5
	Lot estimate	50.06 hr			

B. A hemispherical blasting cap is to be molded from 85-S ceramic body. The cap has one internal diameter change. The outside diameter and height dimensions are 2.2 and 1.1 respectively. If the molding pressure is 3000 psi and 9680 caps are to be produced, determine the unit estimate, hr/100 units, and lot estimate.

Table Number	Process Description	Table Time	Adjustment Factor	Cycle Minutes	Setup Hours
2.10-S	Obtain production request and ready machine	.5			.5
2.10-1	Set mold and finished parts aside	.13		.13	
2.10-2	Wipe and blow die	.11		.11	

Table Number	Process Description	Table Time	Adjustment Factor	Cycle Minutes	Setup Hours
2.10-3	Reposition mold and wait	.09		.09	
2.10-4	Fill and press	.00		.00	
	Total			.33	.5
	Hr/100 units	.550			
	Lot estimate	53.74 hr			

TABLE 2.10 ISOSATIC MOLDING MACHINES

Setup   0.5 hr

Operation elements in estimating minutes

1. Handle and remove parts and mold   0.13

2. Clean die, wipe and blow

L + 2(Dia) + NDC	Min	L + 2(Dia) + NDC	Min	L + 2(Dia) + NDC	Min
0.5	0.03	5.5	0.12	10.5	0.22
1.0	0.04	6.0	0.13	11.0	0.23
1.5	0.05	6.5	0.14	11.5	0.24
2.0	0.06	7.0	0.15	12.0	0.25
2.5	0.07	7.5	0.16	12.5	0.26
3.0	0.08	8.0	0.17	13.0	0.27
3.5	0.09	8.5	0.18	13.5	0.28
4.0	0.10	9.0	0.19	14.0	0.29
4.5	0.10	9.5	0.20	14.5	0.30
5.0	0.11	10.0	0.21	15.0	0.31

3. Reposition rubber mold and molding

L + 2(Dia)	Min	L + 2(Dia)	Min	L + 2(Dia)	Min
.5	.11	4.5	.09	8.5	.08
1.0	.11	5.0	.09	9.0	.08
1.5	.10	5.5	.09	9.5	.07
2.0	.10	6.0	.09	10.0	.07
2.5	.10	6.5	.09	10.5	.07
3.0	.10	7.0	.08	11.0	.07
3.5	.10	7.5	.08	11.5	.07
4.0	.09	8.0	.08	12.0	.07

PRESSWORK

3.1 Power Shear Machines

DESCRIPTION

The power shear machine is used for shearing sheets of steel, aluminum, brass, etc. It can be powered mechanically or hydraulically. Hold-down plungers are spaced across the bed to prevent movement of the sheet during the guillotine squaring-shear cutting action. In operation, the sheet is advanced on the bed so that the line of cut is under the shear. When the foot treadle or pedal is depressed, the hold-down plungers clamp on the material and the blade cuts progressively across the sheet.

Power-squaring shear machines may have features of thickness adjustment in thousandths, adjustable pressure hold-fingers, side gage, and front-operated back gages. Horsepower (hp) ranges between 5–25 hp. Figure 3.1 has a 60 spm cutting speed, back gage range of 36 in., and front gage range of 55 in. Now, consider estimating data for the variety power shear machines.

ESTIMATING DATA DISCUSSION

Several elements have width and length entry variables, and usually the estimator will jump to the next higher number rather than interpolate. The data are independent of thickness and weight, as this factor is averaged into the data. The data are for one operator. If with heavier gages the estimator concludes two operators are necessary, double the estimate.

Element 2 includes the get, move, and release of the raw material on the shear bed. The shear machine time, Element 4, may be selected from foot-pedal or push-against mechanical electronic probes to actuate the guillotine squaring shear. Following the reverse element, the trim element may be used. The trim element includes one shear machine time. The strips or blanks may be processed by an automatic back stacker, they may be loose behind the machine, or they may be on the front edge, etc., after the final shear. Elements 7, 8, and 9 are for various conditions for removing the strips or blanks.

The edges, if they are to be rolled or deburred, are passed through a timesaver, and Element 11 includes pickup of the blanks, feed the blanks through the roll-edging machine, align, and stack. A surfacing operation is also provided in Table 21.2. Some companies include Element 11 as part of the assigned work to the shear operator.

EXAMPLES

A. Determine the time to shear 37 × 80-in. strips and 4¼ × 37-in. blanks from 48 and 144-in. 14-gage, hot-rolled, pickled, and oiled steel sheets. The requirements are for 200 blanks, and each sheet will provide eighteen 4¼ × 37-in blanks. Make strips from sheet. Strip production estimates are covered in Example A.

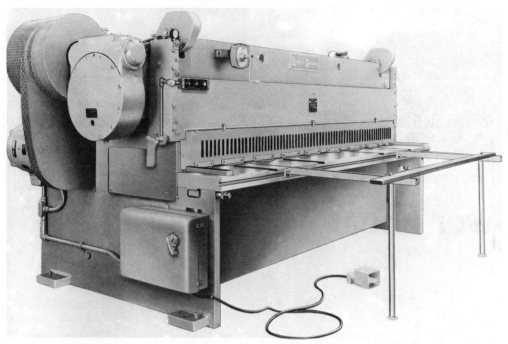

FIGURE 3.1 Power squaring shears with a capacity of 3/16 in. × 4 ft to 3/16 in. × 14 ft in mild steel. *(Lodge & Shipley Company)*

Table Number	Process Description	Table Time	Adjustment Factor	Cycle Minutes	Setup Hours
3.1-S	Setup	.25			.25
3.1-2	Get sheet	.75	1/18	.042	
3.1-3	Trim one edge	.14	1/18	.008	
3.1-5	Reverse 48 × 80 in.	.15	1/18	.008	
3.1-3	Trim	.09	1/18	.005	
3.1-4	Sheer strip, push 37½	.18	1/18	.01	
3.1-7	Remove 37 × 80 from machine	.14	1/18	.008	
3.1-9	Remove scrap, 2 pcs	.004	1/18	.000	
	Unit estimate			.0812	

B. Make 4¼ × 37-in. blanks from 37 × 80-in. 14-gage steel sheets.

Table Number	Process Description	Table Time	Adjustment Factor	Cycle Minutes	Setup Hours
3.1-S	Setup, 18 blanks on a strip	.25			.25
3.1-2	Get strip 37 × 80	.34	1/18	.019	
3.1-3	Trim one edge	.07	1/18	.004	
3.1-4	Push shear, 4¼ in.	.01		.01	
3.1-7	Remove last piece from machine	.07	1/18	.004	
3.1-11	Roll edges using timesaver	.12		.12	
	Total			.157	.25
	Lot estimate for 200 units	.77 hr			

C. A formed hatch cover is made from 5052-H32 aluminum. Raw material is in 48 × 120-in. sheets. The first operation is to shear strips, and each strip will have 9 blanks per strip. The lot quantity is 72 units.

Table Number	Process Description	Table Time	Adjustment Factor	Cycle Minutes	Setup Hours
	Shear strips 9-in. wide × 120 long				

Table Number	Process Description	Table Time	Adjustment Factor	Cycle Minutes	Setup Hours
	5 strips per sheet				
	9 blanks per strip				
	45 parts per sheet				
3.1.S	Setup	0.25			0.25
3.1.1	Start and stop	0.03	1/2	.00	
3.1.2	Get sheet, strip	0.58	1/45	.01	
3.1.3	Trim and square	0.11	1/45	.00	
3.1.4	Shear push < 9 in. 9 blanks per strip	0.04	1/9	.00	
3.1.7	Remove blank	0.04	1/9	.00	
3.1.9	Remove	0.01		.01	
3.1.10	Move material	0.36	1/45	.01	
	Material is in strip				
	Total			0.04	0.25
	Lot hours	.30 hr			

Now consider the shearing of blanks from strip where the strips are completed using the above estimate. There are 9 pieces on a strip.

Table Number	Process Description	Table Time	Adjustment Factor	Cycle Minutes	Setup Hours
	Shear blanks 9 × 12.5				
3.1.S	Setup	0.25			0.25
3.1.1	Start and stop	0.03	1/2	.00	
3.1.2	Get sheet, strip	0.58	1/9	0.06	
3.1.3	Trim and square	0.09	1/9	0.01	
3.1.4	Shear blank 12.5 push	0.05		0.05	
3.1.7	Remove blank	0.04	1/9	.00	
3.1.10	Move material	0.36	1/2	.01	
3.1.10	Move material	0.35	1/2	.00	
	Material is blank				
	9 × 12.5 × 1/8 in.				
	Total estimates			0.14	0.25
	Lot hours	.42			
	Total for job	.72			

D. PN 8871 is shown by Figure 3.2B. Material is 18-ga CRS and is supplied in 4 × 8 96 ft sheets. A quantity of 2500 is required. Output of this operation is strips in 2.750-in. dimension × 48-in. long. Each strip has 21 pc, and each sheet will have 714 pc. The subsequent operation to shear strips is to shear blanks, which is a different operation. <u>The adjustment factor uses various dividers to adjust for the output of one unit.</u> For example, the division of the start and stop time of .03 by 2500 gives a really insignificant time on a per unit basis but is shown for the sake of accuracy. The adjustment by 2500 says that the job is started and stopped only once for the entire job. The output of .03/2500 is insignificant, but to leave it at .03 per unit when it is not required also will excessively inflate the unit time. For jobs which have a large quantity of output, the start and stop might be ignored, but is shown here for completeness. In this estimate we want to find the unit, setup, and the total lot time.

Table Number	Process Description	Table Time	Adjustment Factor	Cycle Minutes	Setup Hours
3.1-S	Setup	0.25			0.25
3.1-1	Start and stop	0.03	1/2500	0.00	
3.1-2	Get sheet, strip	0.42	1/714	0.00	
3.1-3	Trim and square	0.09	1/714	0.00	
3.1-4	Shear push 2.75-in. strip which has 21 pc per strip	0.01	1/21	0.00	
3.1-7	Remove scrap from sheet	0.09	1/714	0.00	

Table Number	Process Description	Table Time	Adjustment Factor	Cycle Minutes	Setup Hours
3.1-9	Remove from back of shear in strips	0.04	1/21	0.00	
3.1-10	Move material	0.36	1/2500	0.00	
3.1-10	Move material	0.35	1/2500	0.00	
	Total			0.01	0.25
	Total lot hours	0.57			

E. The strips of Example D are to be sheared in a second operation referred to as "shear blanks." The incoming raw material is the 2.75 × 48-in. strips and the output of the operation is the 2.75 × 2.29-in. blank, which is ready for a pierce and blanking operation. The lot is 2500 units. Estimate the time for cycle, setup, and the lot. There are 21 pieces on a strip.

Table Number	Process Description	Table Time	Adjustment Factor	Cycle Minutes	Setup Hours
3.1.S	Setup	0.25			0.25
3.1.1	Start and stop	0.03	1/2500	0.00	
3.1.2	Get strip with 21 pc	0.05	1/21	0.002	
3.1.3	Trim and square	0.02	1/21	0.001	
3.1.4	Shear 2.29-in. dim	0.01		0.01	
3.1.9	Remove	0.01		0.01	
3.1.10	Move material	0.36	1/2500	0.00	
	Total			.023	0.25
	Total lot hours	1.23			

TABLE 3.1   POWER SHEAR MACHINES

Setup   .25 hr

Operation elements in estimating minutes

1. Start and stop   .03

2. Get one sheet or strip and load on machine

				Length				
Width	60	72	84	96	108	120	136	144
6			.05	.13	.21	.29	.40	.47
12			.09	.17	.25	.34	.45	.50
20		.06	.14	.23	.31	.39	.51	.56
24		.09	.17	.25	.34	.42	.53	.58
30	.05	.13	.21	.29	.38	.46	.57	.63
36	.09	.17	.25	.34	.42	.50	.61	.67
48	.17	.25	.34	.42	.50	.58	.69	.75

Hoist transport   1.79

3. Trim and square (includes a shear) one sheet or strip

				Length				
Width	60	72	84	96	108	120	136	144
24	.02	.04	.05	.06	.07	.09	.10	.11
36	.04	.05	.06	.07	.09	.10	.12	.12
48	.05	.06	.07	.09	.10	.11	.13	.14

4. Shear:
   1ft on and off: .02

Push Length	5	7	10	13	16	20
Shear Min	.01	.02	.04	.05	.07	.08
Push Length	25	30	35	40	45	50
Shear Min	.12	.14	.16	.18	.22	.24

5. Reverse—turn sheet, strip, or blank, around or over, or rotate side

Width	Length						
	30	60	72	96	108	120	144
12		.04	.06	.09	.11	.13	.16
24	.01	.06	.07	.11	.13	.15	.18
36	.03	.07	.09	.13	.15	.16	.20
48	.05	.09	.11	.15	.16	.18	.22

   Hoist reversal   1.27

6. Relocate—after reversal, use trim element

7. Remove blank or scrap from machine and place on pile or table, size

Width	Length				
	20	30	36	40	48
10		.04	.05	.07	.09
15		.05	.07	.08	.10
20	.04	.07	.08	.09	.12
30	.07	.09	.11	.12	.14

   Hoist aside of sheet, strip, blank, or scrap   1.42

8. Remove piece and drop—50% of 7, blank size

9. Remove—pick up from back of machine, blank or strip size
   - to   6 × 18:   .01
   - to   18 × 24:   .02
   - Above   24:   .04

10. Move material
    - Walk to rear of shear and return   .36
    - Move truck from rear to front   .35
    - Hand truck skid of stock or pans away and return   .88

11. Roll edges on abrasive belt machine

Length	Min	Length	Min
4.0	.03	24.2	.08
7.5	.04	32.7	.10
11.9	.05	43.4	.12
17.3	.06	56.8	.15

3.2 Punch Press Machines (First Operation)

DESCRIPTION

First operations for punch press machines are construed to involve strip or coil stock materials. There are many operations including blanking, perforating, cutting, etc. Part design, type of die, and press selection must be known or visualized by the estimator in estimating this high-speed production process. This operation may occur on a variety of presses. The advantage of an inclined press over a straight-side press, or vice versa, is dictated by governing conditions within the plant. These estimating data focus on the manual or automatic elements, rather than a specific press. A straight-side eccentric-geared press is shown in Figure 3.2A. These presses have large die areas and long stroke with engagement high upon the stroke for drawing. Figure 3.2B has a front-back bolster dimension of 42 in. and a left-right dimension of 72 in.

ESTIMATING DATA DISCUSSION

The setup is 0.8 hr. This includes setup and teardown to a clean bolster plate. Exceptions for additional setup hours are provided.

The elemental times are affected by part design, die, and presses. These factors are either identified for separate selection, or are averaged in the data. The estimating data are for single gang dies. Divide the final estimate by the number of gangs when using a multiple gang die. When one operator tends two or more presses, divide number of estimated man-minutes by the number of machines assigned to the operator. In a progressive die, for instance, each stroke of the press produces one part.

Raw material is either strips or coils. The strips have been previously sheared to width, and lengths can vary up to the length of the sheet stock. Coils are thinner material and length varies with the application. Strip stock calls for operator handling, while coil stock requires operator tending.

Elements 1 and 4, loading and oiling strip(s), can be for one unit or for a bundle, and the unit time depends on the number of pieces per strip and the strips handled. In Element 6, strip advance depends on the type of stop, pins of various kinds, or a sight-stop. If the part advance, which depends on the layout of the part on the strip, exceeds 6 in., additional increments of time can be calculated and included.

Machine time selection for strips depends on several factors. Some or all of the part advance may occur during the ram upstroke. In continuous run, the operator trips the pedal and the press continues to stroke until the material is run or the operator stops the press. This method of operation may be used on strip stock, providing the advance time does not exceed two-thirds of the continuous machine time, and it is believed that the motion pattern is acceptable. If the advance is slower than a continuous press stroke cycle, then intermittent stroking, using either a foot pedal or hand buttons to trip the press, is necessary. In intermittent run, there is a clutch lag time included for engagement. Conditions for intermittent run depend on the opportunity to advance the strip during ram upstroke.

Machine time for automatic coil operation may be the only operational time included in the estimate.

FIGURE 3.2A A 300-ton double crank gap frame press. *(Niagara Machine and Tool Works)*

Several average conditions are given where the minutes exceed the expected direct output from spm. Also, some spm averages, based upon press tonnage, are given. For a specific job, the speed of a press depends upon the length of stroke, nature of operation (blanking vs. drawing), physical properties of the material, and the method and speed of loading and unloading the work.

EXAMPLES

A 6-ft strip has a width of 1.36 in. and a weight of 1.72 lb. The number of pc/ft is about 6, thus the advance is 2 in. The die is visualized to have an open-end stop, and a press machine having an 83 crank rpm is planned for the process. Estimate the job with different assumptions for the following three examples.

A. Assume intermittent machine operation and advance during the nonmachine cycle.

Table Number	Process Description	Table Time	Adjustment Factor	Cycle Minutes	Setup Hours
3.2-1	Load strips	.20	1/36	.006	
3.2-2	Pick up strip, move to die	.035	1/36	.001	
3.2-3	Assemble strip to die	.019	1/36	.001	
3.2-4	Oil strips	.25	1/36	.007	
3.2-5	Trip press, each piece	.006		.006	
3.2-6	Advance strip	.007		.007	
3.2-7	Medium-speed press machine time	.022		.022	
3.2-8	Dispose of scrap in skeleton	.092	1/36	.003	
	Total			.053	

B. Consider the same problem with intermittent run, but die and part conditions allow some of the machine cycle for advance of stock.

Table Number	Process Description	Table Time	Adjustment Factor	Cycle Minutes	Setup Hours
Repeat	Repeat Elements 1–6 from Example A	.027		.027	
3.2-7	Machine time	.017		.017	
3.2-8	Dispose of scrap	.092	1/36	.003	
	Total			.047	

C. Estimate the same job for continuous running.

Table Number	Process Description	Table Time	Adjustment Factor	Cycle Minutes	Setup Hours
Repeat	Repeat Elements 1–4 from Example A	.015		.015	
3.2-5	Trip press	.006	1/36	.0002	
3.2-6	Advance strip, within machine time	0		0	
3.2-7	Continuous run	.017		.017	
3.2-8	Dispose of scrap	.092	1/36	.0026	
	Total			.0348	

D. Refer to Figure 3.2B for a sketch of a sheet metal part. The material for the bracket is 18 ga (.048-in.) CRS and is supplied to the operation in 2.75 × 48-in. strips which were previously sheared. Each strip has 21 pc and the lot is 2500 units. This operation converts the strip to blanks which are also pierced, but not formed. The blanking die will leave the final outside shape. Note in the following Adjustment Factor that 21 is frequently the divisor, which converts a unit of work input to a time for one unit of output.

Table Number	Process Description	Table Time	Adjustment Factor	Cycle Minutes	Setup Hours
3.2-S	Setup for strip stock	0.80			0.80
3.2-S	Setup with knockout	0.05			0.05

Punch Press Machines (First Operation) 99

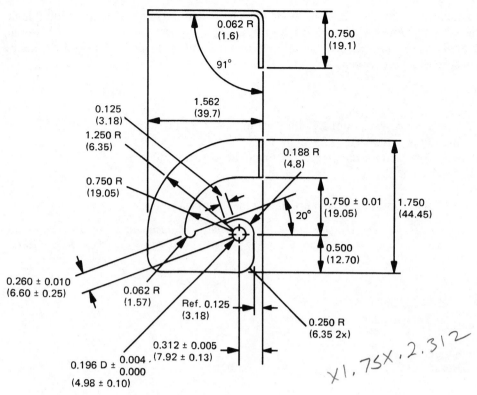

FIGURE 3.2B Part for first-operation punch press equipment.

Table Number	Process Description	Table Time	Adjustment Factor	Cycle Minutes	Setup Hours
3.2-1	Load strip(s); each strip has 21 pc	0.20	1/21	0.01	
3.2-2	Pick up strip	0.04	1/21	0.00	
3.2-3	Insert in die	0.02	1/21	0.00	
3.2-4	Oil strip(s) every 5 strips	0.25	1/105	0.00	
3.2-5	Trip	0.01		0.01	
3.2-6	Advance	0.01		0.01	
3.2-7	Strip stock punch 80 rpm, intermittent	0.02		0.02	
3.2-10	Remove skeleton	0.09	1/21	0.00	—
	Total			0.06	0.85
	Total lot hours	3.47			

TABLE 3.2 PUNCH PRESS MACHINES (FIRST OPERATION)

Basic Setup .80 hr
    Add for features of ejection, knockout, rubber cushion, subplate .05 ea
    Oven, rewind reel, straightener .25 ea

Operation elements in estimating minutes

1. Load strip(s) from skid to feed board, occurrence .20
   Load coil on reel by truck, crane, hoist, occurrence 4-25

2. Pick up strip and move to die .035

3. Insert strip in die .019
   Add'l weight over 2 lb .006/lb
   Add'l for no pins or mechanical stops .004
   Add'l for width over 6in., each 6 in. .016

4. Oil strip(s)	.25
5. Trip press, each piece or first piece only	—
Foot pedal	.006
Hand buttons	.015
6. Advance strip	.007
To push or pull stops	.007
Over pin stop	.017
Over mechanical stop	.028
Add'l per in. of advance over 6	.001

7. Punch
A. Strip stock; each

Speed	Crank rpm	Continuous, min	Intermittent Run, min	Intermittent Run and Partial Advance, min
Low	30	.020	.026	.019
	50	.021	.025	.018
Medium	80	.017	.022	.017
	110	.016	.020	.015
Higher	150	.013	.018	.013
	250	.007	.011	.006
	500	.003	.004	.003

B. Roll feed; punch each

SPM	Min	SPM	Min	SPM	Min
100	.014	200	.008	600	.0017
125	.012	225	.006	700	.0014
150	.011	250	.004	800	.0013
175	.009	500	.002	1000	.001

C. Press tonnage, punch; each

Press Capacity, Tonnage	Min
22	.017
48	.020
75	.018
150	.092
Higher	.11+

Double if fiber, thin rubber, or paper

8. Remove strip skeleton from die, dispose   .092

3.3 Punch Press Machines (Secondary Operations)

DESCRIPTION

Press work estimates can be developed on the source of power (manual, mechanical, hydraulic, pneumatic); ram (single or double acting); design of frame (bench, inclinable, gap, arch, straight-side, horn, pillar); according to method of applying power (crank, cam, eccentric, screw); purpose of press (squaring shears, brake, punching, drawing, extruding, forming, coining); or types of dies. But in this section, two basic types of secondary punch-press operations are considered. The first is the operation which utilizes conventional dies. The second type uses standard punches and dies. Either operation may be single or multiple stroke. In a single-stroke operation, the operator picks up a part, loads the part, trips the press, and removes and places the part aside. In multiple-stroke operations, the operator picks up a part, loads the part, trips the press, relocates the part, trips the press, and continues the relocation and tripping until the operation is completed.

Any die, other than a first-operation die, is considered a second-operation die. It may be the fifth, second, etc., for the specific part, but in die terminology it is considered a second-operation die. Many operations that can be classified as second operations on sheet metal include coining, countersinking, drawing, dinking, embossing, extruding, flattening, forming, necking, notching, piercing, pinching, redrawing, shaving, shearing, sizing, slotting, staking, stamping, swaging, trimming, etc.

One type of press that is useful for both first and secondary operations is shown in Figure 3.3. The mechanical advantage of the box section gives stiff support to the crankshaft at the point of load application. In addition to operator-station palm buttons, plug-in foot switches are available. This press can be inclined to allow rear ejection of parts. Punching holes in thick steel plates or blanking steel with shear or drawing operations usually requires a slow-punch movement to reduce shock and to increase die life. Figure 3.3, a 90-ton OBI press, is available in 72, 60, 46, or 36 spm speeds.

FIGURE 3.3 A 90-ton open-back inclinable press with front-to-back crank shaft. *(Niagara Machine and Tool Works)*

ESTIMATING DATA DISCUSSION

Element 1 is an inclusive element involving the following: stock from skid, box, etc. to bolster plate; load-in die or on punch and die; stroke of press where hands touch buttons; ram down and up; release of buttons; remove part, stack in box, on skid, etc.; and occasional clean out of slugs. A machine stroke constant of .048 standard min is used. The reposition element starts at release of buttons following ram ascent, reach to piece in die or on die and punch buttons, reposition, turn around or over in die, punch buttons, and ram descends. The estimates are one-man, and if two operators are required, the values are doubled. All ferrous and nonferrous materials are included in stock sizes. Distinctions for presses are averaged. While the time to perform the operation varies with type of operation and part, only the blank size has been used as the time driver for location points on part, size, and weight methods of ejection, speed of machine.

EXAMPLES

A. A part has been blanked to dimension 8 × 22½ in., and one secondary operation of piercing holes using a pierce die is planned. A lot quantity of 2820 is required. Find the estimate.

Table Number	Process Description	Table Time	Adjustment Factor	Cycle Minutes	Setup Hours
3.3-S	Setup	.65			.65
3.3-1	Blank $L + W = 32\frac{1}{2}$ and using next higher table value	.23		.23	—
	Total			.23	.65
	Lot estimate	11.46 hr			
	Hr/100	.383			
	Pc/hr	260			

B. A ¼-in. plate is to be notched and pierced in a secondary operation. A lot quantity of 2820 units is planned for a flat blank size of 26 × 32 in. Find the estimates.

Table Number	Process Description	Table Time	Adjustment Factor	Cycle Minutes	Setup Hours
3.3-S	Setup	.65			.65
3.3-1	Blank size of 58 in., but using next higher table entry	.36		.36	
3.3-3	Addition for heavy plate	.08		.08	—
	Total			.44	.65
	Lot estimate	21.33 hr			
	Hr/100	.733			
	Pc/hr	136			

C. Refer to Figure 3.2B for a sketch of an OEM part. Assume that flat blanks have been sheared to size for a blanking and piercing operation. The blanking and piercing operation will give a flat blank ready for the later operation of forming. Assume that for this operation, the size of the material is 2.75 × 2.29 × .048 in., and a lot quantity of 2500 is required. The entry value for this operation is $L + W = 5.04$.

Table Number	Process Description	Table Time	Adjustment Factor	Cycle Minutes	Setup Hours
3.3-S	Setup for pierce and blank die	0.65			0.65
3.3-1	Pierce, blank	0.09		0.09	
	Total			0.09	0.65
	Total lot hours	4.23			

D. Refer to Figure 3.2B and Example C. Assume that the part has been blanked and pierced to size. Now the estimate is for the 91-degree forming operation where the part has a $L + W = 5.14$-in. dimension, which is the same as in Example C. For a forming operation using the identical equipment as in Example C, the estimates are identical.

TABLE 3.3 PUNCH PRESS MACHINES (SECONDARY OPERATIONS)

Setup .65 hr

Operation elements in estimating minutes

1. Pierce, blank, form, emboss, 1 stroke

L + W	Min	L + W	Min	L + W	Min
2.0	.071	24.0	.18	79.4	.45
2.7	.075	25.8	.19	84.0	.48
3.5	.078	27.7	.20	88.9	.50
4.3	.082	29.7	.21	93.9	.53
5.1	.086	31.8	.22	99.2	.55
6.0	.091	34.0	.23	104.8	.58
6.9	.095	36.3	.24	110.7	.61
7.9	.100	38.8	.25	116.8	.64
8.9	.105	41.3	.27	123.3	.67
9.9	.110	44.0	.28	130.1	.70
11.0	.116	46.8	.29	137.2	.74
12.2	.122	49.8	.31	Add'l	.005
13.4	.128	52.9	.32		
14.7	.134	56.1	.34		
16.1	.141	59.6	.36		
17.5	.143	63.2	.37		
19.0	.155	66.9	.39		
20.6	.163	70.9	.41		
22.2	.171	75.1	.43		

Note: Estimate is for one operator. Double if size requires two operators.

2. Reposition part and restroke, each time

L + W	Min
0–10	.06
11–20	.09
21–30	.14
31–60	.16

3. Miscellaneous

Deduct for air eject if small part	0.33
Deduct for toss if medium part	.021
Add for parts that tangle	.018
Two-station die complete 1 part per stroke, add	.021
One-station die complete 1 part each 2 strokes, add	.036
Pry part out of nest, add	.06
For 7-ga or ¼-in. plate:	
2.0–17.5	.05
19.0–52.9	.06
56.6–	.08

3.4 Turret Punch Press Machines

DESCRIPTION

Turret punch presses are especially adapted for the production of flat sheet metal parts having varied hole patterns of many sizes. For some presses, a templet locates the holes, and the hole size is selected from a cylindrical turret containing punches. Other turret punch presses are NC, CNC, or DNC. Tools are located in the turret for instant use.

Figure 3.4 is an example of a 30-ton CNC machine having a 50-in. throat. It handles sheets up to 48×72 in., and there are 32 turret stations. Position and hit speed on this machine is 175 hits/min with 1-in. centers. Slower speeds are necessary for longer distances. Now, consider estimating data for the general class of turret punch press machines.

ESTIMATING DATA DISCUSSION

Setup is related to the number of tool stations. If standard tools are garrisoned permanently in the turret, it may be unnecessary to include this as a tool item for setup. But the number of different punch and die configurations is used as a time driver. Usual elements of setup are also included. Neither templet or tape programming is included, but tape preparation may require 15 to 30 min per tape.

The load and unload time is related to blank size or length plus width ($L + W$). Fingers which clamp the blank may be used. A reposition of the blank is included for operational selection.

Depending on the selection of the turret by the NC tape, Element 2 includes turret rotation and piercing the first hole or hit. Element 3 is used for piercing holes remaining after the first hole. For instance, if ten .250-in. holes were pierced, Element 2 covers the first hole and Element 3 provides for the remaining nine holes. The distance between holes and rotational time between successive punch and die sets are averaged. A hole or opening may not be the result of one hit. Thus, Element 4 is time per hit. Openings, splitting, nibbling, notching, shearing, and punching action may call for hits rather than holes. A faster and slower time are two distinctions that are recognized. The hit time for the faster machine is 0.008 min for typical parts and 0.033 min for slower machines. These values are incorporated into a table allowing for selection.

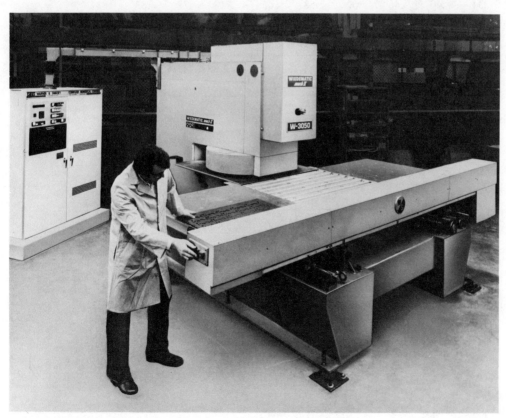

FIGURE 3.4 Turret punch press, 30 ton with 50-in. throat and 32 tool stations. (*Wiedemann Division, Warner & Swasey Company*)

EXAMPLES

A. A sheet metal part, 12.255 × 10.065-in. 7 ga aluminum has seven different sizes of holes, two similar corner-angle notches, and one slot for a total of 49 openings. The slot is odd-shaped, and requires two hits of a specially shaped punch. Each notch requires two hits. Total number of hits is 52, and the number of tool stations is 10. Lot quantity is 63. Using a high-speed machine, find the unit, lot, and the shop estimates.

Table Number	Process Description	Table Time	Adjustment Factor	Cycle Minutes	Setup Hours
3.4-S	Setup for 10 stations	.83			.83
3.4-1	Load and unload blank, $L + W = 22.320$ in.	.34		.34	
3.4-2	First hit of each size, 10	.65		.65	
3.4-4	Remaining hits = 52 − 10 = 42	.40		.40	
	Total			1.39	.83
	Lot estimate	2.29 hr			
	Shop estimate	43 pc/hr			

TABLE 3.4 TURRET PUNCH PRESS MACHINES

Setup

No. of Stations (TOOL STAT.)	1	2	3	4	5	6	7	8	9	10
Hr	.28	.34	.40	.46	.52	.58	.65	.71	.77	.83

No. of Stations	11	12	13	14	15	16	17	18	19	20	Add'l
Hr	.89	.95	1.01	1.07	1.14	1.20	1.26	1.32	1.38	1.44	.061

Operation elements in estimating minutes

1. Handle

$L + W$	Min	$L + W$	Min	$L + W$	Min
8.0	.22	36.6	.42	93	.76
15.5	.27	51.2	.53	121	.87
24.9	.34	69.5	.66	Add'l	.007

   Reposition part end for end, small   .76

2. Rotate turret and punch first hole or hit of each size

No.	Min	No.	Min	No.	Min
1	.07	8	.52	15	.98
2	.13	9	.59	16	1.05
3	.20	10	.65	17	1.11
4	.26	11	.72	18	1.18
5	.33	12	.78	19	1.24
6	.39	13	.85	20	1.31
7	.46	14	.91	Add'l	.065

3. Punch remaining holes

Hole	Min	Hole	Min	Hole	Min
1	.04	18	.65	50	1.80
2	.07	19	.68	55	1.98
3	.11	20	.72	60	2.16
4	.14	22	.79	65	2.34
5	.18	24	.86	70	2.52
6	.22	26	.93	75	2.70
7	.25	28	1.01	80	2.87
8	.29	30	1.08	85	3.05
9	.32	32	1.15	90	3.23
10	.36	34	1.22	95	3.41
11	.40	36	1.29	100	3.59
12	.43	38	1.37	105	3.77
13	.47	40	1.44	110	3.95
14	.50	42	1.51	115	4.13
15	.54	44	1.58	120	4.31
16	.57	46	1.65	125	4.49
17	.61	48	1.72	Add'l	.036

4. Punch remaining hits

Low-Speed Machine

Hits	Min	Hits	Min
1	.03	20	.66
2	.07	30	.99
3	.10	40	1.32
4	.13	50	1.65
5	.17	60	1.98
6	.20	70	2.31
7	.23	80	2.64
8	.26	90	2.97
9	.30	100	3.30
10	.33	Add'l	.033

High-Speed Machine

Hits	Min	Hits	Min
1	.008	20	.16
2	.02	30	.24
3	.02	40	.32
4	.03	50	.40
5	.04	60	.48
6	.05	70	.56
7	.06	80	.64
8	.06	90	.71
9	.07	100	.79
10	.08	Add'l	.008

3.5 Single-Station Punching Machines

DESCRIPTION

Punching machines have power-driven punch and their location may be by numerical control, stops, or a punched-hole templet. These machines are designed for hole-punching from prototype to medium-run quantities. Tooling changeover is minimal, as the punch holder swings out and a replaceable punch can be inserted. Die removal is simple as well. Finger-gripping of parts is common. If a machine is templet con-

FIGURE 3.5 Hole punching machine with a 30-in. throat and a 30 × 60 in. maximum sheet size. *(Strippit Houdaille)*

trolled, the operator locates a stylus point in each pilot hole, and concurrently the workpiece is positioned under the punch. The ram is automatically tripped, punching the hole. Templet pilot holes can be color-keyed to specific punch sizes, permitting the operator to punch all holes of the same size or shape before changing the punch and die. Front and back gage setting for hole locations are also available.

Figure 3.5 is an example of a hole-punching machine having a throat depth of 30 in., maximum hole diameter of 3½ in. in 12 ga, sheet size of 30 × 60 in., and a maximum of 165 hits/min.

ESTIMATING DATA DISCUSSION

For each size hole the setup is 0.15 hr and includes items of work customarily associated with machine setup and teardown to a neutral machine table. Oftentimes, in successive hole diameters, the only work is a hole diameter change. The subsequent setup is reduced. Run time includes only Element 1, as it involves a pickup blank or strip, punch, and piece aside to stack. For certain materials, such as stainless steel or materials over 3/16-in. thick, or diameters over 3/4 in., oil time may be allowed, as found in Element 2.

EXAMPLES

A. A sheet metal blank, 13-ga hot-rolled, pickled and oiled stock, has a $L + W$ dimension of 27.15 in., and 13 holes are punched. Estimate the lot time for 75 parts and the hr/100 units.

Table Number	Process Description	Table Time	Adjustment Factor	Cycle Minutes	Setup Hours
3.5-S	Setup for 1 hole	.15			.15
3.5-1	Pierce 13 holes in blank of $L + W = 27.15$	.48		.48	—
	Total			.48	.15
	Lot estimate	.75 hr			
	Hr/100 units	0.800			

B. A part has a blank size $L + W = 27.15$ in. The number of holes is seven. There are 75 parts. The operation is a continuation of Example A.

Table Number	Process Description	Table Time	Adjustment Factor	Cycle Minutes	Setup Hours
3.5-S	Setup	.05			.05
3.5-1	Pickup, 7 holes of same size, unload	.29		.29	—
	Total			.29	.05
	Lot	.41			
	Hr/100	.483			

TABLE 3.5 SINGLE-STATION PUNCHING MACHINES

Setup
  1st hole diameter   .15 hr
  Subsequent hole diameter, same part   .05 hr

Operation elements in estimating minutes

1. Pick up blank, punch, blank aside

					Holes						
L + W	1	2	3	4	5	6	7	8	9	10	Add'l
3	.13	.17	.20	.23	.25	.29					.03
3.5	.12	.15	.19	.22	.25	.28	.31				.03
4	.11	.14	.18	.21	.24	.28	.31	.34			.03
6	.10	.13	.17	.20	.23	.26	.30	.33	.36	.40	.03
15	.09	.12	.15	.19	.22	.25	.29	.32	.35	.39	.08
25	.10	.13	.17	.20	.23	.26	.30	.33	.36	.40	.03
30	.11	.14	.18	.20	.22	.25	.29	.32	.35	.39	.03
32	.12	.15	.19	.21	.23	.26	.30	.33	.36	.40	.03
34	.13	.17	.20	.22	.24	.28	.31	.34	.37	.41	.03
36	.14	.18	.21	.24	.28	.31	.34	.37	.41	.44	.03
40	.15	.19	.22	.25	.29	.32	.39	.39	.42	.45	.04
44	.17	.21	.24	.29	.33	.37	.42	.46	.51	.55	.04

Add'l L + W   .015   Add'l hole   .05

2. Oil time

Holes	1	2	3	4	5	6	7	8	9	10	Add'l
Min	.03	.04	.06	.07	.08	.09	.10	.11	.12	.13	.01

3. Reposition 180°, flip, or turn

L + W	25	35	44
Min	.03	.04	.05

3.6 Power Press Brake Machines

DESCRIPTION

Press brakes are used to brake, form, seam, trim, and punch light-gage sheet metal. Pressure capacity of a press brake is established by the material, length of work, thickness of the metal, and radius of the bend. Minimum inside radius of a bend is usually limited to material thickness. Press brakes have short strokes and are generally equipped with an eccentric type of drive mechanism.

Conventional power press brakes may be either hydraulic or mechanical. Hydraulic presses are more popular for larger tonnages. Figure 3.6 is a small hydraulic press brake equipped with numerically controlled programmable back gage and depth stop, which enables an operator to punch in a program of a sequence of different dimensions at various positions on a sheet and the press automatically adjusts to the settings.

ESTIMATING DATA DISCUSSION

Setup hours depend on brake length and number of stops. The length of a 30-in. brake separates the time per length. Setup standards are for conventional machines.

The operation elements are compiled with time as the central column. The data are constructed to require lip length (lip in.) of the folded metal, then dropping down to the sum of the blank width and length (L + W). At this intersection move horizontally to the middle and read min. The first braking elements are composed of the work needed to move the material to and from the machine. Any additional brakes include work to reposition the material and expose a new lip. In a corresponding way, for each additional brake, start by using the next higher lip length, then descend vertically to the next higher blank L + W. At this point, move horizontally to the center for time.

EXAMPLES

A. A sheet metal part having a flat blank size of 16 × 24 in. has one fold of 4 in. along the 16-in. dimension and a second fold of 6 in. along the 24-in. dimension. Standard commercial tolerances are required. A lot of 756 is planned.

FIGURE 3.6 A 100-ton capacity hydraulic press brake with NC programmable backgage and depth stop. *(Niagara Machine & Tool Works)*

Table Number	Process Description	Table Time	Adjustment Factor	Cycle Minutes	Setup Hours
3.6-S	Setup	.30			.30
3.6-1A	4-in. lip and 16 + 24 = 40 in.	.21		.21	
3.6-1B	Reposition and brake 6 in., use the 8-in. lip and 46.3 blank size	.16		.16	—
	Total			.37	.30
	Total lot estimate	4.96 hr			

B. A chassis is folded four times with 3.65-in. lips. Total $L + W = 46.2$ in. A lot of 231 units is required. Find the estimates.

Table Number	Process Description	Table Time	Adjustment Factor	Cycle Minutes	Setup Hours
3.6-S	Setup	.25			.25
3.6-1A	Constant for first lip	.23		.23	
3.6-1B	Additional lips	.16	3	.48	—
	Total			.71	.25
	Lot estimate	2.98			
	Hr/100	1.183			

TABLE 3.6 POWER PRESS BRAKE MACHINES

Setup

	Stops	
Brake Length	1	2
Under 30 in.	.25 hr	.30 hr
Over 30 in.	.30 hr	.35 hr

Operation elements in estimating minutes

1. Brake

A.	First Brake, $L + W$					B.	Add'l Brake, $L + W$			
	Lip In.						Lip In.			
1	2	4	8	16	Min	16	8	4	2	1
2.0					.05					
3.6	3.5				.06					2.9
5.3	5.2				.07				5.5	5.7
9.2	9.1	9.0			.08		10.3	11.5	12.1	12.3
11.5	11.4	11.2	10.9		.09		14.1	15.3	15.9	16.1
14.0	13.9	13.7	13.4		.10		18.3	19.4	20.0	20.3
16.7	16.6	16.5	16.2		.11	20.5	22.9	24.0	24.6	24.9
19.7	19.7	19.5	19.2	18.5	.12	25.5	27.9	29.1	29.7	30.0
23.1	23.0	22.8	22.5	21.9	.13	31.1	33.5	34.7	35.3	35.6
26.7	26.6	26.5	26.1	25.5	.14	37.2	39.6	40.8	41.4	41.7
30.7	30.6	30.5	30.2	29.5	.16	44.0	46.3	47.5	48.1	48.4
35.1	35.1	34.9	34.6	33.9	.17	51.4	53.8	54.9	55.5	55.8
40.0	39.9	39.8	39.4	38.8	.19	59.5	61.9	63.1	63.7	64.0
45.4	45.3	45.1	44.8	44.2	.21	68.5	70.9	72.0	72.6	72.9
51.2	51.2	51.0	50.7	50.0	.23	78.4	80.7	81.9	82.5	82.8
57.7	57.6	57.5	57.2	56.5	.25	89.2	91.6	92.8	93.3	93.6

First Brake, L + W (Cont'd.)						Add'l Brake, L + W (Cont'd.)				
Lip In.						Lip In.				
1	2	4	8	16	Min	16	8	4	2	1
64.8	64.8	64.6	64.3	63.6	.28	101.1	103.5	104.7	105.3	105.6
72.7	72.6	72.4	72.1	71.5	.30	114.3	116.6	117.8	118.4	118.7
81.3	81.2	81.0	80.7	80.1	.34	128.7	131.1	132.3	132.8	133.1
90.8	90.7	90.5	90.2	89.6	.37	144.6	146.9	148.1	148.7	149.0
101.2	101.1	100.9	100.6	100.0	.41	162.0	164.4	165.6	166.2	166.5
112.6	112.6	112.4	112.1	111.4	.45					
125.3	125.5	125.0	124.7	124.1	.49					
139.1	139.0	138.9	138.6	137.9	.54					
154.4	154.3	154.1	153.8	153.2	.59					
171.2	171.1	170.9	170.6	170.0	.65					

Additional lip in., add .00028
Additional L + W, add .0035

Additional lip in., add .0062
Additional L + W, add .0021

Note: When L + W = 100 in. or L = 7 in. for 7 ga, time is for 2 workers.

2. If parts must be stacked, add .03
    For locating part on pins, add .08

3. Miscellaneous
    For 7 ga and ¼-in. plate, add for first brake .09
    Each add'l brake .11

3.7 Jump Shear, Kick Press, and Foot Brake Machines

DESCRIPTION

These machines are light duty and require leg and body motions to effect the shear, pierce, or brake operation. Motorized power is not involved. The stock cannot be oversized to the machine, i.e., stock too thick or long or hard temper to be sheared, pierced, or braked. Machine limitations with respect to the material must be known. These estimating data are between bench and powered sheet-metal machines. A jump shear as shown by Figure 3.7 has a 36-in. maximum shearing width of 16-ga mild steel. Now consider estimating data for the general class of jump shear, kick press, and foot brake machines.

FIGURE 3.7 Foot-operated shear with 36-in. wide capacity in 16-ga mild steel. *(Di-Acro Division, Houdaille Industries, Incorporated)*

ESTIMATING DATA DISCUSSION

The elemental data are the "get and place" type with entry variables as width and length of the sheet, strip, or blank. Times are for single operator, except for sizes in excess of 24 × 48 in. For a sheet 48 × 96 in. to be handled, two operators are suggested and Element 1 time is doubled or 2 × .18 = .36 min. Elements 2.4, and 5 also use Element 1.

EXAMPLES

A. A lot consists of 814 units. Shear a 6 × 48-in. strip from sheet stock 24 × 48 in. in size. Determine unit estimate and pc/hr for blanks 6 × 6 in. There are 4 strips of 6 × 48-in. size for 32 pc/sheet.

Table Number	Process Description	Table Time	Adjustment Factor	Cycle Minutes	Setup Hours
3.7-S	Setup	.4			.4
3.7-1	Pick up sheet, 2 operators	.09	2/32	.006	
3.7-2	Locate and lock, 2 operators, 1 time, 8 pc/strip	.09	2 × 2/8	.045	
3.7-3	Shear, 2 operators, 3 times, 8 pc/strip	.09	2 × 3/8	.068	
3.7-4	Reposition and lock, 2 operators, 2 times, 8 pc/strip	.09	2 × 2/8	.045	
3.7-5	Remove pc, 6 × 48, 2 operators, 4 times	.09	2 × 4/8	.09	
	Total			.18	.4
	pc/hr	333 units			
	Lot estimate	2.84 hr			

B. Shear a 6 × 6-in blank from a strip 6 × 48 in. Determine unit estimate and pc/hr. A lot quantity is 814 units. A strip has 8 units. One operator is used.

Table Number	Process Description	Table Time	Adjustment Factor	Cycle Minutes	Setup Hours
3.7-S	Setup	.1			.1
3.7-1	Pick up strip	.09	⅛	.011	
3.7-2	Locate and lock	.09	⅛	.011	
3.7-3	Shear, push length 6 in.	.12		.12	
3.7-4	Relocate and lock	.09		.09	
3.7-5	Unlock and remove, 1 pc	.06		.06	
3.7-6	Pick up pc	.06		.06	
	Total			.352	.1
	pc/hr	.70 units			
	Hr/100	.587			

C. Pierce 2 holes in the blank 6 × 6 in.

Table Number	Process Description	Table Time	Adjustment Factor	Cycle Minutes	Setup Hours
3.7-S	Setup	.10			.1
3.7-1	Pick up piece	.04		.04	
3.7-3	Pierce hole	.05		.05	
3.7-4	Reposition against stop	.04		.04	
3.7-3	Pierce second hole	.05		.05	
3.7-1	Remove piece	.04		.04	
	Total			.22	.1

Jump Shear, Kick Press, and Foot Brake Machines

D. Find the unit estimate for Example C where a brake of 2 in. is necessary.

Table Number	Process Description	Table Time	Adjustment Factor	Cycle Minutes	Setup Hours
3.7-S	Setup	.10			.1
3.7-1	Handle	.04		.04	
3.7-2	Locate and lock	.04		.04	
3.7-3	Brake	.06		.06	
3.7-5	Remove piece	.04		.04	
	Hr/100	.300		.18	.1

TABLE 3.7 JUMP SHEAR, KICK PRESS, AND FOOT BRAKE MACHINES

Setup
  1st   .4 hr
  Each add'l   .1 hr

Operation elements in estimating minutes

1. Pick up part to machine bed

Width	|				Length					
	3	6	9	12	18	24	36	48	60	72	96
3	.03	.04	.04	.04	.05	.06	.07	.08	.09	.10	.13
6	.03	.04	.05	.05	.06	.07	.08	.09	.10	.12	.14
12			.05	.06	.07	.08	.09	.12	.12	.12	.13
18					.07	.08	.09	.12	.12	.12	.13
24						.09	.10	.09	.09	.12	.15
36							.12	.09	.12	.14	.15
48								.12	.15	.18	.20

2. Locate and lock (use Element 1)

3. Shear or brake or pierce

Length	12	24	36	48
Time	.06	.07	.08	.09

Pierce hole   .05

4. Reposition and lock (use Element 1)
   Unlock and reverse piece (use Element 1)

5. Remove pc (use Element 1)

6. Unlock and remove to table or cart

Width					Length				
	9	12	18	24	36	48	60	72	96
3	.06	.06	.07	.08	.09	.10	.12	.13	.15
6	.06	.07	.08	.09	.10	.12	.13	.14	.16
12	.07	.08	.09	.10	.12	.14	.16	.14	.15
18			.09	.10	.12	.14	.16	.14	.15
24				.12	.13	.12	.12	.14	.17
36					.14	.12	.14	.16	.17
48						.14	.16	.19	.21

3.8 Hand-Operated Brake, Bender, Punch Press, Multiform, Coil Winder, Shear, Straightener, and Roller Machines

DESCRIPTION

This equipment is hand-operated and is classified as bench equipment, although some may be free standing. Like foot-operated equipment, the forces necessary to process thin gage sheet or foil stock are restricted to what an operator is physically able to perform. The equipment is varied. Figure 3.8A shows a hand-operated bender with a capacity of $\frac{1}{16}$ to 1-in. round steel bar. Round, square, hexagonal, channel, flat, and tubing can be handled by this machine. Figure 3.8B is a box finger brake and is rated to 16-ga mild steel. Figure 3.8C is a single-station punch with 4-in. capacity in 16-ga mild steel. Figure 3.8D is a 12-station turret punch press.

ESTIMATING DATA DISCUSSION

Inasmuch as the various pieces of equipment are small, manually operated, sheet-metal bench tools, additional tooling is normally general purpose and immediately available. The setup is 0.2 hr per item of equipment.

Elements for this equipment are described in general terms and are similar to get-place tables developed from time studies. Parts are small and light and are hand held. Small production quantities, where predesigned tooling is the exception rather than the rule, are the jobs that would use these estimating data.

FIGURE 3.8A Hand-operated bender. *(Di-Acro Houdaille)*

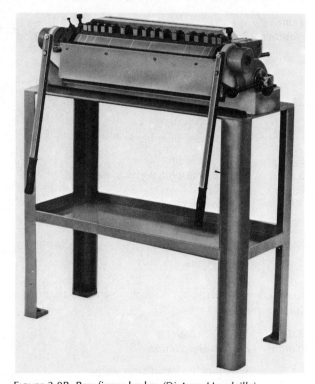

FIGURE 3.8B Box finger brake. *(Di-Acro Houdaille)*

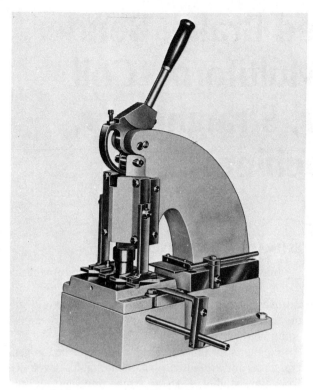

FIGURE 3.8C Single-station punch, hand operated. *(Di-Acro Houdaille)*

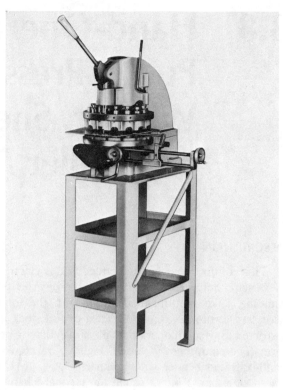

FIGURE 3.8D A 12-station turret punch press. *(Di-Acro Houdaille)*

EXAMPLES

A. Estimate the time to shear a 1.105 × 0.872-in. blank from 0.005-in. thick material, where sheet size is 18 × 24-in. stock. The shear is followed by a multiform operation. Compute the estimate as setup and hr/100 units.

Table Number	Process Description	Table Time	Adjustment Factor	Cycle Minutes	Setup Hours
3.8-S	Shear setup	0.2			.2
3.8-1	Pick up, move, position against stop	0.06		.06	
3.8-3	Shear twice	.02	2	.04	
3.8-4	Reposition for 1 shear	.08		.08	
3.8-5	Piece aside	.02		.02	
	Estimates for shear blank			.20	.2
3.8-S	Multiform setup	0.2			.2
3.8-1	Pick up, transport, position, clamp	0.07		.07	
3.8-3	Multiform	.07		.07	
3.8-5	Piece aside	.02		.02	
	Estimate for multiform			.16	.2
	Total setup	0.4 hr			
	Hr/100 units	.600			

B. A wire spring has 13 turns, ½-in. OD mandrel and two half-circle ends. Find the unit time to make this spring.

Table Number	Process Description	Table Time	Adjustment Factor	Cycle Minutes	Setup Hours
3.8-S	Wire-form and multiform setup	0.4			.4
3.8-2	Advance wire against stop	.03		.03	

Table Number	Process Description	Table Time	Adjustment Factor	Cycle Minutes	Setup Hours
3.8-2	Position mandrel, lock wire, back off, and remove	.51		.51	
3.8-3	Wind wire turns	.03	.0	.30	
3.8-3	Cut wire	.05		.05	
3.8-3	Form one end with multiform	.07		.07	
3.8-4	Reposition for multiform of end	.15		.15	
3.8-5	Piece aside	.02		.02	
	Total			1.13	.4

TABLE 3.8 HAND-OPERATED BRAKE, BENDER, PUNCH PRESS, MULTIFORM, COIL WINDER, SHEAR, STRAIGHTENER, AND ROLLER MACHINES

Setup   .2 hr

Operation elements in estimating minutes

1. Pick up, transport, position, clamp
   - For brake, multiform, sheet roller   .07
   - For bender, punch-press forming   .18

   Pick up transport, position against stop
   - For notch, shear, pierce   .06
   - Pick up, transport   .02

2. Advance strip or wire against stop   .03
   Position mandrel, lock wire, back off, remove wire   .51

3. Process
   - Brake, release   .05
   - Form or bend, release   .19
   - Punch press form   .05
   - Multiform   .07
   - Notch, shear, or pierce   .02
   - Air shear   .01
   - Cut wire   .05
   - Wind wire turns   .03/turn
   - Hand straighten   .25
   - Roll sheet flat or curving   .15

   Position between blocks and hammer blocks   .11

4. Reposition, clamp for each add'l lay   .08
   Reposition, reform, and release
   - Form or bend   .35
   - Punch press   .09
   - Brake   .08
   - Multiform   .15
   - Notch, shear, or pierce   .08

5. Piece aside   .02
   Piece aside, reach for new part   .06

3.9 Nibbling Machines

DESCRIPTION

The nibbling process is used to cut odd shapes in ferrous, nonferrous and nonmetallic materials. Depending on the machine, cuts up to approximately ½-in. thick can be made in mild steel. The nibbling machine has a punch that moves with a rapidly oscillating stroke in a die. A cut is made by moving the material against the pilot of the punch and inching the material along following each punch stroke. Templets or scribe lines are used as guides. Some nibbling machines can fold, bead, louvre, flange, and slot. Sizes of machines are rated by throat depth and the maximum thickness of mild steel that can be cut. An example of a nibbler is given in Figure 3.9.

ESTIMATING DATA DISCUSSION

There are two operation elements: handle part and cut or nibble. The entry variable for handling part is length plus width ($L + W$) of the sheet metal blank. For process time, Element 2, the entry variable is cut or nibble length. Cutting is assumed to be in a straight line while nibbling is curving. The processing times are for soft thin metals.

EXAMPLES

A. Find the unit min to nibble a 17-in. curve and cut a 15-in. straight line on a 5052 H2 aluminum blank which is 20 × 10 in. in size. Lot size is 87 units.

Table Number	Process Description	Table Time	Adjustment Factor	Cycle Minutes	Setup Hours
3.9-S	Setup	.2			.2
3.9-1	Blank size = $L + W$ = 30	.42		.42	

FIGURE 3.9 Nibbler punching machine (*Trumpf America, Incorporated*)

Table Number	Process Description	Table Time	Adjustment Factor	Cycle Minutes	Setup Hours
3.9-1	Reposition, 1 required	.29		.29	
3.9-2	Nibble 17-in. contour	.60		.60	
3.9-2	Cut 10-in. *L*	.17		.17	—
	Total			1.48	.2
	Lot estimate	2.35 hr			

TABLE 3.9 NIBBLING MACHINES

Setup .2 hr

Operation elements in estimating minutes

1. Handle part

L + W	Min	L + W	Min
12.0	.16	31.2	.42
13.2	.18	34.3	.47
14.5	.20	37.8	.51
16.0	.22	41.6	.56
17.6	.24	45.7	.62
19.4	.26	50.3	.68
21.3	.29	55.3	.75
23.4	.32	60.9	.82
25.8	.35	67.0	.91
28.4	.38	Add'l	.0007

Reposition .29

2. Cut or nibble

Cut L	Nibble L	Min	Cut L	Nibble L	Min
2.0	0.9	.03	18.4	9.1	.31
2.5	1.1	.04	20.2	10.1	.34
3.1	1.5	.05	22.3	11.1	.37
3.9	1.8	.07	24.5	12.2	.41
4.9	2.3	.08	26.9	13.4	.45
6.1	2.9	.10	29.6	14.8	.50
7.1	3.5	.12	32.6	16.3	.55
7.8	3.8	.13	35.8	17.9	.60
8.6	4.2	.14	39.4	19.7	.66
9.4	4.6	.16	43.4	21.7	.73
10.4	5.1	.17	47.7	23.9	.80
11.4	5.6	.19	52.5	26.3	.88
12.6	6.2	.21	57.7	28.9	.97
13.8	6.8	.23	63.5	31.8	1.07
15.2	7.5	.26	Add'l		.017
16.7	8.3	.28	Add'l		.033

3.10 Tube Bending Machines

DESCRIPTION

The manufacturing process being estimated is tube bending. Machines used to bend tubes may be hydraulic or manual. The machines are able to bend any kind of metal tube, but are machine limited to the range of diameter. Cold forming is the principle used as temperature is not involved. A mandrel and die fixture for a certain diameter are necessary to match each tube diameter size. For example, a tube with a 2-in. diameter requires a mandrel that fits snuggly inside the tube. A die fixture is needed to fit tightly to the outside of the tube.

These machines can bend tubes of various lengths. Figure 3.10 shows the function of a typical tube bending machine. It is a requirement that there be enough tubing left in the fixture (about 4 in.) so that the tube can still be securely held. There is also a limit as to how close the tangent points can be to each other for the bending process.

The quality of the bent tubing as described in these data is important. It is necessary to avoid wrinkles or bumps on the outside of the tube.

ESTIMATING DATA DISCUSSION

Besides the usual chores for setup, the time includes interchanging the dies from the preceding cycle to dies that are compatible for a tube in the present cycle. If the diameter of the preceding tube is the same, then the setup is zero. If quality requirements demand, a plaster cast of an acceptable tube can be constructed. A time is provided for this work.

The elemental times are most influenced by the degree of bend and the length of each tube. For a manual tube bending machine, Elements 1, 4, 5, and 7 are basically constant since they are not influenced greatly by any time drivers. For Elements 2 and 3, load tube on mandrel and line up tangent points, the time depends on the length of the tube, which is given in inches. Element 6, bend, is influenced by the degree of bend required by the tube. The amount of bend is in degrees. If the specifications of the tube are important, a plaster cast or fixture is provided to check the degree of bend, points of tangency, and length of tube. Here, Element 8, inspect with plaster cast, is considered to be constant.

For the hydraulic tube bender, Element 3, line up tangent points, is not a variable element since stops are provided on the bender, which line up the tangent points automatically. Element 6 is the only variable time element in the cycle and is a function of the amount of bend which is in degrees.

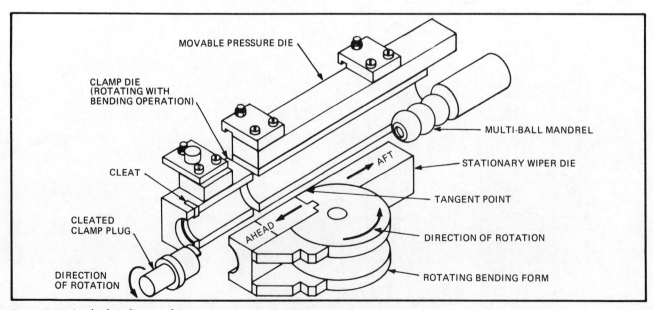

FIGURE 3.10 A tube bending machine.

EXAMPLES

A. Estimate the time to bend a .75-in. OD tube to 85°. The tube is 7.2-in. long. There is an interchange of dies and inspection. Work is on the hydraulic tube bender. Find the unit estimate, hr/100 units, and total setup for 93 units.

Table Number	Process Description	Table Time	Adjustment Factor	Cycle Minutes	Setup Hours
3.10-S	Clean up, get tube etc.	.33			.33
3.10-S	Interchange dies	.56			.56
3.10-S	Inspect	.10			.10
3.10-1	Oil inside of tube	.15		.15	
3.10-2	Put tube on mandrel	.049		.049	
3.10-3	Line up tangent points	.11		.11	
3.10-4	Set/remove clamp die	.13		.13	
3.10-5	Set/remove movable pressure die	.12		.12	
3.10-6A	Bend	.20		.20	
3.10-7	Remove tube from machine	.06		.06	
	Total			.82	.99
	Hr/100 units	1.367			
	Lot estimate	2.26 hr			

B. Estimate the time to bend a tube 14.4-in. long with a 20°-bend. There is no interchange of dies nor any inspection. This job is done on the manual tube bender. The lot quantity is 410 units.

Table Number	Process Description	Table Time	Adjustment Factor	Cycle Minutes	Setup Hours
3.10-S	Clean up, get tubes, etc.	.33			.33
3.10-1	Oil inside of tube	.15		.15	
3.10-2	Put tube on mandrel	.072		.072	
3.10-3	Line up tangent points	.14		.14	
3.10-4	Set/remove clamp die	.13		.13	
3.10-5	Set/remove movable pressure die	.11		.11	
3.10-6A	Bend	.18		.18	
3.10-7	Remove tube from machine	.04		.04	
	Total			.823	.33
	Lot estimate	5.95 hr			

C. Estimate the time to bend a tube 15.6-in. long with a 40.1°-bend. There is no interchange of dies, but there is an inspection of the finished tube. The bend is done with the hydraulic tube bender. Find the unit estimate, hr/100 units, and lot time for 15 units.

Table Number	Process Description	Table Time	Adjustment Factor	Cycle Minutes	Setup Hours
3.10-S	Clean up, get tubes, etc.	.33			.33
3.10-S	Inspection	.10			.10
3.10-1	Oil inside of tube	.15		.15	
3.10-2	Put tube on mandrel	.072		.072	
3.10-3	Line up tangent points	.11		.11	
3.10-4	Set/remove clamp die	.13		.13	
3.10-5	Set/remove movable pressure die	.12		.12	
3.10-6B	Bend	.094		.094	
3.10-7	Remove tube from machine	.058		.058	
3.10-8	Inspect with plaster cast	.02		.02	
	Total			.75	.43
	Hr/100 units	1.250			
	Lot estimates	.62 hr			

D. Estimate the time to bend a tube 1-ft long with a 16.5°-bend. There is no interchange of dies, but there is an inspection of the finished tube. The bend is done with the manual tube bender. Find the lot time for 1850 units.

Table Number	Process Description	Table Time	Adjustment Factor	Cycle Minutes	Setup Hours
3.10-S	Clean up, get tubes, etc.	.33			.33
3.10-S	Inspection	.10			.10
3.10-1	Oil inside of tube	.15		.15	
3.10-2	Put tube on mandrel	.064		.064	
3.10-3	Line up tangent points	.13		.13	
3.10-4	Set/remove clamp die	.13		.13	
3.10-5	Set/remove movable pressure die	.11		.11	
3.10-6	Bend	.15		.15	
3.10-7	Remove tube from machine	.04		.04	
	Total			.77	.43
	Hr/100 units	1.283			
	Lot estimate	24.17 hr			

TABLE 3.10 TUBE BENDING MACHINES

Setup

Clean up, get tubes, basic	.33 hr
Interchange dies, if necessary	.56 hr
Make plaster cast for inspection	.10 hr

Operation elements in estimating minutes

1. Oil inside of tube .15

2. Load tube on mandrel

L (In.)	1.2	2.6	4.0	5.5	8.2	9.2	11.1	13.4	15.8
Min	.033	.037	.040	.044	.049	.053	.059	.064	.072

3. Line up tangent points
   Manual machine

L (In.)	7.2	8.7	10.2	12.0	13.9	16.0
Min	.090	.099	.11	.12	.13	.14

   Hydraulic machine .11

4. Set/remove clamp die .13

5. Set/remove movable pressure die; manual .11
                                   hydraulic .12

6. Bend
   A. Manual machine

Degree of Bend	18	20	22	24	27	30	33
Min	.15	.16	.18	.20	.22	.24	.26
Degree of Bend	36	40	44	50	54	60	
Min	.29	.32	.35	.39	.43	.47	

B. Hydraulic machine

Degree of Bend	32	36	40	43	48	53
Min	.071	.078	.086	.094	.11	.13
Degree of Bend	63	69	76	84	92	
Min	.14	.15	.17	.18	.20	

7. Remove tube from machine; manual .04
   hydraulic .06

8. Inspect tube with plaster cast .02

3.11 Ironworker Machines

DESCRIPTION

The ironworker is a general purpose machine and may be operated by either one or two operators independently. Each may perform a variety of work which includes, shearing, punching, or other similar processes (blanking, notching, piercing, etc.).

The shearing operation involves pushing the strip, barstock, etc., through the machine until it impacts a triggering mechanism, where the bar or sheet is cut and the finished part drops into a waiting bin. Once the stock has been completely sheared, the remaining end piece is removed, and a new piece is prepared for shearing. This process continues until the lot requirement is completed.

Similarly, the punching process involves a handling and positioning time, after which the part is punched. In addition, if several punches are required for a part, a rehandling and repunching time enter the estimating process. After the punch or punches have been performed, the finished part is then tossed into a waiting bin, and the cycle is repeated.

ESTIMATING DATA DISCUSSION

A distinction is provided for shearing and punching. Usually, shearing starts with stock to be separated. Punching deals with individual blanks.

Many of the operation elements require information on the number of pieces that are to be sectioned from raw material. The shear length is the length required for a sheared part.

The estimator selects the elements, keeping in mind that the units for setup are in hours while the cycle elements are in minutes. Notice also that it is important to select the cycle time that is related to one unit of output, since much of the ironworker operations deal with multiples. For example, oil piece is done every 50 pieces or so. The element time to oil is greater than that shown by Element 7, but it has been divided by 50 units. So, this is a requirement every 50 units or so.

EXAMPLES

A. A lot of 1900 parts is to be made 7-in long from 58 strips of metal. Thirty-three parts/strips may be sheared. Find the cycle estimate.

Table Number	Process Description	Table Time	Adjustment Factor	Cycle Minutes	Setup Hours
3.11-1	Slide strip onto rollers	.001		.001	
3.11-2	Push and shear strip	.02		.02	
3.11-3	Remove end piece	.003		.003	
3.11-9	Inspect part	.01		.01	
3.11-4	Settle parts	.005		.005	
	Cycle estimate			.039	

B. A lot of 500 parts 35.5-in. long is to be sheared from 84 strips with 6 parts/strip. Find the unit, hr/100 units, and lot estimate.

Table Number	Process Description	Table Time	Adjustment Factor	Cycle Minutes	Setup Hours
3.11-S	Shearing setup	.30			.3
3.11-1	Slide strip onto rollers	.005		.005	
3.11-9	Position and inspect	.01		.01	
3.11-2	Push and shear strip	.06		.06	
3.11-3	Remove end piece	.01		.01	
3.11-4	Settle parts	.005		.005	—
	Total			.09	.3
	Hr/100 units	.150			
	Lot estimate	1.05 hr			

C. A quantity of 2000 parts is to be punched once per part. Part length equals 35.5 in. Find the unit estimate.

Table Number	Process Description	Table Time	Adjustment Factor	Cycle Minutes	Setup Hours
3.11-5	Pick up stock	.05		.05	
3.11-7	Lubrication time	.03		.03	
3.11-6	Positioning	.05		.05	
3.11-7	Punch	.02		.02	
3.11-9	Inspection	.01		.01	
3.11-8	Drop finished part	.02		.02	
3.11-4	Settle parts	.005		.005	
	Total			.185	

D. Five hundred 28-in. long parts are to be punched twice. Find the unit, hr/100 units, shop and lot estimate for 181 pieces.

Table Number	Process Description	Table Time	Adjustment Factor	Cycle Minutes	Setup Hours
3.11-S	Punching setup	.15			.15
3.11-5	Pick up stock	.05		.05	
3.11-7	Lubrication	.03		.03	
3.11-6	Positioning	.02		.02	
3.11-7	Punch	.02		.02	
3.11-6	Repositioning	.03		.03	
3.11-7	Punch	.02		.02	
3.11-9	Inspection (measurement)	.01		.01	
3.11-8	Drop finished part in bin	.02		.02	
3.11-4	Settle parts	.005		.005	—
	Total			.205	.15
	Hr/100 units	.342			
	Shop estimate	292 pc/hr			
	Lot estimate	1.86 hr			

TABLE 3.11 IRONWORKER MACHINES

Setup
    Shearing operation   .30 hr
    Punching operation   .15 hr
Operation elements in estimating minutes

1. Slide bar or strip onto rollers for shear   .05/bar

No. of Pieces from Bar	1	5	10	20	50
Prorated Min	.05	.01	.005	.003	.001

2. Push and shear; part drops into bin

Push L, In.	10	20	30	40
Prorated Min	.02	.03	.04	.06

3. Remove end piece, shear

No. of Pieces from Bar	1	3	10	30	60
Prorated Min	.03	.01	.003	.001	.0005

4. Settle sheared parts in bin   .20

Push L, In.	10	20	30	40
Prorated Min	.005	.007	.01	.005

   Settle punched parts in bin   .005

5. Pick up single part for punching, blanking, notching, etc.   .05

6. Position single part for operation   .02
   Reposition for extra strike   .03

7. Punch, notch, etc.   .02
   Lubricate strip or part, prorated over 50 punches   .03

8. Drop finished part in bin   .02

9. Inspect or count part   .01

MARKING

4.1 Marking Bench and Machines

DESCRIPTION

Permanent impressions can be made by forcing a high-speed steel-lettered die into a metal surface. These cold impressions are used for marking, numbering, graduating, embossing, knurling, and tag or nameplate making. But these estimating data refer to metal marking. Machines for metal marking are hydraulic, mechanical, or pneumatic powered. Impact, rolling, or cold forming are some of the cold-working principles employed. Even though metal is the primary object of these data, plastics, leather, etc. can be similarly marked. Figure 4.1 is an example of a high production metal-marker machine. Production of this machine is 2500 tubes per hr. Hand marking machines are small enough to sit on a table in front of the operator.

ESTIMATING DATA DISCUSSION

Entry variables for Elements 1, 2, or 3 are part size: small, medium, or large. Small parts are handled easily, no dimension exceeds 3 in., and weight does not exceed 1/5 lb. Medium parts have no dimension or weight that exceed 9 in. and 3 lb respectively, while large parts are limited to 20 in. and 8 lb.

For manual operation, Element 1 is always required. If semi- or fully automatic, knowledge of loading is required. The target for the part in Element 2 is either universal or fitted nest.

The marking or machine time is given by Element 5. For machines that have a lever which is rolled and manipulated by the operator, Element 5 will be used.

A ram action that is vertical and uses pressure to force the marking die into the surface adopts Element 6. A round surface is covered by Element 7.

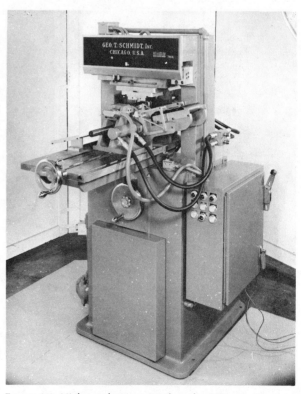

FIGURE 4.1 High-production metal marker. (*George T. Schmidt, Incorporated*)

Marking Machines 127

EXAMPLES

A. A medium-size part is loaded in a fitted nest and a pressure-stamping machine is used. Find the lot time for 1800 parts.

Table Number	Process Description	Table Time	Adjustment Factor	Cycle Minutes	Setup Hours
4.1.S	Setup	.15			.15
4.1-1	Get and aside part	.050		.50	
4.1-2	Load in fitted nest	.044		.044	
4.1-6	Pressure stamp	.038		.038	
	Total			.132	.15
	Lot estimate		4.11		

TABLE 4.1   MARKING MACHINES

Setup .15 hr

Operation elements in estimating minutes

		Part Size		
		Small	Medium	Large
1.	Get and aside part	.035	.050	.11
2.	Load and unload part in:	.022	.028	.035
	universal nest	.038	.044	.050
	fitted nest	.030	.042	.048
3.	Place part against stop			
4.	Clamp with C			
	Install and remove			.22
	Tighten or loosen			.05
	Quick-acting vise			.04
5.	Manual rotation of handle to stamp			
	Length of stamp < .7 in.			.006
	Length of stamp ≥ .7 in.			.011
6.	Pressure stamp or mark, machine			.038
7.	Machine roll stamp			
	Length of stamp, 0-3 in.			.036
	Length of stamp, > 3 in.			.066
8.	Hand-pressure ink mark			
	Area being marked,	in.^{2}		min.
		.4		.05
		.9		.06
		2.0		.07
9.	Inspect			
	L + W + H of part			
		2.8		.027
		3.5		.05
		4.3		.08
10.	Wash ink pad			.095

4.2 Screen Printing Bench and Machines

DESCRIPTION

The apparatus for manual screen printing consists of a base equipped with gages or jigs for registry of the part to be screened. The base is hinged with a frame which can be raised and lowered. The frame is often counterweighted. A silk screen or other material is held taut within the frame. After a piece has been located, the frame is lowered and paint, previously applied on the screen, is wiped across using a rubber squeegee. This forces a paint film through the mesh that is not blanked off by the stencil onto the surface of the work beneath. The frame is raised; the piece is removed, and it is replaced by a second. The operation is then repeated.

In semiautomatic screen printing, the manual work of screen manipulation has been removed, leaving loading and unloading for the operator. Figure 4.2 is an example of a semiautomatic machine. A variable speed drive with 400 to 1140 impressions/hr can be maintained. For this model, materials ranging from printed circuits to fabrics to metals $3/4$ in. in thickness can be accommodated. The printer can be adapted to conveyorized or turntable operation. Integrated automatic and conveyorized oven units are also available. Infrared heating elements are used to dry the ink.

ESTIMATING DATA DISCUSSION

A fixed screen is one mounted on the silk-screen bench while a free screen is portable and placed on the part by hand. The setup and the cycle operation separates the estimating data on this basis.

The first operation is the initial screening, as the subsequent operation requires less work. Entry variables are first or following operation, fixed or free screen, part box size (= L + W + H) and the length of the squeegee pass for the silk screen area. The time is found at the intersection of these entry variables, and is total for handling and silk screening. Following the first screen for an operation, additional screening for that operation would use the additional screen at the bottom of the table. Element 1 is for manual screen printing of fabricated parts. Element 3 deals with 12×18 in. printed circuit boards.

EXAMPLES

A. A part, L + W + H = 15.7 in., has one bench-mounted screening operation followed by a portable screen on a different area. The screen lengths are 12 in. and 5 in. Determine lot time for 118 units.

FIGURE 4.2 Screen printing press. *(Lawson Printing & Drying Machine Company)*

Table Number	Process Description	Table Time	Adjustment Factor	Cycle Minutes	Setup Hours
4.2-S	Fixed screen and 1 free screen, .2 + .10 + .05	.25			.25
4.2-1	First operation, $L + W + H = 15.7$, fixed screen at screen length of 12 in.	.43		.43	
4.2	First operation, free screen at 5 in.	.49		.49	
	Total			.92	.25
	Lot time	2.06 hr			

B. The part previously silk screened in Example A above is to have additional screening. Operational facts are 1 fixed screen of 7 in. and 2 free screens of 4 and 9 in. Estimate pc/hr and lot time. Quantity = 496.

Table Number	Process Description	Table Time	Adjustment Factor	Cycle Minutes	Setup Hours
4.2-S	Fixed screen and 2 additional portable screens	.54			.54
4.2-1	Second operation, fixed screen at L of 7 in., $L + W + H = 15.7$	.22		.22	
4.2-1	Free screen, second operation, $L = 4$ in.	.30		.30	
	Total			.52	.54
	Shop estimate	115 pc/hr			
	Lot estimate	4.84 hr			

TABLE 4.2 SCREEN PRINTING BENCH AND MACHINES

Setup

Screen machine	.25 hr
Fixed screen bench	.2 hr
Free screen	.10 + .05 hr/screen

Operation elements in estimating minutes

1. Manual screen printing

Part		$L + W + H$		Length of Screened Area							
First Operation		Following Operations									
Fixed	Free	Fixed	Free	3	5	6	7	9	12	14	18
3		11		.16	.18	.18	.19	.21	.23	.26	.29
		13		.17	.18	.19	.20	.22	.25	.27	.30
6		15		.18	.19	.20	.21	.23	.26	.28	.31
9		16		.19	.20	.21	.22	.25	.27	.29	.32
10		19	9	.20	.21	.22	.23	.26	.28	.30	.33
11	6	22	14	.22	.23	.25	.26	.28	.30	.32	.36
11		24	18	.25	.26	.27	.28	.30	.32	.35	.38
12		26	21	.27	.28	.29	.30	.32	.35	.37	.40
13	11	29	26	.32	.33	.35	.36	.38	.40	.42	.46
15	13	32	31	.36	.37	.38	.39	.41	.43	.46	.49
17	14	33	32	.40	.41	.42	.43	.46	.48	.50	.53
19	15	35	34	.43	.45	.46	.44	.49	.51	.53	.57
20	18	36	36	.48	.49	.50	.51	.53	.56	.58	.61
21	21	37	37	.52	.53	.54	.56	.58	.60	.62	.66
23	22	39	39	.58	.59	.60	.61	.63	.60	.68	.71

Part		L + W + H		Length of Screened Area							
First Operation		Following Operations									
Fixed	Free	Fixed	Free	3	5	6	7	9	12	14	18
25	25	40	40	.63	.64	.66	.67	.69	.71	.77	.77
26	27			.70	.71	.72	.77	.76	.78	.80	.84
29	32			.72	.78	.78	.80	.82	.85	.87	.90
33	35			.85	.86	.86	.88	.90	.92	.95	.98
36	38			.93	.95	.96	.97	.98	1.01	1.04	1.07
39				1.04	1.05	1.06	1.07	1.09	1.10	1.13	1.15
40	41			1.15	1.16	1.17	1.18	1.20	1.23	1.25	1.28
43	43			1.26	1.27	1.28	1.29	1.31	1.34	1.36	1.39
Add'l				.15	.16	.17	.18	.20	.22	.24	.27

    Hook and unhook park   .27
    Rubber stamp   .10

2. Semiautomatic screen printing
    Rotary turntable
    5 × 5-in. area, metal part   .06

    Conveyor handling and printing

Impressions/hr	200	500	1000	1500
Min/unit	.30	.12	.06	.04

4.3 Laser Marking Machines

DESCRIPTION

Laser marking is a means of noncontact pressure marking useful for a variety of nontransparent materials such as synthetics and rubber, metal, and diamond. After the continuous-wave laser radiation exits a laser head, and becomes an enlarged beam, mirrors controlled by galvanometers focus the beam on the workpiece. At the point of focus, the laser vaporizes the material of the workpiece, and as it is scanned in the X and Y directions it traces out a trench to engrave whatever character, symbol, or pattern is established by the computer program.

On some machines, software is provided on tape and is loaded into random-access read/write memory using a tape reader that is part of the system. Once loaded, the program may be modified from the control panel. Figure 4.3 shows a machine capable of 4 mil dia focused spot size. Engraving fields up to 3½ × 3½ in. are available.

ESTIMATING DATA DISCUSSION

Setup times allow for average programming, and if the program is saved on any storage medium, the amount is reduced to 0.1 hr.

Part handling is associated with part size. Very small implies parts are difficult to control. Although they can be easily handled in handfuls, tweezers are sometimes used. Small parts are easy to manage with the fingers. No dimension exceeds 3 in. and weight does not exceed ¼ lb. Medium-sized parts are easily handled by one hand; maximum dimension is 9 in.

A door has to be opened to allow entry of the part. Following the marking, the door has to be open for removal. A time of 0.04 min allows for both actions.

Laser marking depends upon character height. Distinctions for flourishes, strokes, and various character differences are averaged. Material is considered hard. Count the number of different height characters to find a time.

FIGURE 4.3 Laser marking machine. *(JEC Lasers, Inc.)*

EXAMPLES

A. A fragile transistor has "2N" and "4856" laser marked on top of the metal cover unit. The unit is very small and is loaded into a nest. The characters are ⅛-in. high. Find the lot time for a new order where the quantity is 75 units.

Table Number	Process Description	Table Time	Adjustment Factor	Cycle Minutes	Setup Hours
4.3-S	Setup program and initialize	.3			.3
4.3-2	Open and close door	.04		.04	
4.3-2	Start and stop	.02		.02	
4.3-1A	Get and aside part, stack	.042		.042	
4.3-1C	Place into nest	.033		.033	
4.3-3	Laser mark 6 characters, ⅛ in. high	.30		.30	—
	Total			.435	.3
	Lot estimate	.90 hr			

B. A machine tool dial is laser marked. It has 360 marks on the circumference. Every major 5° has ⅜-in. long marks for a total of 36. Each minor mark is ⁵⁄₁₆-in. long and there are 324 marks. The dial is loaded on an indexable mandrel. There are ninety-nine ⁵⁄₁₆-in. numerals. Find the unit estimate to make one part. Assume door is opened for each mark, and the part is indexed to position the dial.

Table Number	Process Description	Table Time	Adjustment Factor	Cycle Minutes	Setup Hours
4.3-S	Setup	.3			.3
4.3-2	Open and close door (360 + 99)	.04	459	18.36	
4.3-2	Start and stop, controllable, once	.02		.02	
4.3-1C	Medium part load on indexable mandrel	.045		.045	

Table Number	Process Description	Table Time	Adjustment Factor	Cycle Minutes	Setup Hours
4.3-3	Major lines, 36 at ⅜-in. height 2.83 + (36 − 25) .113	1.24		1.24	
4.3-3	Minor lines, 324 at 3⁄16-in. height 1.40 + (324 − 25) .056	18.14		18.14	
4.3-3	Numbers, 99 at 3⁄16-in. height 1.40 + (99 − 25) .056	5.54		5.54	
4.3-1C	Reposition (360 + 98) .01	.01	458	4.58	—
	Total			47.89	.3
	Lot estimate	1.10 hr			

TABLE 4.3 LASER MARKING MACHINES

Setup

Program and initialize	.3 hr
Preprogrammed	.1 hr

Operation elements in estimating minutes

1. Handling

    A. Get and aside part

Part Size	Toss Aside	Stack Aside
Very Small	.027	.042
Small	.021	.039
Medium	.030	.045

    B. Turn over part

Part Size	Min
Very Small	.009
Small	.012
Medium	.015

    C. Place and position part into nest or against pins

Part Size	Nest	Two Pins
Very Small	.033	.045
Small	.030	.039
Medium	.036	.051

    Reposition part   .01

2. Machine operation

Open and close door	.04
Start and stop	.02

3. Laser mark, min

Character Height (In.)	No. of Characters							
	1	2	3	4	5	6	7	8
⅛	.04	.08	.11	.15	.19	.23	.27	.30
3⁄16	.06	.11	.17	.22	.28	.34	.39	.45
¼	.08	.15	.23	.30	.38	.45	.53	.60
5⁄16	.09	.19	.28	.38	.47	.56	.66	.75
⅜	.11	.23	.34	.45	.57	.68	.79	.90
½	.15	.30	.45	.60	.75	.90	1.05	1.20

Character Height (In.)	No. of Characters							
	9	10	12	14	17	20	25	Add'l
1/8	.34	.38	.46	.53	.65	.76	.95	.038
3/16	.50	.56	.67	.78	.95	1.12	1.40	.056
1/4	.68	.75	.90	1.05	1.28	1.50	1.88	.075
5/16	.85	.94	1.13	1.32	1.60	1.88	2.35	.094
3/8	1.02	1.13	1.36	1.58	1.92	2.26	2.83	.113
1/2	1.35	1.50	1.80	2.10	2.55	3.00	3.75	.150

4.4 Pad Printing Machines

DESCRIPTION

Pad printing (or transfer printing) is a process in which text and other graphic artwork is placed on the surface of a product. The medium used in this process is an epoxy ink. The ink is stored in a reservoir and is allowed to flow across a metal plate that is engraved with the desired artwork. This metal plate is mounted in a fixture called a cliche; a blade then scrapes the excess ink off the plate. At this time, a silicone rubber pad lifts the ink from the plate and transfers it to the product. The epoxy ink dries to the touch within 20 seconds and cures completely within two days. After it is cured, the ink will not chip or rub off. It is resistant to wear and has long life. The printed image is precise and considered by some superior to the print obtained by hotstamping or silkscreening.

The pad printing technique can be used on a variety of materials, including metal, plastic, and glass. The pads, which have a shelf life of approximately 60 days, come in various shapes and sizes depending on the application. The hardness of the pad is also a factor. The more fragile the product, the softer the pad that is used.

Two important factors that influence the quality of the printed image are the amount of compression produced by the pad and the cycle time of the printing process. The compression usually depends on the material on which the artwork is printed. The texture of the material is also a factor. In general, smooth surfaces require less compression and coarse surfaces require more. Cycle time is also important in maintaining an accurate printed image. If the cycle is too slow, the ink will lose its consistency and the image will deteriorate.

The epoxy ink will not adequately adhere to oily or waxy surfaces. Therefore, it is essential that oil and silicone spray be removed from the part surface before printing. If plastic parts are to be printed, no mold sprays should be used in the molding process. Furthermore, care should be taken to avoid getting dust or other contaminants on the parts prior to printing. The printing environment should be free of dust because the silicone pad builds up a static charge that attracts dust and other particles. Because of this, periodic cleaning of the pad and static charge removal are necessary. Exposing the ink reservoir to the air makes it necessary to thin the ink occasionally.

In the printing environment, constant humidity and temperature should be maintained for maximum accuracy and repeatability. The recommended temperature range is 58°F to 90°F. Above 90°F, the ink dries too quickly and below 58°F it is too thick for effective printing. The ideal temperature for the process is about 72°F.

ESTIMATING DATA DISCUSSION

In pad printing, the main factor in setup is the size of the cliche. Large cliches require longer setup times. Setup values can be found in Table 4.4 given in hours.

The operational elements are pick up part and position in fixture, print part, unload and set aside, inspect part, and put in box. These are given in estimating minutes. Elements 4.4-1, pick up part and position in fixture; 4.4-2, print part; 4.4-3, unload and set aside; and 4.4-5, put in box, are allowed once per part on a constant basis. Element 4.4-4, inspect part, should be multiplied by the frequency of inspection. For example, if one part in five is to be inspected, the Adjustment Factor would be 1/5.

The other elements to consider include clean pad, clean blade, thin paint, heat part, and add for parts that tangle. Element 6, clean pad, must be divided by the average number of parts printed between cleanings. This procedure should be followed for clean blade and thin paint. If the part is to be heated to facilitate quicker drying time, heat part must be added once per part. If parts have a tendency to tangle, add for parts that tangle (should be added once per part).

These estimates can be extended to a variety of pad printing machines, provided that the parts do not require special fixtures and are reasonably easy to handle. This extension is possible because most pad printing machines operate in a similar manner.

EXAMPLES

A. Estimate the time required to pad print a polycarbonate part with 79 characters of text. The printing plate requires a large cliche. Inspection of each part is required. The pad should be cleaned after 100 parts. The blade should be cleaned every 800 parts. The paint needs to be thinned every 200 parts. A lot of 3000 will be produced.

Table Number	Process Description	Table Time	Adjustment Factor	Cycle Minutes	Setup Hours
4.4-S	Setup (large cliche)	2.75			2.75
4.4-1	Pick up and position part in fixture	.035		.035	
4.4-2	Print part	.012		.012	
4.4-3	Unload and set aside	.033		.033	
4.4-4	Inspect part	.041		.041	
4.4-5	Put in box	.022		.022	
4.4-6	Clean pad	.47	1/100	.005	
4.4-6	Clean blade	1.26	1/800	.002	
4.4-6	Thin paint	1.08	1/200	.005	
	Total			.155	2.75
	Hr/100	.258			
	Lot estimate	10.50 hr			

B. Estimate the time required to pad print a polycarbonate part with 7 characters of text. The printing plate requires a small cliche. Inspection of each part is required. The pad should be cleaned after 500 parts. The blade should be cleaned every 1000 parts. The paint needs to be thinned every 1000 parts. A lot of 10,000 will be produced. Each part needs to be heated to speed up the drying time.

Table Number	Process Description	Table Time	Adjustment Factor	Cycle Minutes	Setup Hours
4.4-S	Setup (small cliche)	2.00			2.00
4.4-1	Pick up and position part in fixture	.035		.035	
4.4-2	Print part	.012		.012	
4.4-3	Unload and set aside	.033		.033	
4.4-4	Inspect part	.029		.029	
4.4-6	Heat part	.021		.021	
4.4-6	Put in box	.022		.022	
4.4-6	Clean pad	.47	1/500	.002	
4.4-6	Clean blade	1.26	1/1000	.003	
4.4-6	Thin paint	1.08	1/1000	.001	
	Total			.157	2.00
	Hr/100	.262			
	Lot hr	28.18			

C. Estimate the time required to pad print a polycarbonate part with 38 characters of text. The printing plate requires a large cliche. Inspection of each part is required. The pad should be cleaned after 200 parts. The blade should be cleaned every 1000 parts. The paint needs to be thinned every 500 parts. A lot of 5000 will be produced.

Table Number	Process Description	Table Time	Adjustment Factor	Cycle Minutes	Setup Hours
4.4-S	Setup (large cliche)	2.75			2.75
4.4-1	Pick up and position part in fixture	.035		.035	
4.4-2	Print part	.012		.012	
4.4-3	Unload and set aside	.033		.033	
4.4-4	Inspect part	.035		.035	
4.4-5	Put in box	.022		.022	
4.4-6	Clean pad	.47	1/200	.002	
4.4-6	Clean blade	1.26	1/1000	.001	
4.4-6	Thin paint	1.08	1/500	.002	
	Total			.142	2.75
	Hr/100	.237			
	Lot estimate	14.58 hr			

TABLE 4.4 PAD PRINTING MACHINES

Setup

    Small cliche   2.00 hr

    Large cliche   2.75 hr

Operation elements in estimating minutes

1. Pick up part and position in fixture   .035

2. Print part   .012

3. Unload and set aside   .033

4. Inspect part

Number of Characters	Min
16	.029
25	.033
40	.035
55	.038
70	.041

5. Load in box   .022

6. Machine quality (prorate over quantity)
   - Clean pad   .47
   - Clean blade   1.26

7. Miscellaneous
   - Thin paint (prorate over quantity)   1.08
   - Heat part   .021
   - Add for parts that tangle   .030

HOTWORKING

5.1 Forging Machines

DESCRIPTION

These estimating data are intended for hammer-or drop-forging machines. Hammer forging consists of hammering heated metal between flat dies in a steam-or air-powered hammer. Drop forging differs from hammer forging in that closed-impression, rather than open-face, dies are used. The forging is produced by impact or pressure which compels hot and pliable metal to conform to the shape of the die cavity. Repeated blows on the metal gradually change the form, and the number of steps in the process varies according to the size and shape of the part, forging qualities of the metal, tolerances, machine, and crew size. For parts of large or complicated shapes, a preliminary shaping operation using more than one set of dies is possible.

Two principal types of drop-forging hammers are the steam or air hammer, and the gravity drop or board hammer. In the former, the ram, hammer, and upper die half are lifted by steam or air pressure and the force of the blow and the number of blows per min can be controlled by throttling the steam or air. Figure 5.1 is a 5000-lb drop hammer. Now we consider the general class of forging machines and their estimating data.

ESTIMATING DATA DISCUSSION

Setup data are linked to machine rating and while the number of forges, die complexity, and mass are significant factors, these in turn are related to forge size. The setup times are independent of the crew size. For a total time, multiply setup by crew size.

There are three time drivers used for estimating the optional time or weight of the raw billet, diameter multiplied by billet length, and forge machine rating. Crew sizes vary with forge machine rating. For the

FIGURE 5.1 A 5000-lb drop hammer. *(Chambersburg Engineering Company)*

1500-, 3000-, 5000-, and 12,000-lb. machine rating, typical crew sizes are 2, 3, and 4 men. A maximum crew is the hammerman, helper, furnace, and trim press operator. All elemental times are for the hammerman only, and for a total elemental time, the hammerman's time is multiplied by crew size since the hammerman is the controlling factor. Material forgeability is averaged in these data. Furthermore, the number of blows for rolling, fullering, blocking, etc., is not always accurately known, and these data are averaged, although rough and finishing forging is related to the number of blows.

In Element 1, the hammerman places the tongs on billet, which has been placed on the die block by the furnace attendant. Also in Element 1, the hammerman places his tongs on the billet and positions the billet in either the roll and breakdown impression or in the forge impression. Billet weight is the initial weight including tong hold and extra stock for fullering.

In roll and breakdown, the entry variable is diameter by billet length. For instance, a 2-in. billet of 20-in. length would be entered as 40. The hammerman rolls and breaks down the billet to prepare the billet for the forge impression. The helper blows the scale from the dies and oils the dies. Element 3 has the hammerman position the billet in the bend impression, finish forge impression, or, on the flat part of the die and bend, straighten or flatten the billet. It may include placing the billet on the flat part of the die to pop scale.

In reposition billet, the hammerman removes the billet from one impression and positions the billet in another impression. A forge billet either follows a roll and breakdown or does not. In these elements, the hammerman forges the billet while the helper blows the scale from the die and oils the dies as necessary.

In the element "move billet to trimmer, trim and cutoff flash," the hammerman moves the billet to the trimmer and positions the billet on the trimmer, while the operator trims the billet and cuts off the flash. After the billet has been trimmed and the flash removed, the hammerman slides the billet into the tote box while the helper walks to the furnace and places a cold bar in the furnace.

For example, in Element 7, the hammerman removes the billet from the die block and places the billet on the table while the operator returns to the hammer. A tong and hook are used by the hammerman helper and trim press operator to remove the billet from the die block using an overhead moveable hook and tongs. In Element 7, the hammerman picks up the tongs, either from the table or the pail of water near the table, and returns to the hammer.

The hammerman walks to the table and picks up previous billet and returns to the hammer. The operator forges a tong hold and places the billet on the table, and the operator returns to the hammer. In "change tongs," the hammerman holds the billet out and the helper picks up tongs, fastens tongs to extended end, and the operator removes the tongs he is holding and asides tongs, taking the tongs the helper is holding. Helper returns to blow and oil while the operator turns and places billet on die.

The restrike element requires the hammerman to pick up trimmed billet, place billet on die, and strike billet using ram. Operator moves billets to table.

In "trim and restrike," the hammerman takes the billet to the trimmer, trims, returns to the hammer, and restrikes billet, then places billet on table.

EXAMPLES

A. A connecting rod requires heavy stock on one end and lighter stock on the other. Singly, these require fullering and rolling while in multiple-die impression dies, grain flow permitting, these can be nested to eliminate blows. Consider a 1500-lb forging machine making two pieces per platter and a gross weight of 1.42 lb per forging for all multiples. With two parts end to end, we estimate 2 blows for fullering, 2 blows for rollering, 5 blows for blocking, and 3 blows for finishing. The bar Dia and L is approximately 1-in. OD by 1.125-ft long for 3.6 lb. Estimate the times for a hammerman and helper for 817 parts. The adjustment factor considers two operators and two units.

Table Number	Process Description	Table Time	Adjustment Factor	Cycle Minutes	Setup Hours
5.1-S	Setup	1.0			1.0
5.1-1A	Tongs on billet and position	.08	½	.16	
5.1-2	Roll and breakdown, 1 × 1.125 ft = 12 ⅛ in.	.25	½	.25	
5.1-3A	Reposition and pop scale	.06	½	.06	
5.1-4A	Reposition and rough forge	.21	½	.21	
5.1-4B	Reposition, finish forge, 3 blows	.12	5	.60	
5.1-5	Move billet to trimmer and trim, 2 times	.13	2 × ½	.26	

Table Number	Process Description	Table Time	Adjustment Factor	Cycle Minutes	Setup Hours
5.1-5	Operator returns to hammer from trimmer	.10	2 × ½	.10	—
	Estimate for hammerman for multiple parts			1.62	1.0
	Estimate for hammerman and helper	3.24 min			
	Unit estimate	1.62 min			
	Lot estimate	23.06 hr			

B. A flange forging, C-1020 material, 1½-in. round, cut 7%6 in. will give 4 cuts per bar. Nine hammer blows (3 in roller, 2 in blocker, 3 in the finisher, and 1 cutoff) is required to forge the part. This would result in 36 hammer blows per bar. A 3000-lb forging press and a two-man crew is anticipated. Find the crew hr/100 units. This estimate will be calculated for 4 pc per bar. Quantity is 723.

Table Number	Process Description	Table Time	Adjustment Factor	Cycle Minutes	Setup Hours
5.1-S	Setup	1.5			1.5
5.1-6	Operator picks up billet, forges tong hold	.28	½	.14	
5.1-1A	Position billet, 4 lb	.08	½	.04	
5.1-2	Roll and breakdown, $Dia \times$ billet $L = 11.3$ in.	.25	½	.125	
5.1-4G	Forge billet	.60	½	.30	
5.1-6	Trim	.24	½	.12	—
	Estimate for bar			.725	1.5
	Unit estimate for hammerman	.725			
	Hr/100 for crew	2.42			
	Lot estimate	18.97 hr			

C. A 6½-in. OD bar, 11-in. long, AISI 8822 material, is used for a heavy-duty gear shaft. Two hubs are forged from the material, match tolerances are commercial, and a long run is planned, but quantities are within one die set life. Crew size for a 12,000-lb rated forge machine is 4 operators. Find lot time for 375 units. Each bar weighs 103 lb.

Table Number	Process Description	Table Time	Adjustment Factor	Cycle Minutes	Setup Hours
5.1-S	Setup	3.0	2		6.0
5.1-7C	Load billet with tong and hook	.18	4	.72	
5.1-7E	Operator picks up tongs and returns to billet	.09	4	.36	
5.1-7D	Forges tong hold	.74	4	2.96	
5.1-2	Roll and breakdown, $Dia \times$ billet $L = 71.5$	.63	4	2.52	
5.1-4H	Forge billet, 2 hubs	1.46	2 × 4	11.62	
5.1-6	Operator places part in tote box and returns	.11	4	.44	—
	Total			18.62	6.0
	Crew estimate	18.62			
	Lot estimate	122.37 hr			

TABLE 5.1 FORGING MACHINES

Setup
1500-lb hammer	1 hr
3000-lb hammer	1.5 hr
5000-lb hammer	2 hr
12,000-lb hammer	3 hr

Operation elements in estimating minutes

1. Handling and position

   A. Tongs on billet .10

Billet Weight	5000 lb 12,000 lb. Hammer
4	.08
11	.10
22	.13
33	.16
39	.18

   Tongs on billet and position 1500-lb, 3000-lb hammer .10

   B. Position billet on die to make 2 parts, 3000-lb hammer .10
   Helper moves billet to die block 1500-lb hammer .10
   Position billet

Weight	Min
6	.08
31	.10
58	.13
97	.16

2. Roll and breakdown
   $Dia \times$ billet L

	Hammer lb			
1500	3000*	5000	12,000	Min
1	1			.06
5	5	1		.13
11	8	8		.19
17	12	15	3	.25
21	14	23	18	.31
27	18	30	35	.38
37	25	46	71	.50
	33		98	.63
	39			.75
	46			.88

   *2 impression die

3. Reposition

   A. Reposition, bend, flatten, pop scale, or straighten 1500-lb hammer .06

   B. Reposition billet, 5000- and 12,000-lb hammer

Weight	Min
5	.06
35	.08
65	.10
95	.12
124	.14

C. 3000-lb hammer
    Reposition billet, flop billet 180° on die .09
    Pop scale or flatten billet .09
    Place billet on edge roll and draw as required .15
    Reposition and flatten billet or pop scale .11

4. Reposition and forge

   A. Reposition and rough forge (follows bend or flatten 1 blow) 1500-lb hammer

Dia × Billet L	Min
4	.11
6	.14
9	.16
12	.19
15	.21

   B. Reposition, finish forge, turn billet over, strike 1 blow, 1500-lb hammer .12

   C. Reposition and forge, follow roll and breakdown, 1500-lb hammer

Dia × Billet L	Min
3	.13
5	.19
8	.28
10	.34

   D. Rough forge, 3000-lb hammer

Dia × Billet L	Min
3	.10
5	.15
7	.23
12	.34
17	.51
26	.75

   E. Forge billet, follow roll and breakdown, 3000-lb hammer

Dia × Billet L	Dia < 3	Dia ≥ 3
5	.15	.35
10	.31	.55
15	.46	.75
20	.64	.95
25	.80	1.20

   F. Reposition and forge, no breakdown, 3000-lb hammer

Dia × Billet L	Min
4	.45
10	.56
16	.70
23	.88

G. Reposition and finish forge

Dia × Billet L	1500-lb Hammer	3000-lb Hammer
3	.11	.23
5	.16	.29
7	.20	.38
10	.26	.46
13		.60
17		.74
22		.91

H. Forge billet, 5000- and 12,000-lb hammer, no breakdown and 1 impression

Dia × L	$2 \leq Dia < 5$	$2 \leq Dia < 5, L \geq 10$	$5 \leq Dia \leq 6$	$6 < Dia \leq 7$
5	.49			
9	.54			
16	.60		.70	1.01
18	.63		.88	1.03
24	.69	.23	1.09	1.06
34		.39	1.44	1.14
44		.35	1.79	1.21
58		.44		1.33
76		.55		1.46
98		.69		1.63
126		.86		

I. Forge billet, follow roll and breakdown, 5000-lb hammer

Dia × Billet L	Min
2	.25
7	.31
15	.39
24	.49
32	.61
46	.76

J. Forge billet, 12,000-lb hammer

Dia × Billet L	1 Impression after Breakdown
6	.13
8	.16
11	.20
13	.25
16	.30
21	.39
26	.48
35	.60
44	.75
55	.93
66	1.16
84	1.45

5. Elements for 1500-lb hammer
   Move billet to trimmer and cut off piece .10

Move billet to trimmer and trim	.13
Move billet to trimmer, trim, and cut off flash	.18
Operator returns to hammer from trimmer	.10
Operator returns to hammer, holds tong and throws billet on floor, helper places cold bar in furnace	.13
Helper places billet in furnace	.11
Operator places billet in tote box and returns to hammer	.11
Helper strikes knockout pin	.09
Restrike	.13

6. Elements for 3000-lb hammer

Helper removes billet from die block, operator picks up tongs and returns to hammer	.08
Operator places billet on table, returns to hammer	.08
Operator returns to hammer and positions with tongs	.09
Operator and helper move billet to trimmer, trim and punch out center, and replace pin	.30
Operator turns billet 180°, takes billet to trimmer, positions, trims and cuts off flash	.21
Operator moves billet to trimmer, trims and returns to hammer, helper positions billet and straightens billet on second press, returns to hammer	.24
Operator picks up trimmed part, positions part in die, restrikes	.15
Operator takes billet to trimmer, trims and returns to hammer, places part in die, restrikes to reforge or straighten, removes part to box	.23
Operator places part in tote box and returns to hammer	.11
Operator picks up billet, forges tong hold, places billet on table, and returns to hammer	.28
Changes tongs and places tongs on billet	.11
Cool dies	.10
Helper strikes knockout pin	.13

7A. Elements for 5000-lb and 12,000-lb hammer

Operator moves billet to table

Weight	9	29	39	70
Min	.08	.12	.13	.15

B. Operator moves billet to table and returns to hammer

Weight	10	30	60	88
Min	.10	.13	.16	.20

C. Helper and trim press operator remove billet from die block with tong and hook

Weight	10	56	120
Min	.11	.14	.18

D. Operator picks up billet, forges tong hold, places billet on table, and returns to hammer

Weight	9	28	47	67
Min	.38	.48	.59	.74

E.
Operator returns to hammer from table	.09
Changes tongs and places tongs on billet	.13
Restrike	.26
Trim and restrike	.31

Forging Machines 143

5.1

5.2 Explosive Forging Machines

DESCRIPTION

In this hot forging process steel billets automatically feed to an induction coil for heating. They are then placed in the die and the hammer is released and pounds the billet. The forged piece is removed, placed in a press and trimmed. Figure 5.2 illustrates a typical work station layout. The forging hammer is propelled by expanding nitrogen.

ESTIMATING DATA DISCUSSION

Data for setup in hours are given in Table 5.2. In general, one shift is allowed for die changing. Time for preheating the die depends on the size. Also included in setup is inspection of the first part which entails lab analysis of the stresses and okaying of the dimensions.

Elements 1 and 2 are constant minutes per piece for part handling, namely grabbing the hot billet with tongs, positioning in the die, and starting the explosive hammer cycle. Element 3 is a variable entry for hammer and aside time. The time depends on the finished thickness of the forged part. Element 4 is constant time for trimming. If no trimming is necessary this time should be added to the delay found in Element 5. The delay is a function of the hammer time and heating time of the billet. Notice the inverse nature of Element 5. During the delay activities such as lubricating the die and checking the feed may be performed. As the process is generally hot, heavy, and loud, two operators are used. The data are for two operators.

EXAMPLES

A. A lot of 200 gear spurs is to be forged. The finished thickness is 1 in. The part is to be trimmed. Estimate lot time.

Table Number	Process Description	Table Time	Adjustment Factor	Cycle Minutes	Setup Hours
5.2-S	Change and heat die	10.			10.
5.2-S	Inspection	.5			.5
5.2-1	Grab	.17		.17	
5.2-2	Position and start	.11		.11	
5.2-3	Hammer (1 In.)	.42		.42	
5.2-4	Trim	.18		.18	
5.2-5	Wait	.15		.15	
	Total			1.03	10.5
	Hr/unit	0.017			
	Lot time	13.93 hr			

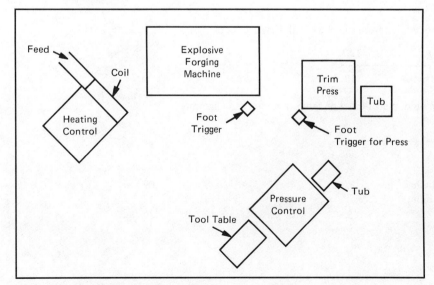

FIGURE 5.2 A typical work station layout.

B. A part will be run where the finished thickness will be approximately 2 in. after which it will be trimmed and a hole punched simultaneously. Estimate the operation time.

Table Number	Process Description	Table Time	Adjustment Factor	Cycle Minutes	Setup Hours
5.2-1	Grab	.17		.17	
5.2-2	Position and start	.11		.11	
5.2-3	Hammer (2 in.)	.19		.19	
5.2-4	Trim (no extra for punch)	.18		.18	
5.2-5	Wait	.36		.36	
	Total			1.01	

C. The trimming of the gear spur in Example A is to wait until machining. Re-estimate operation time.

Table Number	Process Description	Table Time	Adjustment Factor	Cycle Minutes	Setup Hours
5.2-1	Grab	.17		.17	
5.2-2	Position	.11		.11	
5.2-3	Hammer	.42		.42	
5.2-5	Wait (no trim: 0.15)	.33		.33	
	Total			1.04	

D. A lot of 2732 units is to be estimated. The part is halved in the trim press, but the multiple is forged together. Finished thickness is ¾ in.

Table Number	Process Description	Table Time	Adjustment Factor	Cycle Minutes	Setup Hours
5.2-S	Change die	8.			8.
5.2-S	First part inspection	1.			1.
5.2-1	Grab billet	.18	½	.09	
5.2-2	Position and start	.11	½	.055	
5.2-3	Hammer and part aside to trim	.48	½	.24	
5.2-4	Trim	.18	½	.09	
5.2-5	Wait after trim use, .48	.09	½	.045	
	Total			.52	9.
	Hr/100	.867			
	Lot estimate	32.68 hr			

TABLE 5.2 EXPLOSIVE FORGING MACHINES

Setup
    Change die   8–12 hr
    First part inspection   5–2 hr

Operation elements in estimating minutes

1. Grab billet   .18

2. Position and start   .11

3. Hammer and part aside to trim

Finished Thk	Min	Finished Thk	Min
½	.54	1½	.31
⅝	.51	1⅝	.27
¾	.48	1¾	.24
⅞	.45	1⅞	.22
1	.42	2	.19
1⅛	.39		
1¼	.36		
1⅜	.33		

4. Trim .18

5. Wait after trim

Hammer Min	Min	Hammer Min	Min
.54	.04	.33	.23
.51	.07	.31	.25
.48	.09	.27	.28
.45	.12	.24	.31
.42	.15	.22	.33
.39	.17	.19	.36
.36	.20		

Wait, no trim; add to above time .18

TURNING

6.1 Engine Lathes

DESCRIPTION

The engine lathe is an all-purpose turning machine tool. It cuts cylindrical forms with a single-point cutter moving parallel to the axis of rotation of the work. These estimating data are principally for the manufacturing classification of engine lathes, where numerical control is not a primary feature. If tracer machines or numerical control are involved, we choose to interpret these machines as requiring special treatment, and other estimating data scattered throughout the turning section can be adopted. A typical medium-size engine lathe is shown in Figure 6.1A. Now, consider estimating data for the general class of engine lathes produced by many manufacturers.

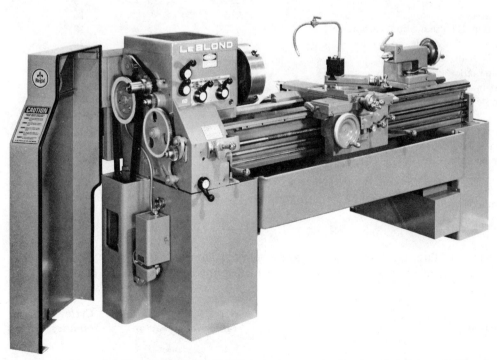

FIGURE 6.1A Manual shift engine lathe with 15 ½-in. swing over bed and carriage wings. *(Le Blond)*

ESTIMATING DATA DISCUSSION

Setup is composed of four elements. The basic time includes typical chores and has the entry variables of holding device and no. of tools. If a tool is to be used for multiple cuts of two or more dimensions, time is provided for additional dimensions. Makepiece can be included in setup, if that is company policy. Time for cut adjustment is provided for a critical tolerance.

The cycle time is arranged into five major groups. Element 1 deals with handling. This handling also includes disposal. If the part is heavy, large, or awkward, two hands may be required, which is covered by the term *large*. One-handed handling is either undesignated, or it is *small*. A variety of elements are described in Element 4; while the listing is extensive, the estimator may have to decompose the element into 2/3 and 1/3 ratios for elements that may not appear because of the wide adaptability of engine lathes. The first part of the element is the two-thirds contribution of the time.

The time for turning and allied machining, turning tool replacement, and part inspection are found in other tables.

EXAMPLES

A. An irregular-shaped valve consisting of SAE 4130 steel casting has several machine elements. The part requires two hands for loading because of its irregularity. The schedule of the cuts is given by the description below. Find the unit estimate for the cycle.

Table Number	Process Description	Table Time	Adjustment Factor	Cycle Minutes	Setup Hours
6.1-1	Load and unload in 4 jaw chuck	.30		.30	
6.1-4	Position cross-slide tool, set for depth advance tool to work, engage feed	.09	2	.18	
11.1-2	Make face cut, 2.81-in. OD, 1.81-in. ID, length of cut = .5 in., rough pass with carbide	.12	.5	.06	
6.1-4	Retract and advance cross slide	.09		.09	
11.1-2	Finish cut, 2.81-in. OD, 1.81-in. ID, length of cut = .5 in., finish pass with carbide	.28	.5	.14	
6.1-4	Change tool, set for proper dimension	.40		.40	
6.1-4	Advance tool to work, engage feed, and retract	.12		.12	
11.1-3	Trial cut, 1.570-in. ID, clean up surface by rough bore, use medium carbon steel	.52	.25	.13	
22.1-3	Mike dimension	.16		.16	
11-1-3	Finish bore 1.570-in. ID, use medium carbon steel	.72	.25	.18	
6.1-4	Retract tool for rough and finish and dial tool	.24		.24	
6.1-4	Position to bore 1.81-in. ID	.16		.16	
6.1-4	Hand dial tool to mark	.06		.06	
11.3-3	Rough bore 1.81-in. OD, 1.65-in. long, HSS tool point, use medium carbon steel	.60	1.65	.99	
4.3-3	Finish bore	.83	1.65	1.37	
6.1-4	Retract tool for rough and finish boring	.18		.18	
6.1-5	Turn coolant on and off, 6 times	.03	6	.18	
6.1-4	Turn spindle on and off, 7 times	.06	7	.42	
6.1-5	Air-clean part	.10		.10	
11.1-7	Break edges	.16		.16	
11.4-1	Tool life, 3.07 min of cutting time	.07		.07	—
	Total cycle estimate			5.69	

B. A bell-body has an internal thread, 2.75-in. in length, and a nominal diameter of .875 in. The part which is brass can be handled with one hand, and is loaded in a collet. A lot has 65 units. Find the lot time.

Table Number	Process Description	Table Time	Adjustment Factor	Cycle Minutes	Setup Hours
6.1-S1	Collet, 2 tool	.46			.46
6.1-1	Load collet	.13		.13	

Table Number	Process Description	Table Time	Adjustment Factor	Cycle Minutes	Setup Hours
6.1-4	Start and stop spindle	.06		.06	
6.1-4	Advance tail stock, lock, unlock, return	.18		.18	
	Drill for hole to be threaded, 2.75 + .39 = 3.14 in. for lead and drill distances				
11.2-4	Use power drilling, brass .075-in. ID	.10	3.14	.31	
6.1-4	Advance and retract carriage	.11		.11	
11.1-5	Threading, 7 threads per in	.18	2.75	.50	
6.1-4	Reposition tool, set depth, engage feed for 3 passes	.16	3	.48	
6.1-4	Clear tool	.09		.09	
6.1-5	Blow threads clean	.05		.05	
11.1-7	Break 1 edge internal	.06		.06	—
	Total			1.97	.46
	Lot estimate	2.59 hr			

C. The part "small cap" is partially described by the dimensioned sketch of Figure 6.1B. This stainless steel forging stock has been previously machined and the engine lathe is being used to do one cut, a counterbore on the left inside which is .187 in. Other machining is done elsewhere. For this operation there is a special collet tooling for clamping the end. For the 181 units in the lot, find the cycle, setup, and total lot hours.

Table Number	Process Description	Table Time	Adjustment Factor	Cycle Minutes	Setup Hours
6.1-S1	Basic setup, 1 tool	0.34			0.34
6.1-S2	Multiple cuts	0.06			0.06
6.1-S3	Make piece	0.01	2		0.02
6.1-S4	Adjust cut tolerance	0.08			0.08
6.1-1	Handle	0.13		0.13	
6.1-4	Start, stop	0.06		0.06	
6.1-4	Splash shield	0.07		0.07	
6.1-4	Engage feed	0.03	2	0.06	
6.1-4	Advance, retract, cross slide	0.09		0.09	
6.1-4	Cross slide to dial	0.16		0.16	
11.1-3	Bore, rough .187-in. dia with carbide, .05-in. depth of cut	0.02	.05	0.00	
11.1-3	Bore finish .187-in. dia with carbide	0.03	.05	0.00	

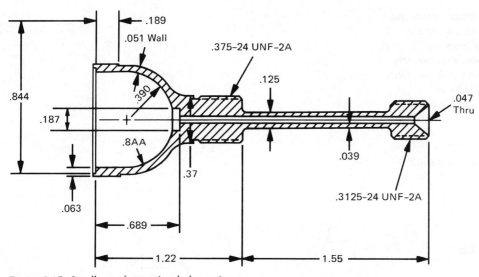

FIGURE 6.1B Small part for engine lathe estimate.

Table Number	Process Description	Table Time	Adjustment Factor	Cycle Minutes	Setup Hours
1i.1-7	Break edges	0.06		0.06	
11.4-1	Turn tool life	0.00		0.00	
	Total			0.63	0.50
	Total lot hours	2.41			

TABLE 6.1   ENGINE LATHES

Setup in estimating hours

S1. Basic time

Holding Device	No. of Tools		
	1	2	3
Collet	.34	.46	.57
Chuck	48	.57	.69
Fixture	.52	.63	.75

S2. Multiple cuts, no. of cuts per tool   .06/occurrence

S3. Make piece (optional): 2 × hr/unit

S4. Adjust cut

Tolerance	Over .01	.0031 .010	.0011 .0030	.0001 .0010
Hr	.03	.08	.12	.15

Operation elements in estimating minutes

1. Pick up part, move and place, pick up and lay aside
   - Load and unload collet or air chuck,
     - Small .13
     - Large .20
   - Load and unload T-wrench chuck,
     - Small .18
     - Large .30
   - Load fixture, clamp and unclamp, unload,
     - Small .20
     - Large .40
   - Load, unload arbor without arbor nut .39
   - Load, unload arbor with arbor nut .60
   - For each add'l pc on arbor add .11
   - Load and unload collet with tweezers .20
   - Load and unload collet with split bushings .15
   - Load and unload plate fixture .51
   - Load and unload expanding arbor .10
   - Place between and remove from centers,
     - Small .12
     - Large .68
   - Center to work and lock tail stock .08
   - Jib load part
   - Chuck jaws 5.30
   - Fixture 4.29

2. Clamp and unclamp
   - Put on and remove dog,
     - Small .12
     - Large .20

Press pc on and off mandrel	.21
Hammer pc to seat or loosen	.05

3. Part manipulation
   Unload, turn end for end, reload T-wrench chuck,

Small	.20
Large	.35

   Unload, turn end for end in collet,

Small	.05
Large	.10

4. Engine lathe machine operation

Start and stop spindle	.06
Reverse spindle	.04
Raise and lower splash shield	.07
Dial tool post to mark	.06
Engage and disengage feed (cross-slide or carriage)	.03
Change feed direction	.03
Advance and retract cross-slide or tool	.09
Advance and retract cross-slide to dial reading	.16
Lock and unlock carriage	.07
Advance and retract carriage	.11
Advance carriage to dial indicator, lock, unlock, retract carriage	.10
Advance and retract tailstock	.15
Advance tailstock, lock, unlock, and return tailstock	.18
Dwell	.05
Advance and retract compound	.13
Advance compound to dial reading and return	.17
Cut in or out half nut lever	.04
Change tool in tool post, set for proper position	.40

5. Machining

Turn, bore, form, cutoff, thread, start drill	Use Table 11.1
Drill, ream, counterbore	Use Table 11.2

6. Part inspection   Use Table 22.2

6.2 Turret Lathes

DESCRIPTION

A principal characteristic of turret lathes is the consecutive arrangement of tools. With a hexagon or octagon turret and a cross-slide turret, many possibilities exist for prearranging the selection of tools. Controls for these machines are restricted to manual or electric, and numerical control turret lathe machines are found elsewhere. Two figures are shown which illustrate a ram and saddle type machine. In Figure 6.2A, a 4½-in. capcacity ram type is shown. Usually, ram-type turret lathes are faster than saddle types. Figure 6.2B shows a saddle type which has the advantage of heavier cuts. Power rapid traverse is available for hex and square turrets. Preselection of speeds and feeds are options provided by various manufacturers. Now, we consider the estimating for the general class of turret lathe machines.

ESTIMATING DATA DISCUSSION

In determining the setup value, the basic time given by Element 1 has the no. of tools as the entry variable. A duplex tool using one turret, or a combi-

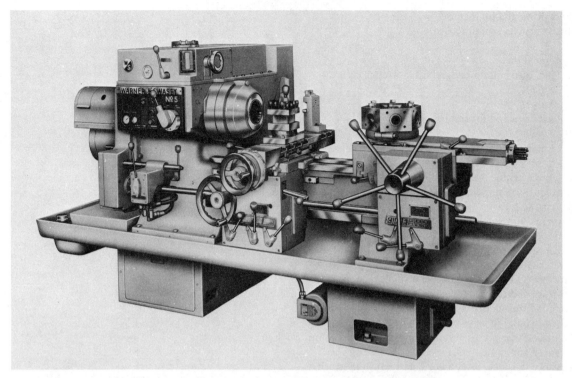

FIGURE 6.2A Universal ram-type turret lathe with a 4½-in. capacity. *(Warner and Swasey)*

nation-turning box tool would count as two tools. Setup Element 2 accounts for different types of holding equipment, or special machine-mounted tools, or gages have times selected from Element 3. Should there be critical tolerances, Element 5 is applied.

The three types of turret lathes are identified as small, medium, and large. Even though the variety that really exists exceeds this classification, we have limited the variations. The machine distinctions are refined with elements suitable for a particular machine

FIGURE 6.2B Saddle type universal turret lathes. *(Jones & Lamson)*

tool. The types are classified with respect to maximum collet diameter as small (1 in. or less), medium (more than 1 in. to 3 in.), and large (more than 3 in. to 6 in.). Turret lathes come in many sizes from ½-in. to 10-in. diameters. Nor do we categorize between ram or saddle machines, but some elements deal with these distinctions. Because of the varied machining operations performed on the turret lathe, tables can also serve for identical operations performed on other and more specialized machines.

Turret lathes will permit the functioning of several tools simultaneously. It is possible to have a hexagonal-turret tool perform a drilling element at the same time a cross-slide tool turns or faces. The longer of these simultaneous operations is selected. It may be possible to gage or break edges during machine cuts.

Element 1 deals with many variations of handling. Principally, the holder may be a collet, chuck, or fixture. Various opportunities for clamping are suggested.

Element 3 deals with machine operations. Major distinctions are made for the turret, cross-slide, and carriage. Element 4 is a consolidation of the various elements and can simplify element selection. This element is for routine parts. The consolidated sets usually include start and stop, advance turret and feed stock, retract, index, advance, and engage feed. Additional position oil line and splash guards are included. If elements other than these are anticipated, they may be added; conversely, elements can be deducted to facilitate ease of usage.

Elements 5, 6, and 7 are found using other tables. These machining tables do not encourage the adjustment of feeds and speeds to those actually available on the selected machine for the job. While the estimator can correct to actual rpm, this practice is time consuming and may not improve accuracy. See the discussion on this technique given by the machining tables.

EXAMPLES

A. An SAE 1020 thread adapter is completely machined on a large-size turret lathe. The schedule of the turrets is as follows: (1) combination stock stop and start drill, (2) drill, (3) boring bar, (4) reamer mounted in floating holder, (5) recessing tool using a quick-acting slide tool, (6) tap mounted in square turret position 1. The stock diameter is 3 in., and additional machining data are stated below. Find the lot time for a quantity of 315. Develop elements rather than use the consolidated handling elements.

Table Number	Process Description	Table Time	Adjustment Factor	Cycle Minutes	Setup Hours
6.2-S1	Seven tools for setup	3.0			3.0
6.2-S2	Bar stock	.2			.2
6.2-S4	Make piece 2 × .11	.11	2		.22
6.2-S5	Tolerance checking for .001 in. tolerance	.08			.08
	Setup total				3.40
6.2-3	Start and stop spindle	.09		.09	
6.2-1	Open and close spindle	.10		.10	
6.2-3	Advance turret and feed stock to 6 in.	.29		.29	
6.2-3	Index duplex holder for start drill	.07		.07	
6.2-3	Advance turret and engage feed	.05		.05	
11.1-6	Start drill for 1¼-in. tap drill dia	.85		.85	
6.2-3	Return turret, index, advance, and engage feed, for 5 turrets	.16	5	.80	
11.2-4	Drill 1¾₄-in. hole, 4-in. long + .25 lead	.23	4.25	.98	
11.1-3	Bore 2.5-in. long × 1.65-in. OD, HSS, rough	.31	2.5	.78	
11.2-5	Ream 1.65-in. OD	.27	2.5	.68	
11.1-3	Machine recess 1¾-in. ID, ¼-in. long, use finish bore time,	.43	.25	.11	
6.2-3	Quick-acting tool holder up and down	.03		.03	
11.2-3	Tap 1¼–7 NC, 1.25-in. long	.32	1.25	.40	
6.2-3	Reverse spindle	.15		.15	
11.1-7	Break edges	.08		.08	
11.1-4	Cut off 3-in. OD, length of cut = .68 in., using carbide	.26	.68	.18	
	Total			5.91	3.40
	Hr/unit	.098			
	Lot time	34.5 hr			

B. A cast iron air-starting cam is partially machined on the turret lathe. Drilling, facing, and boring are the machining elements. The part weighs 3 lb and is irregular. Holding is by chuck. Turret positions are (1) center drill, (2) drill, (4) boring, and square turret position (1) faces. Use an elemental approach to find unit and setup time. A medium-size turret lathe will be used with a two-jaw chuck.

Table Number	Process Description	Table Time	Adjustment Factor	Cycle Minutes	Setup Hours
6.2-S1	Three hex and 1 square turret tools	2.0			2.0
6.2-S2	Chuck	.1			.1
6.2-S4	Make piece, (not allowed)				
6.2-S5	Tolerances, .001	.08			.08
6.2-1	Get, load in chuck with T-wrench	.37		.37	
6.2-3	Start and stop	.09		.09	
6.2-3	Advance turret, engage feed	.05		.05	
11.1-6	Center drill for $1^{19}/_{32}$-in. hole	.25		.25	
6.2-3	Return, index and advance	.07		.07	
11.2-3	Drill $1^{19}/_{32}$-in. hole, $2^{3}/_{8}$-in. long	.33	2.37	.78	
6.2-3	Return and skip index for turret	.07		.07	
6.2-3	Change speed	.04		.04	
6.2-3	Face end using square turret, advance and engage, carbide tool	.07		.07	
11.2-2	Face end $2^{1}/_{4}$-in. length of cut, $2^{3}/_{8}$-in. OD	.07	2.25	.16	
6.2-3	Advance turret, engage feed, hex no. 4	.05		.05	
6.2-3	Bore 1.501-in. hole, $2^{1}/_{4}$-in. long, carbide tooling,	.06	2.25	.14	
6.2-3	Return turret to hex position 1, estimate	.05		.05	
11.2	File burrs during boring	0		0	
22.1-3	Gage part during drill hole time	0		0	
	Total			2.19	2.18

TABLE 6.2 TURRET LATHES

Setup in estimating hours

S1. No. of Tools	1	2	3	4	5	6	7	8	Add'l
Hr	1.09	1.27	1.45	1.63	1.82	2.00	2.17	2.35	.15

S2. Holding tool
    Collet   .2
    Chuck   .1
    Bar stock   .2
    Face plate, fixture   .4

S3. Miscellaneous
    Dial indicator   .2
    Cross-slide indicator stop   .2
    Install carriage bed stop   .2
    Large manual chuck   .4
    Bore soft jaws   .3
    Tap chasers   .3
    Box tool rollers   .1

S4. Make-piece (optional): $2 \times$ hr/1 unit

S5. Adjust cut:

Tolerance	Over .01	.0031 .010	.0011 .0030
Hr	.03	.08	.12

Operation elements in estimating minutes

1. Handling

	Type of Turret Lathe		
	Small	Medium	Large
Get, load, close collet, open collet, and unload part	.13	.19	.25
Get, load in air chuck, clamp, unclamp, and unload		.21	.36
Get, load in chuck with T-wrench, clamp, unclamp, and unload		.37	.50
Get, load in fixture, clamp, unclamp, and unload		.99	1.15
Catch cutoff part and unload	.03	.03	
Open and close collet, feed stock by hand	.12	.14	
Unclamp, unload, turn end-for-end, load, and clamp in collet	.07	.12	.21
Unclamp, unload, turn end-for-end, load, and clamp in chuck		.12	.25
Unclamp, unload, turn end-for-end, load, and clamp in with T-wrench in chuck		.75	.85
Open and close collet			.10
Relieve stress (before a finish bore, etc.)			1.00
Jib load in chuck, unload, 0–24 in.			2.15
Jib load in chuck or fixture, 24 + in.			3.15

2. Cleaning and lubricating

	Type of Turret Lathe		
	Small	Medium	Large
Position and clear coolant line	.04	.04	.04
Air-clean collet	.05	.05	.16
Air-clean part	.06	.06	.13
Air-clean fixture or chuck		.07	.18
Place and remove oil guard		.08	.10

3. Turret lathe machine operation

	Type of Turret Lathe		
	Small	Medium	Large
Start and stop spindle	.05	.05	.09
Speed, change range or high–low	.02	.04	.05
Reverse spindle and change back	.03		.15
Shift gears			.20
Advance turret and feed stock			
to 3 in.	.03	.11	
to 6 in.	.04	.15	
to 9 in.	.04	.18	
by hand		.12	
Turret. Advance and return	.03	.04	
Advance turret and engage feed	.05	.05	.05
Engage and disengage feed	.03		.08
Return	.07		
Return, index, and advance	.03	.07	.15
Return, index, advance, and engage feed	.06	.08	.16
Return and skip index	.04	.07	.07
Index, stop roll	.02		
Lock and unlock carriage or cross-slide	.07		.09
Return and index		.04	

	Type of Turret Lathe		
	Small	Medium	Large
Engage feed		.02	.03
Disengage feed		.01	
Index and back index			.07
Unlock and advance			.08
Unlock saddle, retract, index, advance, engage feed			.22
Advance and lock			.10
Change feed			.03
Cross-slide. Advance			.09
Advance and engage feed		.07	.10
Advance to dial stop and lock			.20
Lock and unlock		.05	.06
Return			.07
Unlock and return			.10
Disengage feed and return			.08
Index square turret, 1 station		.04	.07
Advance and return	.03	.13	
Index cross-roll stop			.04
Advance, engage feed, and return		.14	.17
Advance cross-slide to stop, lock, unlock, and return		.26	
Change feed direction, engage		.05	
Carriage. Engage feed			.03
Lock and unlock			.07
Advance			.04
Advance and engage feed		.11	.11
Advance to dial indicator and lock			.11
Return			.04
Unlock and return		.06	.07
Return, index stop roll			.10
Return, index stop roll, and advance		.14	.16
Return, index, advance to indicator stop, and lock			.20
Reset self-opening die or tap	.03	.04	.04
Index duplex holder		.07	.07
Move slide holder up and down		.08	.13
Move quick-acting holder up and down		.05	.08
Engage lead screw	.06		
Dwell	.03	.03	.03

4. Consolidated handling element

   A. Bar stock elements, small and medium turret lathes, min

| No. Turrets | 1 | 2 | 3 | 4 | 5 | 6 | 1 | 2 | 3 | 4 | 5 | 6 |
No. Cross-Slides	1	1	1	1	1	1	2	2	2	2	2	2
Small	.27	.33	.39	.45	.51	.57	.30	.36	.42	.48	.54	.60
Medium	.41	.49	.57	.65	.73	.81	.54	.62	.70	.78	.86	.94

   B. Collet elements, small and medium turret lathes, min

| No. Turrets | 1 | 2 | 3 | 4 | 5 | 6 | 1 | 2 | 3 | 4 | 5 | 6 |
No. Cross-Slides	0	0	0	0	0	1	1	1	1	1	1	1
Small	.28	.34	.40	.56	.52	.58	.31	.37	.43	.49	.55	.61
Medium	.44	.52	.60	.68	.76	.84	.58	.72	.86	1.00	1.14	1.28

| No. Turrets | 1 | 2 | 3 | 4 | 5 | 6 |
No. Cross-Slides	2	2	2	2	2	2
Small	.34	.40	.46	.52	.58	.64
Medium	.72	.86	1.00	1.14	1.28	1.42

C. Chuck elements, medium turret lathes, min

No. Turrets	1	2	3	4	5	6	1	2	3	4	5	6
No. Cross-Slides	0	0	0	0	0	0	1	1	1	1	1	1
Medium	.46	.54	.62	.70	.78	.86	.60	.74	.88	1.02	1.16	1.30

No. Turrets	1	2	3	4	5	6
No. Cross-Slides	2	2	2	2	2	2
Medium	.74	.88	1.02	1.16	1.30	1.44

D. Bar stock elements, large turret lathes, min

Cross-Slides	No. Turrets					
	1	2	3	4	5	6
1	.55	.71	.87	1.03	1.19	1.35
2	.72	.88	1.04	1.20	1.36	1.52
3	.89	1.05	1.21	1.37	1.53	1.69
4	1.06	1.22	1.38	1.54	1.70	1.86
5	1.23	1.39	1.55	1.71	1.87	2.03

E. Collet elements, large turret lathes, min

Cross-Slides	No. Turrets					
	1	2	3	4	5	6
1	.81	.97	1.13	1.29	1.45	1.61
2	.98	1.14	1.30	1.46	1.62	1.78
3	1.15	1.31	1.47	1.63	1.79	1.95
4	1.32	1.48	1.64	1.80	1.96	2.12
5	1.49	1.65	1.81	1.97	2.13	2.29

F. Chuck castings, large turret lathes, min

Cross-Slides	No. Turrets					
	1	2	3	4	5	6
1	1.06	1.22	1.38	1.54	1.70	1.86
2	1.23	1.39	1.55	1.71	1.87	2.03
3	1.40	1.56	1.72	1.88	2.04	2.20
4	1.57	1.73	1.89	2.05	2.21	2.37
5	1.74	1.90	2.06	2.22	2.38	2.54

5. Machining
   Turn, bore, form, cut off, thread, start drill, and break edges Use Table 11.1
   Drill, counter bore, ream, counter sink, or tap Use Table 11.2

6. Turning tool replacement Use Table 11.4-1

7. Part inspection Use Table 22.1

6.3 Vertical Turret Lathes

DESCRIPTION

A vertical turret lathe resembles a vertical boring mill, but has the additional characteristics of a turret for tool mounting. The VTL includes a rotating horizontal table with the turret mounted to ram on the cross-rail. There may be side heads provided with square turrets. This machine facilitates the mounting, holding, and the machining of large-diameter heavy parts. Only chucking or fixture work is done on a VTL. The main and auxiliary turrets function in the same manner as the hexagon and square turrets on a horizontal lathe. The side head has a rapid traverse and feed independent of the turret, which provides for simultaneous machining adjacent to elements performed by the turret. The machine can be provided with numerical control, manual or automatic operation. Control permits automatic operation of each head, including rate and direction of feed, change in spindle feed, indexing of turret, starting, and stopping. Simultaneous cutting by the overhead turret or the side head is possible. Tool changer with ram is another option. Figure 6.3 is a 36-in. VTL configured with a five-station turret.

ESTIMATING DATA DISCUSSION

The setup basic time is a required constant. It includes punch in/out, instructions and study, install and remove tape, clean chips, adjust rail height, assemble dial indicator for indicating fixture, and reverse.

In Element 2, various holding devices are considered. A jib and sling hoist to an eyebolt on the 3-jaw chuck or fixture, position on machine table, indicator (fixture only), secure to table, clean, and reverse work elements. Round and standard jaws are secured by positioning T-nuts in the slot of the machine table, positioned, secured, and removed. Standard jaws require two cap screws. Locator pins have the T-nut positioned in a machine table slot, secured, turned selected height pin into locator pin base, and reverse. Spring

FIGURE 6.3 Vertical turret lathe with side turret and NC control. *(Bullard)*

loaded jacks have T-nut positioned in machine table slot, position, secure with one cap screw, and adjust jack to piece, and reverse. The work is similar for an adjustable jack.

Element 3 deals with tool installation. Tool installation involves cleaning turret face, securing tool block or bar to machine turret, positioning tool holder in tool block or bar, and reverse.

Should the manuscript instructions require inspection, work involves obtaining mike or other gage, checking piece, and aside to storage.

Consider the description of several cycle-time elements. Handling includes these general items of work: remove chips from fixture using air hose, brush, or rag; get sling(s), eyebolts, attach, position, and aside hoist; place chip guards in position on machine; and reverse. Clamping, Element 2, involves turning and positioning of clamp, obtaining wrench and tightening and reverse.

In Element 3, the program stop involves removing chips, obtaining tools, and inspection.

If the part has received heavy cutting, and is unable to be moved or inspected, the cooling involves flooding by coolant, and approximations are provided in Element 3.

EXAMPLES

A. A heavy high-temperature alloy part is to be turned on the vertical turret lathe. One main bore, one step bore, two faces, and outside turning are performed on the forging. An NC vertical turret lathe is planned. Find the lot time for 74 parts. The machining schedule is given by the list of elements. Cutting velocity from Table 11.1 is 40 fpm. Use cutting approach with formulas rather than tables.

Table Number	Process Description	Table Time	Adjustment Factor	Cycle Minutes	Setup Hours
6.3-S1	Basic time	.35			.35
6.3-S2	Round jaws	.14			.14
6.3-S3	Six tools	.12	6		.72
6.3-1	Jib load to chuck and unload	7.18		7.18	
6.3-2	Clamp	1.38		1.38	
6.3-3	Start program and back two turrets	.38		.38	
11.1-2	Face 26.5-in. OD rough with carbide $$\text{min/in.} = \frac{26.5\pi}{12 \times .021 \times 40} = 8.26 \text{ and length of cut} = 2.1 \text{ in.}$$	8.26	2.1	17.35	
11.1-3	Bore step 18.1-in. ID rough with carbide $$\text{min/in.} = \frac{18.1\pi}{12 \times .021 \times 40} = 5.64 \text{ and length of cut} = 1.3 \text{ in.}$$	5.64	1.3	7.33	
11.1-2	Finish face 26.5-in. OD with carbide $$\text{min/in.} = \frac{26.5\pi}{12 \times .015 \times 60} = 7.71 \text{ and length of cut} = 2.1 \text{ in.}$$	7.71	2.1	16.19	
6.3-1	Jib turnover and clamp, 5.28 + 1.38	6.66		6.66	
11.1-2	Face 26.5-in. OD at 8.26 min/in. for length = 4.1, while simultaneously turning 26.5-in. OD for length of 7.1 in. using side tool. Accept longer time of two cuts or	8.26	4.1	33.87	
11.1-3	Counter bore 17.3-in. hole rough with carbide for a length of 1.2 in. $$\text{min/in.} = \frac{17.3\pi}{12 \times .021 \times 40} = 5.39$$	5.39	1.2	6.47	
11.1-3	Finish counterbore, carbide for 1.2 in. $$\text{min/in.} = \frac{17.3\pi}{12 \times .015 \times 60} = 5.03$$	5.03	1.2	6.04	
6.3-3	Cool off piece	3.00		3.00	
6.3-3	Turret manipulation to return during cooling	.00		0	
22.1-3	Inspection for 6 dimensions	.16	6	.96	
11.4-1	Tool life and replacement for 87 min, 87/40 × 36	.79		.79	
	Total			107.60	1.21
	Lot estimate	133.92 hr			

TABLE 6.3 VERTICAL TURRET LATHES

Setup elements in estimating hours

- S1. Basic time .35

- S2. Install and take down
    - Fixture .48
    - Chuck .32
    - Invert jaws (chuck only) .07
    - Round jaws
        - 2 jaws .07
        - 4 jaws .14
    - Standard jaws
        - 2 jaws .23
        - 4 jaws .46
    - Locator pins (3) .25
    - Spring loaded jacks (4) .34
    - Adjustable jack, ea .07
    - Piece clamp, ea .07

- S3. Tool
    - Tool block or flanged bar, ea .12
    - Bar holder, ea .14

- S4. Inspection, per requirement .02

Operation elements in estimating minutes

1. Handling

    Manual load and unload

Weight	25	50
To Table Clamps	3.93	6.28
To Fixture	1.55	2.21
To Chuck	1.95	2.60

    Jib load, unload
    - To fixture 6.03
    - To chuck 7.18

    Overhead crane load and unload
    - To fixture 10.77
    - To chuck 11.43

    Turnover
    - Jib or chuck or fixture 5.28
    - Overhead crane or chuck or fixture 9.44

2. Clamp, unclamp
    - Fixture, per clamp .65
    - Chuck or standard jaws 1.38

3. Machine operation
    - Start, stop
    - Engage feed .16
    - Align piece per program instructions
    - Adjustable jacks with surface gage 3.23
    - Indicate chuck, fixture, and time with rough surface 6.82
    - Indicate chuck, fixture, and true with finish surface 12.30
    - Program stop for inspection dimension per tool 1.03
    - Turret advance, retract, index, or clear

No.	1	2	3	4	5
Time	.19	.38	.57	.76	.95

Ram or side slide, advance, retract, index, ea	.30
Flood piece for cooling	
Light	1.00
Massive	3.00

4. Machining
    Turn, bore, form, thread, start drill, and break edges   Use Table 11.1
    Drill, counterbore, ream, or tap   Use Table 11.2

5. Tool wear and replacement   Use Table 11.4

6. Inspection   Use Table 22.1-3

6.4 Numerically Controlled Turning Lathes

DESCRIPTION

The machines described in this section are turning lathes with numerical control (NC). They are designed for heavy-duty production. Slant and horizontal bed type are varieties within this description. Vertical turret lathes and chucking type horizontal machines are considered elsewhere. There is a variety of configurations for the multiple-slide turrets or tool changer as arranged in various positions on the machine. These turrets are able to hold a total of four to twelve or more tools.

NC lathes, in general, have advantages over automatic turret lathes in that starting, stopping, feeds, speeds, tool indexing or changing, and the tool path are controlled automatically by tape.

Figure 6.4A is a machine that behaves as a turning model if it is supplied with a tailstock. With an 18-in. swing over the bed, and a 3-in-hole through spindle, the number of ID/OD tools depends upon the needs. In this model, tools can be arranged in various ID/OD combinations. Figure 6.4B is a combination lathe and has a versatility for turning and chucking.

ESTIMATING DATA DISCUSSION

The basic setup time is always provided. It includes general machine cleanup, sweep chips, position chip guard, and operator self-cleaning. It also includes remove tape from reader, wind up on two spools, and return to tape container. General cleanup, manuscript caring, and job instructions also are provided for. The NC manuscript or process sheet is verified for the tools, and the first tool is gathered along with the first mike, dogs, and other general or special purpose tooling. The foregoing is included in the basic time. The basic time covers only the first tool. Another element considers any additional tools or mike.

The setup element for adjust tool offset is initiated after the cycle is started, and a trial of approximately ¼ in. mike diameter, and adjust offset are included. At the completion of the job, the operator returns all offsets to zero for the finishing tool.

A time consideration is provided to mike each decimal tolerance which is used once per each decimal tolerance on the shaft. Placing and removing a steady rest with a jib and sling is provided. On some models, the steady rest is on a swing mechanism, and the time to set up the steady rest is the same. The opportunity to load and unload a tailstock depends upon the specific machine. On some machines the tailstock is hinged and may swing down.

Place and remove part includes obtain jib and sling, sling part, hook up, and move to lathe. It also includes wipe centers on both part and lathe, position between centers, advance and lock tailstock, unhook and remove sling, aside jib and sling. Obtain jib and sling, sling part, hook up, unlock and clear tail, move to box or pallet, remove, and aside sling and jib also are included.

The place and remove part with jib element includes obtain jib and sling, sling part, hook up, and move to lathe. Wipe centers on both part and lathe, position between centers, advance and lock tailstock, unhook and remove sling, aside jib, and sling also are included, as are obtain jib and sling, sling part, hookup, unlock and clear tail, move to box or pallet, and remove and aside sling and jib.

Place and remove part with the overhead crane includes notify hooker for first part only, wipe centers on

FIGURE 6.4A A numerically controlled turning lathe. *(Waterbury Farrel, Division of Textron, Inc.)*

FIGURE 6.4B A numerically controlled combination lathe. *(Waterbury Farrel, Division of Textron, Inc.)*

both part and lathe, position between centers, advance and lock tail, remove, and aside sling. After the machining is concluded, it includes notify hooker, obtain steel tape and chalk, locate and mark balance line, aside tape and chalk, sling part while crane is in transit, unlock and clear tail, and guide from between centers.

The element turn part 180° with jib includes obtain jib and sling, sling part, hook up, unlock and clear tail, turn part 180°, wipe centers on both part and lathe, position between centers, advance and lock tail, unhook, remove sling, and aside sling and jib.

The turn part 180° with overhead crane includes notify hooker, obtain steel tape and chalk, locate and mark balance line, aside tape and chalk, obtain sling, sling part while crane is in transit, turn part 180°, wipe centers of both part and lathe, position between centers, advance and lock tail, unhook, and remove and aside sling and jib.

The elements secure and release and dog with one bolt include obtain dog from spindle center and position on part, secure by hand, obtain wrench from tool tray on top of headstock and secure tight, aside wrench, obtain wrench from tool tray and loosen bolt, wrench and hand, aside wrench, remove, and aside dog to spindle center.

If the surface is to be protected, the element guard when dogged on finished surface involves obtain guard from tool tray on top of lathe and place under bolt, and remove and aside guard to tool tray.

Position, open, close, and adjust steady rest includes position to turned spot, slide on machine ways, obtain box wrench from tool tray and secure steady rest to ways with one bolt, close steady rest and secure with wrench, position two bottom rollers to part and secure with wrench, and loosen clamping bolt on side. Position top roller to part and secure with wrench, secure clamping bolt, aside wrench to tool tray and obtain chip shield and place also are included. Also included are: remove chip shield and aside to chip pan, obtain wrench from tool tray, unclamp and retract rollers, release nut and reposition bolt, open steady rest, release rest from ways and slide toward headstock, and aside wrench to tool tray.

This cycle start time allows for a total stop when programmed or necessary.

The stop and start spindle and start feed element accommodates programmed or planned feed stops.

A set speed manually will cover walk to control panel, turn tape control knob and turn toggle to desired position, push start button to return to normal operation, and turn tape control knob to auto.

Set tool from shaft end with length gage, shaft not faced, means that the direct-labor operator will obtain gage from bench and place on turret for a controlled length. This distance is from the tool to the end of the shaft. Turn the incremental adjustment knob. If length of shafts are not consistent, set the length each time shaft is turned or placed in the machine.

Set tool from shaft end with length gage (shaft faced) considers obtain gage from bench and place on turret for measurement of decimal length. This is the distance from the tool to the end of the shaft. Incremental adjustment follows. Set length each time shaft is turned or placed in machine, as length of shafts is not consistent.

Set tool with gage block from finished shoulder implies that the operator obtain gage from bench, position to shaft, position tool to gauge by turning the incremental adjustment knob, and pushing the left or right job button. Aside gage to bench eventually.

Adjust tool, incremental, includes a trial cut and the tool is adjusted for the rough and finish cuts first end, finish cut second end. Afterwards, mike an adjustment may have to be made.

EXAMPLES

A. An 8-in. SAE 4140 shaft is turned for several diameters and lengths. The schedule of *Dia* and *L* is listed below. An NC heavy-turning lathe is used. Determine setup, unit time, and lot time for 10 units. An NC tape is unavailable at the time of the estimate.

Table Number	Process Description	Table Time	Adjustment Factor	Cycle Minutes	Setup Hours
6.4-S1	Basic time	.39			.39
6.4-S2	Check tool magazine for 4 tools	.08			.08
6.4-S4	Check 4 dimensions	.08			.08
6.4-1	Place and remove part with crane between centers	4.39		4.39	
6.4-1	Dog, 1 bolt	.63		.63	
6.4-2	Set tool with gage	.57		.57	
6.4-2	Cycle start	.08		.08	
11.1-2	Rough machine, 7.85-in. OD × 69.60-in. long, carbide, 1 pass	.32	69.60	22.27	
6.4-2	Stop and start spindle, start feed	.11		.11	
6.4-2	Mike and adjust tool	.57		.57	
11.1-2	Turn rough 7.25-in. OD × 54.15-in. long, carbide	.32	54.15	17.33	

Table Number	Process Description	Table Time	Adjustment Factor	Cycle Minutes	Setup Hours
6.4-1	Turn part with crane	2.54		2.54	
6.4-1	Dog, 1 bolt	.63		.63	
6.4-2	Set tool with gage	.57		.57	
6.4-1	Cycle start	.08		.08	
11.1-2	Rough machine 6.95-in. OD × 39.90-in. long	.28	39.90	11.17	
11-1.2	Rough machine 6.45-in. OD × 18.25-in. long	.28	18.25	5.11	
6.4-2	Start and stop spindle, start feed	.11		.11	
6.4-2	Mike and adjust tool	.57		.57	
11.1-2	Finish machine	.65	18.25	11.86	
6.4-1	Turn part 180°, overhead crane	2.54		2.54	
6.4-1	Dog, 1 bolt	.63		.63	
6.4-2	Set tool with gage	.57		.57	
6.4-2	Cycle start	.08		.08	
11.1-2	Finish cut 7.85-in. OD × 15.45-in. long	.75	15.45	11.59	
6.4-2	Start and stop spindle, start feed	.11		.11	
6.4-2	Mike and adjust tool	.57		.57	
11.1-2	Finish machine 6.95-in. OD × 21.65-in.	.65	21.65	14.07	
6.4-4	Indexing of tools, 8 times	.034	8	.27	
8.2-1	Rapid traverse, 237 in., 200 in./min	1.50		1.50	
11.4-1	Tool wear for 93.21 min, 93.21/40 × .22	.51		.51	
	Total			111.02	.55
	Lot time	19.05 hr			

B. An airframe part, No. 50532, is made of 4340 bar stock in the annealed state. The raw material is supplied in 12-ft lengths. A quantity of 7075 is to be estimated using a heavy NC turning lathe. The adjustment factor can be the length of cut or the number of cuts or tools. A facing cut has its length of cut as half of the diameter. Notice in the estimate below that the elements can be in most any order. A partially dimensioned sketch is given by Figure 6.4C.

Table Number	Process Description	Table Time	Adjustment Factor	Cycle Minutes	Setup Hours
6.4-S1	Turn, drill, thread, basic time	.39		.39	
6.2-S1	No. of tools, check	.02		.16	
	Eight tools for hex and cross slide		8		

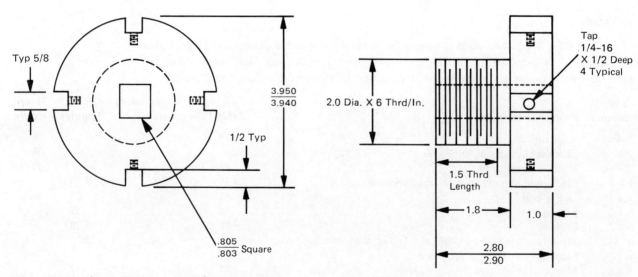

FIGURE 6.4C Airframe part, raw material 4340 annealed.

Table Number	Process Description	Table Time	Adjustment Factor	Cycle Minutes	Setup Hours
6.4-2	Start and stop mach	0.11		0.11	
6.4-1	Open, close collet	0.10		0.10	
6.4-3	Adv turret, feed stock	0.15		0.25	
11.1-2	Face (same as turn)				
	OD = 4, L = 2, carb	0.37	2	0.74	
11.1-2	Finish turn				
	L = 2.9, OD = 4, carb	0.37	2.9	1.07	
11.1-2	Rough turn 4-in. OD	0.16	1.8	0.29	
	L = 1.8 in., one pass	0.29	2	0.58	
	Right turn, two add'l passes from above				
	Three passes to obtain third dia.				
11.1-5	Threading	0.16	1.5	0.24	
	Thread dia = 2				
	Self-opening die	0.24	2	0.48	
11.2-4	Drill $^{25}/_{32}$, 3-in dp for broached hole	0.27	3	0.81	
11.1-4	Form or cutoff				
	OD = 4, L = 2 in.	0.54	2	1.08	
6.4-3	Clean, lubricate				
	Coolant lines	0.13		0.13	
11.1-7	Break edges	0.12		0.12	
6.2-S4	Make piece setup				
	Optional choice	0.10	2		0.20
	Total			6.00	.75
	Lot hours	708.25			

C. Figure 6.4D shows a partially dimensioned part. This gas turbine product part is of AMS 6265 material (similar to AISI 4340 steel) and is forged to oversize for length, but is generally close to 2½-in. OD stock. Several operations have preceded this operation. Notably the right end of the bar has been preliminary turned to provide an accurate holding and locating surface for work on the left end of the drawing. This operation considers the left end of the barstock. The machining elements are shown below and estimated. All tooling is carbide, and the machine is a medium sized NC turning lathe. The lot is 41 units. Notice in the estimate that the machining elements are listed first followed by the handling for the equipment. This is a logical way to approach many machining operations, since the machining is necessary and the handling and setup corresponds to this requirement. Insofar as the values of the estimate are concerned, the ordering of the elements is unimportant as long as they are included. The estimator needs to select the right value for the adjustment factor. In many elements, this selection is based on the distance that the tool is cutting. This estimate finds the cycle, setup, and the lot times.

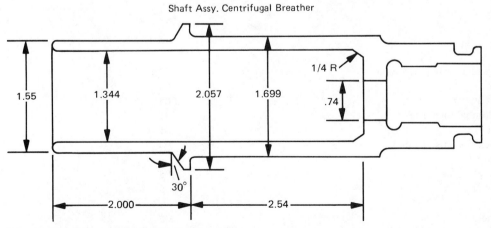

FIGURE 6.4D Airframe part, forging barstock.

Table Number	Process Description	Table Time	Adjustment Factor	Cycle Minutes	Setup Hours
11.1-2	Face left end, rough carbide tooling, 1.25 length, OD bar stock = 2.5 forging	0.09	1.25	0.11	
11.1-2	Face left end finish	0.20	1.25	0.25	
11.1-6	Start drill	0.13		0.13	
11.2-4	Power drilling, ⅝ length = 4.54 in. + drill pt of ⁵⁄₁₆ in.	0.41	4.85	1.99	
11.1-3	Bore first pass, ruf 4.54 in. = Length ID of bore = ⅝ in.	0.04	4.54	0.18	
11.1-3	Bore 2nd pass, ID = about 1 in., rough	0.11	4.54	0.50	
11.1-3	Bore finish to 1.344	0.16	4.54	0.73	
11.1-2	Turn radius, ¼ R inside bore, length = .4 in. Carbide	0.08	.4	0.03	
11.1-2	Turn outside rough 2 ½-in. OD stock About 2.0-in. length for 2.057-in. dim	0.09	2	0.18	
11.1-2	Turn 2.057 finish	0.18	2	0.36	
11.1-2	Turn rough, start at 2.057-in. dim to 1.55-in. dim with finish allowance	0.08	2	0.16	
11.1-2	Turn 1.55-dim fin. and form 30-degree rough	0.14	2	0.28	
11.1-4	Form 30 angle finish	0.27	.125	0.03	
11.1-4	Chamfer right end special, length = ⅛	0.21	.125	0.03	
11.1-7	Break edges, one inside two outside edges	0.12		0.12	
11.4-1	Turn tool life 5.08 min of turning	0.05		0.05	
6.4-1	Handling into collet on premachined right hand of part	0.19		0.19	
6.4-2	Start and stop	0.08	2	0.16	
6.4-3	Blow chips, rake	0.25		0.25	
6.4-4	Index, ad, retract for eight tools,	0.25		0.25	
6.4-4	But use 14 tools	0.03	6	0.18	
6.4-7	Inspection Mike every other pc	0.33		0.33	
6.4-S1	Basic time	0.39			0.39
6.4-S2	Tools	0.02			0.02
6.4-S2	Tools	0.03	5		0.15
6.4-S4	Tolerance	0.02			0.02
6.4-S5	Make piece 2 × hr/unit	0.10	2		0.20
	Total			6.49	0.78
	Total lot hours	5.21			

TABLE 6.4 NUMERICALLY CONTROLLED TURNING LATHES

Setup in estimating hours

- S1. Basic time .... .39
- S2. Tools
    - Check tool magazine for required tools .... .02/tool
    - Additional tool .... .03
    - Additional micrometer .... .005
        - Mike per decimal tolerance .... .01
    - Dog, additional .... .01
    - Adjust tool offset, per finishing tool .... .04
    - Place and remove
        - Steady rest .... .05
        - Driving center .... .06
        - Fixture
            - to spindle .... .05
            - to face plate .... .22

S3. Delete program block .05

S4. Tolerance .02
    Check per decimal tolerance

S5. Miscellaneous elements
    Change insert .03
    Grind tool bit .07
    Check tool or gage .02
    Clean surface plate .03
    Hone tool .02

S6. Make pc (optional): 2 × hr/unit

Operation elements in estimating minutes

1. Handling

	Type of Lathe		
	Small	Medium	Large
Get, load, close collet, open collet, and unload part	.13	.19	.25
Catch cutoff part and unload	.03	.03	
Open and close collet, feed stock by hand	.12	.14	
Unclamp, unload, turn end-for-end, load, and clamp in collet	.07	.12	.21
Open and close collet			.10
Place and remove part between centers using jib			3.23
Place and remove part with overhead crane			4.39
Turn part 180° with jib			1.49
Turn part 180° with overhead crane			2.54
Secure and release dog with one bolt			.63
Secure and release dog with two bolts			1.07
Change dog			.29
Position, open, close, and adjust steady rest			1.70
Protect surface, if finished, when dogged, add'l			.14

2. NC lathe machine operation
    Cycle start .08
    Stop and start spindle, start feed .11
    Set speed manually .16
    Shift gears .25
    Adjust tools
        Set tool from shaft end with length gate
            Shaft end not faced .72
            Shaft end faced .29
        Set tool with gage block from finished shoulder .57
        Adjust tool, incremental .16

3. Cleaning and lubricating, blow chips, rake
    Blow chips, rake .25

	Type of Lathe		
	Small	Medium	Large
Position and clear coolant line (manual)	.04	.04	.04
Air-clean-collet (manual)	.05	.05	.16
Air-clean part (manual)	.06	.06	.13
Place and remove oil guard		.08	.10

Deburr (Use Table 11.1-7)

4. Index, change tool, advance, retract cross slide turret

Tools	1	2	3	4	5	6	7	8	Add'l
Min	.03	.07	.10	.13	.16	.19	.22	.25	.03

5. Machining
    Turn, trim, break edges   Use Table 11.1
    Rapid traverse   Use Table 8.2-1

6. Turning tool replacement   Use Table 11.4-1

7. Part inspection
    Mike   .33
    Vernier caliper   .78
    Groove gage   .77
    Plug gage   .44
    Thread gage   .66
    For other gaging   Use Table 22.1-3

6.5 Numerically Controlled Chucking Lathes

DESCRIPTION

The machines described here are chucking lathes with numerical control. Figure 6.5 shows a typical NC chucking lathe. They are designed for heavy duty production. Internal contouring and external machining are the usual features. Turrets may be slant-bed mounted and have short tool projections. Coolant is provided externally or through boring bars and drills. Basic chuck sizes are from 12 to 36 in., and maximum swings are associated with these sizes. Turrets may be hexagonal or octagonal. Other machine tool designs are also possible. For some machines, the chuck can be removed and bar work is possible. Specific estimating data for universal NC turning lathes are not given, except that the estimator may examine both Sections 6.4 and 6.5 for selection of information. Adaptation of the data from both sections is useful for universal NC turning lathes.

These chucking lathes may have one or more turrets with a capacity of holding from 4 to 12 or more tools. NC chuckers have the advantage of reduced setup data over standard machines because much of the data are supplied to the machine by the NC tape. Repeat orders or smaller lots benefit from a machine having these controls. Machine cutting time can be reduced with multiple cuts, as two or more surfaces may be machined with only one action of the end working

FIGURE 6.5 A numerically controlled chucking lathe. *(Waterbury Farrel, Division of Textron, Inc.)*

hexagon turret, or in combination of a cross-slide and a turret. Two or more axis machines can offer simultaneous coutouring. Chip disposal may be by an end chute. Older models may require stopping the machine and raking.

ESTIMATING DATA DISCUSSION

The basic setup is always provided. It includes tape, manuscript and print handling, and punch in/out. It allows for setting offsets to zero, changing, indexing, or resetting of jaws and their blowing, and adjust coolant lines.

The checking of the tool magazine includes checking of required tools in load position and indicator lights, and provides for new cutting edge, index, or replace insert.

Changing tools involves checking tool catalog, obtaining bar or insert or insert holder, remove holder from magazine, place in holding fixture, remove bar by releasing screws, and place and secure next bar.

Spacer work obtains spacers, T-slot bolts, wipe chuck, place and secure, and aside. The discussion provided in 6.4, Numerically Controlled Turning Lathes, is also appropriate for chucking lathes, although estimating data are different.

The handling elements include a place and remove part with jib where the part exceeds 50 lb. The operator obtains the jib with hook, hooks part from box or pallet, transports to chuck, blows chuck, and wipes part as required. Position part in chuck, secure jaws, adjust coolant, followed by aside jib with hook, close sliding guard, and start cycle. Open sliding guard, and blow chips as required. Obtain jib with hook, hook part, align driver to chuck, release jaws, transport to box or pallet and release.

Place the element and remove part by the hand includes obtain part from box or pallet, transport to chuck, blow chuck, wipe part as required. Position part in chuck, secure jaws, adjust coolant, close sliding guard, start cycle. After a machining element, open sliding guard, and blow chips as required. Align driver to chuck, release jaws, and transport to box or pallet.

The element handling turn part 180° with jib involves open sliding guard and blow chips as required. Obtain jib with hook, hook part, align driver to chuck, release jaws. Transport part to bench, rehook part from opposite end, transport part to chuck. Blow chuck, wipe part as required. Position part in chuck, secure jaws, adjust coolant. Aside jib with hook and finally close sliding guard and start cycle.

If the operator turns the part 180° by hand for a part that is less than 50 lb, the operator will open sliding guard, blow chips as required. Align driver to chuck, release jaws, turn part 180°. Blow chuck, wipe part as required, position part in chuck, secure jaws, adjust coolant, close sliding guard, and finally start cycle.

Seat part element requires obtain shim or feeler gage from pocket or bench, check from part to locator pins, obtain lead hammer, seat part, check with shim or feeler, and finally aside gage and hammer.

Push bar may be required. If it is, the operator will switch to manual control, clear push bar to remove part as required. Place plate on push bar, advance to hold part, resecure jaws, clear, aside plate, and finally return turret to zero.

In checking with a plug gage, the operator will obtain plug gage from bench, assemble to bore, check, remove and aside to bench. A thread gage requires obtain thread gage from bench, turn on, check, turn off and aside to bench.

EXAMPLES

A. An oil-field valve bonnet is machined from a large steel forging. The round forging, 21¼-in. OD × 20-in. long has a forged bore of 8 in. and a total weight of 1700 lb. A series of machined bores and turning are scheduled and is shown below. Both ends are faced and have step bores. Simultaneous machining is used. Some diameters exceed Table 11.1 values, so use the formula min/in. $= \frac{\pi Dia}{12fV}$ (see Estimating Data Discussion for Sec. 11.1). Find unit estimate, hr/10 units, and lot time for 9 units. A large NC chucker will be used.

Table Number	Process Description	Table Time	Adjustment Factor	Cycle Minutes	Setup Hours
6.5-S1	Basic time	.29			.29
6.5-S2	Check tools, 7	.14			.14
6.5-S3	Change two holders	.28			.28
6.5-S6	Adjust tool offsets with gage block for one dimension	.10			.10
6.5-1	Place and remove part with hoist	5.16		5.16	
6.5-1	Seat part	.54		.54	
6.5-2	Cycle start	.08		.08	

Table Number	Process Description	Table Time	Adjustment Factor	Cycle Minutes	Setup Hours
11.1-2	Rough face left end, 21¼-in. OD × 6.63-in. long cut, carbide, 1 pass, $V = 360$ *fpm*, $f = .020$ *ipr*, and min/in. $= \dfrac{\pi \times 21.25}{12 \times .020 \times 360} = 0.77$ Length of cut = 6.63 + .125 = 6.76 in., where .125 in. is safety stock.	.77	6.76	5.20	
11.1-2	Finish face 21¼-in. OD × 6.63-in. long, carbide, $V = 475$ *fpm*, $f = .007$ *ips*, and min/in. $= \dfrac{\pi \times 21.25}{12 \times .008 \times 475} = 1.67$ Length of cut = 6.76 in.	1.67	6.76	11.31	
11.1-3	Rough bore 8-in. ID × 19.5-in. long to 8.4-in. OD	.63	19.5	12.29	
11.1-2	Rough turn 21¼-in. OD to 20¾-in. OD partial length of 10-in. carbide. min/in. $= \dfrac{\pi \times 21.25}{12 \times .020 \times 360} = .77$	.77	10	7.70	
11.1-2	Finish turn 20¾-in. OD	1.67	10	16.70	
6.5-1	Reverse bonnet with hoist into chuck	2.10		2.10	
6.5-1	Seat part	.54		.54	
6.5-2	Cycle start	.08		.08	
11.1-2	Rough face right end, same as above	.77	6.76	5.20	
11.1-2	Finish face right end, same as above	1.67	6.76	11.31	
11.1-2	Rough turn 21¼-in. OD to 19⅛-in. OD dimension. Two passes of approximate .5 in. depth of cut Stock removal each. Length = 10.1 in.				
	First pass at 0.77 in./min	.77	10.1	7.78	
	Second pass, 0.77 in./min	.77	10.1	7.78	
11.1-2	Finish turn 18.93 in. OD, 9⅛-in. long min/in. $= \dfrac{\pi \times 18.93}{12 \times .020 \times 360} = .69$	.69	9.125	6.28	
11.1-3	Finish bore 8.5-in. ID × 19.0 in. length. Carbide. Previously bored from other side to 8.4-in. ID. Table value at 9-in. *Dia*, 19.0 × .98	.98	19.0	18.62	
11.1-3	Finish bore 8.5-in. ID × 19.0 in. length. Carbide. Previously bored from other side to 8.4 in. ID. Table value at 9-in. *Dia*, 19.0 × .98	.98	19.0	18.62	
11.1-3	Counterbore original 8.5-in. ID to 8.618-in. ID. Length = 9.125 in. Finish pass. Carbide.	.98	9.125	8.94	
11.1-2	Finish turn 18.48-in. OD simultaneously to counterbore 8.618-in. ID. Length of 18.48-in. OD is 9.125 in. min/in. = $\dfrac{\pi \times 18.48}{12 \times .020 \times 360} = .67$	.67	9.125	6.11	
	This finish turn time is less than boring time, so it is not added to total				
6.5-3	Deburr, allow 2 times	.26	2	.52	
6.5-3	Blow chips, twice	.25	2	.50	
6.5-4	Index, 13 times	.43		.43	
8.2-1	Rapid traverse, approximately 125 in. at 200 in./min	.63		.63	
11.4-1	Carbide tool replacement, use stainless steel as an approximate material, machining time = 125 min of carbide, .36 min from table for 40 min., so for 125 min	1.13		1.13	
	Total			136.93	.81
	Hr/10 units	22.82			
	Total lot estimate	21.35 hr			

B. A cast brass nozzle will have a few final machining elements performed on a chucker. Assume material as a coppor alloy for turning purposes. The machining schedule is given below. The normal lot release is 410 per month. Determine the unit estimate, hr/100 units, lot time, and the calendar total time for 12 equal lot orders.

Table Number	Process Description	Table Time	Adjustment Factor	Cycle Minutes	Setup Hours
6.5-S1	Basic time	.29			.29
6.5-S1	Check tools, 5	.10			.10
6.5-S3	Change tools, 5	.25			.25
6.5-1	Load and unload by hand	1.32		1.32	
6.5-2	Cycle start	.08		.08	
11.1-2	Face 2½-in. OD to nominal dimension, 1⁵⁄₁₆-in. length of cut, HSS Finish pass,	.14	1⁵⁄₁₆	.18	
11.2-3	Drill ⅞-in. *Dia*, 7½-in. long	.09	7.5	.68	
11.2-6	Tap 1-in. coarse thread, 8 threads 3-in. long	.316	3	.48	
11.1-4	Contour end shape, depth = .5 in., *L* = 3 in., *Dia* = 4 in.	.40	3	1.20	
11.1-4	Contour ¼-in. O-ring groove of 4¼-in. OD. Length of cut = ¼ in. HSS	.25	1.06	.27	
11.1-7	Break edges, 1 inside and 4 outside edges	.16		.16	
6.5-4	Index 5 times	.17		.17	
	Total			4.54	.64
	Hr/100 units	7.567			
	Lot time	31.66 hr			
	Annual hours	397.96 hr			

TABLE 6.5 NUMERICALLY CONTROLLED CHUCKING LATHES

Setup in estimating hours

- S1. Basic time .29
- S2. Check tool magazine for required tool .02 tool
- S3. Change tool
  - Bar by hand .05
  - Bar by hoist .07
  - Holder .14
  - Bushings/sleeve on boring bar .04
- S4. Place and remove locator pins .17
- S5. Place, secure, remove chuck spacers .09
- S6. Adjust tool offset after gage block and checking length .10
  - Set end stops .12
- S7. Bore soft jaws .17
- S8. Check per decimal tolerance .02
- S9. Make pc (optional): 2 × hr/unit

Operation elements in estimating minutes

1. Handling

	Type of Lathe		
	Small	Medium	Large
Get, load in air chuck, clamp, unclamp, and unload		.21	.36
Get, load in chuck with T-wrench, clamp, unclamp, and unload		.37	.50
Get, load in fixture, clamp, unclamp, and unload		.99	1.15
Unclamp, unload, turn end-for-end, load, and clamp in chuck		.12	.25
Unclamp, unload, turn end-for-end, load, and clamp in with T-wrench in chuck		.75	.85
Relieve stress (before a finish bore, etc.)			1.00
Place and remove part, manual, 50 lb. or greater			1.32
Place and remove part, jib, 50 lb. or greater			2.44
Turn or rotate part, manual hoist, 500 lb or greater			5.16
Turn or rotate part, jib			.74

Table Number	Process Description	Table Time	Adjustment Factor	Cycle Minutes	Setup Hours
	Seat part, heavy			2.10	
	Push bar			.54	
	Index part			1.06	
	By hand			.73	
	By hoist			1.26	
	Index fixture by jib			3.50	
2.	NC lathe machine operation				
	Cycle start			.08	
	Program stop, check length, adjust tool offset			1.98	
	Start and stop chuck, start feed			.11	
	Set feed manually			.16	
	Adjust tools			.16	
	Set tools with gagebock			.57	
	Adjust tool, incremental			.16	

3. Cleaning and lubricating, blow chips

	Type of Lathe		
	Small	Medium	Large
Position and clear coolant line (manual)	.04	.04	.04
Air-clean part (manual)	.06	.06	.13
Air-clean fixture or chuck		.07	.18
Place and remove oil guard (manual)		.08	.10
Blow chips, rake			.25
Stamp			.09
Deburr			.26
Deburr (see also Table 11.1-7)			

4. Index, change tool, advance, retract

Tools	1	2	3	4	5	6	7	8
Min	.03	.07	.10	.13	.17	.20	.23	.27
Tools	9	10	11	12	13	14	15	16
Min	.30	.33	.37	.40	.43	.47	.50	.53
Tools	17	18	19	20	Add'l			
Min	.57	.60	.63	.67	.04			

5.	Machining					
	Turn, bore form, thread, start drill, break edges			Use Table 11.1		
	Drill, counterbore, tap			Use Table 11.2		
	Turning tool replacement			Use Table 11.4-1		
	Rapid traverse			Use Table 8.2-1		
6.	Part inspection					
	Mike per dimension				.33	
	Vernier caliper				.78	
	Plug gage				.44	
	Thread gage				.66	
	For other gaging			Use Table 22.1-3		

6.6 Single-Spindle Automatic Screw Machines

DESCRIPTION

The automatic screw machine (ASM) does work similar to the turret lathe. Automatic screw machines can be single or multiple spindle. A multiple spindle may contain 4, 5, or 6 spindles. Automatic screw machines provide controlling movements for the turret and cross-slides to have tools feed into the work at desired feeds, withdrawn, and indexed to the next position. The machine produces parts with little attention from the operator. Many automatic screw machines not only feed in an entire bar of stock, but also are provided with a magazine to have bars fed through the machine automatically.

A 1-1/4-in. collet size automatic screw machine is shown in Figure 6.6A. A choice of 2-speed or 4-speed drive units is an option. Now, consider the general class of single-spindle automatic screw machines.

ESTIMATING DATA DISCUSSION

The setup has a constant time for single-spindle machines. Additionally, time is provided for each tool for standard tolerances. If exceptional tolerances are required, the affected tool is increased by 0.15 hr.

The first four elements are for nonmachining work. These elements may occur during machining time, and when that happens no time is allowed. For instance, turret positioning, indexing, and retracting may be done during a cross-slide element and the maximum time of the two simultaneous choices is added to the estimate.

Estimates for ASM machining time, by custom, are made in revolutions. However, in this book we use a different approach because more estimators will find it clearer. The first step finds the cutting speed (peripheral speed of the work passing the tool) for the material to be cut. Element 5 provides a sample of materials that are popular for screw machine work. Typically, we have selected low-carbon steel, resulfurized (e.g., 1212); medium carbon-steel, resulfurized (e.g., 1145); low-carbon lead steel (12L14); medium- and high-carbon leaded steel (41L30); free-machining stainless steels, annealed (430F); and an average of aluminum and brass bar stock. Entry variables for Element 5 are tool type, cut conditions, and material. The sfm and feed are noted for a selected tool and material.

Screw machines are sometimes estimated using cams which are pre-engineered for a certain part. This method is accurate, but often times a cam design may be unavailable and our method is prior to any cam design.

The next step is to plan a sequence of machining elements and assign them to various turret positions or cross-slides. The spindle speed in rpm can be calculated using the formula:

$$rpm = \frac{12}{\pi}\left(\frac{sfm}{Dia}\right) = 3.82\left(\frac{sfm}{Dia}\right)$$

where Dia is the largest or initial diameter of the feature being machined or tool diameter.

However, the calculated spindle speed may not be practical for all machining elements. Each element should be questioned before using an estimated spindle speed. If the turning cutting diameter is based on

FIGURE 6.6 A 1½-in. automatic screw machine. *(Brown & Sharpe)*

Cutting speeds (sfm) and feeds (ipr) for HSS tools for ASM

Tool	Width (W), Dia, or Depth (D)	ipr Feeds	sfm Free-machining Resulfurized			sfm Annealed stainless steels	sfm Low-carbon Leaded Steels	Aluminum, Brass	
			Low	Medium	High			ipr Feed	sfm
Counterboring		.004	150	120	105	120	160	.002	175
Turning Single point, box		.007	225	190	170	150	235	.006	350
Center drill	1/8 OD	.003	125	115	100	115	135	.003	200
Cutoff tool Angular,	.062 W	.002	140	115	110	120	150	.003	340
Circular, Straight	.125 W	.0025	140	115	110	120	150	.003	340
Threading die		.003	40	30	20	20	40		35
Chaser die			45	35	25	25	45		35
Drill Twist	.02 OD	.0005	125	115	100	115	135	.0005	200
	.04	.001	125	115	100	115	135	.0001	200
	1/16	.001	125	115	100	115	135	.001	200
	3/32	.002	125	115	100	115	135	.002	200
	1/8	.003	125	115	100	115	135	.003	200
	3/16	.004	125	115	100	115	135	.004	200
	1/4	.005	125	115	100	115	135	.007	200
	5/16	.007	125	115	100	115	135	.008	200
	3/8	.008	125	115	100	115	135	.010	200
	1/2	.010	125	115	100	115	135	.012	200
	5/8	.012	125	115	100	115	135	.015	200
Form tool	.125 W	.0025	140	115	110	120	150	.003	340
Circular	.250 W	.003	140	115	110	120	150	.004	340
Knurl; turret, side		.015	150	110	110	110	160	.020	165
		.002	150	110	110	110	160	.004	165
Pointing, facing		.003	225	190	170	150	235	.004	200
Reamer	≤ 1/8	.005	140	115	95	120	150	.007	240
	> 1/8	.008	140	115	95	120	150	.010	240
Tap		.002	60	50	35	40	65	.002	80
Recessing		.002	150	120	105	120	160	.002	175

stock diameter, the corresponding spindle speed may be too high for tapping or threading, unless special speed reducers are used, or the machine has a stepdown velocity option.

More than likely, the calculated rpm will not match available gear-ration rpm's, and the estimator can adjust to the nearest actual rpm and estimating can proceed. Conversely, if it is uncertain which one of several machines will produce the parts, knowledge of the specific rpm will be unknown, and if the improved accuracy resulting from the rpm adjustment is minor, the estimator may want to overlook rpm recalculation, as we do in our problems. Finally, some materials, notably free-machining brass or aluminum, may exceed the available rpm of the machine, and thus the maximum rpm is a logical candidate.

Temporary rpm's are calculated for each machining element, but a practical rpm is chosen that can be effective with all the machining elements (tapping and knurling exempted). The rpm's are found in the following table. If the cutting length is not included, rpm's for various lengths can be added together for a constant feed. For example, at .0005 feed, 1000 + 1500 = 2500 would be the rpm for 1 1/2 in.

Number of revolutions for length of cut and feed

Feed/in.	Length					
	1/32	1/16	1/8	1/4	1/2	1
.0005	63	125	250	500	1000	1500
.0010	31	63	125	250	500	750
.0015	21	42	83	167	333	500
.0020	16	31	63	125	250	375
.0025	13	25	50	100	200	300
.003	10	21	42	83	167	250
.004	8	16	31	63	125	188
.005	6	13	25	50	100	150
.006	5	10	21	42	83	125
.007	4	9	18	36	71	107
.008	4	8	16	31	63	94
.009	3	7	14	28	56	83
.010		6	13	25	50	75
.012		5	10	21	42	63
.015			8	17	33	50
.020			6	13	25	38

One operator may tend 2, 3, 4, or more single-spindle machines depending upon lot quantities. This multiple machine per operator is handled by the Adjustment Factor.

The times are divided by the number of machines per operator. Element 5 is composed of machining for automatic screw machines. This element is used for multiple automatic screw machines as well as single spindle machines. The entry variables are type of machining, diameter, and material. Six popular materials are given, and their specifications have been already stated. Speeds and feeds for these materials and types of machining have been given above. For example, low carbon free machining resulfurized steel has a ipr and sfm of .007 in./rev and 225 sfm. These rates have been used to give the data for Element 5A. There are A through M different types of machining. These tables are used in the following way.

(1) It is necessary to know the material, and then determine the nature of the cut. The largest diameter is the entry variable. For example, if the original diameter is 1.25 in., but material was removed leaving the diameter at 1.07 in., the next diameter would use the latest diameter in entry for the diameter. If there is no specific diameter, the entry then goes to the next higher value for that material. This is keeping with the practice of this data book.

(2) While the customary practice is to adopt rpm's as the usual way to estimate time for the machining elements, this book departs from that practice. Instead, the practice is to enter the min/in. value from the table. The adjustment factor is the length of tool travel for this factor. The length of cut is distance that the tool moves, including safety stock, approach, part length required, and overtravel. Not all of these four factors are necessarily required for tool travel. But once the distance is known, it is entered as the adjustment factor on the estimating form. The multiplication of the two, table time and adjustment factor, is posted to the cycle minutes column.

EXAMPLES

A. CDA 360 brass is used for a bushing 1.245 in. in length for a 1.125-in. OD. The total part length is 1.390 in., including .015-in. facing and .125-in. cutoff. A single spindle machine with a 6-hole turret is used. The machining schedule is given below. Two form elements using the cross-slide are not added since they overlap with turret elements. Run part at 500 sfm or 1534 rpm.

Table Number	Process Description	Table Time	Adjustment Factor	Cycle Minutes	Setup Hours
6.6-S	Setup	3.0			3.0
6.6-S	Tools	.25	8		2.0
6.6-1	Feed stock twice	.006	2	.012	
6.6-2	Index turret	.006		.006	
6.6-5C	Center drill 1/8-in. OD, drill point =.063 + approach (=.030) = .093 in. The drill feeds at 31 revolutions (rev)	.055	.093	.005	
6.6-2	Index turret	.006		.006	
6.6-5G	Step drill through cutoff width, and length of cut = 1.390 in., 625-in. OD, 1.390 ÷ .015 = 93 rev	.055	1.39	.076	
6.6-2	Index turret	.006		.006	
6.6-2	Position recess. The element chamfers the inside edge and is pushed to cut by a cross-slide. The position for recess is estimated to be two indexes	.006	2	.012	
6.6-5N	Recess .015 in. deep at .002, 8 rev,. 625 in.-ID	.56	.015	.008	
6.6-2	Index turret	.006		.006	

Single-Spindle Automatic Screw Machines

Table Number	Process Description	Table Time	Adjustment Factor	Cycle Minutes	Setup Hours
6.6-5L	Ream, flat bottom, and face. A ⅝-in. counterbore will be reamed, flat-bottomed, and the end faced using a combination tool. Total tool travel = .340 in. and .340 ÷ .002 = 170 rev	.068	.340	.023	
6.6-2	A dwell is used for flat bottom, 5 rev	.003		.003	
6.6-5L	Ream .436/.437-in. hole, and travel = .936-in. depth, approach = .014 in., and into cutoff = .010 in. for total of .960-in. Rev = .960 ÷ .010 = 96 rev	.055	.96	.053	
6.6-5E	Cutoff uses a ¾-in. formed *Dia.* Travel from OD to ID = .157 in. for .157 ÷ .003 = 52 rev	.19	.157	.030	
6.6-5I	Front slide forms the 1-in. *Dia* The tool will travel .068-in. + .012-in, approach = .080-in., and .080 ÷ .003 = 26 rev, but is done during step drill operation, and is simultaneous (simo) and not allowed.	0		0	
6.6-5I	Form using rear slide travels a total of .210-in. and at .003 ipr rev = 70, but this is done simo to drilling and reaming operations	0		0	
6.6-2	Clear	.018		.018	
	Total			.264	5.0
	Min = 455 ÷ 1534 = .297 from revs for machining				
	Unit estimate	.264			
	Hr/1000 units	4.400			
	Pc/hr	227			

Note. The calculations for "rev" are not necessary to determine the estimates, but are shown for completeness.

B. A threaded part is made from ⅝-in. hex medium carbon steel. A turning velocity of 917 is determined from 150 sfm. Find pc/hr as if one operator was tending one machine.

Table Number	Process Description	Table Time	Adjustment Factor	Cycle Minutes	Setup Hours
6.6-1	Advance material to bar stop	.006		.006	
6.6-2	Retract, index, and advance turret	.024		.024	
6.6-5K	Point and chamfer, .16-in. of cut and approach, .16 ÷ .003 = 53 rev	.23	.16	.037	
6.6-2	Retract, index, and advance turret	.024		.024	
6.6-5A	Turn thread *Dia*, 1¼-in. length of cut and approach, 1.25 ÷ .004 = 312 rev	.15	1.25	.188	
6.6-2	Retract, index, and advance turret			.024	
6.6-5M	Thread ⁵⁄₁₆-18 on and off, ¾-in. long, .75 × 18 = 13.5 + 3 for lead, rpm = 245	.66	.75	.49	
6.6-2	Clear turret and index	.036		.036	
6.6-3	Advance side tool (.036) simo during turning	0		0	
6.6-5I	Form chamfer, ⅛-in. of cut ⅛ ÷ .003 = 42 rev which is done simo during turning	0		0	
6.6-3	Advance side tool (.012) done simo	0		0	
6.6-5E	Cut off part, ⁵⁄₁₆-in. length of cut, .002 feed = 94 rev	.68	.188	.128	
	Total			.957	
	Unit estimate	.957 min			
	Hr/1000 units	15.9500			
	Pc/hr	62			

TABLE 6.6 SINGLE-SPINDLE AUTOMATIC SCREW MACHINES

Setup
- Single-spindle, basic — 3.00 hr
- Install and adjust tool, ea — .25 hr
- Add'l or fine tolerance, per tool affected — .15 hr

Operation elements in estimating minutes

1. Material advance to stock stop .006
   Load bar in collet .30

2. Turret operation
   Index .006
   Back turret tool from work, index, advance new tool .024
   Back turret tool from work, index to start position, advance cross-slide .036
   Clear turret from work .018
   Dwell .003

3. Cross slide operation
   Advance cross-slide tool to work .012
   Start or stop machine .02

4. Simultaneous elements
   Remove parts from pan 0
   Count 0
   Inspection 0
   Clean chips

5. Automatic screw machining, min/in.

5A. Turning, single point box

						Diameter							
	1/64	1/32	1/8	1/4	1/2	3/4	7/8	1	1.125	1.25	1.5	2	3
Free Machining Resulfurized Steel													
Low carbon	.003	.005	.021	.042	.08	.13	.15	.17	.19	.21	.25	.33	.50
Medium carbon	.003	.006	.025	.049	.10	.15	.17	.20	.22	.25	.30	.39	.59
High carbon	.003	.007	.027	.055	.11	.17	.19	.22	.25	.28	.33	.44	.66
Annealed Stainless Steel	.004	.008	.031	.062	.12	.19	.22	.25	.28	.31	.37	.50	.75
Low Carbon Leaded Steel	.002	.005	.02	.04	.08	.12	.14	.16	.18	.20	.24	.32	.48
Aluminum, Brass	.002	.004	.03	.03	.06	.09	.11	.13	.14	.16	.19	.25	.38

5B. Counterboring

						Diameter							
	1/64	1/32	1/8	1/4	1/2	3/4	7/8	1	1.125	1.25	1.5	2	3
Free Machining Resulfurized Steel													
Low carbon	.01	.01	.05	.11	.22	.33	.38	.44	.49	.55	.65	.87	1.31
Medium carbon steel	.01	.02	.07	.14	.27	.41	.48	.55	.61	.68	.82	1.09	1.64
High carbon steel	.01	.02	.08	.16	.31	.47	.55	.62	.70	.78	.93	1.25	1.87
Annealed Stainless Steel	.01	.02	.07	.14	.27	.41	.48	.55	.61	.68	.82	1.09	1.64
Low Carbon Leaded Steel	.01	.01	.05	.10	.20	.31	.36	.41	.46	.51	.61	.82	1.23
Aluminum, Brass	.01	.01	.05	.09	.19	.30	.34	.38	.43	.46	.55	.75	1.15

5C. Center drill

	Diameter												
	1/64	1/32	1/8	1/4	1/2	3/4	7/8	1	1.125	1.25	1.5	2	3
Free Machining Resulfurized Steel													
Low carbon	.01	.02	.09	.17	.35	.52	.61	.70	.79	.87	1.05	1.40	2.09
Medium carbon	.01	.02	.09	.19	.38	.57	.66	.76	.85	.95	1.14	1.52	2.28
High carbon	.01	.03	.05	.11	.22	.65	.76	.87	.98	1.09	1.31	1.75	2.62
Annealed Stainless Steel	.01	.02	.09	.19	.38	.57	.66	.76	.85	.95	1.14	1.52	2.28
Low Carbon Leaded Steel	.01	.02	.08	.16	.32	.48	.57	.65	.73	.81	.97	1.29	1.94
Aluminum, Brass	.01	.01	.05	.11	.22	.33	.38	.44	.49	.55	.65	.87	1.31

5D. Cutoff tool, angular

	Diameter												
	1/64	1/32	1/8	1/4	1/2	3/4	7/8	1	1.125	1.25	1.5	2	3
Free Machining Resulfurized Steel													
Low carbon	.01	.03	.12	.23	.47	.70	.82	.94	1.05	1.17	1.40	1.87	2.81
Medium carbon	.02	.04	.14	.28	.57	.85	1.00	1.14	1.28	1.42	1.71	2.28	3.41
High carbon	.02	.04	.15	.30	.60	.89	1.04	1.34	1.49	1.49	1.79	2.38	3.57
Annealed Stainless Steel	.02	.03	.14	.27	.55	.82	.95	1.09	1.23	1.36	1.64	2.16	3.27
Low Carbon Leaded Steel	.01	.03	.11	.22	.44	.65	.76	.87	.98	1.09	1.31	1.75	2.62
Aluminum, Brass	.004	.01	.03	.06	.13	.19	.22	.26	.29	.32	.95	.51	.77

5E. Cutoff tool, circular or straight

	Diameter												
	1/64	1/32	1/8	1/4	1/2	3/4	7/8	1	1.125	1.25	1.5	2	3
Free Machining Resulfurized Steel													
Low carbon	.01	.02	.09	.19	.37	.56	.65	.75	.84	.94	1.12	1.50	2.24
Medium carbon	.01	.03	.11	.23	.46	.68	.80	.91	1.02	1.14	1.37	1.82	2.73
High carbon	.01	.03	.12	.24	.48	.71	.83	.95	1.07	1.19	1.43	1.90	2.86
Annealed Stainless Steel	.01	.03	.11	.22	.44	.65	.76	.87	.98	1.09	1.31	1.75	2.62
Low Carbon Leaded Steel	.01	.02	.09	.17	.35	.52	.61	.70	.79	.87	1.05	1.40	2.09
Aluminum, Brass	.004	.01	.03	.06	.13	.19	.22	.26	.29	.32	.39	.51	.77

5F. Threading die

	Diameter												
	1/64	1/32	1/8	1/4	1/2	3/4	7/8	1	1.125	1.25	1.5	2	3
Free Machining Resulfurized Steel													
Low carbon	.03	.07	.27	.55	1.09	1.64	1.91	2.18	2.45	2.73	3.27	4.36	6.55
Medium carbon	.05	.09	.36	.73	1.45	2.18	2.55	2.91	3.27	3.64	4.36	5.82	8.73
High carbon	.07	.14	.55	1.09	2.18	3.27	3.82	4.36	4.91	5.45	6.54	8.73	13.09
Annealed Stainless Steel	.07	.14	.55	1.09	2.18	3.27	3.82	4.36	4.91	5.45	6.54	8.73	13.09
Low Carbon Leaded Steel	.03	.07	.27	.55	1.09	1.64	1.91	2.18	2.45	2.73	3.27	4.36	6.55
Aluminum, Brass	.04	.08	.31	.62	1.25	1.87	2.18	2.49	2.80	3.12	3.74	4.99	7.48

5G. Drill, twist

	Diameter					
	1/64	1/32	1/8	1/4	1/2	5/8
Free Machining Resulfurized Steel						
Low carbon	.065	.065	.087	.11	.11	.18
Medium carbon	.071	.071	.095	.11	.11	.12
High carbon	.082	.082	.11	.13	.13	.14
Annealed Stainless Steel	.071	.071	.10	.11	.11	.12
Low Carbon Leaded Steel	.061	.061	.081	.10	.10	.10
Aluminum, Brass	.041	.041	.055	.047	.055	.055

5H. Form Tool, circular

	Diameter												
	1/64	1/32	1/8	1/4	1/2	3/4	7/8	1	1.125	1.25	1.5	2	3
Free Machining Resulfurized Steel													
Low carbon	.01	.02	.08	.16	.31	.47	.55	.62	.70	.78	.93	1.25	1.87
Medium carbon	.01	.02	.09	.19	.38	.57	.66	.76	.85	.95	1.14	1.52	2.28
High carbon	.01	.02	.10	.20	.40	.59	.69	.79	.89	.99	1.19	1.59	2.38
Annealed Stainless Steel	.01	.02	.09	.18	.36	.55	.64	.73	.82	.91	1.09	1.45	2.18
Low Carbon Leaded Steel	.01	.02	.07	.15	.29	.44	.51	.58	.65	.73	.87	1.16	1.75
Aluminum, Brass	.003	.01	.02	.05	.10	.14	.17	.19	.22	.24	.29	.38	.58

Single-Spindle Automatic Screw Machines

5I. Form Tool

						Diameter							
	1/64	1/32	1/8	1/4	1/2	3/4	7/8	1	1.125	1.25	1.5	2	3
Free Machining Resulfurized Steel													
Low carbon	.01	.02	.09	.19	.37	.56	.65	.75	.84	.94	1.12	1.5	2.24
Medium carbon	.01	.03	.11	.23	.46	.68	.80	.91	1.02	1.14	1.37	1.82	2.73
High carbon	.01	.03	.12	.24	.48	.71	.83	.95	1.07	1.19	1.43	1.90	2.86
Annealed Stainless Steel	.01	.03	.11	.22	.44	.65	.76	.87	.98	1.09	1.31	1.75	2.62
Low Carbon Leaded Steel	.01	.02	.09	.17	.35	.52	.61	.70	.79	.87	1.05	1.40	2.09
Aluminum, Brass	.004	.01	.03	.06	.13	.19	.22	.26	.29	.32	.39	.51	.77

5J. Knurl; side

						Diameter							
	1/64	1/32	1/8	1/4	1/2	3/4	7/8	1	1.125	1.25	1.5	2	3
Free Machining Resulfurized Steel													
Low carbon	.01	.03	.11	.22	.44	.65	.76	.87	.98	1.09	1.31	1.75	2.62
Medium carbon	.02	.04	.15	.30	.60	.89	1.04	1.19	1.34	1.49	1.70	2.38	3.57
High carbon	.02	.04	.15	.30	.60	.89	1.04	1.19	1.34	1.49	1.79	2.38	3.57
Annealed Stainless Steel	.02	.04	.15	.30	.60	.89	1.04	1.19	1.34	1.49	1.79	2.38	3.57
Low Carbon Leaded Steel	.01	.03	.10	.20	.41	.61	.72	.82	.92	1.02	1.23	1.64	2.45
Aluminum, Brass	.01	.02	.08	.16	.32	.45	.58	.64	.80	.90	1.15	1.50	2.30

5K. Pointing, facing

						Diameter							
	1/64	1/32	1/8	1/4	1/2	3/4	7/8	1	1.125	1.25	1.5	2	3
Free Machining Resulfurized Steel													
Low carbon	.01	.01	.05	.10	.19	.29	.34	.39	.44	.49	.58	.78	1.16
Medium carbon	.01	.01	.06	.11	.23	.34	.40	.46	.52	.57	.69	.92	1.38
High carbon	.01	.02	.06	.13	.26	.38	.45	.51	.58	.64	.77	1.03	1.54
Annealed Stainless Steel	.01	.02	.07	.15	.29	.44	.51	.58	.65	.73	.87	1.16	1.75
Low Carbon Leaded Steel	.01	.01	.05	.09	.19	.28	.32	.37	.42	.46	.56	.74	1.11
Aluminum, Brass	.01	.01	.04	.08	.16	.25	.29	.33	.37	.41	.49	.65	.98

5L. Reaming

	Diameter					
	1/64	1/32	1/8	1/4	1/2	5/8
Free Machining Resulfurized Steel						
Low carbon	.006	.012	.047	.058	.12	.15
Medium carbon	.007	.014	.057	.071	.14	.18
High carbon	.009	.017	.069	.086	.14	.22
Annealed Stainless Steel	.007	.014	.055	.068	.14	.17
Low Carbon Leaded Steel	.005	.011	.044	.055	.11	.14
Aluminum, Brass	.002	.005	.019	.027	.055	.068

5M. Tapping

	Diameter				
	1/64	1/32	1/8	1/4	1/2
Free Machining Resulfurized Steel					
Low carbon	.03	.07	.27	.55	1.09
Medium carbon	.04	.08	.33	.66	1.31
High carbon	.06	.12	.47	.94	1.87
Annealed Stainless Steel	.05	.10	.41	.82	1.64
Low Carbon Leaded Steel	.03	.06	.25	.50	1.01
Aluminum, Brass	.03	.05	.20	.41	.82

5N. Recessing

	Diameter												
	1/64	1/32	1/8	1/4	1/2	3/4	7/8	1	1.125	1.25	1.5	2	3
Free Machining Resulfurized Steel													
Low carbon	.01	.03	.11	.22	.44	.65	.76	.87	.98	1.09	1.31	1.75	2.62
Medium carbon	.02	.03	.14	.27	.55	.82	.95	1.09	1.23	1.36	1.64	2.18	3.27
High carbon	.02	.04	.16	.31	.62	.94	1.09	1.25	1.40	1.56	1.87	2.49	3.74
Annealed Stianless Steel	.02	.03	.14	.27	.55	.82	.95	1.09	1.23	1.36	1.64	2.18	3.27
Low carbon Leaded Steel	.01	.03	.10	.20	.41	.61	.72	.82	.92	1.02	1.23	1.64	2.45
Aluminum, Brass	.01	.02	.09	.19	.37	.56	.65	.75	.84	.94	1.12	1.50	2.24

6. Tool wear and replacement   Refer to Table 11.4-1

6.7 Multispindle Automatic Screw Machines

DESCRIPTION

The automatic screw machine (ASM) provides rapid production of screw machine parts. The process is similar to that of the turret lathe. The automatic screw machines that are discussed here are multispindle machines. Multispindle ASM contains five or more spindles. (See Figure 6.7.) The machine automatically provides movement by the turret and crossslides to have various tools fed into the waiting stock at certain feeds, withdrawn, and indexed to the next position. The process is very similar to that of a gatling gun which rotates each time it fires a bullet. These machines are self-sufficient and require little direct attention from the operator, other than the loading of the new stock and periodic inspection of parts. These machines feed in the raw bar stock automatically and can take up to a 12 or 20-ft length of bar with a maximum diameter of about 1 in. to several inches. Machine capacities vary.

One operator may tend three or more machines. In the case of multiple machine operation the estimator "de-joints" the direct-labor time by dividing the cycle estimate, which is the time from floor-to-floor for one unit from one machine, by the number of machines tended by the operator. Now consider estimating data for the general class of multispindle automatic screw machines.

ESTIMATING DATA DISCUSSION

Setup effort is broken down into a basic time plus additional time for installation and adjustment of tools. If the estimator believes that tolerance considerations for the specific multiple ASM requires additional time, time is provided for each tool.

Before proceeding with discussion of the elements, it is useful to point out practical steps for estimating jobs that are done on these kinds of machines.

From the part print, note the grade, size and shape of the material, tolerances, micro-finish, and concentricity requirements. Process the part from the end of cutoff piece. Plan the sequence of elements starting with internal tool slide followed by the external tool slide; then cross slide elements position by position.

Determine the tool travel required for the tool slide, cross slides, auxiliary slides, recessing cross travel, and the number of threads to be cut. Sometimes tool travel is composed of safety stock, approach, part length, and over travel. From page 174 determine cutting rate. Check surface speeds on outside diameter, drills, and taps. Also select drilling and drill speed ratios. Information about Element 5 is given in Equipment 6.5.

The position requiring the most time is the time of the job. The first four elements in the element listing are for non-machining time. These elements sometimes occur during machining time, and if this happens no time is allowed.

Estimates for the machining time for the automatic screw machine are customarily made in revolutions. But this book provides rates.

Machining times are obtained from Element 6.6-5. This element is similar for multiple spindles. That table uses information such as type of material; cutting method such as drill, turn, etc.; and *Dia*. Once these inputs are known, the table value is posted on the "Table Time" column. The time value is a rate, or min/in. This factor must be multiplied by the length

FIGURE 6.7 A five-spindle automatic screw machine. *(Davenport Machine Tool Division, Dover Corporation)*

of cut which includes safety stock, approach, drawing requirement, and overtravel length. This book does not use rpm's to establish time, although it is customary in the automatic screw machine industry. We found it simpler to use rates rather than rpm's and that practice is standard throughout this book.

The machining elements are sequenced and assigned to various spindles. The spindle speed in rpm is calculated using

$$\text{rpm} = \frac{12}{\pi} \frac{(sfm)}{(Dia)} = 3.82 \frac{(sfm)}{(Dia)}$$

where Dia is the largest or initial diameter of the bar stock being machined or the tool diameter.

Usually, the calculated rpm will not match the gear-ratio rpm's. The corresponding spindle speed for cutting may be too high for drilling, tapping, or threading so the estimator may need to adjust the rpm to the nearest actual rpm and then he or she can proceed with the estimating.

Temporary rpm's are calculated for each machining element, but a practical rpm is chosen that can be effective with all of the machining elements, except tapping and knurling. The calculated rpm's are shown in the discussion for single spindle ASM. For multi-spindle work, the longest machine element establishes the unit time. Other spindle rpm's being less are not the limiting production requirement. For the automatic screw machine, the independent variable is the number of revolutions that each machining element needs to perform its function.

EXAMPLES

A. 12L14 steel is used to make a part .422 in. in length for a .445-in. OD. A five-spindle machine is used. Stock size of ½ in. is required. The rpm is 2500. Find the estimate. Assume one operator per machine.

Table Number	Process Description	Table Time	Adjustment Factor	Cycle Minutes	Setup Hours
6.7-S	Setup, 9 tools	10.05			10.05
6.7-1	Load bar stock .15 × 5 bars/320 units per bar	.00		0	
6.7-1	Tap bar into position .10 × 5/320	.00		0	
6.7-2	One index time	.05		.05	
6.6-5C	Position 1: Spot drill .125 Dia for .125-in. depth rev = ⅛ ÷ .0012 = 100 rev. Knurl, narrow ground 1 roll knurl. rev = 33.	.081	.125	(.010)	
6.6-5I	Position 2: Drill ³⁄₁₆-in. Dia halfway with main tool. Rough form .445-in. Dia with cross slide. Drill distance = .245 in. ÷ .0024 = 100 rev				
	Rough form .420 ÷ .0025 = 168 rev, use form	.35	.42	(.15)	
6.6-5I	Position 3: Drill past cutoff ³⁄₁₆-in. Dia. Finish form .445-in. Dia with cross slide. Drill distance = .25 in. ÷ .0024 = 100 rev				
	Finish form .445 OD. .420 ÷ .0024 = 175 rev	.35	.42	(.15)	
6.6-5L	Position 4: Ream drill ¹³⁄₆₄ in. thru. ¹³⁄₆₄ ÷ .0045 = 45 rev	.055	.42	(.023)	
6.6-5	Position 5: Pickoff and burr .005 = 100 rev				
6.6-5	Cutoff .445-in. OD. .445 ÷ .0025 = 178 rev	.35	.445	.155	
	Longest machining element position 5, 178 rev	.155			
	Machine time	.155 min			
	Unit estimate	.205 min			
	Pc/hr	292			

B. A part is made from ¼-in. brass on a five-position machine. The part is .45-in. long and .235 in. in diameter. A turning velocity of 1528 rpm is determined from 100 sfm. Find the unit estimate and pieces per hour. Assume one operator per three machines. Find the hr/1000 units where one operator tends three ASM's and the lot time for 10,000 parts.

Table Number	Process Description	Table Time	Adjustment Factor	Cycle Minutes	Setup Hours
6.7-S	Setup, 7 tools	9.15			9.15
6.7-2	Index	.05		.05	
6.7-3	Position stock	.00			
6.7-5I	Position 1: Spot drill .125 Dia ⅛ ÷ .0010 = 125 rev. Rough form .235-in. Dia with cross slide L.O.C. = .40 in. rev = .40 ÷ .0025 = 160 rev. Form is maximum of position.	.06	.25	(.016)	

6.6-5G	Position 2: Knurl, narrow ground 1 roll knurl rev = 33. Drill thru ⅛ *Dia.* Drill dist. = .45 in. ÷ .0025 = 200 rev. maximum	.055	.51	.028	
6.6-5L	Position 3: Ream drill 9/64 in. thru 9/64 ÷ .0045 = 32 rev	.027	9/64	(.004)	
6.6-5I	Position 4: Finish form .235-in. *Dia* with cross slide L.O.C. = .40 ÷ .0024 = 167 rev	.06	.05	(.003)	
6.6-5E	Position 5: Cutoff .235-in OD. .235 ÷ .0020 = 118 rev	.064	.12	(.008)	
	Longest machining element position 2, 200 rev	.028			
	Machine time	.028 min			
	Unit estimate	.078 min			
	Pc/hr	767			
	Net unit estimate	.026 min			
	Hr/1000 units	.433 hr			
	Lot time	13.48 hr			

C. A .375-in. part is made from 12L14 steel on a five-position machine. The part is .275 in. in diameter. The turning velocity is 2500 rpm. Find the unit estimate. One operator tends four machines. Find the net unit estimate. Find the lot hours to charge for direct-labor cost for 50,000 parts.

Table Number	*Process Description*	*Table Time*	*Adjustment Factor*	*Cycle Minutes*	*Setup Hours*
6.7-S	Setup, 7 tools	9.15			9.15
6.7-2	Index	.05		.05	
6.6-5I	Position 1: Rough form .275 *Dia* with cross slide L.O.C. = .25 in. rev = .25 ÷ .0025 = 100 rpm				
	Drill halfway drill distance = .1875 in. ÷ .0025 = 100 rpm	.35	.25	.09	
6.6-5I	Position 2: Finish form .275 *Dia* with cross slide L.O.C. = .25 in. rev = .25 ÷ .0024 = 104 rpm				
	Ream ⅛ in. halfway ⅛ ÷ .0045 = 28 rev	.35	.25	.09	
6.6-5J	Position 3: Knurl, narrow ground, 1 roll knurl rev. = 33	.20	1/16	(.013)	
6.65I	Position 4: Finish form .255-in. OD. L.O.C. = .05 in. .05 ÷ .0024 = 21 rev	.35	.05	(.018)	
6.6-5E	Position 5: Cutoff .275-in. OD. .275 ÷ .0025 = 110 rev, maximum	.17	.14	(.023)	
	Longest machining element position 5, 110 rev				
	Machine time	.09 min			
	Unit estimate	.14 min			
	Pc/hr	428			
	Net unit estimate	.035 min			
	Lot estimate for order of 50,000	38.32 hr			

D. A multispindle screw machine part is made from 12L14 steel. The part is .40 in. in diameter and .375 in. in length. The rpm is 2000. Find the unit estimate as a floor-to-floor estimate.

Table Number	*Process Description*	*Table Time*	*Adjustment Factor*	*Cycle Minutes*	*Setup Hours*
6.7-S	Setup, 8 tools	9.60			9.6
6.7-2	Index	.05		.05	
6.6-5I	Position 1: Spot drill .125 in. with .125 *Dia* rev = .125 ÷ .0012 = 100 rev				
	Rough form. 40-in. *Dia* with cross slide L.O.C. = .35 35 ÷ .0025 = 140 rev, maximum for position	.13	.35	(.045)	
6.6-5G	Position 2: Drill thru ⅛-in. *Dia* L.O.C. = .375 .0025 = rev				
	Finish form .40 *Dia* with cross slide .35 ÷ .0024 = 146 rev				
	Maximum position.	.081	.375	(0.30)	
6.6-5L	Position 3: Ream drill ⅛ thru. ⅛ ÷ .0045 = 28				
	Finish form .38-in. *Dia* with cross slide .05 ÷ .0025 = 20 rev	.044	.375	(.017)	
6.6-5	Position 4: Pickoff and burr .005 = 100 rev				
6.6-5E	Position 5: Cutoff .40 in. OD. .40 ÷ .0025 = 160 rpm	.35	.20	.070	
	Machine time	.070 min			
	Unit estimate, floor-to-floor time	.12 min			

E. Brass is used as a material for a screw machine part. The part diameter is .45-in. OD and is .375 in. in length. Turning velocity is 1500 rpm. Find the lot time to produce 1000, 5000, 10,000 units. Assume one operator per machine.

Table Number	Process Description	Table Time	Adjustment Factor	Cycle Minutes	Setup Hours
6.7-S	Setup, 7 tools	9.15			9.15
6.7-2	Index	.05		.05	
6.6-5I	Position 1: Spot drill .125 Dia ⅛ in. ⅛ ÷ .0010 = 125 rev. Rough form 45-in. OD with cross slide L.O.C. = .35 in. .35 ÷ .0025 = 140 rev	.13	.35	.045	
6.6-5I	Position 2: Drill halfway L.O.C. = .1875 in. .0025 = 75 rev. Finish form .45-in. Dia L.O.C. = .35 in. .35 ÷ .0024 = 146 rev	.13	.35	.045	
6.6-5I	Position 3: Finish form .43-in. OD L.O.C. = .10 in. .10 ÷ .0015 = 67 rev	.13	.10	(.013)	
6.6-5L	Position 4: Ream drill ³⁄₆₄ in. halfway ³⁄₆₄ ÷ .0045 = 32 rev	.027	.187	(.005)	
6.6-5	Position 5: Cutoff .45 in. OD .45 ÷ .0025 = 180 rev, maximum	.13	.225	.029	
	Machine time	.045 min			
	Unit estimate	.095 min			
	Pc/hr	631			
	Lot time for 1000 units	10.73			
	Lot time for 5000 units	17.07			
	Lot time for 10,000 units	24.98			

TABLE 6.7 MULTISPINDLE AUTOMATIC SCREW MACHINES

Setup

Multiple spindle, basic	6.00 hr
Install and adjust tool, ea	.45 hr
Add'l or fine tolerance, ea	.15 hr
Six–eight cams, entirely changed	10–12 hr
Changing gearing only	4–6 hr
Changing gearing partially for family of parts	2–3 hr

Operation elements in estimating minutes

1. Pick up bar, load in spindle .15
   Tap bar into position .10

2. Index .05

3. Operation
   Start or stop macine .00
   Position oil guards, turn on oil .00

4. Simultaneous elements
   Inspection .00
   Remove parts .00
   Clean chips .00

5. Automatic Screw Machining (refer to Table 6.6-5)

6. Tool wear and replacement (refer to Table 11.4-1)

MILLING MACHINES

7.1 Milling Machines Setup

ESTIMATING DATA DISCUSSION

These data are for setup of milling machine operations. The units are estimating hours. There are three elements: 1) basic time; 2) make part time; and 3) tolerance checking. Entry variables are the holding device and the type of milling cut and cutter for Element 1. Several entry choices are available. While not all setups are covered, the estimator can adapt the list as necessary to other setup requirements.

Element 2 deals with the making of a part. The inclusion of this element in the setup is optional with the estimator. Certainly, the making of one or several parts is necessary for the completion of a setup. The time to make one or several parts is also included in the unit estimate or hours per lot. There is the chance of "double estimating." The time required to complete the setup may be longer than the entire production run. Also, if the first part is scrapped and additional material is provided for this eventuality, the "make time" is a reasonable inclusion. If the element will be included, the estimator determines the operation estimate first. Then it is converted to hours per unit (hr/unit). This value is mutliplied by 2 for the usual case of requiring one part to complete a setup. If the hours per unit is relatively large, its importance to the setup is tangible. Multipliers other than the suggested 2, such as 1, 3, or 4, may be chosen by the estimator as conditions warrant.

Element 3 is concerned with exceptional tolerance requirements of a setup. This would exclude "cut and try," but include "measure" for critical dimensions. Commercial milling practices are included in basic time, Element 1.

Examples showing the use of these data are given with the type of milling machine.

TABLE 7.1 MILLING MACHINES SETUP

Setup in estimating hours

S1. Basic time

Type of Milling Cut and Cutter / Holding Device	Cutter Used in Prior Operation	Plane Surface (End, Shell, Face)	Shoulder Cut (End or Shell)	Slot Profile (End or Shell)	Saw Slab, Form (Plain)	Slot (Peripheral)	Straddle Mill (2 Cutters)	Three-Cutter Gang
Holding device in prior operation. (Dial table only.)	.15	.55	.60	.90	.55	1.00	1.35	1.60

Type of Milling Cut and Cutter / Holding Device	Cutter Used in Prior Operation	Plane Surface (End, Shell, Face)	Shoulder Cut (End or Shell)	Slot Profile (End or Shell)	Saw Slab, Form (Plain)	Slot (Peripheral)	Straddle Mill (2 Cutters)	Three-Cutter Gang
Holding device in prior operation is re-setup.	.30	.70	.75	1.00	.70	1.15	1.45	1.75
Small part. (Placed manually.)	.50	.90	1.00	1.25	.90	1.40	1.70	2.00
Large part. (Hoist required.)	.60	1.00	1.05	1.30	1.00	1.45	1.75	2.05
Vise.	.70	1.15	1.15	1.45	1.15	1.65	1.90	2.20
Table setup. (Part length ≤ 12 in.)	.90	1.30	1.35	1.40	1.20	1.60	1.90	2.15
Table setup. (Part length > 12 in.)	1.05	1.45	1.50	1.55	1.45	1.70	2.00	2.30
Collet or chuck. (No index.)	.50	.90	1.00	1.25	.90	1.40	1.70	2.00
Collet or chuck. (With index.)	.85	1.25	1.30	1.55	1.25	1.70	2.00	2.30
Index device in prior operation. (Part relocated.)	.35	.75	.85	1.10	.75	1.25	1.55	1.85
Index. (Device with tail stock.)	1.05	1.50	1.55	1.80	1.50	1.95	2.25	2.55
Angle plate.	.80	1.20	1.25	1.55	1.20	1.65	2.00	2.25

S2. Make piece (optional): $2 \times$ hr/unit

S3. Tolerance checking:

Tolerance	Hr
1/64 –.011	.03
.01 –.0051	.05
.005–.0031	.08
.003–.0011	.10
.001–	.13

7.2 Knee and Column Milling Machines

DESCRIPTION

Knee and column machines are utility milling machines. The name is derived from the column-shaped main frame and the knee-shaped projection from the column, which supports the saddle and work-holding table. The machines have general adjustments of their moving members. Three styles are included: plain (horizontal arbor), universal, and vertical.

The plain milling machine is usually used in jobbing shops and tool rooms. It may be adapted to quantity production with attachments and fixtures. The main elements of this machine are the column, spindle, overarm, knee, saddle, and table. Figure 7.2 is an example of the plain machine. It has a 60 × 13.7-in. table with 16 speeds that range from 25–1500 rpm and 16 feeds ranging from ½ to 60 ipm. Controls are located in front and back.

The vertical knee and column machine has the spindle vertical, or placed parallel to the column face, and at right angles to the top surface of the table. The vertical knee and column machine is adapted to operations with end and face mills, profiling interior and exterior surfaces, milling dies and metal molds, and for locating and boring holes in jigs and fixtures.

The universal milling machine resembles the plain type. The chief difference is that the table is supported and carried by the housing, which is swiveled in a horitzontal plane. The universal milling machine can mill helices in addition to those operations of the plain machine. Now consider the following estimating data for the general class of knee and column milling machines.

ESTIMATING DATA DISCUSSION

Data provided in Table 7.2 include operating elements only. Element 1 deals with the basic part handling. Entry variables are weight and holding devices. As part weight increases, its distance and holding container is expected to vary, consistent with quantity, of course. There are four major categories of holding devices. Element 1 allows for the operator's movement time to and from the machine and part and grasping and releasing time. However, the tabular values in Element 1 do not include clamping.

The listed elements in 1 include the necessary movements to accomplish the action. For instance, "seat with mallet" includes reaching and grasping the mallet, and letting go of the mallet in its nearby resting spot after seating. Similarly, "wipe off parallels" involves the use of a rag; its inclusion in the element is implied. The "load and unload part in collet, angle plate, cup arbor, chuck and dividing head" does include the necessary clamping action. These last loading and unloading times are a compilation of the more complete work, thus allowing the estimator a choice of how the element is applied.

If the handling work is done during a machine cut, a portion of the element can be deducted. For example, from ⅓ to ½ can be deducted if loading is partially compensated by the machine cut. If the weight is less than 1 lb and a knee and column machine is used, the values should be considered a necessary threshold value. While the estimator can interpolate tabular values for greater precision, the better practice is to apply the higher value in estimating.

The clamp and unclamp elements are provided in Element 2. With knowledge of what is expected, the estimator selects the appropriate matching work. The time values are for both clamp and unclamp. If loos-

FIGURE 7.2 Plain general purpose milling machine. (*Cincinnati Milacron*)

ening and tightening will occur during the machine-controlled rapid retract time, the values are reduced 25%. The necessary wrenches are assumed to be on the machine table or not more than an arm's reach from the operator.

In Element 3, the convention for rotate or turn part in vise, does not include the opening or closing. It does, however, include fixture and part manipulation. For the listed elements in 3, the work is inclusive.

Element 4 lists those elements suitable to knee and column milling operation. Some are single purpose, start or stop spindle, while others are consolidated, and use the word "and" to include the inclusive property. The "start" element, for example, is used in several elements in various ways. Depending upon the need, a particular one is selected. Whenever the spindle, saddle, knee, or table is moved, an average distance is figured. Thus, the rapid traverse or rapid retract rate of the machine is assumed. An alternate approach is to use individual elements and determine the distance for rapid traverse and retract, then use the specific machine traverse rate for that distance. This approach, while accurate, is more time consuming. (The following examples show how it is done.) In some plain milling operations only one elemental value may be needed, while complicated vertical milling operations may require a combination of elements to give machine operation. Some operating elements may be done during machine operation and would not be entirely required. The automatic or manual machine operations depend upon the specific machine being estimated, but all elements are provided.

Element 5 deals with cleaning, lubricating, and wiping. Type of tool, air hose, brush, oil hose, rag, or hand and area are the entry variables.

EXAMPLES

A. A rocker shaft is constructed of aluminum. Four parts are cast together and an operation is necessary to separate each. While sawing the parts, the two outside ends can be milled. Thus, three saws and two side mills can be gauged on an arbor. For milling the two ends the job requires two standard high-speed steel side mills, 8-in. diameter, ¾-in. wide. To mill apart the four pieces of the valve rocker shaft requires three 8-in. diameter, 5/16-in. wide staggered teeth standard HSS cutter. All cutters are set for right-gang cut. A horizontal arbor knee and column milling machine is available. The hand-clamping fixture must be provided with four clamping points to prevent the four brackets from shifting the fixture as they are milled apart. Two quick-acting clamps control each final part. Depth and length of cut are 1¼ in. each. The part weighs 13.2 lb. Find the unit and lot estimate for 680 units.

Table Number	Process Description	Table Time	Adjustment Factor	Cycle Minutes	Setup Hours
7.1-S1	Setup 5-gang cutter, 2.35 + 2 × .30	2.95			2.95
7.1-S2	Make part, 2 × hr/1 = 2 × .008	.008	2		.016
7.1-S3	Tolerance, not required				
7.2-1	Load and unload part	.24	¼	.06	
7.2-1	Seat with mallet	.11	¼	.028	
7.2-2	Clamp and unclamp with quick-acting clamps	.04	2/4	.02	
7.2-4	Start and stop spindle	.06	¼	.015	
7.2-4	Trip lever, rapid traverse ready to cut, retract for clearance and stop	.30	¼	.06	
	Length of cut = cutter approach, safety stock + over travel = 3.46 + .14 + 0 = 3.60 in.				
11.3-3	Mill 3.60 in. Check to find longest time for sawing or side milling.	.07	3.60/4	.063	
7.2-5	Position mist	.05	¼	.011	
	Unload 4 pieces at 3.3 lb each or	.13		.13	
7.2-5	Brush away chips	.14	¼	.035	
7.2-5	Wipe locators with rag	.15	¼	.038	
11.4-2	Cutter replacement, not allowed			0	
22.1-3	Part inspection, not allowed			0	
	Total			.46	2.97
	Unit estimate	.46 min			
	Hr/100 units	.767			
	Lot estimate	8.18 hr			

B. An 18-in. long SAE 1020 steel bar requires a shell mill operation, 2 × 2 in. along one edge. A vertical knee and column machine will be used. Simple table clamping is required. Each part weighs 93 lb. A similar previous operation is used. A 4-in. HSS shell mill will be considered. Find unit and lot time for 40 parts.

Table Number	Process Decription	Table Time	Adjustment Factor	Cycle Minutes	Setup Hours
7.1-S1	Holding device in prior operation is resetup	.70			.70
7.1-S2	Make part time, 2 × .12	.24			.24
7.1-S3	Tolerance, not allowed				
7.2-1	Load and unload, hoist	1.44		1.44	
7.2-2	Allen screws	.21	4	.84	
7.2-4	Start and stop spindle, table	.06		.06	
	Approach, safety allowance + over travel = .27 + .11 = .38 in.				
	Total distance milled = 18.38				
11.3-1	Mill 18.38 in. distance	.24	18.38	4.41	
8.2-1	Rapid travel 6 in., retract 24 in., at 300 in./min	.08		.08	
7.2-5	Start coolant	.05		.05	
7.2-5	Blow off table	.15		.15	
11.4-2	Tool life and replacement	.30		.30	—
	Total			7.33	.94
	Lot estimate	5.83 hr			

TABLE 7.2 KNEE AND COLUMN MILLING MACHINES

Setup   See Table 7.1

Operation elements in estimating minutes

1. Pick up part, move and place; pick up and lay aside

	Weight						
Holding Device	1.0	2.5	5	10	15	25	35
Open vise; parallels; simple fixture without pins or stops	.08	.10	.13	.16	.19	.20	.22
Simple fixture with locating pins or blocks; between centers; V-block with stops	.08	.11	.14	.18	.21	.23	.25
Complex fixture with pins or blocks to fit piece; chuck or complex clamps	.09	.12	.16	.20	.24	.26	.27
Blind fixture with stops or pins; index plate with alignment; or align to mark or dial on fixture	.10	.13	.17	.22	.27	.29	.32

Seat with mallet	.11
Wipe off parallels	.26
Unwrap and wrap pc for protection	.14
Straighten several loose pcs in fixture or vise	.02/ea
Check squareness and tap with mallet to align	.51
Place gage on surface and remove	.19
Level piece	.18
Assemble split bushing or insert for hold; disassemble	.16
Pry part out	.06
Use feeler or shim for locating	.18

Load and unload in collet	.24
Load and unload part to angle plate	1.01
Load and unload part in cup arbor	.48
Load and unload part in dividing head and tail stock	.39

2. Clamp and unclamp

Vise, ¼	.05
Vise, 2 turns	.12
Small quick clamp	.04
Medium quick clamp	.05
Large quick clamp	.06
Small cam clamp, 90°–180° throw	.02
Medium cam clamp, 90°–180° throw	.03
Large cam clamp	.04
Hand (star) wheel, 2 turns	.09
Thumb wheel or screw, 2 turns	.07
Position U-clamp/ea	.08
Allen screw (2 finger turns, 2 wrench turns)/ea	.18
Hex nut/ea	.21
Air cylinder: 1 in., .06; 3 in.,	.07
C-washer, on and off	.10

3. Part or part and fixture or attachment manipulation

   A.

Rotate, Flip	Weight			
	10	20	30	45
90°	.05	.07	.09	.11
180°	.06	.09	.11	.14

   B. Trunnion fixture

Rotate	90°	180°	270°
Min	.11	.13	.16

   C.

Turn Part in Vise	Weight				
	5	10	20	30	45
Min	.06	.07	.09	.13	.15

   D.
Pin and unpin	.09
Turn part end-for-end in dividing head and tail stock	.25
Turn table, lock and unlock	.14
Turn table, tune per 3°	.01
Index cup arbor to 90°	.16
In collet end-for-end	.09
Between centers, end-for-end	.34

4. Knee and column milling machine operation

Start or stop spindle	.03
Start and stop spindle	.06
Reverse spindle	.04
Spindle clamp, lock and unlock	.05
Table clamp, lock and unlock	.07
Cross slide, lock and unlock	.14
Knee, lock and unlock	.25
Start, advance work to cutter feed	.12
Trip lever, rapid traverse ready to cut, after cut retract for clearance and stop	.30
Advance table transversly to stop and lock	.17
Unlock table, return transversly	.10
Raise, lower spindle, clamp, unlock	.21

  Change feed to speed
    Single lever   .04
    Double lever   .07
  Engage clutch   .03
  Adjust table dial to mark (horizontal)   .50
  Adjust table dial to mark (vertical)   .32
  Adjust cross feed dial to mark (vertical)   .29
  Adjust spindle dial to mark (vertical)   .29
  Adjust cutter or table to micrometer depth of cut   .20
  Back work from cutter and stop machine
    Lever hand feed   .05
    Screw hand feed   .10
  Boring tool, adjust   .38
  Tool change for horizontal machine   4.90
  Tool change for vertical machine   2.80

5. Cleaning, lubricating or wiping
  Clean with air hose
    Small area 6 × 6-in. or under   .07
    72 × 12-in. area   .12
    12 × 12 to 12 × 24-in. area   .15
    12 × 24 to 18 × 48-in. area   .18
    Small fixture   .11
    Large fixture   .20
    Partial T-slots, add'l   .15
    Blow lube and chips from machined recess   .13
  Brush chips
    Small area and fixture   .12
    Medium area, fixture to 12 × 12 in.   .14
    Large area, fixture to 12 × 24 in.   .20
    Complex area and fixture   .26
    Empty chips from inside of container   .11
  Lubricate tool or piece
    Oil tool or piece with brush   .07
    Adjust coolant or mist hose   .05
    Place or remove splash guard   .08
    Wash part   .06
  Clean area, holding device
    Wipe locators with fingers   .04
    Wipe 6 × 6-in. area by hand   .05
    Wipe V-block or centers or attachment with rag   .15
    Wipe vise with rag   .12

6. Machining
  Face, side, slot, form, stradde, end, or saw milling   See Table 11.3
  Drill, counter bore, ream or tap   See Table 11.2

7. Milling cutter replacement   See Table 11.4-2

8. Part inspection   See Table 22.1-3

9. Rapid traverse   See Table 8.2-1

7.3 Bed Milling Machines

DESCRIPTION

Bed milling machines are manufacturing type machines used primarily for quantity production of identical parts. Using face and shell end mills, and arbor-mounted cutters singly or in combination, these machines can be set up for a variety of milling operations. These machines have a fixed-table support or bed. They can be further subdivided into plain-milling, with one horizontal spindle, or duplex, with two horizontal spindles located on opposite sides of the table. The headstock, together with the bed, forms the main frame of the machine. The spindle carrier slides on the vertical ways of the headstock and encloses the spindle, mounted in a quill, which provides cross-wise adjustment necessary for setting up the machine. The table guided in ways is provided on the bed. It moves longitudinally, or rotates, at right angles to the axis of rotation of the spindle. The working surface of the table has a series of T-slots. (See Fig. 7.3.)

In general, bed-type milling machines are semi- or fully-automatic and of simple, but sturdy, construction. There are several distinctive features: automatic cycle of approach of cutter and work relative to each other; rapid movement during the noncutting part of the cycle; and selective spindle stops and speeds. Some cross and vertical adjustments are operated manually or semi-automatically depending upon the manufacturer. After the machine is set up, the operator is required only to load and unload the machine and to start the automatic cycle, which may be controlled by pre-set switches. Numerical control features can be added. Now consider estimating data for the general class of bed milling machines.

FIGURE 7.3 Bed milling machine with two horizontal milling cutters. *(White-Sundstrand Machine Tool, Incorporated)*

ESTIMATING DATA DISCUSSION

Data for setup are provided in Table 7.1. Most of the operation elements are also listed in Table 7.2. Because of basic machine differences, additional elements are listed for bed milling. For many values the elements are described similarly, but the times are different. Whenever elements are the same, refer to the estimating data discussion for Tables 7.1 and 7.2.

Entry variables for load and unload are weight and type of holding device. If the part requires a crane hoist, three choices are: trunnion and L-hooks, 3-claw chain, and plain chain. A trunnion has two projecting cylinders to provide a hook point for lifting and rotation. The hoist time is for both load and unload part.

EXAMPLES

A. Estimate the finish mill operation of the top of a cylinder block. The block is 170–197 Bhn, and depth of cut is 1/16 in. The width of cut varies between 2⅝ in. minimum and 7½ in. maximum due to core openings. The length of the surface to be milled is 31¼ in. Locating points are two crankshaft bearings at opposed ends. A 9-in. diameter 20-tooth carbide face mill for right-hand cutting is necessary to overlap on the 7½-in. width of surface. The weight of the casting requires a crane hoist. The fixture has two sliding and tapered plugs and two loading rails. Determine the unit and shop estimate for a lot quantity of 675 units.

Table Number	Process Description	Table Time	Adjustment Factor	Cycle Minutes	Setup Hours
7.3-1	Hook, hoist, load, and unload	1.44		1.44	
7.3-1	Seat with mallet	.11	2	.22	
7.3-2	Air cylinder	.07		.07	
8.2-1	Move table to approach at rapid traverse rate, 3.5 in. at 200 in./min	.02		.02	
	Cutter approach, safety stock, and overtravel at end of cut = 2 + .17 + 2 = 4.17. Total length at feed = 4.17 + 31.25 = 25.42 in.				
11.3-2	Face mill. 35.42 in. (Use 8-in. diameter.)	.19	35.42	6.73	
8.2-1	Table travel after ending feed and rapid traverse back = 10 + 47 = 57 in.	.30		.30	
7.3-4	Change feed	.03		.03	
7.3-5	Blow fixture	.20		.20	
7.3-5	Adjust coolant	.05		.05	
7.3-5	Wipe locators	.05		.05	
11.4-2	Milling cutter replacement	1.00		1.00	
22.1-3	Part inspection (during cutting time)			0	—
	Total			10.11	0
	Shop estimate	5.9 pc/hr			

B. Large hexagon nuts, alloy steel, 250–300 Bhn, have six surfaces to be milled. The depth of cut is ¼ in. Width of cut is 3 7/16 in. along one of the six sides of the nut. The length of cut is 1 11/16 in. Locating is off a preturned hole. The six surfaces of the nut can be straddle milled, two at a time, and indexed twice. Cutters are side type with inserted blades of sintered carbide of a grade suitable for milling alloy steel. The cutters are 8 in. in diameter and the number of teeth is standard. Two surfaces are plunged in conjunction with indexing. The nut is placed and removed from the indexing fixture by hand and the fixture is hand operated. One turn of the index crank is equivalent to 60° rotation of the nut and brings a new pair of surfaces into position. A clearance of 12 in. is necesaary for safe loading. The bed-type milling machine will have an automatic table cycle controlled by trip dogs. A machine with specific performance capability is selected by the estimator. Find hr/100 units and shop time for 125 units.

Table Number	Process Description	Table Time	Adjustment Factor	Cycle Minutes	Setup Hours
7.1-S1	Straddle mill with expanding plug	2.00			2.00
7.1-S2	Make parts, not allowed				0

Bed Milling Machines 195

Table Number	Process Description	Table Time	Adjustment Factor	Cycle Minutes	Setup Hours
7.1-S3	Commercial inspection, not allowed				0
7.3-1	Load and unload	.22		.22	
7.3-1	Medium quick clamp	.05		.05	
7.3-4	Start and stop spindle	.06		.06	
8.2-1	Table travel to beginning of feed + travel at end of feed = 12 + 12 = 24 at rapid traverse rate of 150 in./min	.13		.13	
11.3-3	Safety stock + approach + overtravel = .14 + .53 + .53 = 1.20 in. Straddle mill (1.20 + 1¹¹⁄₁₆ = 2.89)	.25	2.89	.72	
7.3-3	Index two times for new surfaces	.05	2	.10	
11.3-3	Straddle mill two additional times	.72	2	1.44	
	Total			2.70	2.00
	Hr/100 units	4.500			
	Shop time	5.63 hr			

C. A part has been previously machined. Notice in Figure 6.4C where a turret lathe and the broaching operations have been completed. Next is the bed milling operation of the four ⅝-in. slots. The material is 4340 annealed steel, and four slots are required for a quantity of 7075 units.

Table Number	Process Description	Table Time	Adjustment Factor	Cycle Minutes	Setup Hours
	Mill 4 slots				
7.2-S1	Basic	1.55			1.55
11.3-3	Slot, straddle	0.39	2.75	1.07	
	Approach, safety, design, overtravel length = 2.75 in.				
11.3-3	Add'l three slots	1.07	3	3.21	
7.3-1	Pick up, move	0.17		0.17	
7.3-2	Clamp, unclamp	0.06		0.06	
7.3-3A	Manipulation	0.05	4	0.20	
	Rotate four times for slots				
7.3-4	Machine operation	0.12		0.12	
7.3-5	Clean, lubricate	0.12		0.12	
11.4-2C	Tool wear, 4.28 min of machining with carbide	0.56		0.56	
	Total			5.51	1.55
	Lot hours	651.57			

TABLE 7.3 BED MILLING MACHINES

Setup See Table 7.1

Operation elements in estimating minutes

1. Pick up part, move and place; pick up and lay aside

	Weight						
Holding Device	5	10	15	25	35	45	60
Open vise; parallels; simple fixture without pins or stops	.13	.16	.19	.20	.22	.24	.26
Simple fixture with locating pins or blocks; between centers; V-blocks with stops	.14	.18	.21	.23	.25	.28	.31
Complex fixture with pins or blocks to fit piece; chuck or complex clamps	.16	.20	.24	.26	.27	.31	.35
Blind fixture with stops or pins; index plate with alignment; or align to mark or dial on fixture	.17	.22	.27	.29	.32	.37	.40

Position, hook hoist to parts on pallet, hoist part to holding device, aside hoist; and remove part to pallet using:

Trunnion, L-hooks, or tongs	1.44
3-claw chain	1.55
Plain chain	1.89
Place cardboard, per layer, on skid	.12
Seat with mallet	.11
Wipe off parallels	.26
Straighten several loose parts in fixture or vise	.02
Check squareness and tap with mallet to align	.51
Level part	.18
Load and unload part to angle plate	1.01
Pry part out	.06
Use feeler or shim for locating	.18

2. Clamp and unclamp

Vise, ¼ turn	.05
Vise, 2 turns	.12
Medium V-quick clamp	.05
Large quick clamp	.06
Medium cam clamp, 90°–180° throw	.03
Large cam clamp	.04
Hand (star) wheel, 2 turns	.09
Thumb wheel or screw	.07
Position U-clamp/ea	.08
Allen screw (2 finger turns, 2 wrench turns)/ea	.18
Hex nut/ea	.21
Air cylinder: 1 in., .06; 3 in. .07	
Operate foot-air clamp	.02

3. Part or part and fixture or attachment manipulation

A.

Rotate, Flip	Weight				
	10	20	30	45	60
90°	.05	.07	.09	.11	.13
180°	.06	.09	.11	.14	.16

B. Trunnion fixture

Rotate	90°	180°	270°
Min	.11	.13	.16

C.

Turn Part in Vise	Weight					
	5	10	20	30	45	60
Min	.06	.07	.09	.13	.15	.21

4. Bed milling machine operation

Start or stop spindle	.03
Start and stop spindle	.06
Reverse spindle	.04
Spindle clamp, lock and unlock	.05
Start, advance work to cutter feed; after cut, retract for clearance and automatic stop	.12
Advance table transversly to stop and lock	.17
Unlock table, return transversly away from work	.10
Change feed or speed	.03
Adjust table dial to mark	.25

5. Cleaning, lubricating, or wiping

Clean with air hose	.07

Bed Milling Machines

Small area 6 × 6-in. or under	.07
12 × 12-in. area	.12
12 × 12-in. to 12 × 24-in. area	.15
12 × 24-in. to 18 × 48-in. area	.18
Small fixture	.11
Large fixture	.20
Partial T-slots, add'l	.15
Blow lube and chips from machined recess, add'l	.13
Brush chips	
Small area and fixture	.12
Medium area, fixture to 12 × 12 in.	.14
Large area, fixture to 12 × 24 in.	.20
Complex area and fixture	.26
Lubricate tool or piece	
Oil tool or piece with brush	.07
Adjust coolant or mist hose	.05
Clean area, holding device	
Wipe locators with fingers	.04
Wipe 6 × 6-in. area by hand	.05
Wipe V-block of centers or attachment with rag	.15
Wipe vise with rag	.12

6. Machining
   - Face, side slot, form, straddle, end or saw milling — See Table 11.3
   - Drill, counterbore, or ream — See Table 11.2

7. Milling cutter replacement — See Table 11.4-2

8. Part inspection — See Table 22.1-3

9. Rapid traverse — See Table 8.2-1

7.4 Vertical-Spindle Ram-Type Milling Machines

DESCRIPTION

The vertical-spindle ram-type milling machine design includes a knee and column support arrangement for the table. A saddle supporting the table provides in-and-out motions with respect to the column. The motor is mounted on the tool head supported by the overarm. A feature of this machine is the ram. The ram is a sliding element of the machine's overarm and enables the spindle to move via hand crank or power, in and out, parallel to the movement of the saddle. Ram-type vertical mills can incorporate other motions. The entire ram overarm can pivot about the main upright axis to describe an arc over the worktable. Another motion is the ability to tilt the spindle axis off of vertical, either right or left, and forward or aft.

The machines operate at low horsepower and have a toolroom or light milling orientation. End mills are a common form of tooling but formed cutters for slotting and dovetailing and shell-type cutters for milling flat surfaces are suitable. These machines can also handle boring and tapping operations.

The turret model shown in Figure 7.4 has a 4 hp variable speed motor, 58-in. table length, 15-in. cross travel and a 15-in. vertical capacity.

ESTIMATING DATA DISCUSSION

Data for setup are listed in Table 7.1, and most of the operation elements are also listed in Table 7.2. Because of machine differences, additional elements are provided for vertical-spindle ram-type milling machines. In some cases, the elements are described similarly, but the times are different.

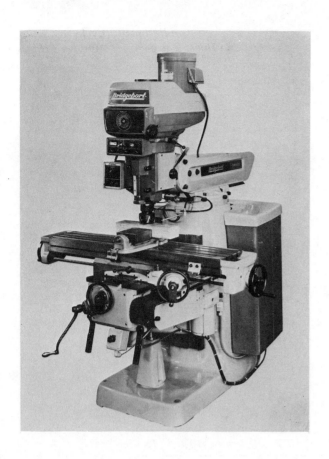

FIGURE 7.4 Vertical-spindle turret-type milling machine. (*Bridgeport Machines*)

EXAMPLES

A. An aluminum cast head is end milled for a company called SoHo. The bored hole is .750-in. OD and 2-in. deep. The depth of cut is 1/32 in. Length of cut is 2 in. The head weighs 6.28 lb. Fixture mounting is critical and tolerance checking is to be added to the estimate. Find the unit estimate, hr/unit, and the lot estimate for 87 units.

Table Number	Process Description	Table Time	Adjustment Factor	Cycle Minutes	Setup Hours
7.1-S1	Table setup	1.45			1.45
7.1-S2	Make part, 2 × unit estimate	.02			.04
7.1-S3	Tolerance check, .001-in. depth tolerance	.13			.13
7.4-1	Load and unload part	.20		.20	
7.4-2	Clamp, 2 places	.09	2	.18	
7.4-2	Pry part, add'l	.06		.06	
7.4-4	Start and stop boring head	.06		.06	
8.2-1	Rapid traverse ready to cut and retract, 100 in./min for 8 in.	.08	2	.16	
11.3-4	End mill opening ¾-in. cutter, HSS	.01	2	.02	
7.4-5	Adjust mist	.05		.05	
7.4-5	Blow chips	.15		.15	
11.4-2	Tool wear out allowance	0		0	
22.1-3	Part inspection with depth micrometer, .001-in. tolerance	.29		.29	
	Total			1.17	1.82
	Hr/unit	.020			
	Lot estimate	3.317 hr			

B. An aluminum cast head has a manifold area faced milled. Surface area is 8.25 × 2 in. overall with various cored areas. The head weighs 6.28 lb and tolerances are not critical. A previous fixture is used. An HSS 4-in. shell mill is used. Find pc/hr for the job.

Vertical-Spindle Ram-Type Milling Machines

Table Number	Process Description	Table Time	Adjustment Factor	Cycle Minutes	Setup Hours
7.1-S1	Re-setup of previous operation	.75			.75
7.1-S2	Make part, not allowed	0			0
7.1-S3	Tolerance check, not allowed	0			0
7.4-1	Load and unload part	.20		.20	
7.4-2	2 star wheels	.09	2	.18	
7.4-4	Start and advance work to cutter feed	.12		.12	
	Cutter approach, safety stock allowance + overtravel = .27 + .11 = .38. Shell mill .38 + 8.25 = 8.63 in.				
11.1-3	Shell mill	.01	8.63	.09	
7.4-5	Adjust mist hose	.05		.05	
7.4-5	Blow chips	.15		.15	
	Total			.79	.75
	Pc/hr		76		

TABLE 7.4 VERTICAL-SPINDLE RAM-TYPE MILLING MACHINES

Setup   See Table 7.1

Operation elements in estimating minutes

1. Pick up part, move and place; pick up and lay aside

Holding Device	Weight				
	1.0	2.5	5	10	15
Open vise; parallels; simple fixture without pins or stops	.08	.10	.13	.16	.19
Simple fixture with locating pins or blocks; between centers; V-blocks with stops	.08	.11	.14	.18	.21
Complex fixture with pins or blocks to fit piece; chuck or complex clamps	.09	.12	.16	.20	.24
Blind fixture with stops or pins; index plate with alignment; or align to mark or dial on fixture	.10	.13	.17	.22	.27

Seat with mallet	.11
Wipe off parallels	.26
Check squareness and tap with mallet to align	.51
Place gage on surface and remove	.19
Level part	.18
Assemble split bushing or insert for holding and disassemble	.16
Load and unload part in collet	.24
Load and unload part to angle plate	1.01
Load and unload part in cup arbor	.48
Load and unload part in dividing head and tail stock	.39
Pry part out	.06
Use feeler or shim for locating	.18

2. Clamp and unclamp

Vise, ¼ turn	.05
Vise, 2 turns	.12
Small quick clamp	.04
Medium quick clamp	.05
Small cam clamp, 90°–180° throw	.02
Medium cam clamp, 90°–180° throw	.03
Hand (star) wheel, 2 turns	.09

    Thumb wheel or screw, 2 turns .07
    Position U-clamp leach .08
    Allen screw (2 finger turns, 2 wrench turns)/ea .18
    Hex nut/ea .21
    Air cylinder: 1 in., .06; 3 in., .07

3. Part or part and fixture or attachment manipulation

    A.

Rotate, Flip	Weight		
	10	20	30
90°	.05	.07	.09
180°	.06	.09	.11

    B. Trunnion fixture

Rotate	90°	180°	270°
Min	.11	.13	.16

    C. Pin, unpin .09

Turn Part in Vise	Weight			
	5	10	20	30
Min	.06	.07	.09	.13

    Advance rotary table to work and return to load position. Each 4.5° .01
    Place and remove spacer block .05

4. Turret milling machine operation
    Start or stop spindle .03
    Start and stop spindle .06
    Reverse spindle .04
    Ram or headstock swivel clamp, lock and unlock .19
    Spindle clamp, lock and unlock .05
    Table clamp, lock and unlock .07
    Cross-slide, lock and unlock .14
    Start and advance work to cutter feed .12
    Trip lever, rapid traverse ready to cut, after cut retract for clearance and
        stop .16
    Jog table transversely to stop and lock .17
    Unlock, jog table transversely .10
    Raise, lower spindle, clamp, unlock .21
    Change feed or speed .05
    Adjust table dial to mark (horizontal) .40
    Adjust table dial to mark (vertical) .28
    Adjust cross feed dial to mark .26
    Adjust cutter or table to micrometer depth of cut .20
    Lower spindle 1 in. .03
    Lower spindle 1 in. and lock .04
    Raise spindle 1 in. .02
    Unlock and raise spindle .04
    Advance table and engage feed .10
    Return table .08

5. Cleaning, lubricating or wiping
    Clean with air hose
        Small area, 6 × 6-in. or under .07
        12 × 12-in. area .12
        12 × 12-in. to 12 × 24-in. area .15
        Small fixture .11
        Partial T-slots, add'l .15
        Blow lube and chips from machined recess, add'l .13

Brush chips
    Small area and fixture .12
    Medium area, fixture to 12 × 12 in. .14
Lube tool or piece with brush .07
Adjust coolant or mist hose .05
Clean area, holding device
    Wipe locators with fingers .04
    Wipe 6 × 6-in. area, by hand .05
    Wipe V-block or centers or attachment with rag .15
    Wipe vise with rag .12

6. Machining
    Face or end milling   See Table 11.3
    Drill, counterbore, ream, or tap   See Table 11.2

7. Milling cutter replacement   See Table 11.4-2

8. Part inspection   See Table 22.1-3

7.5 Router Milling Machines

DESCRIPTION

Routers use high speeds, often in excess of 10,000 rpm, to face mill, profile, pocket, cut, or drill nonferrous metals and plastics. Routers can be classified as radial, profile, and shaper. The radial-type has a moveable head with the cutter in a vertical position mounted on hinged arms which allow motion over the table. A profile router has the cutter mounted in a rotating-spindle, fixed-position head stock. In this machine, profiling and pocketing is controlled by the movement of a pattern over a guide pin, while depth is controlled by vertical movement of head. In the shaper-type, the cutter head is mounted under the machine table and projects above surface. A guide is fastened to the table for positioning of the workpiece.

The router shown in Figure 7.5 has push-button controls, allows for interchanging heads, guide pin lock and release, handwheel for vertical table adjustment, and an airfoot pedal for lowering and raising the router head.

ESTIMATING DATA DISCUSSION

Data for setup are listed in Table 7.1, and most of the operation elements are also listed in Table 7.2. Because of machine differences, additional elements are provided for router machines. In some cases the elements are described similarly, but times may differ. For a description of the elements, see the estimating data discussion in section 7.2.

FIGURE 7.5 Milling router. (*Onsrud Division, Danly Machine Corporation*)

EXAMPLES

A. An aluminum alloy is routed in a pocket operation. A two-flute end mill, ¾-in. diameter, is used to pocket an inside step. The shape of the pocket is controlled by movement of pattern over a guide pin. The aluminum part has holes drilled and reamed through the part for locating on the pattern. Two small quick clamps hold the part against the plate. The pocket is 21-in. long and the part weighs 2.4 lb. Find the unit estimate. There ae 903 units in a lot.

Table Number	Process Description	Table Time	Adjustment Factor	Cycle Minutes	Setup Hours
7.1-S1	Table setup	1.40			1.40
7.5-1	Load and unload	.11		.11	
7.5-2	Two small quick-acting clamps	.04	2	.08	
7.5-4	Start and stop machine	.05		.05	
7.5-4	Lower and raise head 1 in.	.10		.10	
11.3-5	End mill with ¾-in. HSS 21 in.	.01	21	.21	
7.5-5	Blow off table top	.12		.12	
	Total			.67	1.40
	Unit estimate	.67 min			
	Lot estimate	11.48 hr			

B. A hydraulic automatic transmission valve body is face milled with a carbide 4-tooth 4-in. cutter. The body is 6-in. wide requiring two passes 8-in. long. The holding fixture is a vise with handle bars allowing the fixture and part to be hand fed under the cutter. A stock of ¹⁄₁₆ in. is removed. The body weighs 12.3 lb. Estimate the hourly production.

Table Number	Process Description	Table Time	Adjustment Factor	Cycle Minutes	Setup Hours
7.1-S1	Vise with end-mill setup	1.45			1.45
7.5-1	Load and unoad part	.19		.19	
7.5-2	Two quick-acting clamps	.04	2	.08	
7.5-4	Start and stop spindle	.05		.05	
7.5-5	Adjust mist hose	.05		.05	
	Cutter approach + safety distance = .68 + .11 = .79 in. Total length per pass = 8.79				
11.3-1	Face mill 8.79 in two passes	.01	8.79 × 2	.18	
7.5-5	Blow off chips	.15		.15	
	Total			.70	1.45
	Pc/hr	86			

C. Four .0625-in. plastic strips are clamped together and fed along a fence mounted to a table. Each 24-in. strip has several holes predrilled allowing location on a sliding fixture element with 2 pins. A 1-in. end mill, 2 flute, is used to edge the four parts. Find the lot esimate for 1250 units.

Table Number	Process Description	Table Time	Adjustment Factor	Cycle Minutes	Setup Hours
7.1-S1	Table setup with end mill	1.55			1.55
7.5-1	Load and unload 4 parts	.08		.08	
7.5-2	Allen screws	.18	½	.09	
7.5-4	Start and stop spindle	.05	¼	.01	
11.3-4	Side mill 24 in. long plastic part	.05	24/4	.30	
7.5-5	Blow plastic chips	.28	¼	.07	
	Total			.55	1.55
	Four strips, floor-to-floor	2.21			
	Lot estimate	13.06 hr			

Router Milling Machines

TABLE 7.5 ROUTER MILLING MACHINES

Setup See Table 7.1

Operation elements in estimating minutes

1. Pick up part, move and place; pick up and lay aside

Holding Device	Weight				
	1.0	2.5	5	10	15
Open vise; parallels, simple fixture without pins or stops	.08	.10	.13	.16	.19
Simple fixture with locating pins or blocks; between centers; V-blocks with stops	.08	.11	.14	.18	.21

Straighten several loose parts in fixture or vise	.02
Pry part out	.06
Unwrap and wrap part for protection	.12

2. Clamp and unclamp

Vise, ¼ turn	.05
Vise, 2 turns	.12
Small quick clamp	.04
Small cam clamp, 90°–180° throw	.02
Hand (star) wheel, 2 turns	.09
Position U-clamp	.08
Allen screw, (2 finger turns, 2 wrench turns)	.18
Hex nut/ea (2 finger turns, 2 wrench turns)	.21
Air cylinder 1-in. throw	.06

3. Part or part and fixture or attachment manipulation

   A.

Rotate, Flip	Weight		
	10	20	30
90°	.05	.07	.09
180°	.06	.09	.11

   B. Trunnion fixture

Rotate	90°	180°	270°
Min	.11	.13	.16

   Pin, unpin .09

   C.

Turn Part in Vise	Weight			
	5	10	20	30
Min	.06	.07	.09	.13

4. Router milling machine operation

Start and stop spindle	.05
Raise or lower head and cutter, 1 in.	.10
Handwheel adjustment of table, 1 in.	.30
Guide pin lock and release	.25
Change spindle speed	.14
Set depth stops	.35
Tilt table and lock	.50
Adjust table fence	.60
Brake air spindle	.05

5. Cleaning, lubricating or wiping
    Clean with air hose
        Small area 6 × 6-in. or under .07
        12 × 12-in. area .12
        12 × 12-in. to 12 × 24-in. area .15
        12 × 24-in. to 36 × 48-in. table .28
        Add'l for excessive aluminum or plastic chips .15
        Small fixture .11
    Blow lube and chips from machined recess, add'l .13
    Brush chips from small area including fixture .12
    Adjust mist hose .05
    Wipe locators with fingers .04

6. Machining
    Face, end or engrave milling   See Table 11.3
    Drill   See Table 11.2

7. Milling cutter replacement   See Table 11.4-2

8. Part inspection   See Table 22.1-3

7.6 Special Milling Machines

DESCRIPTION

The family of milling machine types is diversified and a great variety of models is available. To provide estimating data becomes cumbersome due to these distinctions. Thus, we have identified milling machines as standard and special. The standard milling machines are described elsewhere. The special machines are not divided into subgroups. Trade names abound, such as duplicators, panto mills, contour,

FIGURE 7.6 Hydro-tel milling machine with switchable computer numerical control and universal electronic tracer. *(Cincinnati Milacron)*

planer, engravers, hydrotel, etc. Our purpose is to show estimating data for these special machines as a unit.

Figure 7.6 is a 40-in. Hydro-Tel milling machine with switchable computer NC and a universal electronic tracer. Three milling heads are found in this machine. Notice that this machine has a traveling planer bed. A follower is shown suspended from the crossrail on the right side.

ESTIMATING DATA DISCUSSION

Data for setup are listed in Table 7.1, and most of the operation elements are also listed in Table 7.2. Because of machine differences, additional elements are provided for special milling machines. In some cases the elements are described similarly, but the times may differ. For a description of the elements, see 7.2.

EXAMPLES

A. Keyways are milled simultaneously on two steel shafts placed abreast on the table of a rise-and-fall milling machine. The keyway is ¼-in. wide × 11-in. long × .145-in. deep in a 1-in. OD soft-steel bar stock. The two shafts are placed in V-blocks, side by side against end stops, and are clamped with two hand clamps. Two HSS staggered tooth slotting cutters are spaced at a required distance on an arbor. The rise and fall feature engages the cutter into the bar stock with a minimum approach. Cutter size is 3-in. dia. Find lot size for 34 units.

Table Number	Process Description	Table Time	Adjustment Factor	Cycle Minutes	Setup Hours
7.1-S1	Table setup with straddle mill	1.90			1.90
7.1-S2	Make part time	.061	2		.12
7.1-S3	Inspection, not allowed				0
7.6-1	Load 2.5 lb stock in V-blocks	.14		.14	
7.6-2	Small cam clamp	.02	2/2	.04	
7.6-4	Start and stop spindle	.06	½	.03	
7.6-5	Adjust lubricant	.05	½	.025	
	Rapid traverse cutter to work + safety. Stock = .83 + .11 = .94				
11.3-3	Mill 11.94 in. of stock	.55	11.94/2	3.29	
8.2-1	Rapid retract 12 in. at 80 in./min	.15	½	.08	
7.5-5	Blow off chips	.18	½	.09	
	Total			3.69	2.02
	Two-part estimate	7.39			
	Lot estimate	4.10 hr			

B. A pocket and a step operation is planned for a milling machine using a duplicating arm mechanism. The size of the reduction is 2 to 1. The end mill is ⅛ in. in diameter, 2 flute HSS. A slot ⅛-in. in width is milled ⅛-in. deep inside the edge of an aluminum extrusion 9.2-in. long. The step is 8.6 in. in length, and 1/16-in. wide and deep. Find the unit estimate. There are 1318 units in the lot.

Table Number	Process Description	Table Time	Adjustment Factor	Cycle Minutes	Setup Hours
7.1-S1	Table setup with part length, 12 in. for end mill	1.60			1.60
7.1-S2	Make part	.038	2		.08
7.6-5	Clean nest with air	.15		.15	
7.6-1	Load and unload, 3.1 lb part	.13		.13	
7.6-2	2 thumbscrews	.07	2	.14	
7.6-5	Lubricate piece	.07		.07	
7.6-4	Start and stop spindle	.06		.06	
7.6-4	Position ⅛-in. tool to work	.20		.20	
11.3-4	Mill ⅛-in.	.03	9.2	.28	
7.6-5	Blow chips	.15		.15	
7.6-4	Start and stop spindle	.06		.06	
9.3-4	Key-chuck tool change	.30		.30	
7.6-4	Position 1/16-in. tool to work	.20		.20	
7.6-4	Change speed	.07		.07	

11.3-4	Mill 1/16-in.		.04	8.6	.34	
7.6-5	Blow chips		.15		.15	
	Total				2.30	1.68
	Lot estimate		52.20 hr			

TABLE 7.6 SPECIAL MILLING MACHINES

Setup   See Table 7.1

Operation elements in estimating minutes

1. Pick up part, move and place; pick up and lay aside

			Weight				
Holding Device	5	10	15	25	35	45	60
Open vise, parallels; simple fixture without pins or stops	.13	.16	.19	.20	.22	.24	.26
Simple fixture with locating pins or blocks; between centers; V-blocks with stops	.14	.18	.21	.23	.25	.28	.31
Complex fixture with pins or blocks to fit piece; chuck or complex clamps	.16	.20	.24	.26	.27	.31	.35
Blind fixture with stops or pins; index plate with alignment; or align to mark or dial on fixture	.17	.22	.27	.29	.32	.37	.40

Position hook hoist to part on pallet, hoist, part to holding device, aside hoist; and remove part to pallet using trunnion, L-hooks, or tongs	1.44
3-claw chain	1.55
Plain chain	1.89
Seat with mallet	.11
Wipe off parallels	.26
Unwrap and wrap part for protection	.24
Straighten several loose parts in fixture or vise	.02
Check squareness and tap with mallet to align	.51
Place gage on surface and remove	.19
Level part	.18
Assemble split bushing or insert for hold and disassemble	.16
Load and unload part in collet	.24
Load and unload part in angle plate	1.01
Load and unload part in cup arbor	.48
Load and unload part in dividing head and tail stock	.39
Place on index head and tighten nut	.06
Pry part out	.06
Use feeler or shim for locating	.18
For engraver machine, slide small part to stops	.04
For engraved piece, fill character, clean off excess, per character	.01

2. Clamp and unclamp

Vise, 1/4 turn	.05
Vise, 2 turns	.12
Small quick clamp	.04
Medium quick clamp	.05
Large quick clamp	.06
Small cam clamp, 90°–180° throw	.02
Medium cam clamp, 90°–180° throw	.03
Large cam clamp	.04
Hand (star) wheel, 2 turns	.09
Thumb wheel or screw, 2 turns	.07

Position U-clamp/ea	.08
Allen screw (2 finger turns, 2 wrench turns)/ea	.18
Hex nut/ea	.21
Air cylinder: 1-in. throw, .06; 3-in. throw	.07

3. Part or part and fixture or attachment manipulation

   A.

	Weight				
Rotate, Flip	10	20	30	45	60
90°	.05	.07	.09	.11	.13
180°	.06	.09	.11	.14	.16

   B. Trunnion fixture

Rotate	90°	180°	270°
Min	.11	.13	.16

   Pin, unpin   .09

   C.

	Weight					
Turn Part in Vise	5	10	20	30	45	60
Min	.06	.07	.09	.13	.15	.21

4. Special milling machine operation

Start or stop spindle	.03
Start and stop spindle	.06
Reverse spindle	.04
Spindle clamp, lock and unlock, raise	.24
Table clamp, lock and unlock	.08
Cross-slide, lock and unlock	.15
Swivel head, lock and unlock	.22
Start, advance work to cutter feed	.14
Advance table transversely to stop and lock	.19
Unlock table, return transversely	.12
Change feed or speed	.07
Adjust dial to mark	.35
Adjust cutter or table to micrometer depth of cut	.20
For each manual shift of tool	.06
For each manual shift of stylus on pattern	.18

5. Cleaning, lubricating or wiping

Clean with air hose	
Small area 6 × 6-in. or under	.07
12 × 12-in. area	.12
12 × 12-in. to 12 × 24-in. area	.15
12 × 24-in. to 18 × 48-in. area	.18
Small fixture	.11
Large fixture	.20
Partial T-slots, add'l	.15
Blow lube and chips from machined recess, add'l	.13
Brush chips	
Small area and fixture	.12
Medium area, fixture to 12 × 12 in.	.14
Large area, fixture to 12 × 24 in.	.20
Complex area and fixture	.26
Lubricate tool or piece	
Oil tool or piece with brush	.07
Adjust coolant or mist hose	.05
Clean area, holding device	
Wipe locators with fingers	.04

Wipe 6 × 6-in. area by hand	.05
Wipe V-block or centers or attachment with rag	.15
Wipe vise with rag	.12

6. Machining

Face, side, slot, form straddle, end, saw, or engrave milling	See Table 11.3
Drill, counterbore, ream, or tap	See Table 11.2

7. Milling cutter replacement — See Table 11.4-2

8. Part inspection — See Table 22.1-3

7.7 Hand Milling Machines

DESCRIPTION

Hand millers are a class of manually-fed machines where the table is controlled by the operator. The operator may use a hand-feed lever or a wheel. Customarily, the spindles are horizontal, and arbor-mounted cutters are common. The tables are about 30 in. in length, and width is approximately 10 in. T-slots are available for mounting light fixtures and attachments. Parts are loaded manually, and their size is small.

Figure 7.7 shows two milling machines where hand-level controls feeds to head, saddle and table motion. Other models may have motorized feeds.

FIGURE 7.7 Hand-lever control feeding of small milling machines. *(Barker Milling Machine Company)*

ESTIMATING DATA DISCUSSION

Data for setup are listed in Table 7.1, and most of the operation elements are also described in 7.2. Because of machine differences, additional elements are provided for hand miller machines. In some cases the elements are described similarly, but times differ.

EXAMPLES

A. Alloy steel rods are splined milled on two ends. Shafts, 14.65-in. long and ½ in. in diameter, weigh .8 lb. Inserted into a collet and indexed by a dividing head, each end has four splines milled a length of 1.215 in. by a specially-formed cutter 4 in. in diameter. However, it is similar to a plain cutter. Tolerances are commercial. Lot size is 125 units. Find unit and lot estimate.

Table Number	Process Description	Table Time	Adjustment Factor	Cycle Minutes	Setup Hours
7.1-S1	Setup collet with indexing	1.25			1.25
7.1-S2	Make part, 2 × unit estimate	.18			.18
7.1-S3	Tolerance check, not allowed	0			0
7.7-1	Load and unload part in collet	.24		.24	
7.7-4	Start and stop spindle	.06		.06	
7.7-4	Advance, ready to cut, retract	.09		.09	
11.3-3	Slot 1.215 in.	.97	1.215	1.17	
7.7-3A	Rotate collet	.04		.04	
Repeat	Repeat 3 above elements for 3 more splines, 3 × (.09 + 1.17 + .04)	3.90		3.90	
7.7-3	Turn end-for-end in collet	.09		.09	
Repeat	Repeat advance, slot and rotate	1.30	4	5.20	
	Total			10.79	1.43
	Lot estimate	23.91 hr			

B. An aluminum extrusion having an irregular area but cutting length of .6 in. is sawed by a hand milling operation. The extrusion is 20 in. in length and pushed against a stop, followed by the sawing operation. For a large quantity, find hr/1000 units. The extrusion weighs 6.1 lb., and the cutter is a standard. Each extrusion yields 27 parts.

Table Number	Process Description	Table Time	Adjustment Factor	Cycle Minutes	Setup Hours
7.7-1	Load extrusion	.14	1/27	.005	
7.7-2	Quick clamp	.04	1/27	.001	
7.7-4	Start and stop spindle	.06	1/27	.002	
Estimate	Push against stop	.02	1/27	.001	
11.3-5	Saw aluminum	.06	.6	.04	
Repeat	Repeat push and saw element 26 times	.06		.06	
7.7-1	Toss aside scrap	.03	1/27	.001	
7.7-5	Blow scrap	.15	1/27	.006	
	Total			.167	—
	Extrusion estimate	4.51 min			
	Hr/1000 units	2.783			

TABLE 7.7 HAND MILLING MACHINES

Setup   See Table 7.1

Operation elements in estimating minutes

1. Pick up part, move and place; pick up and lay aside

Holding Device	Weight			
	½	1	2½	5
Open vise; parallels; simple fixture without pins or stops	.06	.08	.10	.13
Simple fixture with locating pins or blocks; between centers; V-blocks with stops	.07	.08	.11	.14

Seat with mallet	.11
Wipe off parallels	.26
Straighten several loose parts in fixture or vise	.02
Check squareness and tap with mallet to align	.51
Load and unload part in collet	.24
Load and unload part in dividing head and tailstock	.39
Load and unload part in universal chuck	.32
Pry part out	.06
Use feeler of shim for locating	.18
Air eject, deduct	.04
Toss aside, deduct	.03

2. Clamp and unclamp

Vise, ¼ turn	.05
Small quick clamp	.04
Small cam clamp. 90°–180° throw	.02
Hand (star) wheel, 2 turns	.09
Thumb wheel or screw, 2 turns	.07
Position U-clamp	.08
Allen screw (2 finger turns, 2 wrench turns)	.18
Hex nut/ea	.21
Air cylinder, 1-in. throw	.06

3. Part or part and fixture manipulation

   A.

Rotate, Flip	Weight		
	5	10	20
90°	.04	.05	.07
180°	.05	.06	.09

   B. Trunnion fixture

Rotate	90°	180°	270°
Min	.11	.13	.16

   Pin, unpin .09

   C.

Turn Part in Vise	Weight		
	5	10	20
Min	.06	.07	.09

In collet, end-for-end	.09
Turn part, end-for-end, in dividing head and tail stock	.25

4. Hand miller machine operation

Start or stop spindle	.03
Start and stop spindle	.06
Reverse spindle	.04
Spindle clamp, lock and unlock	.05
Table clamp, lock and unlock	.07
Trip lever, advance ready to cut; after cut, retract for clearance and stop	.09

Change speed or screw feed	.07
Adjust cutter or table to micrometer depth of cut	.20
Back work from cutter and stop machine	.06

5. Cleaning, lubricating or wiping
   Clean with air hose

6 × 6-in. area or under	.07
12 × 12-in. area or under	.12
12 × 12-in. to 12 × 24-in. area	.15
Fixture in vise	.11
Partial T-slots, add'l	.08
Brush area and fixture	.12
Adjust coolant or mist hose	.05
Wipe locators with fingers	.04
Wipe 6 × 6-in. area with hand	.05
Wipe V-block or centers or attachment with rag	.15

6. Machining
   Side, slot, form, straddle or saw milling   See Table 11.3

7. Milling cutter replacement   See Table 11.4-2

8. Part inspection   See Table 22.1-3

MACHINING CENTERS

8.1 Machining Centers

DESCRIPTION

Early design placed control units on existing machine tool structures to achieve numerical control. As experience expanded, the machine center evolved and a distinctive class of machine tools developed with generally two- to five-axis capability. Pallets for off-machine loading, shuttles, and a tool holder that can accommodate many tools are only a few of the special features available. While machining centers have a variety of configurations, they are basically chip removal structures with NC control. These data consider this class of machine similar to a general purpose mill able to mill, drill, bore, face, spot, counterbore, route, and more. Part positioning is a function of rotation, indexing, and table traversing at various rates. Figure 8.1 is a four-axis model that has a rotary table.

ESTIMATING DATA DISCUSSION

These data are for the estimating of machining center setup and the units are estimating hr. The values include the build-up and removal from the machine table and control unit parts, fixture, clamps, tape, etc. required to produce the part.

Job preparation includes these elements: punch in/out, receive instructions, tape, drawing, position tape in reader, get blocking, parallels, fixture, pins or other tooling, position machine to proper coordinates on X–Y axis using pin or piece as a "0" point, measure for clamp heights, adjust control boxes, start machine, check hole location on first piece and check drawing, position hoist or crane, and after lot is run, sweep off chips and clean area, and remove tape from machine.

Setup Element 2 uses entry variables of number of clamps and clamp height. This work may include using parallels, filler blocks, step blocks, straps, positioning and tightening threaded studs, etc. Teardown includes similar work except in reverse order.

Pin stop is for any dowel pin to locate piece or fixture as indicated by drawing or grid sketch. Element includes study grid sketch, mark hole for pin location, clean, and teardown.

Locator plug or fixture consists of obtaining plug or fixture and positioning on table. Plug or fixture is bolted to table, and includes teardown.

Adapter ring fits on holding fixture and is attached by threaded bolts. Set tooling involves obtain, assemble tool to adapter, load tool and adapter in tool matrix, and teardown. Some tools require miking for dimension or adjusting for dimension after trial cut.

Element 4 provides for the study of grid sketch for pin location, clean out holes, position dowel pins, align angle plate, tighten, and teardown. Element 5 deals with the optional inclusion of "make part." Certainly, it is necessary to finish a part before the setup is concluded, but this time is also included in the cycle time. Element 6 consists of exceptional tolerance requirements of a setup.

In cycle Element 1, there is a distinction of a single- or double-table layout. The double layout uses two set-

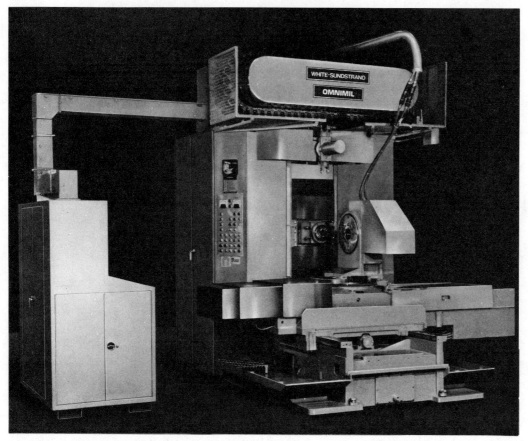

Figure 8.1 Four-axis machining center. *(White-Sundstrand)*

ups on the same table, or a second setup is on the pallet arrangement. The single layout involves loading, machining one piece to completion, and unload followed by another piece in the same fixture. Double layout consists of two setups on the grid table for machining one piece to completion, load second piece ready for machining during period of first part machining, machine second piece and simultaneously remove previous piece, and prepare next piece for loading.

Entry variables for using Table 8.1 are pin stops or locator plugs, weight, and part configuration. For weight under 45 lb, the work includes move piece from skid to grid table, align piece to pin stops or locator plug, and reset tape. After machining, remove piece to skid. An occasional mallet blow is included for seating or loosening. When weight exceeds 45 lb, a jib crane is used, and a part configuration is identified for entry. A part may be mostly circular and a 50-in. diameter is used to separate the selection of a time value. A bottom-area perimeter of 35 in. is used for selection when the configuration is mostly rectangular or square. In clamp and unclamp, the work involves the positioning of an end clamp over piece and step-block, tighten, and aside wrench, done in reverse for teardown.

Element 4 involves a selection of machining-center operations. For automatic tool changes see Table 8.2 which is used for several tape-drive machines having automatic tool selection. A manual tool change is possible at the quill and a choice is provided.

Element 5, for practical purposes, can be done during the NC cycle, although location and fixture cleaning is done off the NC tape. The element that indicates piece involves locating to the origin point with a dial indicator, clear spindle of previous tool, position table, position spindle with correct tool to location, and check coordinates.

EXAMPLES

A. A front-cover aluminum part is machined on a NC machining center. The part is loaded one at a time on a horizontal table. Ten tools are required and twenty position changes are necessary to machine the cover. Clamping in T-slots will be used. A tape is unavailable at the time of the estimate. The machining schedule is provided below. Find the unit and shop estimate.

Table Number	Process Description	Table Time	Adjustment Factor	Cycle Minutes	Setup Hours
8.1-4	Start machine	.10		.10	
8.1-1	Load to pin stops	1.03		1.03	
8.1-2	Tighten 3 clamps	.50	3	1.50	
8.2-1	Rapid traverse to and from part at 300 in./min for 25 in.	.16		.16	
8.2-2	Ten tool changes, magazine rate at .30	.30	10	3.00	
8.1-4	Coordinate changes	.08	20	1.60	
11.2-4	Drill 1 hole, 19/32-in. OD, 507-in. long	.08	.507	.05	
11.2-4	Flat-bottom drill, 1 hole, 19/32-in. OD., .311-in. long	.08	.311	.02	
11.2-7	Ream 1 hole to .609 in., .507-in. long	.05		.05	
11.2-4	Countersink hole	.08	.241	.02	
11.3-4	End mill .062-in. radius to .312-in. dimension, .537-in. long, HSS	.03	1.6	.05	
11.3-4	Counterdrill .33-in. deep, 6 times, ⅝-in. tool	.08	6 × .33	.16	
11.3-4	Step drill .196-in. large OD, 6 times by .612-in. deep	.07	6 × .612	.26	
11.3-4	Countersink 120° by .24-in. diameter by .151-in. deep 6 times.	.07	6 × .24	.10	
11.2-8	Tap 6 holes 19–32 UNF-3B × .435-in, deep	.13	6 × .435	.34	—
	Unit estimate			8.52	
	Shop estimate		7 pc/hr		

TABLE 8.1 MACHINING CENTERS

Setup in estimating hours

S1. Job preparation, constant   .35

S2. Tooling

A. Prepare clamps with locator plug

Clamp Height	No. Clamps					
	1	2	3	4	5	6
1	.05	.10	.15	.20	.25	.30
4	.06	.13	.19	.26	.32	.38
10	.08	.16	.23	.31	.39	.47
16	.09	.17	.26	.34	.43	.51
Add'l	.01	.01	.02	.03	.04	.04

B. Prepare clamps without locator plug

Clamp Height	No. Clamps					
	1	2	3	4	5	6
1	.09	.14	.20	.26	.31	.37
4	.10	.16	.24	.30	.36	.43
10	.11	.18	.26	.33	.41	.49
16	.12	.20	.29	.37	.46	.54
Add'l	.01	.02	.03	.04	.05	.06

C. Pin stop   .02/pin

D. C-clamp   .01/clamp

E. Locator plug or fixture

No. Bolts	0	1	2	3	4
Under 45 lb	.02	.05	.08	.11	.15
Over 45 lb	.06	.09	.13	.16	.19

F. Adapter ring   .06

S3. Set tools
- Drilling .02
- Reaming .04
- Core-spade .04
- Boring .11
- Tapping .04
- Facing .02
- Presetting tool .25
- Load tool in turret or station .17/tool

S4. Angle block
- With gage stop .14
- Without gage stop .13

S5. Make piece (optional): 2 × hr/per 1 unit

S6. Tolerance checking (optional)

Tolerance	Hr
1/64 −.011	.03
.01 −.0051	.05
.005−.0031	.08
.003−.0011	.10
.001−	.13

Operation elements in estimating minutes

1. Pick up part, move and place; pick up and lay aside

   Single-table layout:

Position	Weight		
	To 45 lb	Over 45 lb to 50-in. Dia to 35-in. Perm	Over 45 lb over 50-in. Dia over 35-in. Perm
Align to pin stops	1.03	3.54	5.06
Align on locator plug	2.26	4.07	5.53

   Double-table layout:

Position	Weight		
	To 45 lb	Over 45 lb to 50-in. Dia to 35-in. Perm	Over 45 lb over 50-in. Dia over 35-in. Perm
Align to pin stops	.73	3.10	4.62
Align on locator plug	2.08	3.62	5.09

2. Clamp and unclamp

   Tighten clamp or strap

No.	1	2	3	4	5	6	7
Min	.53	1.06	1.50	2.11	2.64	3.17	3.70

   Set screw or C-clamp   1.19/ea
   Adjustable bottle jack   .36/ea

3. Part and fixture manipulation
   - Shuttle — .90
   - Reposition — 60% of basic time

4. Machining center operation
   - Start machine — .10
   - Rapid traverse — See Table 8.2-1
   - Automatic tool change — See Table 8.2-2
   - Manual tool change — .86
   - Keylock time — .17
   - Spindle on-time — .04
   - Tape reader factor — 1.00
   - Table index
     - Small machine, under 10 hp — .18
     - Large machine — .30
   - Coordinate change — .08

5. Cleaning, lubrication or wiping
   - Clean area with air hose — .18
   - Large fixture, air hose — .20
   - Brush chips, complex area and fixture — .26
   - Lubricate tool or piece — .14
   - Adjust coolant hose — .10
   - Clean and wipe locators or pins — .08

6. Machining
   - Face, side, slot, form, straddle, end — See Table 11.3
   - Drill, counterbore, ream — See Table 11.2
   - Bore — See Table 11.1

7. Milling cutter replacement — See Table 11.4-2

8. Part inspection — See Table 22.1-3
   - Indicate piece — 4.08

8.2 Rapid Travel and Automatic Tool Changer Elements

DESCRIPTION

Many machines have some form of rapid travel of a machine element and automatic tool changing. These data have been collected here because of convenience and the necessity of space reduction. For various reasons, other estimating data may have their own rapid travel and automatic tool changer data. For milling and machining centers, the data are compiled here, however.

ESTIMATING DATA DISCUSSION

The rapid traverse element consists of a pure machine element move. It is one direction only, and does not include any machine dwell, tarry, or stepping motor peculiarity. The entry variables are distance for the table, spindle, headstock, pallet, or other machine element. Velocity is expressed in minutes per inch (min/in.). The body of the table reads minutes per occurrence (min/occurrence). The time does not include provision for feeds or machining. The rapid-travel distance in elemental time value is calculated by assuming a distance from the loading position to the position where the machine feed starts. By applying the time values per inch from the chart, a total time value can be computed for the rapid travel "in." The specified rapid travel per inch is chosen by referring to the machine specification for the machine. The rapid travel "out" element for the estimate is found by applying the same per inch time value as used in the rapid travel "in." element. The total length of machine cut plus the original rapid travel in value is the distance used to compute the value.

Automatic tool changing varies machine to machine. Three situations are provided. "Prepositioned" involves tools that are accessible immediately for changing. Studies suggest that 10 hp is a convenient machine size to make another separation. Tool changing implies that the tool is removed from the spindle and another one inserted, which has been automatically preselected from the tool magazine.

TABLE 8.2 RAPID TRAVEL AND AUTOMATIC TOOL CHANGER ELEMENTS

Operation elements in estimating minutes

1. Rapid traverse of machine table, spindle, headstock or pallet, or other machine element, min/occurrence

Distance	Velocity, In./Min						
	25	50	100	150	200	300	400
1	.04	.02	.01	.01	.01	.003	.003
2	.08	.04	.02	.01	.01	.01	.01
4	.16	.08	.04	.03	.02	.01	.01
6	.24	.12	.06	.04	.03	.02	.02
8	.32	.16	.08	.05	.04	.03	.02
10	.40	.20	.10	.07	.05	.03	.03
15	.60	.30	.15	.10	.08	.05	.04
20	.80	.40	.20	.13	.10	.07	.05
25	1.00	.50	.25	.17	.13	.08	.06
30	1.20	.60	.30	.20	.15	.10	.08
40	1.60	.80	.40	.27	.20	.13	.10
50	2.00	1.00	.50	.33	.25	.17	.13
60	2.40	1.20	.60	.40	.30	.20	.15
100	4.00	2.00	1.00	.66	.50	.34	.26
200	8.00	4.00	2.00	1.33	1.00	.68	.50
300	12.00	6.00	3.00	2.00	1.50	1.00	.75
500	20.00	10.00	5.00	3.33	2.50	1.68	1.25

2. Automatic tool changing, min/occurrence

No. Changes	Prepositioned	10 hp or Less	Greater than 10 hp
1	.15	.30	.50
2	.30	.60	1.00
3	.45	.90	1.50
4	.60	1.20	2.00
5	.75	1.50	2.50
6	.90	1.80	3.00
7	1.05	2.10	3.50
8	1.20	2.40	4.00
9	1.35	2.70	4.50
10	1.50	3.00	5.00
11	1.65	3.30	5.50
12	1.80	3.60	6.00

No. Changes	Prepositioned	10 hp or Less	Greater than 10 hp
13	1.95	3.90	6.50
14	2.10	4.20	7.00
15	2.25	4.50	7.50
16	2.40	4.80	8.00
17	2.55	5.10	8.50
18	2.70	5.40	9.00
19	2.85	5.70	9.50
20	3.00	6.00	10.00
Add'l	.15	.30	.50

DRILLING

9.1 Drilling Machines Setup and Layout

DESCRIPTION

Setup estimating data are provided for the several drilling machines. They are used for job-shop and moderate-run production. For long-term production, setup is often an indirect expense and may not be estimated; it is charged as overhead expense. Time observations were taken for the basic elements in the distinctive machine applications.

In setup, the operator may work partially with indirect labor. This has an influence on the amount of setup. Despite this effect, a setup consists of the following elements.

1. Punch in/punch out
2. Get tools, fixtures, and later return
3. Study drawing and other information
4. Get parts and arrange
5. Setup table, spindles, tooling, NC controls
6. Make part(s), (optional to include)
7. Inspection approval, but wait not included
8. Clean up table when parts are finished

ESTIMATING DATA DISCUSSION

For setup evaluation, there are four major considerations: machine, part configuration, method of tiedown for tooling, and part variables (tolerances, operations required). These variables are often known by the estimator.

In Element 1, a basic setup is provided but exclusive of 6 and 7 above. Additional time beyond the number of spindles provided by the machine (i.e., 1, 2, etc.) is given in Element 5.

For turret machines, NC or otherwise, Element 2 is applied and the entry variable is the number of used turrets. The jigging and fixtures are averaged for the machine.

Multiple spindle or cluster drilling machines have as the entry variable the number of tool spindles. Special tooling for part handling accompanies this machine.

Element 5 provides optional selections such as piece part production and inspection. While piece part production of one, two, or three or more parts is necessary to effect a setup, the time for these setup units is also given in the operation elements. Inclusion in the setup is an optional choice by the estimator. Similarly, first part inspection may or may not be required for all parts. Exceptions may be critical-tolerance parts, parts with a history of being difficult, or expensive materials.

Element 6 is the setup and making of a drill templet. It is not included in any of the above elements. It includes study print; get equipment; figure hole locations; make templet blank; and locate, mark and drill holes.

Making of sheet metal templets for punching,

notching, and blanking operations is estimated by Element 7. It includes study print; clean blank and paint; and scribe 2 lines center punch; and swing circle for each hole. For a notch or brake line, a line is scratched. Part is sheared to final size prior to this work.

Elements 8 and 9 are the layout of locations and a centerpunch for each hole on a part itself. It may be desirable to consider this work as an operation, since one or several parts may be produced in this fashion.

EXAMPLES

These data are demonstrated with the drilling machine tables.

TABLE 9.1 DRILLING MACHINES SETUP AND LAYOUT

Setup in estimating hours

S1. Sensitive drill press and upright drilling machines

Jig or Fixture	No. of Spindles					
	1	2	3	4	5	6
On table or vise	.17	.26	.34	.41	.47	.54
Plate or sandwich	.19	.28	.36	.43	.48	.56
Box or collet	.24	.33	.41	.47	.54	.61
Air chuck, collet	.20	.29	.37	.44	.50	.57
Parallels, V-block	.19	.27	.35	.42	.48	.55

S2. Turret drilling machines

No. Turrets	1	2	3	4	5	6	7	8
Jig, Fixture, Vise, Rails	.25	.33	.40	.47	.55	.62	.69	.76

S3. Cluster spindle machine

No. Tool Spindles	1	2	3	4	5	6	7	8	9	10	Add'l
Hr	.08	.17	.25	.33	.42	.48	.58	.67	.75	.83	.08

Install jig or fixture on table, clean, align	1.00
Critical depth	.10/hole

S4. Radial drill press machine

On table or vise or floor plates	.28
Angle plate	.45
Fixture	.40
Crane loading	.05

S5. Miscellaneous elements

Piece production; 2 × unit time	Optional
First part inspection	Optional
Constant	.03
Location coordinate	.02
Hole size	.003
Add'l hole sizes above number of spindles	.06/ea
Tap or magic chuck attachment	.02/ea
Depth on hole	.05/spindle

S6. Lay out and make drill templet

A.

No. Lines	1	2	3	4	5	6	7	8	9	10	Add'l
Hr	.29	.31	.33	.35	.37	.39	.41	.43	.45	.47	.02

B. Lay out flat sheet and make sheet metal templet

No. Holes	1	2	3	4	5	6	7	8	9	10	Add'l
Hr	.04	.06	.07	.08	.10	.11	.12	.13	.15	.16	.02

C. Lay out flat sheet for notches, brake lines, angles on templet

No. Lines	1	2	3	4	5	6	7	8	9	10	Add'l
Hr	.01	.02	.03	.04	.05	.06	.07	.08	.09	.10	.01

D. Lay out flat parts and machine center-punch hole locations, per occurrence

No. Holes	1	2	3	4	5	6	7	8	9	10	Add'l
Hr	.08	.10	.11	.13	.17	.18	.21	.23	.25	.27	.02

E. Lay out irregular shaped parts and machine center-punch hole locations, per occurrence

No. Lines	1	2	3	4	5	6	7	8	9	10	Add'l
Hr	.13	.15	.18	.19	.22	.24	.27	.28	.31	.34	.03

9.2 ~~Sensitive~~ MANUAL Drill Press Machines

DESCRIPTION

The sensitive drilling machine is a high-speed machine of simple construction similar to the upright drill press. Machines of this type are hand-fed, usually by means of a rack and pinion or involute spline drive on the sleeve holding the rotating spindle. These drills may be driven directly by a motor, belt, friction disk, etc. The machines may have from one to eight spindles. Gearing or belt cones may provide from 4 to 16 speeds.

In Figure 9.2, the motor is an enclosed direct drive, 4-speed motor. Speeds can be changed without stopping the motor.

ESTIMATING DATA DISCUSSION

Element 1 comprises the basic load and unload of the part. Entry variables are weight and type of holding device. Obviously, as the part becomes heavier or lighter, the type of handling is changed and this factor is evaluated in the data. If the part is disposed by a toss-aside, a deduction is provided. If additional parts are multiple-loaded, time for full handling is given for each part.

Once the part is loaded, Element 2 provides for tighten and untighten, and several conditions are specified. Element 3 deals with jib or part and jib manipulation.

FIGURE 9.2 Four-spindle sensitive drilling machine. (Leland-Gifford Company, Division of Fayscott, Incorporated)

The operation of the sensitive drill press is given in Element 4, and a variety of work is described. Once the part is loaded and clamped, it is positioned under the drill. Following the last hole, it is moved away from the spindle for part removal. Thus one hole less is required for hole-to-hole positioning. The selection of these elements depends upon whether the drill press is a single- or gang-spindle machine.

Element 5 deals with part, jig, or fixture and machine cleaning and lubricating by several methods.

In Element 6, the manual drilling elements are used as the sensitive drill press machine does not have power feed.

EXAMPLES

A. A 3-in. long, ½-in. OD low-carbon steel pin has a ¹⁄₁₆-in. OD cross-hole drilled through. A jig is used. Tolerances are commercial. The part weighs .2 lb. Find the min/unit.

Table Number	Process Description	Table Time	Adjustment Factor	Cycle Minutes	Setup Hours
9.2-1A	Pick up part and lay aside	.05		.05	
9.2-2	Two thumb screws	.06	2	.12	
9.3-4A	Raise and lower spindle	.02		.02	
11.2-1	Drill 1 hole, ¹⁄₁₆-in. OD × ½ in. deep	.14		.14	
9.2-5A	Blow jig clean	.06		.06	
9.2-5A	Oil tool	.05		.05	—
	Total			.44	

B. An aluminum fitting, 2¾ in. in diameter, has 16 pilot holes, 0.098 in. in diameter, for a distance of ⁵⁄₁₆ in. The part weights less than 1 lb. A fixture with locating pins is used. Estimate the small lot unit time, hr/100 units, and shop time.

Table Number	Process Description	Table Time	Adjustment Factor	Cycle Minutes	Setup Hours
9.2-1A	Pick up part and lay aside	.08		.08	
9.2-2	Quick-acting clamp	.05		.05	
9.2-4C	Position work under drill	.04		.04	
11.2-1	Drill 16 holes	.10	16	1.60	
9.2-4D	Raise tool, position hole to hole, 15 times	.99		.99	
9.2-4C	Clear drill from work	.04		.04	
9.2-5A	Blow jig, medium area	.11		.11	
22.1-3	No checking time for pilot holes	0		0	
11.4-3	No drill life and replacement	0		0	—
	Total			2.91	
	Hr/100 units	4.85			
	Shop time	20.60 pc/hr			

C. A two-spindle sensitive drill press is used to drill and tap an irregular steel casting. One spindle is used for drilling, and the second is used for tapping. One ¼-20 hole is drilled and tapped for a ½-in. depth while two ⅜-in. holes are drilled only. A standard drill jig is used. Weight is 8.2 lb. One reposition of tool and part is necessary. Find the lot time for 18 parts.

Table Number	Process Description	Table Time	Adjustment Factor	Cycle Minutes	Setup Hours
9.1-S1	Setup two spindles	.33			.33
9.2-1A	Pick up part and lay aside	.18		.18	
9.2-2	Toggle clamp	.05		.05	
9.2-4A	Raise and lower spindle for first hole	.02		.02	
9.2-4A	Add'l spindle operator reposition	.01		.01	
9.2-4B	Change tool, quick acting collet	.09		.09	
11.2-1	Drill ¼-in. holes, ½ in. deep	.72		.72	

Table Number	Process Description	Table Time	Adjustment Factor	Cycle Minutes	Setup Hours
11.2-1	Drill two ⅜-in. holes, ½ in. deep	.92	2	1.84	
9.2-3A	Reposition tool and part	.02		.02	
9.2-4D	Position hole to hole, 4 times	.26		.26	
11.2-8	Tap ¼-20	.15	½	.08	
9.2-5A	Brush chips for medium area	.24		.24	
9.2-5B	Blow out chips from casting cavity	.11		.11	
11.4-3	Tool wear out	.05		.05	—
	Total			3.67	.33
	Lot estimate	1.43 hr			

TABLE 9.2 SENSITIVE DRILL PRESS MACHINES

Setup See Table 9.1

Operation elements in estimating minutes

1A. Pick up part, move and place, pick up and lay aside

Holding Device	Weight					
	.5	1.0	2.5	5.0	10.0	15.0
Table, parallels, open vise, simple fixture	.05	.07	.09	.12	.14	.17
V-block, table between hold-down clamps, fixture with locating pins or blocks, between centers	.06	.08	.10	.13	.16	.19
Chuck, complex clamping, or fixture with locating pins or blocks to fit part	.07	.08	.11	.14	.18	.22
Blind fixture with locating pins or blocks or stops, index plate with alignment, or align to mark in fixture	.07	.09	.12	.15	.20	.24

B. Deduct for toss aside .02

C. For additional part for handling in Element 1

Weight	2	5	10	15
Min	.05	.07	.08	.10

D. Pry or tap part out .06

2. Clamp and unclamp
  - Vise, ¼ turn .05
  - Vise, full turn .08
  - Toggle or quick-acting clamp .05
  - Cam clamp .03
  - Star wheel .07
  - Thumb screw .06
  - Air cylinder .05
  - U-clamp with hex nut .26
  - C-clamp with hex nut .32
  - C-clamp with thumb screw .21
  - Strap clamp with hex nut .31
  - Add'l nut .05

3. Part or part and jib manipulation

   A. Rotate part or part and fixture

	Weight			
Rotate, Flip	10	20	30	45
90°	.01	.02	.03	.04
180°	.02	.04	.05	.07

   B. Trunnion fixture

Rotate	90°	180°	270°
Min	.07	.08	.10

   Pin, unpin .07
   Turn part in vise .10

4A. Sensitive drill press machine operation

   Start or stop spindle, clutch type .03
   Start or stop spindle, button or pull type .01
   Raise and lower spindle .02
   Spindle clamp, lock and unlock .05
   Bushing, place and remove .11
   Bushing, shift hole-to-hole .08
   Change pulley belt speed .30
   Set depth pointer or dial marker .04
   Table clamp, lock and unlock .07
   Rotate table on column .05
   Walk back to first spindle (for progessive spindles) .01

   B. Change Tools

Change Tools	1	2	3	4	5	Add'l
Quick-change collets	.09	.18	.27	.45	.45	.09
Keyed-drill chuck	.30	.60	.90	1.20	1.50	.30
Taper shank with drift	.35	.70	1.05	1.40	1.75	.35

   C. Position work under drill, adjust coolant, advance drill to work, align to work or bushing, raise spindle after machining

Weight	5	10	30	45
Min	.04	.05	.06	.07

   D. Raise tool, position work hole-to-hole, advance tool to work

Holes	Min	Holes	Min	Holes	Min	Holes	Min
1	.07	6	.40	11	.73	16	1.06
2	.13	7	.46	12	.79	17	1.12
3	.20	8	.53	13	.86	18	1.19
4	.26	9	.59	14	.92	19	1.25
5	.33	10	.66	15	.99	Add'l	.06

5. Cleaning, lubricating, or wiping

   A. Clean with air hose
      Small area up to 6 × 6 .06
      Medium area to 12 × 12 .11
      Large area to 12 × 24 .14
      Small fixture .10
      Complex large fixture .18
   Brush chips
      Small area up to 6 × 6 .11
      Medium area to 12 × 12 .13
      Large area to 12 × 24 .18
      Fixture .24

Lubricate tool or piece
    Oil tool or piece with brush   .05
    Dip tool in compound   .02
Adjust coolant hose   .04
Clean area, holding device, or hole
    Wipe locators with fingers   .03
    Wipe 6 × 6-in. area with hand   .04
    Wipe V-block or centers with rag   .11
    Wipe vise with rag   .10

B. Blow lube and chips from hole

Holes	1	2	3	4	5	6	8	9	10	Add'l
Min	.06	.08	.11	.13	.15	.17	.21	.24	.26	.02

6. Machining
    Drill, counterbore, ream, or tap   See Table 11.2

7. Drill replacement   See Table 11.4-3

8. Part inspection   See Table 22.1-3

9.3 Upright Drilling Machines

DESCRIPTION

Upright drilling machines have power feeding mechanisms (and non-powered, as well) for the rotating drill. They are designed for heavier work than the sensitive drilling machines. The column is box-type, and it is more rigid than a round column machine. When the uprights can be grouped on a single table, it is called a gang drill. Tapping can be handled as well.

In the model shown in Figure 9.3, the maximum capacity is 7½ hp. Eight quick-change gears are available and the machine can be ganged in 2 to 6 spindles. Various gear speeds are available, anywhere from 32 to 708 for gear feeds ranging from .055 to .085 ipr.

Now consider estimating data for the broad class of upright drilling machines.

ESTIMATING DATA DISCUSSION

Many elements are common in the upright drilling machine and the sensitive drill press machine.

Entry variables for Element 1 are weight and type of holding device. But, the weight range is higher than was found in Table 9.2 as these machines do drilling of larger parts, although small work is possible. Element 2 is for tighten and untighten and is per occurrence.

Some of the elements of machine operation depend upon the particular machine being estimated. The selection of drill, counterbore, ream, or tap will use power feeds of Table 11.2.

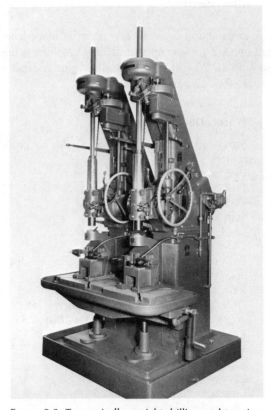

FIGURE 9.3 Two-spindle upright drilling and tapping machine with 7½ hp capacity. *(Barnes Drill)*

EXAMPLES

A. An aluminum casting has four holes drilled. Part weight is 25 lb, and a fixture with locating pins and 2 star clamps are required. The drill schedule is given below. The part is not considered critical in inspection. A 4-spindle machine is used. Find the unit estimate showing a fuller extension of the elements.

Table Number	Process Description	Table Time	Adjustment Factor	Cycle Minutes	Setup Hours
9.1-S1	Box jig and three active spindles	.41			.41
9.1-S5	Piece production, first-part inspection	Not allowed			0
9.3-4A	Start and stop, push button	.01		.01	
9.3-1A	Fixture with locating pins	.23		.23	
9.3-2	2 star clamps	.09	2	.18	
9.3-4A	Position under drill, first hole	.09		.09	
11.2-4	Drill one ⁵⁄₁₆-OD hole × ¾-in. deep (L includes lead and break through)	.07	.75	.05	
9.3-4C	Position to next hole, advance drill	.11		.11	
11.2-4	Drill 2nd ⁵⁄₁₆-OD hole × ¼ in. deep	.07	.25	.02	
9.3-4C	Raise tool into clear, and reposition, advance tool	.11		.11	
9.3-5A	Blow chips, small area	.06		.06	
9.3-3A	Tumble jig, 45 lb or less	.07		.07	
9.3-4B	Install and remove a slip bushing	.08		.08	
11.2-4	Drill ¼ in. through ⁵⁄₁₆-in. deep	.05		.05	
9.3-4C	Raise drill, position hole-to-hole, advance drill	.11		.11	
11.2-4	Drill ⅝-in. hole to depth of 1½ in.	.10	1.5	.15	
9.3-4B	Raise drill into clear, move work into clear	.09		.09	
9.3-5A	Blow jig clean to receive next load	.10		.10	
9.3-5A	Blow table clean to receive next load	.11		.11	
22.1-3	Unit time to check part	Not allowed		0	
11.4-3	Unit time to sharpen drills, .01 + .01	.02		.02	
	Total			1.64	.41
	Unit estimate	1.64 min			

B. A large stainless steel valve stem is loaded between centers, and a drilling cap is clamped on. A 2-spindle upright drilling machine is used to drill and ream 4 holes and drill and tap 3 holes. Weight is 12 lb, and the part is expected to cause some difficulty. Find the lot estimate for 163 parts.

Table Number	Process Description	Table Time	Adjustment Factor	Cycle Minutes	Setup Hours
9.1-S1	Setup 2 spindle, 12 lb, estimate as value not given for between centers	.30			.30
9.1-S5	Piece production	Not allowed			0
9.1-S5	First-part inspection for 4 reamed holes, .03 + 4 × .02 + 4 × .003	.12			.12
9.3-4A	Start and stop	.03		.03	
9.3-1A	Load and unload part between centers	.19		.19	
9.3-2	2 U-clamps	.26	2	.52	
9.3-4B	Position under drill, adjust coolant, advance drill to work, first hole	.09		.09	
9.3-4C	Raise tool, hole-to-hole, advance tool, 13 times	.11	13	1.43	
11.2-4	Drill four .730-in. holes × 2.62-in. deep, lead and break through = .39 × ⅔ = .26, total L = 2.62 + .26 = 2.83	.49	4 × 2.88	5.65	
11.2-7	Ream 4 holes .75-OD	.55	4 × 2.62	5.76	
11.2-4	Tap drill three ²¹⁄₆₄ in.-holes, 1.5-in. deep (1.5 + ⅔ × .18) in.	.39	3 × 1.62	1.90	
11.2-8	Tap ⅜ = 24, 1.5-in. deep	.30	3 × 1.5	1.35	
9.3-4B	Change 3 tools on one spindle, 3 times	.09	3	.27	
9.3-4A	Change speed on one spindle, 3 times	.02	3	.06	
9.3-4A	Change feed, 3 times	.04	3	.12	
9.3-4B	Clear drill after drilling and tapping	.09		.09	

Table Number	Process Description	Table Time	Adjustment Factor	Cycle Minutes	Setup Hours
9.3-3A	Roll stem over, 3 times	.04	3	.12	
9.3-5A	Lubricate drill, average of 4 times	.05	4	.20	
9.3-5A	Lubricate tap	.05	3	.15	
9.3-5A	Blow jig and piece clean, .11 + .10	.21		.21	
22.1-3	Unit time to check piece, 4 holes with .002-OD tolerance, go, no-go	.16	4 × ⅕	.13	
11.4-3	Drill sharpen allowance				
	.730-in. drill, 5.65 min	.44		.44	
	.750-in. reamer, 5.76 min	.44		.44	
	²¹⁄₆₄-in. drill, 1.90 min	.17		.17	
	⅜-24 tap, 1.35 min	.17		.17	—
	Total			19.49	.42
	Lot estimate	53.37 hr			

TABLE 9.3 UPRIGHT DRILLING MACHINES

Setup   See Table 9.1

Operation elements in estimating minutes

1A. Pick up part, move, and place; pick up and lay aside

Holding Device	Weight					
	5	10	15	25	35	45
Table, parallels, open vise, simple fixture	.12	.14	.17	.18	.20	.22
V-block, table between hold-down clamps, fixture with locating pins or blocks, between centers	.13	.16	.19	.21	.23	.25
Chuck, complex clamping, or fixture with locating pins or blocks to fit part	.14	.18	.22	.23	.26	.28
Blind fixture with locating pins or blocks or stops, index plate with alignment, or align to mark on fixture	.15	.20	.24	.27	.28	.33

B. Deduct for toss aside   .03

C. For additional part

Weight	5	10	15	30
Min	.07	.08	.10	.13

D. Pry or tap part out   .07

2. Clamp and unclamp
    Vise, ¼ turn   .06
    Vise, full turn   .09
    Toggle or quick-acting clamp   .06
    Cam clamp   .04
    Star wheel   .09
    Thumb screw   .09
    Air cylinder   .05
    U-clamp with hex nut   .26
    C-clamp with hex nut   .32
    Clamp with thumb screw   .21
    Strap clamp with hex nut   .31
    Add'l nut   .08
    Operate foot air clamp   .02

3. Part or part and jig manipulation

   A. Rotate part or part and fixture

	Weight				
Rotate, Flip	10	20	30	45	60
90°	.01	.02	.03	.04	.05
180°	.02	.04	.05	.07	.08

   B. Trunnion fixture

Rotate	90°	180°	270°
Min	.07	.08	.10

   Pin, unpin .07
   Turn part in vise .10

4. Upright drilling machine operation

   A. Start or stop spindle, clutch type .03
   Start or stop spindle, button or pull type .01
   Reverse spindle, friction clutch .02
   Reverse spindle, positive clutch .01
   Reverse spindle, motor clutch .01
   Raise or lower spindle .03
   Spindle clamp, lock and unlock .05
   Change speed, single lever .02
   Change speed, double lever .05
   Change feed, single lever .04
   Change feed, double lever .06
   Engage feed, clutch type .01
   Engage feed, hand-feed lever type .01
   Set depth pointed or dial marker .04
   Spindle clamp, lock and unlock .05
   Table clamp, lock and unlock .07
   Walk back to first spindle .01

B. Change tools	1	2	3	4	5	Add'l
Quick-change collets	.09	.18	.27	.36	.45	.09
Keyed-drill chuck	.30	.60	.90	1.20	1.50	.30
Taper shank with drift	.35	.70	1.05	1.40	1.75	.35

   Bushing, place and remove: .11, shift hole-to-hole .08
   Position work under drill, adjust coolant, advance drill to work, first hole, or reverse .09

   C. Raise tool, position work hole-to-hole, advance tool to work:

Holes	Min	Holes	Min	Holes	Min
1	.11	5	.55	9	.99
2	.22	6	.66	10	1.10
3	.33	7	.77	11	1.21
4	.44	8	.88	Add'l	.11

5. Cleaning, lubricating, or wiping

   A.
   Clean area with air hose
      Small area up to 6 × 6 .06
      Medium area to 12 × 12 .11
      Area to 12 × 24 .14
      Large area to 18 × 36 .25
      Small fixture .10

Complex large fixture	.18
Brush chips	
Small area up to 6 × 6	.11
Medium area to 12 × 12	.13
Area to 12 × 24	.18
Large area to 18 × 36	.31
Fixture	.24
Lubricate tool or piece	
Oil tool or piece with brush	.05
Dip tool in compound	.02
Adjust coolant hose	.04
Clean area, holding device or hole	
Wipe locators with fingers	.03
Wipe 6 × 6-in. area with hand	.04
Wipe V-blocks or centers with rag	.11
Wipe vise with rag	.10

B. Blow lube and chips from hole

Holes	1	2	3	4	5	6	7	8	9	10	Add'l
Min	.06	.08	.11	.13	.15	.17	.19	.21	.24	.26	.02

6. Machining
    Drill, counterbore, ream, tap   See Table 11.2

7. Drill replacement   See Table 11.4-3

8. Part inspection   See Table 22.1-3

9.4 Turret Drilling Machines

DESCRIPTION

The outstanding characteristic here is the turret. Anywhere from six- to ten-spindle turret head drilling machines are available. Control of the turret, meaning "index, advance, and after drilling, retracting" can be manual or automatically controlled. These machines, if NC or hydraulically or stepped controlled, provide directional positioning in the plane of the table. Z-axis feed and depth control are available. A DC motor provides individual, per spindle, infinitely variable feed rates for the Z-axis. The feed rates, rapid approach points, and depth stops are adjustable. A separate cam for each spindle selects the point at which the tool goes from rapid approach to the desired preset feed rate.

Numerical control is a feature frequently found with these machines. The NC often uses 1-in. wide, 8-channel tape using word address and variable length block format. The positioning control is two-axis simultaneous open loop.

Figure 9.4 is an example of an all-electric, automatic, NC turret machine designed for production, drilling, tapping, and boring. The table size is 24 × 40 in., and it is a bed-type allowing for heavy shuttle tables.

ESTIMATING DATA DISCUSSION

While elements are common in all drilling machines, distinctions in time are required in turret drilling machines because of differences in the mode of machine control, whether it has manual, semiautomatic, or numerical control. Additionally, table size and the convenience or inconvenience of NC controls affect handling.

Element 1 deals with handling and intends to move the part from a tote box, or skid, to the table or holding device, and then return it. For a flat stack item there is a deduction to be made from the table values, while for additional nonstacked parts, there is a lower handling time.

Once the part is loaded in or on the holding device, several methods of clamping and unclamping are listed in Element 2. Part or part and jig manipulation

FIGURE 9.4 NC turret drill press. *(Burgmaster)*

has entry variables: weight, degree rotation, and type of fixture. Elements 1, 2, and 3 are usually under non-NC machine control and are done during operator control of the machine.

Machine operation is considered by Element 4. Change tool deals with removing and inserting tool in a turret station holder during a stop in the cycle. If a drilling machine has two turret heads, movement of the table is considered. The element raise tool, position to new hole, and advance does not include any drilling time and Table 11.2 is used. For sensitive turret drilling machines, Elements 1 and 2 are used. For powered turret drilling machines, the other elements of Table 11.2 are considered. Element 5 considers cleaning, lubricating, or wiping of the table, tool, or fixture.

EXAMPLES

A. An aluminum flat part is stacked drilled in 37 locations. Each part weights 2.7 lb and is ⅛-in. thick and loaded against a rail clamp-type fixture. The stack height is ⅝-in. thick. Four turret stations are used in the NC turret drilling machine. Locations and hole sizes are not critical. A lot quantity of 600 parts is to be estimated for unit, hr/100 units, and pc/hr.

Table Number	Process Description	Table Time	Adjustment Factor	Cycle Minutes	Setup Hours
9.1-S1	Turret drilling machine setup	.47			.47
9.1-S1	Piece production, first-part inspection	Not allowed			0
9.4-4A	Start tape	.02	1/7	.003	
9.4-1A	Load on table, between hold downs, .13, but deduct for flat part .03 = .10, 7 parts are loaded	.10		.10	
9.4-2	2 clamp screws	.21	2/7	.06	
8.2-1	Rapid traverse at 200 ipm, 8 in.	.04	1/7	.006	
9.4-4C	Index 2 blank turret stations	.06	1/7	.009	
11.2-4	Drill twelve .187-in. holes, distance = .10 + .875 = .975 in., drill time for 12 holes	.09	12/7 × .975	.15	
11.2-4	Drill three .4375-in. holes, distance = .24 + .875 = 1.12-in., drill time for 3 holes	.09	3/7 × 1.12	.043	
11.2-4	Drill seven .516-in. holes, distance = .28 + .875 = 1.16-in., drill time for 7 holes	.08	7/7 × 1.16	.093	
11.2-4	Drill fifteen 1/8-in. holes, distance = .07 + .875 = .95 in., drill time for 15 holes	.07	15/7 × .95	.143	
9.4-4B	Position hole-to-hole 36 times	.06	36/7	.31	
11.4-3	Drill replacement				
	1/4 in. or less, drill time = 1.05 + .99	.15	1/7	.021	
	1/4 + to 1/2 in., drill time = .3	.02	1/7	.003	
	1/2 + to 3/4 in., drill time = .65	.03	1/7	.004	
	Total			.945	.47
	Stack estimate	6.615 min			
	Hr/100 units	1.575			
	Shop estimate	63 pc/hr			

B. A gray cast iron 125 psi 4-in. tee casting has a 6-hole bolt circle drilled and tapped on 3 flange faces. The turret drilling machine has the fixture positioned and mounted on the table. Two turrets are used. Location and hole size are ordinary. The tee casting weighs 18 lb. The fixture is indexed twice to present 2 other flanges. Find the unit estimate.

Table Number	Process Description	Table Time	Adjustment Factor	Cycle Minutes	Setup Hours
9.1-S2	Turret drilling machine setup	.33			.33
9.1-S5	Tap attachment for turret	.02			.02
9.4-1A	Load and unload to indexing fixture with alignment	.27		.27	
9.4-2	2 cam clamps	.03	2	.06	
9.4-4A	Start tape	.02		.02	
8.2-1	Rapid traverse 8-in. to first position	.04		.04	
11.2-4	Drill 1-in. hole, 6 locations on 1 flange, distance = .53 + 1.25 = 1.78 in.	.27	6 × 1.78	2.88	
9.4-4C	Index 1 station	.03		.03	
11:2-6	Tap 6 holes, 8 threads/in.	.32	6 × 1.25	2.40	
9.4-4B	Raise tool, position hole-to-hole, advance tool to work for 12 drilled and tapped holes	.66		.66	
	Index empty turrets in above element				
8.2-1	Rapid traverse 200 ipm 8 in. to clear turret for indexing of fixture and tape stops	.04		.04	
9.4-3A	Index fixture to present new flange	.05		.05	
Repeat	Machine second flange face using above 8 elements	6.22		6.22	
Repeat	Machine third flange	6.22		6.22	
9.4-5A	Blow off table	.25		.25	
11.4-3	Drill replacement				
	1-in. drill, 2.88 × 3 = 8.64, .25 + .20	.45		.45	
	1-in. tap, 2.40 × 3 = 7.20, .25 + .15	.40		.40	
	Total			19.79	.35

C. An airframe part is to be turret drilled using an NC drill press. A picture of the part is given by the partially dimensioned picture in Figure 6.4C. The lot quantity is 7075 units. The adjustment factor may be number of flips to manipulate the part and tooling or the depth of the hole, or the number of holes. A drill length is composed of the hole depth, safety stock, and the overtravel of the drill point. The overtravel of the drill point is about one-half of the diameter. Find the estimates.

Table Number	Process Description	Table Time	Adjustment Factor	Cycle Minutes	Setup Hours
	Drill and tap 4 holes				
9.1-S2	Turret machines	0.33			0.33
9.4-1A	Pick Up, move	0.15		0.15	
9.4-2	Clamp, unclamp	0.05		0.05	
9.4-3A	Manipulation; 4 flips for holes on opposite sides	0.01	4	0.04	
9.4-4C	Turret	0.72		0.72	
	Drill and tap with two different turret; rotation is for two holes, drill, and tap				
11.2-4	Drill 4 holes, depth of hole .625 in	0.20	.625	0.13	
	3 add'l holes	0.13	3	0.39	
11.2-8	Tap 4 holes				
	Depth of tap, .5	0.19	.5 × 4	0.038	
11.4-3	Drill tool life	0.10		0.10	
	Drilling time less than .90 min				
	Total			1.96	0.33
	Lot hours	231.45			

D. A die cast part is to be NC drilled on a machine similar to Figure 9.4. The part is described by Figure 2.9C. The part has been deflashed and deburred. Quantity is 500. Find the estimates. The adjustment factor is the length of tool travel and may include the safety stock, design requirement for depth, and overtravel. The adjustment factor may also be the number of holes for that size. Material is magnesium.

Table Number	Process Description	Table Time	Adjustment Factor	Cycle Minutes	Setup Hours
	Numerical control drill press operation				
11.2-7	Power reaming 2.85-in. hole cleanup, .09 in. of metal	0.10	.09	0.01	
11.2-4	Power drilling, .19 in. wall, 8 holes	0.07	.19 × 8	0.01	
11.2-3	Countersinking	0.03	8	0.24	
11.2-8	Tapping 40 NC thread for 8 holes	0.12	.19 × 8	0.02	
11.2-4	Drill .175-hole 2.57-in. deep, 4 bodyholes	0.07	2.57 × 4	0.18	
11.2-4	Drill #50 holes, 4× through wall .09 in.	0.07	.09 × 4	0.01	
9.4-1A	Pick up, move	0.08		0.08	
9.4-2	Clamp, unclamp	0.05		0.05	
9.4-3A	Manipulation for 3 surfaces by flipping tool on table top	0.01	3	0.03	
9.4-4A	Machine operation; start, stop for flip	0.02	3	0.06	
9.4-4B	Position table for 19 holes	1.05		1.05	
9.4-4B	13 add'l holes	0.06	13	0.78	
9.4-4C	Turret index for 6 tools	0.54		0.54	
9.4-5A	Clean area, medium	0.11		0.11	
9.4-5A	Clean, lubricate	0.05		0.05	
9.4-5B	Blow from 8 tap holes	0.21		0.21	
9.1-S2	Turret machines	0.62			0.62
9.1-S5	Inspection	0.03			0.03
9.1-S5	Miscellaneous hole size	0.003	3		0.01
9.1-S5	Depth on hole	0.05	4		0.20
	Total			4.24	0.86
	Lot hours	36.19 hr			

TABLE 9.4 TURRET DRILLING MACHINES

Setup See Table 9.1

Operation elements in estimating minutes

1A. Pick up part, move, and place; pick up and lay aside.

Holding Device	Weight					
	1.0	2.5	5.0	10	15	25
Table, parallels, open vise, simple fixture	.07	.09	.12	.14	.17	.18
V-block, table between hold-down clamps, fixture with locating pins on blocks, between centers	.08	.10	.13	.16	1.9	.21
Chuck, complex clamping or fixture with locating pins or blocks to fit part	.08	.11	.14	.18	.22	.23
Blind fixture with locating pins or blocks or stops, index plate with alignment, or align to mark on fixture	.09	.12	.15	.20	.24	.27

B. Deduct for flat stack parts
   Deduct for toss aside   .03

C. For additional part (nonstacked)   .02

Weight	2	5	10	15	25
Min	.05	.07	.08	.10	.12

D. Pry or tap part out   .07

2. Clamping and unclamping
   - Vise, ¼ turn   .05
   - Vise, full turn   .08
   - Toggle or quick-acting clamp   .05
   - Cam clamp   .03
   - Star wheel   .07
   - Thumb screw   .06
   - Air cylinder   .05
   - U-clamp with hex nut   .26
   - C-clamp with hex nut   .32
   - C-clamp with thumb screw   .21
   - Strap clamp with hex nut   .31
   - Add'l nut   .05
   - Operate foot air clamp   .02

3. Part or part and jig manipulation
   A. Rotate part or part and fixture

Rotate, Flip	Weight			
	10	20	30	45
90°	.01	.02	.03	.04
180°	.02	.04	.05	.07

B. Trunnion fixture

Rotate	90°	100°	270°
Min	.07	.08	.10

Pin, unpin	.07
Turn part in vise	.10
Index of fixture	.15/stop

Rotation of rotary table: 90°, .07; 180°, .12

4. Turret drilling machine operation

   A. Change tool .06
   Move part turret to turret, for multiple turret model .06
   Semiautomatic job table .01/in.
   Start machine or start tape .02

   B. Raise tool, position work to new X–Y coordinate hole, advance tool to work

Holes	Min	Holes	Min	Holes	Min	Holes	Min
1	.06	6	.33	11	.61	16	.88
2	.11	7	.39	12	.66	17	.94
3	.17	8	.44	13	.72	18	.99
4	.22	9	.50	14	.77	19	1.05
5	.28	10	.55	15	.83	Add'l	.06

   C.

Turret	1	2	3	4	5	6	7	8
Index	.03	.06	.09	.12	.15	.18	.21	.24
Retract, index, advance tool	.09	.18	.27	.32	.45	.54	.63	.72

5. Cleaning lubricating or wiping

   A. Clean area with air hose
       Small area up to 6 × 6 .06
       Medium area up to 12 × 12 .11
       Area up to 12 × 24 .14
       Large area up to 18 × 36 .25
       Small fixture .10
       Complex large fixture .18
   Brush chips
       Small area up to 6 × 6 .11
       Medium area up to 12 × 12 .13
       Area up to 12 × 24 .18
       Large area up to 18 × 36 .31
       Fixture .24
   Lubricate tool or piece
       Oil tool or piece with brush .05
       Dip tool in compound .02
       Adjust coolant hose or mist .04
   Clean area, holding device or hole
       Wipe locators with fingers .03
       Wipe 6 × 6-in. area with hand .04
       Wipe V-blocks or centers with rag .11
       Wipe vise with rag .10

B. Blow lube and chips from hole

Holes	1	2	3	4	5	6	7	8	9	10	Add'l
Min	.06	.08	.11	.13	.15	.17	.19	.21	.24	.26	.02

6. Machining
     Drill, counterbore, ream, tap   See Table 11.2

7. Rapid traverse   See Table 8.2-1

8. Drill replacement   See Table 11.4-3

9. Part inspection   See Table 22.1-3

9.5 Cluster Drilling Machines

DESCRIPTION

Sometimes called multi-spindle drilling, this method can have a special drilling attachment which fits to the spindle quill. Multi-spindle drilling machines are also available. The attachment spindles are driven through universal joints and telescoping splined shafts. Various speeds for drilling, reaming, and tapping can be adapted in the headstock by pick-off change gears. Slip spindles handle either drilling or tapping. Cams control rapid traverse and feed length.

A cluster spindle machine shown in Figure 9.5 has a maximum feedstroke of 18 in. and feeding pressure of 16,500 lb. The range of feeding is 1–20 ipm. The number, speed, and feed of the drills depend upon the material being drilled and the particular model.

ESTIMATING DATA DISCUSSION

Handling is listed by Element 1 and is explained under sensitive drilling machines. Element 2 covers clamping and unclamping and is per occurrence. If a rotary table is used, Element 3 would be used. The operation of the multi-spindle machine is treated in Element 4. Rapid-traverse of the headstock or table is found in Table 8.2. The selection of the various elements in Element 4 depends on the automatic cycles incorporated in the machine. Inasmuch as several drills, taps, or reamers are working simultaneously, machine time is different for one tool working, and mutli-spindle machine time is given in Table 11.2, Element 5. Drill replacement and part inspection are calculated. They depend upon cutting time, hole location, and size tolerance.

EXAMPLES

A. A plate approximately 4-in. wide × 8-in. long and ⅜-in. thick is commercial yellow brass. Drill ten ⁵⁄₁₆-in. holes through 2 plates loaded in a simple locating fixture, but the fixture has locating bushings. Tolerances are commercial and operating checking allowance is not required. Find unit time.

Table Number	Process Description	Table Time	Adjustment Factor	Cycle Minutes	Setup Hours
9.1-S3	Setup multiple-spindle drilling attachment	.83			.83
9.1-S1	Install fixture on table	1.00			1.00
9.5-1A	Plate weighs 3.70 lb and is loaded to simple fixture	.12	½	.06	
9.5-1B	Second plate	.07	½	.035	
9.5-2	3 thumb screws	.18	½	.09	
9.5-4	Position under drills and advance drills to work	.08	½	.04	
9.5-4	Start machine, stop	.02	½	.01	

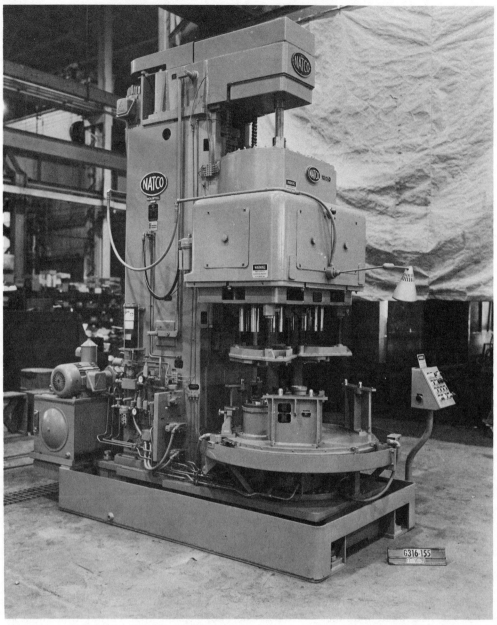

FIGURE 9.5 Multi-spindle drilling machine outfitted with rotary work table and part holding fixture. *(National Automatic Tool Company, Incorporated)*

Table Number	Process Description	Table Time	Adjustment Factor	Cycle Minutes	Setup Hours
11.2-5	Drill 10 holes (.75 in. length of cut)	.12	½	.06	
9.5-4	Position drills into clear and move jig into clear	.08	½	.04	
9.5-5A	Clean jig to receive next load	.10	½	.05	
9.5-5A	Clean table to receive next load	.14	½	.07	
11.4-3	Drill replacement Ten ⁹⁄₁₆-in. drills, 1.20 min of drilling	.10	½	.05	
	Total			.505	1.83
	Stack estimate	1.01 min			
	Hr/100	.842			

B. A turned part is placed on a V-type holding fixture. The fixture holds 4 parts. The parts are manually transferred from station to station being relocated radially in each position. Parts are automatically clamped by spring-loaded pressure pads mounted below the slip spindle plate. A hanging bushing plate is located by 2 guide pins. Material is CRS and weighs under 3 lb ea. A total of 7 holes are simultaneously drilled in the 4 parts with a head-feed multiple-spindle drilling machine. Find unit time and pc/hr.

Table Number	Process Description	Table Time	Adjustment Factor	Cycle Minutes	Setup Hours
9.1-S3	Multiple-spindle drilling machine setup	.58			.58
9.1-S3	Install fixture on table	1.00			1.00
9.1-S3	3 holes have critical depth	.30			.30
9.5-1A	Remove last piece, and load first piece	.12		.12	
9.5-1B	Transfer 3 pieces to next station	.07	3	.21	
9.5-2	Clamping automatic, foot air clamp	.02		.02	
Estimate	Palm button start and automatic return	.04		.04	
9.5-5B	Start oiling with coolant hose	.14		.14	
8.2-1	Head rapid traverse 8 in. at 150 ipm	.05		.05	
11.2-5	Drill 7 holes, maximum depth 1⅛ in.	.47	1.125	.53	
9.5-5A	Blow off fixture	.18		.18	
11.4-3	Drill replacement				
	1⅛-in. diameter, .53 min	.10		.10	
	2¼-in. diameter, 1.06 min	.26		.26	
	4.27-in. diameter, 2.12 min	.26		.26	
22.1-3	Part inspection				
	Go, no-go 4 holes, .005-in. tolerance	.15	⅘	.12	
8.2-1	Head rapid traverse 8 in. up	.05		.05	
	Total	1		2.08	1.88
	Shop estimate	31 pc/hr			

TABLE 9.5 CLUSTER DRILLING MACHINES

Setup See Table 9.1

Operation elements in estimating minutes

1A. Pick up part, move, and place; pick up and lay aside

	Weight					
Holding Device	1.0	2.5	5.0	10	15	25
Table, parallels, open vise, simple fixture	.07	.09	.12	.14	.17	.18
V-block, table between hold-down clamps, fixture with locating pins or blocks, between centers	.08	.10	.13	.16	.19	.21
Chuck, complex clamping or fixture with locating pins or blocks to fit part	.08	.11	.14	.18	.22	.23
Blind fixture with locating pins or blocks or stops, index plate with alignment, or align to mark on fixture	.09	.12	.15	.20	.24	.27

B. For additional part

Weight	2	5	10	15	25
Min	.05	.07	.08	.10	.12

C. Pry or tap part out .07
Dial or magazine part .05

2. Clamp and unclamp, ea
    Toggle or quick-acting clamp .05
    Cam clamp .03
    Star wheel .07
    Thumb screw .06
    Air cylinder .05
    U-clamp with hex nut .26
    C-clamp with hex nut .32
    C-clamp with thumb screw .21
    Strap clamp with hex nut .31
    Add'l nut .05
    Operate foot air clamp .02

3. Part or part and jig manipulation
    Rotate part and fixture horizontally .06
    Pin and unpin .07
    Rotation of rotary table, 90° .07; 180° .12
    Index fixture 90° .05

4. Cluster machine operation
    Start or stop machine, push button .01
    Start or stop machine, palm-operated push button .04
    Raise or lower table, clamp and unclamp .12
    Raise or lower headstock .10
    Foot operated treadle switch .02
    Shift lever for 2-speed control .06
    Position drill into clear and move jig to clear .08

5. Cleaning, lubricating, or wiping

  A.
    Clean area with air hose
      Small area up to 6 × 6 .06
      Medium area up to 12 × 12 .11
      Area up to 12 × 24 .14
      Large area up to 18 × 36 .25
      Small fixture .10
      Complex large fixture .18
    Brush chips
      Small area up to 6 × 6 .11
      Medium area up to 12 × 12 .13
      Area up to 12 × 24 .18
      Large area up to 18 × 36 .31
      Fixture .24

  B. Lubricate tools or piece

Oil Tools	1	2	3	4	5	6	Add'l
Min	.05	.09	.12	.14	.15	.16	.01

    Wipe locators with fingers .03

6. Drill, counterbore, ream, or tap   See Table 11.2

7. Rapid traverse   See Table 8.2-1

8. Drill replacement   See Table 11.4-3

9. Part inspection   See Table 22.1-3

9.6 Radial Drilling Machines

DESCRIPTION

The radial drilling machine is suitable for large work where it may not be feasible to move the unit around if several holes are to be drilled. The vertical column supports an arm which carries the drilling head. The arm may be swung around to any position over the workbed, and the drilling head has radial adjustment along this arm. Drilling in the vertical plane is common but on some machines the head may be swiveled to drill holes at various angles in a vertical plane.

Figure 9.6 is an example of a 3-ft radial drilling machine with a 2½-in. drill capacity in mild steel, 12 spindle speeds, and 9 power feeds.

FIGURE 9.6 A 3-ft radial drill press. *(Clausing Corporation)*

ESTIMATING DATA DISCUSSION

Handling for radial drill work is similarly arranged as in other drilling machines although time values may differ. Alignment requirements of rough or irregular castings and forgings, shimming, and jacking are additional for Element 1.

Clamping and unclamping, Element 2, are similar to other drilling machines except for the differences in time.

Radial drilling machine operation, Element 4, is composed of elements which have been recombined to allow for easy use. For instance, raise spindle (.05), move arm and head hole-to-hole, and lower spindle (.05) have been combined with an average distance between holes and the number of holes. Other elements have been joined to facilitate speed.

EXAMPLES

A. A 78-in long × 21 in.-wide × 12 in.-thick annealed aluminum plate is drilled with six ½-in. holes and twenty-six 1½-in. holes. Locating bump pins are used with screwdown clamps, and a plate fixture is overlaid having holes for bushings. Five parts are to be estimated for lot time.

Table Number	Process Description	Table Time	Adjustment Factor	Cycle Minutes	Setup Hours
9.1-S4	Fixture on floor plates	.28			.28
9.1-S4	Crane loading	.05			.05
9.6-1A	Load and unload part with hoist	.95		.95	
9.6-1B	Align part	.27		.27	
9.6-1B	Shim low points, estimate 2	.57	2	1.14	
9.6-2	Clamp using planer-type T-bolts, 6 necessary	.34	6	2.04	
9.6-2	Plier assist	.07	6	.42	
9.6-1A	Load plate fixture over aluminum stock, 60-16, crane loading	.95		.95	
9.6-2	Lock fixture to plate, 6 C-clamps	.15	6	.90	
9.6-4A	Pull radial arm from clear position	.10		.10	
9.6-4A	Change speed lever, 2 times	.04	2	.08	
9.6-4A	Change feed lever, 2 times	.03	2	.06	
9.6-4A	Start and stop machine	.14		.14	
9.6-4B	Install tool, 2 tools	.09	2	.18	

Table Number	Process Description	Table Time	Adjustment Factor	Cycle Minutes	Setup Hours
9.6-4C	Raise drill, move head hole-to-hole, and lower drill for six ½-in. holes, average distance apart 10 in.	1.16		1.16	
9.6-4E	Bushing, place and remove for ½-in. hole	.14	6	.84	
11.2-4	Drill six ½-in. holes, 2-in. deep, lead and break through = .26, distance = 2.26	.09	6 × 2.26	1.22	
9.6-4C	Raise drill, move arm and head hole-to-hole, and lower drill for twenty-six 1½-in. holes with 5-in. average distance apart, 1.63 + 16 × .16	4.19		4.19	
9.6-4E	Bushing, place and remove	.14	26	3.64	
11.2-4	Drill twenty-six 1½-in. holes, drill point, and breakthrough = .79, total distance = 2.70	.10	26 × 2.79	7.25	
9.6-5	Oil tool, each time	.09	32	2.88	
9.6-5	Clean floor plates	.25	2	.50	
9.6-5	Blow off fixture	.21	2	.42	
9.6-5	Blow off chips off part, every 5th hole	.25	32 × ⅕	1.60	
9.6-4A	Push arm into clear position	.08		.08	
11.4-3	Drill replacement				
	½-in. drill, 1.22 min	.07		.07	
	1½-in. drill, 7.25 min	.21		.21	
22.1-3	Hole size inspection				
	½-in. size, .005-in tolerance, go/no-go 2 end	.15	⅕	.03	
	1½-in. size, fractional tolerance	No time		0	
	Total			31.32	.33
	Lot estimate	2.94 hr			

B. Estimate the part shown in Figure 9.6. The cast iron part weighs 14.2 lb and one 2-in. hole is drilled ¾ in. through, followed by a reaming operation. Find unit time.

Table Number	Process Description	Table Time	Adjustment Factor	Cycle Minutes	Setup Hours
9.6-1A	Load part onto side-hole locator	.22		.22	
9.6-1B	Insert and remove back locating pin	.32		.32	
9.6-2	Tighten U-clamp from hole, star wheel	.18		.18	
9.6-4B	Lower drill from clear position	.05		.05	
9.6-4A	Change speed lever	.04		.04	
9.6-4E	Bushing, place and remove	.11		.11	
11.2-4	Drill 2-in. hole, lead and breakthrough = 1.05 in., L = 1.80 in., drill	.44	1.8	.79	
9.6-4B	Raise drill 3 in.	.05		.05	
9.6-4E	Bushing, place and remove	.11		.11	
9.6-4A	Change speed and feed lever	.07		.07	
9.6-4B	Lower drill 3 in.	.05		.05	
11.2-7	Ream hole	.75	.3	.23	
9.6-4B	Raise reamer 3 in.	.05		.05	
9.6-5	Blow off fixture	.21		.21	
11.4-3	Tool replacement				
	2-in. drill, .79 min	.05		.05	
	2-in. reamer, .23 min	.03		.03	
22.1-3	Part inspection				
	1-in., .003-in. tolerance hole, plug gage	.15	⅕	.03	
	Total			2.59	

TABLE 9.6   RADIAL DRILLING MACHINES

Setup Use Table 9.1

Operation elements in estimating minutes

1A. Pick up part, move, and place; pick up and lay aside.

Holding Device	Weight					
	15	25	35	45	60	Hoist
Floor plates, parallels, table, open vise, simple fixture	.22	.23	.25	.26	.28	.95
V-block, table between hold-down clamps, fixture with locating pins or blocks	.24	.26	.27	.30	.33	1.05
Chuck, complex clamping, or fixture with locating pins, or block to fit part	.26	.28	.30	.33	.36	1.11
Blind fixture with locating pins or blocks or stop, or align to make on fixture	.20	.31	.34	.38	.42	1.23

  B. Add for irregular casting or forging   1.00
     Align part with rule or level   .27
     Insert and remove locating pin   .32
     Align part on table or floor plates with square   .82
     Shim part   .57
     Set jack   .17

2. Clamp and unclamp
    Vise, full turn, hammer to seal   .70
    Toggle or quick-acting clamp   .07
    Cam lever   .10
    Star wheel   .18
    Thumb screw   .08
    Air cylinder   .08
    U-clamp with hex nut   .28
    C-clamp with hex nut, small, 15, large   .34
    C-clamp with thumb screw   .23
    Strap clamp with hex nut   .33
    Add'l nut   .23
    Plier assist   .07

3. Part or part and jig manipulation

   A. Turn part in vise

Weight	15	35	60
Min	.10	.17	.26

   B. Tumble jig

Weight	15	35	60
Min	.08	.14	.18

C. Turn part over on table

Weight	15	35	60
Min	.07	.14	.21

4. Radial drilling machine operation

   A. Pull radial arm from clear position .10
   Push away radial arm to clear position .08
   Change speed lever .04
   Stop, start, or reverse spindle lever .04
   Change feed lever .03
   Start or stop machine .14
   Lock or unlock radial arm or spindle .03

Change Tools	1	2	3	4	5	Add'l
Quick-change	.09	.18	.27	.36	.45	.09

   B. Install tool in quick-change chuck, move arm from clear into position, advance tool to work .24
   Raise spindle, (average of 3 in.) .05

   C. Raise spindle, move arm and head hole-to-hole, and lower spindle, given distance and hole

	Holes										
Distance	1	2	3	4	5	6	7	8	9	10	Add'l
1–3	.15	.30	.46	.61	.76	.91	1.06	1.22	1.37	1.52	.15
4–9	.16	.33	.49	.65	.82	.98	1.14	1.30	1.47	1.63	.16
10–19	.19	.39	.58	.77	.97	1.17	1.35	1.54	1.74	1.93	.19
20–29	.21	.42	.63	.84	1.05	1.26	1.47	1.68	1.89	2.10	.21
Add'l in.											.01

   D. Raise spindle, change tool, change feed and speed, position arm and head hole-to-hole, lower spindle

	Holes										
Distance	1	2	3	4	5	6	7	8	9	10	Add'l
1–3	.31	.62	.94	1.25	1.56	1.87	2.18	2.50	2.81	3.12	.31
4–9	.32	.65	.97	1.29	1.62	1.94	2.26	2.58	2.91	3.23	.32
10–19	.35	.71	1.06	1.41	1.77	2.12	2.47	2.82	3.18	3.53	.35
20–29	.37	.74	1.11	1.48	1.85	2.22	2.59	2.96	3.33	3.70	.37
Add'l in.											.01

   E. Raise spindle, change tools, exchange slip bushings, and lower spindle (no reposition)

Bushings	1	2	3	4	5	6	7	8	10	Add'l
Min	.33	.67	1.00	1.34	1.67	2.00	2.34	2.67	3.34	.33

   Bushing, place and remove .11
   Bushing, shift hole-to-hole .14

   F. Move radial arm or headstock hole-to-hole

Distance	1–3	4–5	10–19	20–29	Add'l
Min	.04	.05	.08	.10	.01

5. Cleaning, lubricating, or wiping
   Clean area with air hose
   Medium area up to 12 × 12 .11
   Area up to 12 × 24 .14
   Large area up to 18 × 36 .25

Fixture	.21
Brush chips	
Area up to 12 × 12	.15
Large area up to 18 × 36	.31
Fixture	.27
Vise	.18
Lubricate tool or piece	
Oil tool	.09
Adjust coolant hose	.06
6. Machining	
Drill, counterbore, ream, tap	See Table 11.2
7. Rapid traverse	See Table 8.2-1
8. Drill replacement	See Table 11.4-3
9. Part inspection	See Table 22.1-3

BORING

10.1 Horizontal Milling, Drilling, and Boring Machines

DESCRIPTION

The horizontal milling machine, also known as the horizontal bar, keeps the work stationary while the tool is rotated. It is adapted to the machining of horizontal holes or milling of surfaces. The horizontal spindle for holding the tool is supported and controlled by the headstock. The headstock can be moved vertically. The table has longitudinal and crosswise movements. A rotary table can be mounted. Instead of a table, floor plates may be used, and the column is then mounted on ways for movement along the floor plates. Configurations can include palleted auxiliary tables, tool changing, and work-handling carousel pallet shuttles, preselected positioning, automatic slope machining, etc. Now consider estimating data for the general class of horizontal bars. (See Figure 10.1A.)

ESTIMATING DATA DISCUSSION

The basic time is required and includes job preparation chores such as punch in/out, study print and process sheet, and remove chips. Elements 1 through 9 are expressed in setup hours per occurrence (hr/occurrence).

Element 2 includes: get angle plate, transport to machine, secure with 2 strap clamps, move spindle to angle plate and align angle plate with spindle, and reverse the work.

Element 2 involves a hoist hookup, transport to machine table or floor ways, position, unhook, secure with 3 or 4 strap clamps, and reverse the work. Element 3 involves a move—either by hand or power—of a fixture, secure with straps or bolts, align fixture to spindle, and reverse the work. Element 4 involves the installation and removal of a block in the floor plates using the overhead crane. The installation of a rotary table, turntable, index table, Element 5, includes the hoist pickup and move, and securing the rotary table to the machine table. It also includes the positioning of a rotary table on an index table and reverse the work.

Element 6 involves screws (before loading parallel blocks), shims, or jack screws (before loading piece), and reverse the work. The inspector checks process sheet for machining operation and drawings for inspections, picks up measuring tools, and inspects the requested dimensions.

Now consider the cycle elements. The manual load and unload includes walking to pallet, pickup, transport to table and position, and reverse the work. A jib or crane is similar except a hoist and sling is used. Entry variable is weight. A bridge or bay crane may be required for heavier parts. Element 2 involves the work to secure and release part. Various holding devices are suggested. The bottle jack is used for leveling piece. A double stud strap or hairpin is clamped to the T-studs. Washers are included.

Element 3 is used to zero in machine. Work includes: study of the machining location, determine coordinates, and use first tool and align piece with spindle to X and Y coordinates per location. Finally, the

FIGURE 10.1A Table type horizontal milling, drilling, and boring machine. *(Giddings and Lewis Machine Tool Company)*

readout is set. The dial indicate element includes work to indicate piece, table, or fixture, and lock headstock and/or table, and reverse the work. The work included with the tables involves retracting tool away, loosening nuts to unlock table, turn table, relock, and position spindle to piece for machining.

Element 4 involves much of the same work except for different tools. One example is select tool and position into spindle, retract spindle, remove tool from spindle with drift key or draw bar, and clean off tool. A stub bar for a first bore will include position adapter to spindle for solid tools, remove solid tool adapter before assembling stub bar, pick up stub bar, position in spindle and position to piece, and retract bar, remove bar from spindle and set aside. An in-line bar for a fixture involves (a) assemble clutch adapter to spindle, (b) position headstock for spindle to accept boring bar, lock, (c) pick up bar with hoist and position bar through fixture and piece and unhook hoist, (d) turn fixture 90° to spindle and secure index table, (e) and reverse the work.

Element 5 deals with assembling of cutter to various tools, securing, and reversing the work. It can include the assembly of a cutter to slot in bar and securing, and reversing.

Element 6 is for advance and retract a tool from the cut. Thus, this element includes the necessary manipulation of levers to engage and disengage the tool to the cut.

For a fraction tolerance, the operator sets the depth of cutting tool and aligns to piece and retracts the tool. For a decimal tolerance, set depth of cutting tool and advance to cut, pick up gage and gage depth and retract cutting tool. The trial portion of this work depends upon the tolerance. The decimal tolerance is given more trials expressed in the time. Distinctions are made for a single- or double-edge cutter.

Element 7 is the weighed average of moving a table, saddle, headstock, column, spindle, or combination to a readout coordinate. Thus, a rapid traverse time is covered by this time rather than Table 8.2.

EXAMPLES

A complicated gear box having box dimensions of 29 × 40 × 17 has many line bores, counter bores, backfaces, taps, and drilled holes. The material is gray cast iron, ferritic

ASTM A 48, class 20, and weighs about 715 lb. For a part of the work, find typical element patterns, and show the time required.

A. Load and unload the work.

Table Number	Process Description	Table Time	Adjustment Factor	Cycle Minutes	Setup Hours
10.1-1	Crane load	8.59		8.59	
10.1-2	Strap clamps, 4	5.84		5.84	
	Total for element to load and strap			14.43	

B. Center drill, drill and tap two 4½ NC holes.

Table Number	Process Description	Table Time	Adjustment Factor	Cycle Minutes	Setup Hours
10.1-4	Position and remove	.57	3	1.71	
10.1-3	Start and stop	.40	3	1.20	
10.1-3	Change feed, speed	.34	2	.68	
10.1-6	Advance and retract tool	.18	3	.54	
10.1-6	Center drill	.49		.49	
11.2-4	Drill 1$^{25}/_{32}$-Dia × 2-in deep	.44	2	.88	
11.2-8	Tap	.38	2	.76	
				6.26	

C. Core drill and line bore 2.75-in. hole inside of casting, decimal hole.

Table Number	Process Description	Table Time	Adjustment Factor	Cycle Minutes	Setup Hours
10.1-4	Position and remove tool	.57		.57	
10.1-3	Start and stop	.40	2	.80	
10.1-3	Change feed and speed	.34	2	.68	
10.1-5	Position and remove cutter	2.15		2.15	
10.1-6	Advance and retract core drill	.18		.18	
11.2-4	Core drill 2.55-in. hole, 2-in. deep	.51	2	1.02	
11.1-3	Rough bore	.10	2	.20	
10.1-6	Advance, retract	1.48		1.48	
10.1-5	Position and remove boring cutters	2.15		2.15	
11.1-3	Finish bore, .13 × 2, carbide	.26		.26	
10.1-6	Advance and retract in-line bar	2.63		2.63	
				12.12	

D. A welded tube has 24 holes drilled on a HBM. A partial dimensioned sketch is shown by Figure 10.1B. The part is loaded on the table. The operation spot drills and then follows with a $^9/_{16}$-in. hole. There are 24 holes. Special clamping is available for this lot of 68 units. Find the unit, setup, and lot estimate.

Table Number	Process Description	Table Time	Adjustment Factor	Cycle Minutes	Setup Hours
	Drill 24 $^9/_{16}$-in. holes with HBM				
10.1.1	Handling with jib crane of welded tube	7.22		7.22	
10.1.2	Clamp, unclamp 4 strap clamps	5.84		5.84	
10.1.4	Install, remove start drill 24 times; in and out of draw bar	0.57	24	13.68	
10.1.S1	Base time setup	0.31			0.31
10.1.S3	Fixture strap hold downs	0.42			0.42
10.1.S9	Piece production	1.00			1.00
11.1.6	Start drill	0.05	24	1.20	
10.1.3	Machine operation; change feed for start drill operation	0.16	24	3.84	
10.1.6	Advance, retract for 24 start drills	0.18	24	4.32	
10.1.4	Install, remove drill $^9/_{16}$-in. size	0.57	24	13.68	
10.1.6	Advance, retract drill	0.18	24	4.32	

Horizontal Milling, Drilling, and Boring Machines

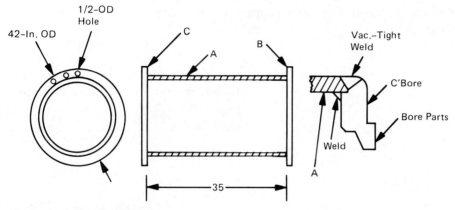

FIGURE 10.1B Part for HBM.

Table Number	Process Description	Table Time	Adjustment Factor	Cycle Minutes	Setup Hours
11.2.4	Power drilling 24 holes at .06 min per hole	0.06	24	1.44	
10.1.7	Travel hole to hole about 10 in. around bolt circle	0.10	24	2.40	
11.4.3	Drill tool life for drill and start drill	0.15		0.15	
	Total			58.09	1.73
	Lot estimate	67.57 hr			

TABLE 10.1 HORIZONTAL MILLING, DRILLING, AND BORING MACHINES

Setup in estimating hours

S1. Base time   .31

S2. Angle plate
    Angle plate, load manually   .25
    Second angle plate   .22
    Crane lift or hoist lift of angle plate
        3 hold downs   .37
        4 hold downs   .44
    Second angle plate
        3 hold downs   .34
        4 hold downs   .41

S3. Fixture load and unload

	Strap Hold Downs			Bolt Hold Downs		
	2	3	4	2	3	4
Hand Carry	.27	.35	.42	.17	.20	.22
Hoist Lift	.36	.44	.51	.26	.29	.31

S4. Center plug
    Less than 6 in., .02; more than 6 in.   .07

S5. Block or table
    Floor plate block   .27
    Install rotary table   .49

S6. Prepare strap clamp for holding piece   .10
    Prepare stop block for holding piece   .05

	Prepare bolt, slot block, set screw or	.01
	adjustable for setting piece	.02
S7.	Select sleeve, assemble to tool and disassemble	.01
S8.	Inspection	
	Fractional tolerance	.01
	Decimal tolerance	.02
S9.	Piece production (optional), 2 × hr per 1 unit	

Operation elements in estimating minutes

1. Handling
   Load and unload manually 1.74

   Jib or crane

Weight	500	1000	2000
Load, Unload	7.22	8.59	9.21
Reposition	4.33	5.15	5.53

2. Clamp and unclamp piece

Number	1	2	3	4	Add'l
Strap clamp	1.46	2.92	4.38	5.84	1.46
Set screw or C-clamp	1.19	2.37	3.56	4.74	1.19
Adjust bottle jack	.36	.73	1.08	1.44	.36
Double stud strap	7.93	15.87	23.79	31.72	7.93

3. Machine operation, ea occurrence
   - Change speed .... .18
   - Change feed .... .16
   - Start and stop .... .40
   - Readout per location 2.79
   - Dial indicate 5.34
   - Manually index rotary table 4.31
   - Power-turn rotary table 1.70

4. Install and remove tool, ea
   - Center, drill, or tap .... .57
   - Spot face or back face .... .87
   - Core drill or ream 1.31
   - Stub bar for first bore 2.05
   - Stub bar for bore, ream, or face 1.35
   - Stub bar for back face 3.15
   - Stub bar for milling 2.61
   - In-line bar through piece 3.94
   - In-line bar through fixture 19.47
   - In-line bar and relocate in fixture 18.73

5. Install and remove cutter, ea occurrence
   - Stub bar with chamfer and spot face 1.01
   - Stub bar bore and ream 1.52
   - Stub bar with back face 2.53
   - Bar in-line for chamfer or ream 1.73
   - Bar in-line for bore, spot face, or back face 2.15

6. Advance and retract tool, ea occurrence
    - Center drill, drill, tap, or end mill ... .18
    - Core drill or reamer ... .60
    - Spot face and back face ... .73
    - Spot face and back face, decimal tolerance ... 1.82
    - Face mill
        - Fraction tolerance ... 1.20
        - Finish decimal tolerance ... 2.69
        - Same depth, different location ... .70
    - Stub bar
        - Single edge cutter, fractional tolerance ... 1.40
        - Single edge cutter, decimal tolerance ... 6.19
        - Double edge cutter, fractional tolerance ... .44
        - Double edge cutter, decimal tolerance ... 1.97
    - Bar, in-line
        - Single edge cutter, fractional tolerance ... 1.48
        - Single edge cutter, decimal tolerance ... 7.01
        - Double edge, fractional tolerance ... .52
        - Double edge, decimal tolerance ... 2.63

7. Machine travel, ea occurrence

Distance	Min	Distance	Min	Distance	Min
10	.10	30	.30	85	.85
12	.12	35	.35	90	.90
14	.14	40	.40	95	.95
16	.16	45	.45	100	1.00
18	.18	50	.50	105	1.05
20	.20	55	.55	110	1.10
22	.22	60	.60	120	1.20
24	.24	65	.65	130	1.30
26	.26	70	.70	140	1.40
28	.28	75	.75	150	1.50
30	.30	80	.80	Add'l	.01

8. Machining
    - Bore, thread, start drill ... See Table 11.1
    - Drill, counterbore, ream or tap ... See Table 11.2
    - Face mill ... See Table 11.3

9. Tool life and replacement ... See Table 11.4

10.2 Boring and Facing Machines

DESCRIPTION

Boring and facing machines are designed for close tolerance boring. Machines have a minimum of two boring heads and can handle boring, facing, and turning elements. They have hydraulically-operated tables and are semiautomatic. The sequence is controlled by adjustable stop dogs. (See Figure 10.2.)

ESTIMATING DATA DISCUSSION

Entry variables for setup are the number of boring heads and tolerance. The handling elements depend upon part size. A small part may be handled with one hand. Medium parts do not weigh over 3 lb; large parts do not exceed 8 lb. In these data, heavy parts are 15 lb or less. These four categories are for fixture loading.

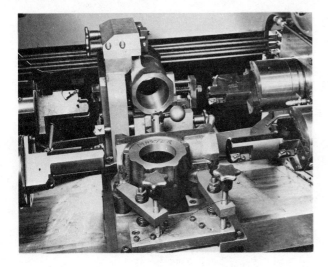

Loading and unloading to an angle plate or machinist vise are additional methods. A variety of devices are provided for in clamping. If a cam lever is used to tighten, the time also includes unclamp. Similarly, a pry part includes the work to obtain and set aside a screwdriver.

FIGURE 10.2 Boring and facing machine working on a hydraulic power component. *(Ex-Cell-O Corporation.)*

EXAMPLES

A. Ductile-iron power-steering gear bodies are to be bored and faced square to the bore. The castings are to be processed on a 4-spindle machine. The part is rough bored, counterbored, and chamfered. The main bore is machined in the front station and reloaded into the rear station for simultaneous small bore machining. The machining schedule is given below. Find the unit time and lot time for a quantity of 1800.

Table Number	Process Description	Table Time	Adjustment Factor	Cycle Minutes	Setup Hours
10.2-S	4 boring heads	1.65			1.65
10.2-S	Exceptional tolerances, 4 heads	.80			.80
10.2-1	Load part, get, dispose	.17		.17	
10.2-2	Clamping plates, 2	.10	2	.20	
10.2-2	Hand knob, 2	.08	2	.16	
10.2-2	Cam lever	.05		.05	
10.2-3	Blow off fixture	.07		.07	
10.2-3	Air blow part	.12		.12	
10.2-4	Machine operation	.02		.02	
10.2-4	Longitudinal distance	.18		.18	
11.1-3	Rough bore 10 in.-long, 4-in. OD, carbide .21 × (10 + .25)	2.15		2.15	
	Counterbore 4.5 in., ½-in. long in front station, .27 (.5 + .25) = .20				
11.1-2	Face 5 in.-OD × ½-in. long, .19 × −.10 (but this simo to rough boring operation)				
Repeat	Load, clamp, unclamp, unload for back station handling	.97		.97	
11.1-3	Rough machine cross bores simultaneously to front station machine time, no time	0		0	
	Total			4.09	2.45
	Lot time	125.15 hr			

TABLE 10.2   BORING AND FACING MACHINES

Setup
    First boring head .60 hr
    Each add'l boring head .35 hr
    Exceptional tolerance per head .20 hr

Operation elements in estimating minutes

1. Handling part

	Small	Medium	Large	Heavy
Get and dispose	.04	.06	.10	.14
Load and unload	.04	.05	.07	.10

Angle plate, load and unload, clamp, unclamp	1.20
Vise for up to 5 lb.	.40
5–25 lb	.60

2. Clamp and unclamp

Cam lever	.05
Position and remove swing or slide clamp	.04
Place and remove clamping plate	.10
Place and remove nut or bolt	.04
Operate air clamp	.03
Tighten and loosen nut using wrench	.17
Tighten and loosen hand knob or wing nut	.08
Pry part out of fixture	.06

3. Cleaning

Blow off fixture	.07
Wash part	.08
Air blow part, small, .06; large	.12

4. Machine operation

Engage feed	.02
Longitudinal feed, rapid traverse to work and return, per cut	.09
Cross feed	.09/in.

5. Machining

Face	See Table 11.3
Drill	See Table 11.2
Bore	See Table 11.1

6. Tool wear and replacement — See Table 11.4

7. Part inspection — See Table 22.1-3

MACHINING and TOOL REPLACEMENT

11.1 Turn, Bore, Form, Cutoff, Thread, Start Drill, and Break Edges

DESCRIPTION

These data are for pure machine time and are appropriate for various metal-cutting machines. Their location within the databook is centralized, thus encouraging ease of application. While primarily to be used with turning machines, other data related to milling and drilling are nearby. Some machines have their machine times integral with handling in a single table such as sawing, hobbing, etc.

ESTIMATING DATA DISCUSSION

These data are used whenever the machine is in a cutting-feed mode. A length of cut is composed of the drawing length plus a safety distance, approach, and overtravel. The entry variable for Element 1 is the diameter of the feature being machined. Element 1 is called safety stock, and provides for that small distance when there is no chips or cutting from the point of conclusion of rapid traverse velocity to cutting. This small length is affected by age of the machine and part tolerance. The total length of cutting is used to multiply the minutes per inch for various types of machining. The length of cut is multiplied by the factor given in Element 2.

Several factors affect the time to turn stock. Diameter, material or part, tool geometry, overhang, material, and sharpness, feed, depth of cut, machine stiffness, and horsepower are essential considerations. Despite the difficulty of precisely defining these machinability factors, the estimator is required to pick a time value in advance of production. Published information and general recommendations often provide ranges for surface-cutting velocity and feed from 300 to 500%, and the estimator is to pick a value. In this book, however, specific time values, minutes per inch (min/in.), are provided for a variety of diameters, materials, and machine cuts.

The factor is "minutes per inch" rather than inch per minute because it is more convenient for estimating. The practice of "min/in." is followed throughout this book. The guide value is multiplied by length of cut resulting in minutes for the machining element.

The length of cut is composed of four factors: safety stock, approach, length for metal removal as determined from the drawing or process plan, and overtravel. The significance of these factors will vary from case to case. Usually, the safety stock is not of great concern if tolerances are close. Approach and overtravel are more important for drilling and milling metal removal. If the process calls for multiple passes, the lengths and dimensions will change. These changes should be understood. The length of cut is entered in the adjustment factor column. If for example, the rate of metal removal is .17 min/in. and the length of cut (LOC) is 14.25 in., the cycle minutes would show 2.42 min.

The length of cut depends if the tool is traveling transversely or crosswise to the stock. If the cutting action is facing or grooving or cutoff, for example, the action of the tool is "plunging" and the length of cut is the distance that the tool moves into the stock, usually the linear distance from the start to the stopping point required by the design. For facing and cutoff, the length of cut is usually a little more than half of the diameter of the stock being parted.

Grooving or forming gives a length of cut starting from the surface of the material to the final depth of penetration. Longitudinal turning, boring, or threading is the length along the bar or casting, for example.

Some companies follow a procedure to match the estimated rpm to available machine rpm's. While this may be useful, the step may not be worth the effort. The table values provided in this data-book are keyed to significant diameters; and many modern machines provide infinite rpm drive cutters. This matching results in negligible improvement over a number of estimates. We recommend using table values, or the next highest value if a specific diameter is unavailable.

Entry variables for Element 2, turn, are tool material, part material and bar diameter. Time values are min/in. and with the length of cut found, the element time is the arithmetic product. The top and lower value are for rough and finish turning. For a two-pass plan of rough and finish, the two values may be added. For a roughing depth of cut in excess of 0.3 in., increase time by 25%. Cutting speed and feed values for selected materials are as follows:

		Turning			
		High Speed Steel		Carbide	
Material	Bhn	Rough	Finish	Rough	Finish
Low-carbon steel, free machining	170	(160, .015)	(210, .007)	(500, .020)	(700, .007)
Low-carbon steel	150	(120, .015)	(160, .007)	(400, .020)	(600, .007)
Medium-carbon steel	215	(190, .015)	(125, .007)	(325, .020)	(400, .007)
Stainless steel, 300	170	(100, .015)	(125, .007)	(300, .015)	(400, .007)
Stainless steel, 400	195	(150, .015)	(160, .007)	(350, .015)	(350, .007)
Steel castings, forging	225	(95, .015)	(125, .007)	(360, .020)	(475, .007)
Cast iron, gray	150	(145, .015)	(185, .007)	(500, .020)	(675, .010)
Aluminum	100	(600, .015)	(800, .007)	(800, .020)	(1200, .010)
Copper alloys		(275, .015)	(350, .007)	(550, .020)	(650, .007)
Plastic		(300, .010)	(300, .005)	(500, .010)	(500, .005)

$$\text{min/in.} = \frac{\pi \, Dia}{12 f V} \text{ where } Dia = \text{diameter, in.}$$

f = feed in./rev.

V = cutting velocity, fpm.

For instance, for free-machining low-carbon steel with an HSS rough speed of 160 and a 0.015 ipr, the 3-in. diameter calculates to give a min/in. of 0.33. Obviously other values for V and f can be substituted and differing values computed.

Cutting velocities and feeds, other than those given above, for some harder materials are as follows for carbide tools:

Material	Roughing	Finishing
Hastelloy	40, .021	75, .007
Inconel	80, .021	160, .007
Rene 41	35, .021	85, .007
J-1570	40, .021	100, .007

Thus, if a 3-in. OD bar of hastelloy was to be turned, the value of V and f could be used, or a ratio of $\frac{.33 \times 160 \times .015}{40 \times .021}$ to give 0.94 min/in. The idea of forming a ratio of the products of $V \times f$ multiplied by the tabular value is quick.

Entry values for boring, Element 3, are tool and part material and diameter of the feature being machined. Basic turning velocities and feeds were reduced for boring. The diameter is chosen before stock removal; and the practice of selecting the next-higher diameter from the table is preferred, especially if interpolation between neighboring values leads to negligible results over a number of estimates.

The basic turning velocities and feeds were reduced for forming and cutoff, Element 4. The maximum diameter of the feature to be formed is the entry variable, and the length of cut is measured as the radial distance moved of the form cutter.

The entry variables for single-point threading, Element 5, are no. of threads per in. and material. The value given by the intersection is multiplied by the product of the length of thread by nominal diameter. Data used in the construction of the table are the following:

Material	Cutting Velocity	Passes
Steel (mild, free-machining)	25	5
Steel (hard, stainless)	10	8

Material	Cutting Velocity	Passes
Cast iron	25	6
Brass	30	3
Aluminum	30	4
Plastic	50	5

The estimating formula is (no. of passes) × (threads per in.) ÷ 3.82 V and the factor in the table is multiplied by the product of L and Dia. For example, assume a threaded Dia of 6 in., L = 1¼ in., and there are 5 threads per in. for an aluminum casting. The time = .17 × 6 × 1.25 = 1.28 min. The time factor includes only threading, as positioning, set depth, etc., are found in other tables. In self-opening die threading, the entry variables are similar to 5, and the factor is multiplied by the L × Dia. The diameter is the nominal size of the thread. The L is full distance of the thread. Tapping time is found in Table 11.2.

The time to start-drill is related to the diameter of the hole being drilled and material. Break edges include the pickup of the file, etc., deburr corner of rotating edges, and lay tool aside. The entry variables are number of inside and outside diameters.

EXAMPLE

A. Notice that Figure 11.1 shows a partially dimensioned sketch. In this example we wish to machine the barstock, low carbon steel, free machining grade. No handling is considered in this example. Handling, or the machine and the operator time, is provided by other equipment information. The purpose of this example is to describe how these data are applied. To use the data, it is necessary to know the tool and part material. We are assuming carbide tool material. The next procedure is to find the particular kind of cutting action. Entry into the table assumes the next higher diameter, if the precise diameter is not provided. Always go to the next higher values. As the material is removed, as it is with this example, the diameter changes, and table entry points are changed. Find the time to machine the part. Do not include handling or machine operation.

Table Number	Process Description	Table Time	Adjustment Factor	Cycle Minutes	Setup Hours
11.1-2	Face end of the barstock; length of cut = half of bar Dia; facing is same as turning; carbide tooling; facing is a finish pass	0.13	1.25	0.16	
11.1-2	Turn ouside Dia of 2.5 in.; carbide tool reducing Dia to 2.15; adjustment factor includes distance of cutoff tool	0.07	5.25	0.37	
11.1-2	Turn 2.15-in. Dia and the 2.25 Dia column chosen; next Dia = 2.057 in.	0.06	5.25	0.31	
11.1-2	Turn finished Dim.	0.12	5.25	0.63	
11.1-4	Form the groove; length of cut = .25; special ground cutter	0.20	.25	0.05	
11.1-4	Cutoff the bar, length of cut = ½ of the remaining bar Dia or 1.035 in., which also includes minor additional stock	0.20	1.035	0.21	
	Total			1.73	0.00

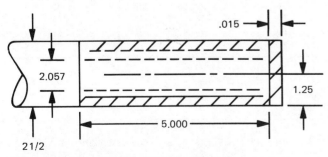

FIGURE 11.1 Partially dimensioned sketch.

TABLE 11.1 TURN, BORE, FORM, CUTOFF, THREAD, START DRILL, AND BREAK EDGES

Setup See other tables

Operation elements in estimating minutes

1. Safety stock, min/occurrence

Dia	¾	1	1¾	2
Min	.01	.03	.05	.09 max

2. Turn or face, min/in.
   High-speed steel

							Diameter												
Material	¼	½	¾	1	1¼	1½	1¾	2	2¼	2½	2¾	3	4	5	6	7	8	9	10
Low-carbon steel, free-machining	.03	.05	.08	.11	.14	.16	.19	.22	.25	.27	.30	.33	.44	.55	.65	.76	.87	.98	1.09
	.04	.09	.13	.18	.22	.27	.31	.30	.40	.45	.49	.53	.71	.89	1.07	1.25	1.42	1.60	1.78
Low-carbon steel	.04	.07	.11	.15	.18	.22	.25	.29	.33	.36	.40	.44	.58	.73	.87	1.02	1.16	1.31	1.45
	.06	.12	.18	.23	.29	.35	.41	.47	.53	.58	.64	.70	.93	1.17	1.40	1.64	1.87	2.10	2.34
Medium-carbon steel	.05	.10	.15	.19	.24	.29	.34	.39	.44	.48	.53	.58	.78	.97	1.16	1.36	1.55	1.75	1.94
	.07	.15	.22	.30	.37	.45	.52	.60	.67	.75	.82	.90	1.20	1.50	1.80	2.09	2.39	2.69	2.99
Stainless steel, 300	.04	.09	.13	.17	.22	.26	.31	.35	.39	.44	.48	.52	.70	.87	1.05	1.22	1.40	1.57	1.75
	.07	.15	.22	.30	.37	.45	.52	.60	.67	.75	.82	.90	1.20	1.50	1.80	2.09	2.39	2.69	2.99
Stainless steel, 400	.03	.06	.09	.12	.15	1.7	.20	.23	.26	.29	.32	.35	.47	.58	.70	.81	.93	1.05	1.16
	.06	.12	.18	.23	.29	.35	.41	.47	.53	.58	.64	.70	.93	1.17	1.40	1.64	1.87	2.10	2.34
Steel castings, forgings	.05	.09	.14	.18	.23	.28	.32	.37	.41	.46	.51	.55	.73	.92	1.10	1.29	1.47	1.65	1.84
	.07	.15	.22	.30	.37	.45	.52	.60	.67	.75	.82	.90	1.20	1.50	1.80	2.09	2.39	2.69	2.99
Cast iron, gray	.03	.06	.09	.12	.15	.18	.21	.24	.27	.30	.33	.36	.48	.60	.72	.84	.96	1.08	1.20
	.05	.10	.15	.20	.25	.30	.35	.40	.45	.51	.56	.61	.81	1.01	1.21	1.42	1.62	1.82	2.02
Aluminum	.01	.01	.02	.03	.04	.04	.05	.06	.07	.07	.08	.09	.12	.15	.17	.20	.23	.26	.29
	.01	.02	.04	.05	.06	.07	.08	.09	.11	.12	.13	.14	.19	.23	.28	.33	.37	.42	.47
Copper alloy	.02	.03	.05	.06	.08	.10	.11	.13	.14	.16	.17	.19	.25	.32	.38	.44	.51	.57	.63
	.03	.05	.08	.11	.13	.16	.19	.21	.24	.27	.29	.32	.43	.53	.64	.75	.85	.96	1.07
Plastics	.02	.04	.07	.09	.11	.13	.15	.17	.20	.22	.24	.26	.35	.44	.52	.61	.70	.79	.87
	.04	.09	.13	.17	.22	.26	.31	.35	.39	.44	.48	.52	.70	.87	1.05	1.22	1.40	1.57	1.75

Carbide

							Diameter												
Material	¼	½	¾	1	1¼	1½	1¾	2	2¼	2½	2¾	3	4	5	6	7	8	9	10
Low-carbon steel, free-machining	.01	.01	.02	.03	.03	.04	.05	.05	.06	.07	.07	.08	.10	.13	.16	.18	.21	.24	.26
	.01	.03	.04	.05	.07	.08	.09	.11	.12	.13	.15	.16	.21	.27	.32	.37	.43	.48	.53
Low-carbon steel	.01	.02	.02	.03	.04	.05	.06	.07	.07	.08	.09	.10	.13	.16	.20	.23	.26	.29	.33
	.02	.03	.05	.06	.08	.09	.11	.12	.14	.16	.17	.19	.25	.31	.37	.44	.50	.56	.62
Medium-carbon steel	.01	.02	.03	.04	.05	.06	.07	.08	.09	.10	.11	.12	.16	.20	.24	.28	.32	.36	.40
	.02	.05	.07	.09	.12	.14	.16	.19	.21	.23	.26	.28	.37	.47	.56	.65	.75	.84	.93
Stainless steel, 300 (ROUGH / FINISHING)	.01	.03	.04	.06	.07	.09	.10	.12	.13	.15	.16	.17	(.23)	.29	.35	.41	.47	.52	.58
	.02	.05	.07	.09	.12	.14	.15	.19	.21	.23	.26	.28	(.37)	.47	.56	.65	.75	.84	.93
Stainless steel, 400	.02	.03	.05	.06	.08	.09	.11	.12	.14	.16	.17	.19	.25	.31	.37	.44	.50	.56	.62
	.03	.07	.10	.13	.17	.20	.23	.27	.30	.33	.37	.40	.53	.67	.80	.93	1.07	1.20	1.34
Steel castings, forgings	.01	.02	.03	.04	.05	.05	.06	.07	.08	.09	.10	.11	.15	.18	.22	.25	.29	.33	.36
	.02	.04	.06	.08	.10	.12	.14	.16	.18	.20	.22	.24	.31	.39	.47	.55	.63	.71	.79
Cast iron, gray	.01	.01	.02	.03	.03	.04	.05	.05	.06	.07	.07	.08	.10	.13	.16	.18	.21	.24	.26
	.01	.02	.03	.04	.05	.06	.07	.08	.09	.10	.11	.12	.16	.19	.23	.27	.31	.35	.39
Aluminum	.01	.01	.01	.02	.02	.02	.03	.03	.04	.04	.04	.05	.07	.08	.10	.11	.13	.15	.16
	.01	.01	.02	.02	.03	.03	.04	.04	.05	.05	.06	.07	.09	.11	.13	.15	.17	.20	.22

Material	Diameter																		
	¼	½	¾	1	1¼	1½	1¾	2	2¼	2½	2¾	3	4	5	6	7	8	9	10
Copper alloys	.01	.01	.02	.02	.03	.04	.04	.05	.05	.06	.07	.07	.10	.12	.14	.17	.19	.21	.24
	.01	.03	.04	.06	.07	.09	.10	.12	.13	.14	.16	.17	.23	.29	.35	.40	.46	.52	.58
Plastics	.01	.03	.04	.05	.07	.08	.09	.10	.12	.13	.14	.16	.21	.26	.31	.37	.42	.47	.52
	.03	.05	.08	.10	.13	.16	.18	.21	.24	.26	.29	.31	.42	.52	.63	.73	.84	.94	1.05

3. Bore, min/in.
   High speed steel

Material	Diameter																		
	¼	½	¾	1	1¼	1½	1¾	2	2¼	2½	2¾	3	4	5	6	7	8	9	10
Low-carbon steel, free-machining	.04	.09	.13	.18	.22	.27	.31	.36	.40	.45	.49	.53	.71	.89	1.07	1.25	1.42	1.60	1.78
	.06	.12	.18	.25	.31	.37	.43	.49	.55	.62	.68	.74	.99	1.23	1.48	1.73	1.97	2.22	2.46
Low-carbon steel	.06	.12	.18	.23	.29	.35	.41	.47	.53	.58	.64	.70	.93	1.17	1.40	1.64	1.87	2.10	2.34
	.08	.16	.24	.32	.40	.49	.57	.65	.73	.81	.89	.97	1.29	1.62	1.94	2.26	2.59	2.91	3.24
Medium-carbon steel	.07	.15	.22	.30	.37	.45	.52	.60	.67	.75	.82	.90	1.20	1.50	1.80	2.09	2.39	2.69	2.99
	.10	.21	.31	.41	.52	.62	.72	.83	.93	1.04	1.14	1.24	1.66	2.07	2.48	2.90	3.31	3.73	4.14
Stainless steel, 300	.07	.15	.22	.30	.37	.45	.52	.60	.67	.75	.82	.90	1.20	1.50	1.80	2.09	2.39	2.69	2.99
	.10	.21	.31	.41	.52	.62	.72	.83	.93	1.04	1.14	1.24	1.66	2.07	2.48	2.90	3.31	3.73	4.14
Stainless steel, 400	.06	.12	.18	.23	.29	.35	.41	.47	.53	.58	.64	.70	.93	1.17	1.40	1.64	1.87	2.10	2.34
	.08	.16	.24	.32	.40	.49	.57	.65	.73	.81	.89	.97	1.29	1.62	1.94	2.26	2.59	2.91	3.24
Steel casting, forgings	.07	.15	.22	.30	.37	.45	.52	.60	.67	.75	.82	.90	1.20	1.50	1.80	2.09	2.39	2.69	2.99
	.10	.21	.31	.41	.52	.62	.72	.83	.93	1.04	1.14	1.24	1.66	2.07	2.48	2.90	3.31	3.73	4.14
Cast iron, gray	.05	.10	.15	.20	.25	.30	.35	.40	.45	.51	.56	.61	.81	1.01	1.21	1.42	1.62	1.82	2.02
	.07	.14	.21	.28	.35	.42	.49	.56	.63	.70	.77	.84	1.12	1.40	1.68	1.96	2.24	2.52	2.80
Aluminum	.01	.02	.04	.05	.06	.07	.08	.09	.11	.12	.13	.14	.19	.23	.28	.33	.37	.42	.47
	.02	.03	.05	.06	.08	.10	.11	.13	.15	.16	.18	.19	.26	.32	.39	.45	.52	.58	.65
Copper alloys	.03	.05	.08	.11	.13	.16	.19	.21	.24	.27	.29	.32	.43	.53	.64	.75	.85	.96	1.07
	.04	.07	.11	.15	.18	.22	.26	.30	.33	.37	.41	.44	.59	.75	.89	1.04	1.18	1.33	1.48
Plastics	.04	.09	.13	.17	.22	.26	.31	.35	.39	.44	.48	.52	.70	.87	1.05	1.22	1.40	1.57	1.75
	.06	.12	.18	.24	.30	.36	.42	.48	.54	.60	.66	.72	.97	1.21	1.45	1.69	1.93	2.17	2.42

Carbide

Material	Diameter																		
	¼	½	¾	1	1¼	1½	1¾	2	2¼	2½	2¾	3	4	5	6	7	8	9	10
Low-carbon steel, free-machining	.01	.03	.04	.05	.07	.08	.09	.11	.12	.13	.15	.16	.21	.27	.32	.37	.43	.48	.53
	.02	.04	.06	.07	.09	.11	.13	.15	.17	.18	.20	.22	.30	.37	.44	.52	.59	.67	.74
Low-carbon steel	.02	.03	.05	.06	.08	.09	.11	.12	.14	.16	.17	.19	.25	.31	.37	.44	.50	.56	.62
	.02	.04	.06	.09	.11	.13	.15	.17	.19	.22	.24	.26	.35	.43	.52	.60	.69	.78	.86
Medium-carbon steel	.02	.05	.07	.09	.12	.14	.14	.16	.21	.23	.26	.28	.37	.47	.56	.65	.75	.84	.93
	.03	.06	.10	.13	.16	.19	.23	.26	.29	.32	.36	.29	.52	.65	.78	.91	1.04	1.16	1.29
Stainless steel, 300	.02	.05	.07	.09	.12	.14	.16	.19	.21	.23	.26	.28	.37	.47	.56	.65	.75	.84	.93
	.03	.06	.10	.13	.16	.19	.23	.26	.29	.32	.36	.39	.52	.65	.78	.91	1.04	1.16	1.29
Stainless steel, 400	.03	.05	.08	.11	.13	.16	.19	.21	.24	.27	.29	.32	.43	.53	.64	.75	.85	.96	1.07
	.04	.07	.11	.15	.18	.22	.26	.30	.33	.37	.41	.44	.59	.74	.89	1.04	1.18	1.33	1.48
Steel castings, forgings	.02	.04	.06	.08	.10	.12	.14	.16	.18	.20	.22	.24	.31	.39	.47	.55	.63	.71	.79
	.03	.05	.08	.11	.14	.16	.19	.22	.25	.27	.30	.33	.44	.54	.65	.76	.87	.98	1.09
Cast iron, gray	.01	.02	.03	.04	.05	.06	.07	.08	.09	.10	.11	.12	.16	.19	.23	.27	.31	.35	.39
	.01	.03	.04	.05	.07	.08	.09	.11	.12	.13	.15	.16	.21	.27	.32	.38	.43	.48	.54
Aluminum	.01	.01	.02	.02	.03	.03	.04	.04	.05	.05	.06	.07	.09	.11	.13	.15	.17	.20	.22
	.01	.02	.02	.03	.04	.05	.05	.06	.07	.08	.08	.09	.12	.15	.18	.21	.24	.27	.30
Copper alloys	.01	.03	.04	.06	.07	.09	.10	.12	.13	.14	.16	.23	.29	.35	.40	.46	.48	.52	.58
	.02	.04	.06	.08	.10	.12	.14	.16	.18	.20	.22	.32	.40	.48	.56	.64	.70	.72	.80
Plastics	.03	.05	.08	.10	.13	.16	.18	.21	.24	.26	.29	.31	.42	.52	.63	.73	.84	.94	1.05
	.04	.07	.11	.14	.18	.22	.25	.29	.33	.36	.40	.43	.58	.72	.87	1.01	1.16	1.30	1.45

4. Form or cutoff, min/in.
   High speed steel

Material	¼	½	¾	1	1¼	1½	1¾	2	2¼	2½	2¾	3	4	5	6	7	8	9	10
Low-carbon steel, free-machining	.09	.18	.27	.36	.45	.55	.64	.73	.82	.91	1.00	1.09	1.45	1.82	2.18	2.55	2.91	3.27	3.64
Low-carbon steel	.12	.24	.36	.48	.61	.73	.85	.97	1.09	1.21	1.33	1.45	1.94	2.42	2.91	3.39	3.88	4.36	4.85
Medium-carbon steel	.16	.32	.48	.65	.81	.97	1.13	1.29	1.45	1.62	1.78	1.94	2.59	3.23	3.88	4.52	5.17	5.82	6.46
Stainless steel, 300	.15	.29	.44	.58	.73	.87	1.02	1.16	1.31	1.45	1.60	1.75	2.33	2.91	3.49	4.07	4.65	5.24	5.82
Stainless steel, 400	.10	.19	.29	.39	.48	.58	.68	.78	.87	.97	1.07	1.16	1.55	1.94	2.33	2.71	3.10	3.49	3.88
Steel castings, forgings	.15	.31	.46	.61	.77	.92	1.07	1.22	1.38	1.53	1.68	1.84	2.45	3.06	3.67	4.29	4.90	5.51	6.12
Cast iron, gray	.10	.20	.30	.40	.50	.60	.70	.80	.90	1.00	1.10	1.20	1.60	2.01	2.41	2.81	3.21	3.61	4.01
Aluminum	.02	.05	.07	.10	.12	.15	.17	.19	.22	.24	.27	.29	.39	.48	.58	.68	.78	.87	.97
Copper alloys	.05	.11	.16	.21	.26	.32	.37	.42	.48	.53	.58	.63	.85	1.06	1.27	1.48	1.69	1.90	2.12
Plastics	.07	.15	.22	.29	.36	.44	.51	.58	.65	.73	.80	.87	1.16	1.45	1.75	2.04	2.33	2.62	2.91

Carbide

Material	¼	½	¾	1	1¼	1½	1¾	2	2¼	2½	2¾	3	4	5	6	7	8	9	10
Low-carbon steel, free-machining	.02	.04	.07	.09	.11	.13	.15	.17	.20	.22	.24	.26	.35	.44	.52	.61	.70	.79	.87
Low-carbon steel	.03	.05	.08	.11	.14	.16	.19	.22	.25	.27	.30	.33	.44	.55	.65	.76	.87	.98	1.098
Medium-carbon steel	.03	.07	.10	.13	.17	.20	.23	.27	.30	.34	.37	.40	.54	.67	.81	.94	1.07	1.21	1.34
Stainless steel, 300	.05	.10	.15	.19	.24	.29	.34	.39	.44	.48	.53	.58	.78	.97	1.16	1.36	1.55	1.75	1.94
Stainless steel, 400	.04	.08	.12	.17	.21	.25	.29	.33	.37	.42	.46	.50	.66	.83	1.00	1.16	1.33	1.50	1.66
Steel castings, forging	.03	.06	.09	.12	15	.18	.21	24	.27	.30	.33	.36	.48	.61	.73	.85	.97	1.09	1.21
Cast iron, gray	.02	.04	.07	.09	.11	.13	.15	.17	.20	.22	.24	.26	.35	.44	.52	.61	.70	.79	.87
Aluminum	.01	.03	.04	.05	.07	.08	.10	.11	.12	.14	.15	.16	.22	.27	.33	.38	.44	.49	.55
Copper alloys	.02	.04	.06	.08	.10	.12	.14	.16	.18	.20	.22	.32	.40	.48	.56	.64	.68	.71	.79
Plastics	.04	.09	.13	.17	.22	.26	.31	.35	.39	.44	.48	.52	.70	.87	1.05	1.22	1.40	1.57	1.75

5. Threading

Threading, single-point; factor × L × Dia = min

Material	4	5	6	7	8	9	10	11	12	14	16	18	20	24	28	32
Steel (mild), free-machining	.21	.26	.31	.37	.42	.47	.52	.58	.63	.73	.84	.94	1.05	1.26	1.47	1.68
Steel (hard), stainless	.84	1.05	1.26	1.47	1.69	1.88	2.09	2.30	2.51	2.93	3.35	3.77	4.19	5.03	5.86	6.70
Cast iron	.25	.31	.38	.44	.50	.57	.63	.69	.75	.88	1.01	1.13	1.26			
Brass	.10	.13	1.6	.18	.21	.24	.26	.29	.31	.37	.42	.47	.52	.63	.73	.84
Aluminum	.14	.17	.21	.24	.28	.31	.35	.38	.42	.49	.56	.63	.70	.84	.98	1.12
Plastic	.10	.13	.16	.18	.21	.24	.26	.29	.31	.37	.42	.47	.52	.63	.73	.84

Threading self-opening die; factor × L × Dia = min

Material	4	5	6	7	8	9	10	11	12	14	16	18	20	24	28	32
Steel (mild), free-machining	.04	.05	.06	.07	.08	.09	.10	.12	.13	.15	.17	.19	.21	.25	.29	.34
Steel (hard), stainless	.10	.13	.16	.18	.21	.24	.26	.29	.31	.37	.42	.47	.52	.63	.73	.84
Cast iron	.04	.05	.06	.07	.08	.09	.10	.12	.13	.15	.17	.19	.21	.25	.29	.34
Brass, Aluminum	.03	.04	.05	.06	.07	.08	.09	.10	.10	.12	.14	.16	.17	.21	.24	.28
Plastic	.02	.03	.03	.04	.04	.05	.05	.06	.06	.07	.08	.09	.10	.13	.15	.17

6. Start drill, min[1]

   Steel

Dia	¼	5/16	½	1	2	3
Min	.02	.03	.05	.11	.35	.66

   Steel (medium)
   Cast iron

Dia	¼	5/16	½	¾	1	1½	3
Min	.03	.04	.07	.12	.14	.25	.83

   Steel (alloy)
   Forgings
   Stainless

Dia	¼	3/8	5/8	1	2	3
Min	.04	.07	.13	.22	.70	1.31

   Steel (high tensile)
   Titanium

Dia	¼	3/8	85/8	1	2
Min	.07	.12	.21	.37	1.16

   Aluminum, brass
   magnesium, plastic

Dia	¼	3/8	½	1	1½	2	3
Min	.01	.02	.05	.05	.08	.14	.26

   [1] Dia is for hole to be drilled.

7. Break edges, min

No. of Inside Dia	No. of Outside Diameters									
	0	1	2	3	4	5	6	7	8	9
0	0	.06	.11	.14	.17	.19	.22	.24	.26	.29
1	.06	.08	.12	.16	.16	.20	.23	.25	.28	.30
2	.08	.11	.14	.18	.20	.23	.25	.28	.30	.31
3	.10	.12	.16	.19	.22	.24	.26	.29	.31	.32
4	.11	.13	.17	.20	.23	.25	.28	.29	.32	.34
5	.12	.14	.18	.22	.24	.26	.29	.31	.34	.35

11.2 Drill, Ream, Counterbore, Countersink, Tap, and Microdrill

DESCRIPTION

These data are for pure machine time and are appropriate for operations involving drilling, counterboring, spotfacing, countersinking, reaming, and tapping, especially with various drilling, turning, and milling machines. The location of this information is centralized to facilitate its use.

Not all drilling machining times are included in this section. But machining for gun drilling, core drilling, taper reaming, and pipe threading can be judged and estimated using a similar context from these and other drilling data. Notice Figure 11.2A for a description of drills.

ESTIMATING DATA DISCUSSION

For Element 1, the entry variables are depth of material to be drilled, diameter of drill, and material. The drilling is sensitive, i.e., manual, using a sensitive drill press. For sensitive drilling no approach or breakthrough distance is added to the stock depth; use the depth only. The time is min per hole. Multiply the hole time by the number of holes for that size. The materials are soft steel, hard steels implying stainless grades and tool steels, brass implying brass, cast and soft bronzes, and aluminum. To expedite table usage, the upper values may be selected to avoid tabular interpolation.

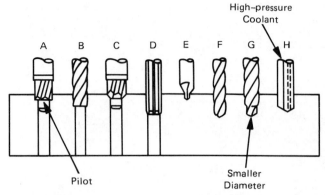

FIGURE 11.2A Types of drilling tools: (A) counterbore; (B) core drilling; (C) piloted countersink; (D) reaming; (E) center drill or start drill; (F) drill; (G) step drill; (H) gun drill.

Naturally not all drill sizes are shown by the tables. But more importantly, it is not necessary since the time distinctions between the various sizes are unimportant. We discourage the practice of interpolating the time values between table entries. Instead the estimator should use the next higher value in the table. If this practice is followed greater consistency and less paper supporting documentation will be required for estimating.

In sensitive reaming, Element 2, which differs from power reaming, only length and ream diameter are evaluated. Material distinctions were overlooked, since they are averaged jointly for the data.

For Element 4, entry variables are drill diameter and five classes of raw material. The basic formula for:

$$\text{min/in.} = \frac{\pi \, Dia}{12 Vf},$$ where V = surface cutting velocity, ft/min, and f = in. per rev

The min/in. value from Elements 4, 6, 7, and 8 require that they be multiplied by L, or length, which may be longer than the depth shown by the drawing. For drilling length, we add for drill point, and the following table is used.

The tabular values provide drill point and breakthrough length as related to drill diameter. Using the 118° included angle of the drill point, the approach and break-through distance is given above. For example, with a .5-in. drill, the value of .26 in. is added

Drill-point lead and breakthrough in.

Drill Dia	Distance	Drill Dia	Distance	Drill Dia	Distance
.019	.01	.416	.22	1⅛	.59
.038	.02	.454	.24	1¼	.66
.057	.03	.5	.26	1½	.79
.076	.04	.53	.28	1⅝	.86
.095	.05	.57	.30	1¾	.93
.113	.06	.61	.32	2	1.05
.132	.07	.64	.34	2⅛	1.12
.151	.08	.68	.36	2¼	1.19
.170	.09	.72	.38	2⅜	1.26
.189	.10	.75	.39	2½	1.32
.208	.11	.79	.42	2⅝	1.39
.227	.12	.83	.44	2¾	1.45
.25	.13	.87	.46	2⅞	1.52
.265	.14	.91	.48	3	1.59
.284	.15	.95	.50	3⅛	1.65
.303	.16	.98	.52	3¼	1.72
.322	.17	1.00	.53	3⅜	1.78
.340	.18	1.02	.54	3½	1.85
.359	.19	1.06	.56		
.378	.20	1.10	.58	Add'l	.53

to the drilling length. For stainless steel, the drill point is flatter, so multiply the tabular value by 0.6. Whenever power drilling time is computed, add to the design drill distance, the distance for drill-point lead and breakthrough. Each drilling rate, *i.e.*, min/in., is multiplied by this total length.

Despite the realization that power drilling is done for a variety of machines where rpm may be fixed rather than infinitely variable, the data are associated directly to velocity without an rpm correction, as rpm adjustment is not worth the effort in estimating. The minimum diameter is ¼ in. as smaller diameters have the same min/in. due to the compensating feed and diameter in the formula.

Material	Velocity
Aluminum and alloys	250
Brass and bronze (ordinary)	250
Die castings (zinc base)	250
Iron, cast (soft)	80
Magnesium and alloy	250
Plastics or similar material	250
Steel	
Mild, .2 to .3 carbon	100
Hard, .4 to .5 carbon	80
Forgings	50
High tensile (heat treated)	30
Stainless steel	50
Titanium alloys	30

Feeds used for Element 4 are as follows:

Dia	Feed	Dia	Feed	Dia	Feed
¹⁄₁₆	.001	⅜	.005	1½	.015
⅛	.002	½	.006	1¾	.015
⁵⁄₃₂	.0025	⅝	.008	2	.015
³⁄₁₆	.003	¾	.008	2½	.016
¼	.004	1	.012	3	.018
⁵⁄₁₆	.0045	1¼	.014		

These speeds and feeds for Element 4 are based upon high-speed steel, and if carbide is used, the rates can be decreased 15 to 30%. If necessary, the estimator can adjust the values for min/in. for other requirements. Remember to multiply the min/in. by the length or depth of the hole. This length may include an addition for point and breakthrough as shown by the above table. For deep holes, raise and lower spindles to relieve chips as necessary.

The Element 5 entry variable for multi-spindle or cluster power-feed drilling is metal length and two types of material. In this type of drilling, several different diameters may be simultaneously working, thus, maximum metal length is selected as the entry variable rather than a drill diameter. No drill point and breakthrough distance is added to metal length, as that consideration is a part of the information used to construct the table.

In counterbore machining, an end-cutting tool having two or more cutting edges is used in producing holes of more than one single diameter. If the counterbore depth is shallow, the operation may be spot facing. Speeds and feeds are slightly less than power drilling to allow for chip discharge caused by the packing during end cutting. These values can be used for counterdrilling and countersinking. This work is considered powered.

In reaming, Element 7, we estimate the reaming element at slower fpm but higher feed rates than drills of corresponding diameter under power-feeding conditions. For high-speed steel reamers, the following conditions were used to construct the tabular entry.

	Feed ipr				
	Steel (Mild)	Steel (Medium)	Steel (Alloy)	Steel (High Tensile)	Aluminum, Brass, Mag.
Dia \ fpm	75	85	40	30	140
⅛	.005	.004	.004	.003	.006
½	.010	.007	.007	.005	.012
1	.016	.012	.012	.008	.020
2	.026	.020	.020	.013	.032
2½	.035	.028	.028	.018	.043
3	.045	.035	.035	.023	.056

Solid and carbide-tipped reamers may be operated at speeds up to 150% of HSS reamers. This work is considered powered. The table may be used for tapered reaming as well as straight reaming.

In Element 8, tapping, the entry variables are no. of threads and material. Feed is not specified for tapping since it is dependent upon thread pitch. The time is for tap, dwell, and reverse out. If an automatic reverse is not available, a time to stop and start the machine and reverse spindle speed is necessary. The times are for min/in. of depth.

Microdrilling is the name for miniature-hole machining, and the diameter is generally considered to be .020 in. or less. However, data for small holes are extremely variable. Depth-to-diameter ratios are high as well. Tool life and clearing of chips present special problems. Specialized microdrilling machines may be hand-fed sensitive units. In automatic operations, a spindle head is fed forward for a predetermined distance and retracted quickly; then the cycle is repeated. Some production reports indicate that drills need regrinding after every 150 holes, although in one instance, several thousand holes were drilled in half-hard brass with a single HSS drill. Carbide and HSS steels are both used, but the data makes no distinction for drill material differences. The element data averages drill material and types of machine, either hand sensitive or automatic.

EXAMPLE

A. A stainless steel part is shown in Figure 11.2B. In this example we want to drill, ream, countersink, spotface, and tap several holes of various lengths and number. Handling and machine manipulation is not demonstrated by this example (see other examples, especially in the drilling equipment). But we do assume that the surface will be spot drilled prior to power drilling, and bushings are not available to guide the drill. The part shown has been milled previously to drilling. The quantity is 217 units. Find the unit, setup estimates, and the total lot hours. The adjustment factor is used for the length of cut and the number of holes. The length of cut will include a safety stock (which is that distance where the drill stops in the rapid traverse mode of machine velocity and begins the machine feeding mode), metal distance required for the hole, and overtravel that may be necessary for the drill point.

Table Number	Process Description	Table Time	Adjustment Factor	Cycle Minutes	Setup Hours
11.1-6	Start drill for NC machine without jig	0.03	7	0.21	
11.2-4	Drill 3 A holes ½; drill length = 2 + .13 = 2.13	0.44	2.13 × 3	2.81	
11.2-4	Drill 2 A holes; length of cut = .5 + .13 = .63 in.	0.44	.63 × 2	0.55	
11.2-4	Tap drill hole for ⅜-in. tap; LOC = 1.1	0.39	1.1	0.43	
11.2-8	Tap ⅜—13 hole LOC = 1 in.	0.39	1	0.39	
11.2-4	Drill 7/64-in. hole LOC = 1.13 in.	0.33	1.13	0.37	
11.2-7	Ream .1299-in. hole; use the next higher time value; LOC = 1 in.	0.47	1	0.47	
11.2-3	Countersinking 3/16-in. C hole	0.03		0.03	
11.2-6	C-sink .255-in. OD C hole	0.61	.105	0.06	
	Total			5.34	0.00
	Total lot hours	19.31			

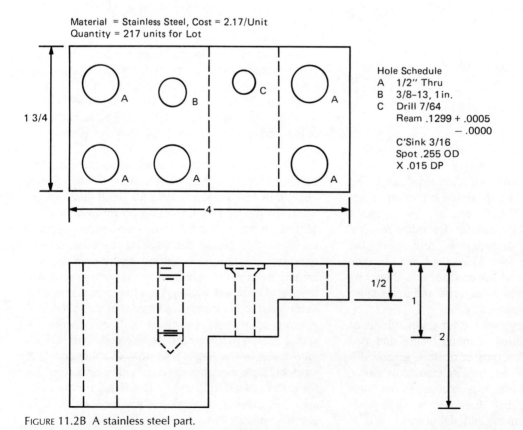

FIGURE 11.2B A stainless steel part.

TABLE 11.2 DRILL, REAM, COUNTERBORE, COUNTER-SINK, TAP, AND MICRODRILL

Setup See other tables

Operation elements in estimating minutes

1. ~~Sensitive~~ MANUAL drilling, min per hole

Depth	Material	.093	.125	.156	.187	.25	.375	.5	.675	.75	1.0
.093	Soft steel					.08	.25	.42			
	Hard steel	.03	.08	.12	.17	.27	.47	66			
	Brass	.03	.05	.06	.08	.11	.18	.24	.33		
	Aluminum	.05	.05	.06	.07	.08	.11	.14	.18	.19	.25
.156	Soft steel					.10	2.7	.44			
	Hard steel	.06	.11	.16	.21	.31	.50	.70			
	Brass	.05	.07	.09	1.0	.13	.20	.26	.36		
	Aluminum	.07	.08	.09	.09	.11	.13	.16	.20	.22	.27
.187	Soft steel					.13	.30	.46			
	Hard steel	.10	.15	.20	.25	.35	.54	.73			
	Brass	.07	.09	.11	.12	.16	.22	.29	.38		
	Aluminum	.10	.10	.11	.12	.13	.16	.19	.23	.24	.30
.25	Soft steel				.09	.17	.34	.51			
	Hard steel	.18	.23	.28	.32	.42	.62	.81			
	Brass	.12	.14	.15	.17	.20	.27	.33	.42		
	Aluminum	.14	.15	.16	.17	.18	.21	.24	.28	.29	.35
.312	Soft steel			.09	.13	.22	.39	.55			
	Hard steel	.25	.30	.35	.40	.50	.69	.89			
	Brass	.16	.18	.20	.21	.24	.31	.39	.47		
	Aluminum	.19	.20	.21	.21	.23	.26	.28	.32	.34	.40
.375	Soft steel		.09	.14	.18	.26	.43	.60			
	Hard steel	.33	.38	.43	.47	.57	.77	.96			
	Brass	.21	.22	.24	.26	.29	.35	.42	.51		
	Aluminum	.24	.25	.26	.26	.28	.31	.33	.37	.39	.45
.5	Soft steel	.14	.18	.23	.27	.35	.52	.69			
	Hard steel	.48	.53	.58	.63	.72	.92	1.11			
	Brass	.30	.31	.33	.34	.38	.44	.52	.60		
	Aluminum	.34	.35	.35	.36	.38	.40	.43	.47	.49	.54
.675	Aluminum	.48	.48	.49	.50	.51	.54	.57	.61	.62	.68
.75	Aluminum	.54	.54	.55	.56	.57	.60	.63	.67	.68	.74
1.00	Aluminum	.73	.74	.74	.75	.77	.79	.82	.86	.88	.93

2. ~~Sensitive~~ MANUAL reaming, min per hole

Depth	.1	.2	.3	.4	.5	.6	.7
.05			.04	.06	.08	.10	.12
.1		.02	.04	.07	.09	.11	.13
.2		.03	.05	.08	.10	.12	.14
.3		.04	.06	.09	.11	.13	.15
.4	.03	.05	.07	.10	.12	.14	.16
.5	.04	.06	.08	.11	.13	.15	.17
.6	.05	.07	.09	.12	.14	.16	.18

Ream Dia (column header over .1–.7)

	Ream Dia						
Depth	.1	.2	.3	.4	.5	.6	.7
.7	.06	.08	.10	.13	.15	.17	.19
.8	.07	.09	.11	.14	.16	.18	.20
.9	.08	.10	.12	.15	.17	.19	.21
1.0	.09	.11	.13	.16	.18	.20	.22

3. ~~Sensitive~~ MANUAL or power countersinking .03/ea

4. Power drilling, min/in.

Drill Dia	Steel (Mild)	Steel (Medium), Cast Iron	Steel (Alloy), Forgings, Stainless	Steel (High Tensile), Titanium	Aluminum, Brass, Magnesium, Plastic
¼	.16	.20	.33	.55	.07
5/16	.18	.23	.36	.61	.07
⅜	.20	.25	.39	.65	.08
½	.22	.27	.44	.73	.09
⅝	.20	.26	.41	.68	.08
¾	.25	.31	.49	.82	.10
1	.22	.27	.44	.73	.09
1¼	.23	.29	.47	.78	.09
1½	.26	.33	.52	.87	.10
1¾	.31	.38	.61	.102	.12
2	.35	.44	.70	1.16	.14
2½	.41	.51	.82	1.36	.16
3	.44	.55	.87	1.45	.17

5. Cluster power drilling, min/occurrence

Metal Depth	Steel (Mild)	Soft Nonferrous
.2	.28	.08
.3	.30	.08
.4	.33	.09
.5	.36	.10
.6	.38	.10
.7	.40	.11
.8	.43	.12
.9	.45	.12
1.0	.47	.13

6. Counterboring, spotfacing, min/in.

Dia	Steel (Mild)	Iron, Steel (Medium)	Steel (Hard)
¼	.24	.31	.55
⅜	.27	.35	.61
½	.29	.37	.65
¾	.36	.47	.82
1	.36	.47	.82
1½	.55	.70	1.23
2	.58	.75	1.31

7. Power reaming, min/in.

Ream Dia	Steel (Mild)	Steel (Medium), Cast Iron	Steel (Alloy), Forgings, Stainless	Steel (High Tensile), Titanium	Aluminum, Brass, Magnesium
⅛	.09	.10	.20	.36	.04
½	.17	.22	.47	.87	.08
1	.22	.26	.55	1.09	.09
2	.27	.31	.65	1.34	.12
2½	.25	.27	.58	1.21	.11
3	.23	.26	.56	1.14	.10

8. Tapping, min/in.

Threads/In.	Steel	Stainless Steel	Aluminum	Brass	Plastic
56	.24		.13	.12	.14
48	.18		.13	.11	.12
40	.16	.38	.12	.10	.11
32	.18	.33	.13	.11	.12
24	.17	.30	.14	.10	.13
20	.15	.30	.12	.10	.12
16	.19	.33	.15	.11	.14
13	.23	.39	.18	.13	.15
11	.26	.39	.18	.13	.16
10	.32	.48	.21	.16	
9	.32	.48	.21	.16	
8	.32	.48	.21	.16	
7	.32	.48	.22	.16	
6	.34	.51	.22	.17	
5	.34	.51	.23	.17	
4	.38	.57	.24	.19	

9. Microdrilling, min per hole

Brass

	Dia	.005	.020	.040
Depth	.10	.003	.025	
	.20	.066	.050	.036
	.30		.075	.053
	.40			.070

Cast iron

	Dia	.020	.040
Depth	.10	.050	.036
	.20	.100	.073
	.30		.100
	.40		.145

Stainless Steel

	Dia	.040
Depth	.10	.05
	.20	.10

11.3 Face, Side, Slot, Form, Straddle, End, Saw, and Engrave Milling

DESCRIPTION

These data are for pure machine cutting time and are suitable for milling operations. Their locations are centralized, thus facilitating their greatest use.

Not all rotating-tool machining times are included in this section, particularly those more closely identified with turning, drilling, hobbing, gear cutting, etc.

ESTIMATING DATA DISCUSSION

The length of cut is the total distance the machine is operating at the cutting feed. The length is found by adding the safety, approach, blueprint length of cut and overtravel, if required. This distance is multiplied by the timer per in. specified for the type and material of cutter and material of part.

The approach for side milling recognizes that when the cutter first touches the material it is not at full depth. Once the vertical centerline of the cutter is over the initial line of material, the approach distance is determined. Entry variables for periphery cutting are diameter of cutter and depth of cut values for approach or overtravel are given below for peripheral cutters.

The effects of approach and overtravel are illustrated by Figure 11.3A, where the objects are non-proportional in size, but intend to convey the principle.

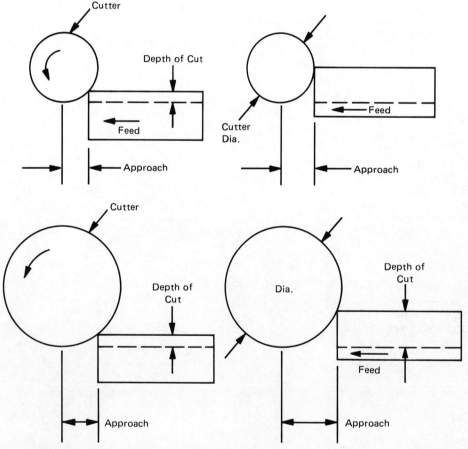

FIGURE 11.3A Side view of cutter and block.

Cutter approach or overtravel—helical, side, saw and key slot cutter, in.

Dia of Cutter, in.	Depth of cut								
	1/16	1/8	1/4	1/2	3/4	1	2	3	4
1¼	.27	.37	.50	.61					
1½	.30	.41	.56	.71	.75				
1¾	.32	.45	.61	.79	.87				
2	.35	.48	.64	.87	.97	1.00			
2½	.39	.54	.71	1.00	1.15	.122			
2¾	.41	.57	.79	1.06	1.22	1.32			
3	.43	.60	.83	1.12	1.30	1.41			
3¼	.45	.63	.87	1.17	1.37	1.50			
3½	.46	.65	.91	1.22	1.44	1.58			
3¾	.48	.67	.93	1.28	1.50	1.66			
4	.49	.70	.97	1.32	1.56	1.73	2.00		
4¼	.51	.72	1.00	1.37	1.62	1.80	2.12		
4½	.53	.74	1.03	1.41	1.67	1.87	2.24		
4¾	.54	.76	1.06	1.46	1.73	1.93	2.35		
5	.56	.78	1.09	1.50	1.79	2.00	2.45		
5½	.58	.82	1.14	1.58	1.89	2.12	2.64		
6	.61	.86	1.20	1.66	1.98	2.24	2.83	3.00	
7	.65	.93	1.30	1.80	2.17	2.45	3.16	3.46	
8	.71	.99	1.39	1.94	2.33	2.65	3.46	3.87	4.00
9	.75	1.05	1.48	2.06	2.49	2.83	3.74	4.24	4.47
10	.79	1.11	1.56	2.18	2.63	3.00	4.00	4.58	4.90
11	.83	1.17	1.64	2.29	2.77	3.16	4.24	4.89	5.29
12	.86	1.22	1.72	2.40	2.91	3.32	4.47	5.19	5.65

These side views show two cutter diameters and two depths of cut. These contrasting sketches point out the relationships for approach and overtravel effects caused by milling cutters. These figures apply only to helical, side, saw, and key slot type of cutters. Notice that for any depth of cut, the larger diameter cutters require a greater approach before full depth of cut is possible. Approach is the distance before the cutter is at the full depth of cut. The table gives the approach or overtravel distance that is added to the safety and drawing length. Overtravel may be zero or minimal when compared to the approach for these types of cutters, because the cutter no longer touches the material surface once the centerline of the cutter is past the ending point of the surface. So for helical, side, saw, and key slot type of cutters, overtravel may not be required.

For any diameter of cutter and depth of cut for that pass of the cutter, the table is used in the following way. The estimator will specify the diameter of the cutter, say 6-in. OD, and from the depth of cut for that pass, say 1 in., find from the table that the approach distance will be 2.24 in. which is added to the distance for the safety and part requirements.

A different approach or overtravel is necessary for a face or shell end mill approaching a flat surface. Entry variables are cutter diameter and width of work piece. Figure 11.3B shows the relationships for approach and overtravel for face, shell, and end milling type of cutters. Notice the top view where the objects are shown disproportionately to demonstrate principle. A cutter is required to traverse a length and return. The width of cut for the part is less than two cutter diameters. For a finish surface, the path length would be equal to safety stock at station no. 1, plus half of the cutter diameter, plus the part length L_d, plus half of the cutter diameter to station no. 2, plus the distance to station no. 3, plus the part length L_d and safety stock to station no. 4, plus the distance returning to the starting station no. 1. If the surface is considered rough, and a finish milling pass will follow, it may not be necessary to remove the cutter from the workpiece at station no. 2 and 3. Instead the cutter will not leave the surface of the part for rough cutting. Another case where data are not given is for milling a flat in a round piece of stock. A formula for this condition is $L = 2(Dia \cdot d - d^2)^{1/2}$ where L = length of cut, Dia = diameter of work piece, and d = depth of cut.

Cutter approach or overtravel—face, shell and end milling cutters, in.

Dia of Cutter, in.	Width of Cut										
	1	2	3	4	5	6	7	8	9	10	12
1	.50										
2	.27	1.00									
3	.09	.38	1.50								
4		.27	.68	2.00							
5		.21	.50	1.00	2.50						
6		.17	.40	.76	1.34	3.00					
7			.34	.63	1.05	1.70	3.50				
8				.53	.88	1.35	2.06	4.00			
10					.76	1.00	1.43	2.00	2.82	5.00	
12						.80	1.03	1.53	2.03	2.68	6.00
14							.88	1.26	1.64	2.10	3.38
16								1.07	1.38	1.76	2.71
18									1.20	1.52	2.29
20											2.00

A safety length is added for a short distance before the cutter enters the material.

Safety distance, in.

Cutter Dia	4	4 + to 8	8 + to 12	12 +
Distance, in.	.11	.14	.17	.20

Figure 11.3C shows the types of cutters that the data covers.

Elements 1 through 6 are the machining times for milling machines and machining centers. Their application depends upon the selection of milling machine initially. The machine times are related to the in. of milling or the no. of characters. The machine times have been determined using practical chip load per tooth, and periphery cutting velocity for either high-speed steel (HSS) or carbide types of cutters. The relationship is

$$\text{min/in} = \frac{\pi Dia}{12 f_t n V}$$

where Dia = cutter Dia, in.,
f_t = chip load per tooth, in. per tooth,
n = number of teeth in cutter,
V = cutter speed, fpm.

Standard cutter sizes and teeth were adopted. Obviously, differing variables using other approaches to determine cutting times are possible but many of their effects tend to offset each other. V is chosen to allow most any practical depth of cut and f_t will allow a com-

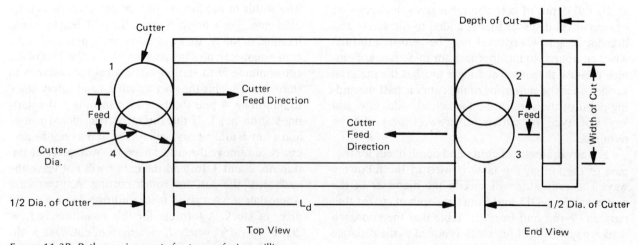

FIGURE 11.3B Path requirements for top surfacing milling.

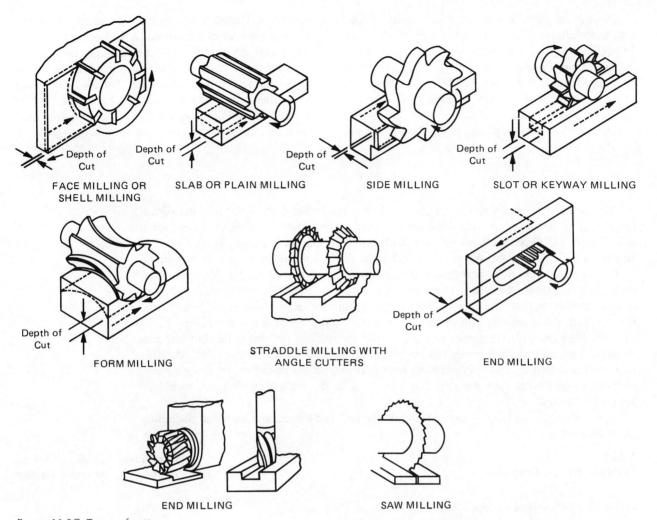

FIGURE 11.3C Types of cutters.

mercial surface finish. The hardnesses for the selected materials are as follows:

Material	Bhn
Low-carbon steel, free-machining	170
Low-carbon steel	150
Medium-carbon steel	215
Stainless steel, austenitic 300	170
Stainless steel, martensite 400	196
Steel castings	230
Aluminum	

While there are thousands of materials, these are a good representation. For other difficult materials the estimator makes comparative estimates.

The tables do not indicate the rpm that might be used. Sometimes the cutting velocity, tool, etc., are matched to the particular machine. For example, the cutting velocity may lead to an rpm that is unavailable on the machine. While a downward adjustment can be made to accommodate, it is not recommended at the time of estimate because rpm adjustment is only one of many factors that is difficult to foresee. Nor are distinctions made for cutting tool grades, lubrication, horsepower required, vertical or horizontal cutting set-ups in these data. These refinements, while oftentimes necessary, are too involved at the time of estimating. Thus, we choose min/in. of cut because that is easiest to use. On balance, this approach provides for consistency, ease, and accuracy.

Element 1, face milling, has the cutter axis perpendicular to the surface, and cuts along the periphery of individual teeth and the surface of the face cutting edges. The machined surface per cutter pass may equal the whole width of the face of the cutter or any fraction of the diameter. Thus, the entry variables are cutter and work piece material and cutter diameter. Overlap between adjacent passes is an assumption the estimator makes. For nonrigid setups, or parts prone to chatter, increase time by 30%.

Element 2, side milling, is different than Element 3 even though the cutter may be identical. Standard side-milling cutters are used for both applications. Convex or concave milling would use Element 3.

In end milling, Element 4, if the ratio of load length to diameter exceeds 3:1, increase the time 25%. (For roughing end mills, reduce the time 15%.) A four-fluted standard end mill is adopted. Other specialized end mill styles, while not true end mills, such as a T-slot or dovetail, will use this element.

Element 5, saw milling, adopts standard metal slitting saws. Distinctions whether the slitting saws are plain or have side chip clearance with or without staggered teeth do not influence the data very much.

Element 6, engraving, includes machine time and reposition tool to next character. The element would adopt the end-milling principles, but geometry, no. of flutes, and tool pointing are different. Engraving data are based upon time-study observations.

EXAMPLES

A. Estimate the part given by Figure 11.3D for machining only. Refer to other equipment data for handling and machine manipulation. These data, 11.3, are for machining only. Figure 11.3D requires two different milling cuts, or the three equally spaced scallops and the 2 × ½ × ½-in. slot. We are considering the scallops in this operation. It is assumed that previous operations have prepared the part for the scallop milling. As with most manufacturing operations the variety of processing methods is large. The milling of the scallops is no exception to that opportunity. Two different approaches are considered here. In the first method, the machining can be with an end mill, and the length of cut will be ¼ in. In this approach the end mill moves or plunges into the surface. Following this cut, the machine, tooling, and operator present two other surfaces for the same cutting action. Machine manipulation, tooling, and operator time requirements are not estimated by this example, and other equipment data, notably that presented in the milling equipment, would be adopted and used.

In this example we have 41 units. Estimate the unit and lot requirements for machining the scallops only.

Table Number	Process Description	Table Time	Adjustment Factor	Cycle Minutes	Setup Hours
	First method:				
11.3-4	End mill three scallops where LOC = ¼ in. per pass; carbide tooling, 2-in. OD cutter	0.06	.25 × 3	0.045	—
	Total			0.045	0.00

Note that the data did not provide a data entry for a 2-in. cutter and that we used the top value as provided which was the 1-in. carbide cutter machining rate. The rules for table lookup as given in the introduction encourage estimator judgment in cases where the *AM Cost Estimator* does not provide data. Normally the advice suggests that the top value be used, or that the estimator make extrapolative judgment.

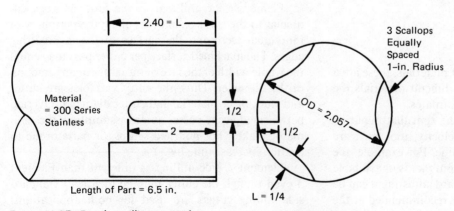

FIGURE 11.3D Part for milling example.

Table Number	Process Description	Table Time	Adjustment Factor	Cycle Minutes	Setup Hours
11.3-4	Second method: End mill scallops by feeding down 2.4 in., three passes for 2-in. carbide cutter	0.06	2.4 × 3	0.43	—
	Total			0.43	0.00

Note that the direction of the milling feed was changed from that given in the first method. An end mill can cut both on the bottom of the cutter as well as the periphery of the cutter.

B. Estimate the slot for machining time only. Assume that the part has been prepared by previous operations. Other time for operator loading, machine manipulation, tooling, etc., would be taken from other equipment data tables. The material is stainless steel grade 300 series.

Table Number	Process Description	Table Time	Adjustment Factor	Cycle Minutes	Setup Hours
11.3-4	End mill slot 2 × ½ × ½ in. with carbide mill ½-in. OD with LOC = 2 in., difficulty increases estimate by 25%	0.04	2 × 1.25	0.10	
	Total			0.10	

Note that the adjustment factor column is used to increase the time for the element time for machine if the estimator has other understanding of the job on the equipment that might be used by the company.

TABLE 11.3 FACE, SIDE, SLOT, FORM, STRADDLE, END, SAW, AND ENGRAVE MILLING

Setup   **In other tables**

Operation elements in estimating minutes

1. Face, plain, or shell milling, min/in.

	HSS Dia					Carbide Dia				
Material	2	3	4	5	6	3	4	5	6	8
Low-carbon steel, free-machining	.08	.09	.10	.11	.12	.03	.04	.04	.05	.05
Low-carbon steel	.18	.22	.24	.26	.27	.08	.11	.10	.12	.13
Medium-carbon steel	.27	.33	.36	.39	.41	.12	.16	.15	.18	.19
Stainless steel, 300	.55	.65	.73	.78	.82	.13	.17	.16	.19	.20
Stainless steel, 400	.36	.44	.48	.52	.55	.12	.16	.15	.18	.19
Steel castings, forgings	.36	.44	.48	.52	.55	.13	.17	.16	.19	.20
Aluminum	.01	.01	.01	.02	.02	.01	.01	.01	.01	.01
Copper alloys	.03	.03	.04	.04	.04	.02	.02	.02	.02	.02
Plastics	.05	.06	.06	.07	.07	.03	.03	.03	.03	.03
Cast iron	.31	.37	.42	.45	.47	.12	.16	.15	.18	.19

2. Side milling, min/in.

Material	HSS Dia						Carbide Dia				
	2	3	4	5	6	8	3	4	5	6	8
Low-carbon steel, free-machining	.07	.10	.11	.13	.14	.16	.05	.07	.06	.06	.07
Low-carbon steel	.14	.18	.22	.24	.26	.30	.10	.14	.13	.12	.15
Medium-carbon steel	.31	.41	.48	.55	.59	.67	.17	.23	.22	.21	.25
Stainless steel, 300	.31	.41	.48	.55	.59	.67	.17	.23	.22	.21	.25
Stainless steel, 400	.37	.49	.58	.65	.71	.81	.17	.23	.22	.21	.25
Steel castings, forgings	.93	1.23	1.45	1.64	.178	2.01	.30	.40	.37	.36	.43
Aluminum	.01	.02	.02	.02	.02	.03	.01	.01	.01	.01	.01
Copper alloys	.07	.09	.10	.12	.13	.14	.06	.08	.08	.08	.09
Plastics	.04	.05	.06	.07	.08	.09	.04	.04	.04	.04	.04
Cast iron	.50	.65	.78	.87	.95	1.40	.22	.29	.27	.26	.32

3. Slot, form and straddle milling, min/in.

Material	HSS Dia						Carbide Dia				
	2	3	4	5	6	8	3	4	5	6	8
Low-carbon steel, free-machining	.10	.14	.17	.19	.21	.24	.04	.09	.08	.10	.11
Low-carbon steel	.42	.55	.65	.73	.79	.90	.14	.18	.17	.20	.22
Medium-carbon steel	.62	.82	.97	1.09	1.19	1.34	.26	.35	.33	.39	.42
Stainless steel, 300	.62	.82	.97	1.09	1.19	1.34	.26	.35	.33	.39	.42
Stainless steel, 400	.75	.98	1.16	1.31	1.43	1.61	.26	.35	.33	.39	.42
Steel castings, forgings	1.25	1.64	1.94	2.18	2.38	2.69	.30	.40	.37	.45	.48
Aluminum	.03	.04	.05	.05	.06	.07	.02	.02	.02	.02	.02
Copper alloys	.13	.18	.21	.23	.25	.29	.08	.11	.10	.13	.13
Plastics	.08	.11	.13	.15	.16	.18	.05	.08	.08	.10	.13
Cast iron	1.00	1.31	1.55	1.75	1.90	2.15	.38	.51	.47	.57	.61

4. End milling, min/in.

Material	HSS Dia				Carbide Dia			
	⅛	¼	½	1	⅛	¼	½	1
Low-carbon steel, free-machining	.10	.10	.06	.10	.03	.03	.02	.02
Low-carbon steel	.18	.18	.12	.18	.05	.05	.03	.04
Medium-carbon steel	.68	.55	.27	.36	.11	.14	.07	.07
Stainless steel, 300	.41	.55	.33	.36	.07	.07	.04	.06
Stainless steel, 400	.55	.65	.65	.36	.07	.07	.04	.06
Steel castings, forgings	.82	1.09	.65	.55	.12	.15	.07	.09
Aluminum	.03	.03	.01	.01	.01	.01	.01	.01
Copper alloys	.16	.18	.13	.12	.04	.04	.02	.03
Plastics	.05	.05	.04	.05	.03	.03	.02	.02
Cast iron	.74	.74	.40	.48	.12	.14	.07	.08

5. Saw milling, min/in.

	HSS Dia, Blade < ⅛ in.					HSS Dia, Blade ≥ ⅛ in.				
Material	2½	3	4	5	6	2½	3	4	5	6
Low-carbon steel, free-machining	.27	.31	.34	.38	.44	.14	.15	.17	.19	.22
Low-carbon steel,	.52	.58	.65	.73	.83	.26	.29	.32	.36	.42
Medium carbon steel	1.30	1.45	1.62	1.82	2.08	.78	.87	.97	1.09	1.25
Stainless steel, 300	1.30	1.45	1.62	1.82	2.08	.78	.87	.97	1.09	1.25
Stainless steel, 400	1.56	1.75	1.94	2.18	2.49	.93	1.05	1.16	1.31	1.50
Steel castings, forgings	1.95	2.18	2.42	2.73	3.12	1.17	1.31	1.45	1.64	1.87
Aluminum	.08	.09	.10	.11	.12	.04	.04	.05	.05	.06
Copper alloys	.17	.19	.21	.23	.27	.08	.09	.10	.12	.13
Plastics	.17	.19	.22	.24	.28	.10	.12	.13	.15	.17
Cast iron	1.56	1.75	1.94	2.18	2.49	.93	1.05	1.40	1.31	1.50

6. Engraving, min per character

	Height of Character							
Material	⅛	5/32	3/16	¼	5/16	⅜	7/16	½
Steel	.15	.16	.17	.18	.19	.19	.21	.22
Brass	.13	.14	.15	.16	.17	.18	.19	.19
Aluminum	.12	.13	.14	.15	.15	.16	.17	.18
Plastic	.07	.07	.08	.08	.09	.10	.12	.12

11.4 Tool Life and Replacement

DESCRIPTION

Cutting tools become dull as usage continues and their effectiveness drops. At some point in time it is necessary to resharpen, index, or replace the tool. The span of usefulness, from sharpness to dullness, is defined as tool life. The quantity of tool life varies with the quality of the cutting materials operating in different materials as well as with many other variables.

When tools become dull, direct-labor time is required to replace them with sharp ones. Not all tools are sharpened by the operator, such as reamers, milling cutters, and gear hobs. While drills can be sharpened, the sharpening can also be done by the tool crib, and this is a discretionary selection by estimators. The time required to keep cutting tools operable is prorated over units produced during the tool life. The following data estimate that time.

These data provide a time to prorate the removal, also possibly resharpening, and replacement of the tool in the machine. The prorated time is added to the operation estimate. Operations related to turning, milling, hobbing, and drilling are historically handled by tool life and replacement as provided here.

ESTIMATING DATA DISCUSSION

The assumptions for tool life and replacement for turning elements are:

Material	HSS	Carbide
Soft steel	160 min	425 min
Medium steel, cast iron	120 min	360 min
Hard steel, stainless	90 min	225 min

The time to replace a turning, boring, etc., type of tool is 2 min. Entry variables are cutting time, type of material, and tool. Inasmuch as there may be several or more turning tools, the cutting time is total machining time for the operation, and the minutes are the prorated life for several tools for one unit. Even though turning operations may include drilling, simply sum all the machining time as the entry variable. Thus, expediency is used for this estimate because of the involved nature of tool life.

Element 2 deals with milling tool life. The first subpart provides an estimate whenever the tool edge is an indexable carbide point. Entry variables are cutting time and part-material. The basic time to index and qualify an insert point is 1.00 min.

In the second subpart of Element 2, the time, as read from the chart, is the prorated unit time for removal and replacement of small tools. For this element, small tools are those having shanks for insertion in the end of the milling machine spindle, those having a center hole for arbor mounting, or those having the cutter back recessed for a single-bolt. Single-bolt mounted face cutters are 6-in. or less. The basic time is 3 min.

Large tools are those face cutters having a multiple-bolt mounting, or face cutter 8-in. or greater. These sizes have a bolt-circle and not single-point mounting. Large tool replacement time is 10 min.

The tool life, in min, of the tool for Element 2 is as follows:

Material	HSS	Carbide
Soft steel	45 min	120 min
Medium steel	30 min	90 min
Cast iron	30 min	90 min
Stainless steel	15 min	60 min

The Taylor tool life equation for cast iron is $VT^{0.12} =$ 225 and $VT^{0.43} = 3000$, for HSS and carbide respectively.

Element 3 deals with tool life for drills and drill-type tools. Entry variables are drill dia, total drill min corresponding to the diameters and part material. The time to remove drill, regrind, and replace is 3 min, and it is this time that is prorated to the job. The tool life for the variables is given below.

Dia (in.)	Stainless, Cast Steel, Forgings, Tool Steels	Steel (Medium, Hard)	Steel (Mild), Cast Iron	Aluminum Brass, Magnesium
¼	25	30	30	60
¼ to ½	35	35	35	75
½ to ¾	40	40	45	90
¾ to 1	40	50	60	105
1 in. +	50	55	60	120

For SAE 1020 steel, if for a ⅝-in. OD drill there is a cycle time of 1.20 min for drilling, then $45/1.20 = 37.5$ parts are drilled, and the resharpen time $= 3/37.5 = 0.08$ min which is about 7% in this example.

EXAMPLES

A. A turret lathe uses 7 tools including start drill, drill, boring bar, reamer, recessing-type tool, tap, and cutoff. The total machine time is 7.62 min. Find the time for tool life to add to the unit estimate.

Table Number	Process Description	Table Time	Adjustment Factor	Cycle Minutes	Setup Hours
11.4-1	Tool life for 7 tools for carbide machining of medium-hardness steel	.04		.04	
	Total			.04	

B. A 9-in face mill made for right-hand cutting and having 20 teeth is used to mill the top of a cylinder block on a vertical knee-and-column milling machine. The cast iron is 170–197 Bhn, length of cut is 31¼ in. The cutting time including cutting, approach, overtravel, and safety stock is 2.19 min. Find the time to add to the unit estimate for indexable inserts, or remove and replace a large tool.

Table Number	Process Description	Table Time	Adjustment Factor	Cycle Minutes	Setup Hours
11.4-2	Unclamp, reposition, qualify, and clamp inserts	.03	20	.60	
	Total			.60	

Table Number	Process Description	Table Time	Adjustment Factor	Cycle Minutes	Setup Hours
11.4-2	Unbolt worn cutter and rebolt new cutter	.28		.28	
	Total			.28	

C. An NC machining center drills 18 holes in a steel forging according to the schedule below. Presuming that the NC tape is interrupted to allow tool changing within the tool storage, what time would be added to the unit estimate?

No.	Dia	Cycle Time	No.	Dia	Cycle Time
3	F	.06	1	1.0	.72
2	25/64	.19	4	.875	.27
2	33/64	.23	2	#45	.08
4	.75	.26			

Table Number	Process Description	Table Time	Adjustment Factor	Cycle Minutes	Setup Hours
11.4-3	Two drills are less than ¼ in., and total cycle time = .34 min	.06		.06	
11.4-3	One drill is less than ½ in., and total cycle time = .38 min	.04		.04	
11.4-3	Two drills are less than ¾ in., and total cycle time = 1.50 min	.15		.15	
11.4-3	One drill is less than 1 in., and total cycle time = 1.08 min	.15		.15	
11.4-3	One drill equals 1 in., and total cycle time = .72 min	.06		.06	
	Total			.46	

The drill life and replacement time is added to the unit estimate for the machine, and the estimate is 11% of the total drilling time.

TABLE 11.4 TOOL LIFE AND REPLACEMENT

Setup   Not applicable

Operation elements in estimating minutes

1. Turning tool life, min per operation

Operation Cutting Time	Soft Steel		Medium Steel, Cast Iron		Stainless Steel	
	HSS	Carbide	HSS	Carbide	HSS	Carbide
.3			.01		.01	
.4	.01		.01		.01	
.5	.01		.01		.01	.004
.6	.01	.003	.01	.003	.01	.01
.7	.01	.003	.01	.004	.02	.01
.8	.01	.004	.01	.004	.02	.01
.9	.01	.004	.02	.01	.02	.01
1.2	.02	.01	.02	.01	.03	.01
1.5	.02	.01	.03	.01	.03	.01
1.8	.02	.01	.03	.01	.04	.02
2.2	.03	.01	.04	.01	.05	.02
2.5	.03	.01	.04	.01	.06	.02
3.0	.04	.01	.05	.02	.07	.03
3.5	.04	.02	.06	.02	.08	.03
4.0	.05	.02	.07	.02	.09	.04
4.5	.06	.02	.08	.03	.10	.04
5.0	.06	.02	.08	.03	.11	.04
5.5	.07	.03	.09	.03	.12	.05
6	.08	.03	.10	.03	.13	.05
7	.09	.03	.12	.04	.16	.06

Operation Cutting Time	Soft Steel		Medium Steel, Cast Iron		Stainless Steel	
	HSS	Carbide	HSS	Carbide	HSS	Carbide
8	.10	.04	.13	.04	.18	.07
9	.11	.04	.15	.05	.20	.08
10	.13	.05	.17	.06	.22	.09
12	.15	.06	.20	.07	.27	.11
14	.18	.07	.23	.08	.31	.12
16	.20	.08	.27	.09	.36	.14
18	.23	.08	.30	.10	.40	.16
20	.25	.09	.33	.11	.44	.18
25	.31	.12	.42	.14	.56	.22
30	.38	.14	.50	.17	.67	.27
35	.44	.16	.58	.19	.78	.31
40	.50	.19	.67	.22	.89	.36

2. Milling tool life, min per operation

   A. Index a throwaway insert, min per insert on tool. Multiply by no. of inserts.

Operation Cutting Time	Soft Steel	Medium Steel, Cast Iron	Stainless Steel
.8	.01	.01	.01
1.0	.01	.01	.02
1.2	.01	.01	.02
1.4	.01	.02	.02
1.5	.01	.02	.03
2.0	.02	.02	.03
2.5	.02	.03	.04
3.0	.03	.03	.05
3.5	.03	.04	.06
4.0	.03	.04	.07
4.5	.04	.05	.08
5	.04	.06	.08
6	.05	.07	.10
7	.06	.08	.12
8	.07	.09	.13
9	.08	.10	.15
10	.08	.11	.17
12	.10	.13	.20
15	.13	.17	.25
20	.17	.22	.33
25	.21	.28	.42

   B. Remove and replace small or shank tool, min per operation

Cutting Time	Soft Steel		Medium Steel, Cast Iron		Stainless Steel	
	HSS	Carbide	HSS	Carbide	HSS	Carbide
.2	.01	.01	.02	.01	.04	.01
.4	.03	.01	.04	.01	.08	.02

Cutting Time	Soft Steel		Medium Steel, Cast Iron		Stainless Steel	
	HSS	Carbide	HSS	Carbide	HSS	Carbide
.6	.04	.02	.06	.02	.12	.03
.8	.05	.02	.08	.03	.16	.04
1.0	.07	.03	.10	.03	.20	.05
1.2	.08	.03	.12	.04	.24	.06
1.4	.09	.04	.14	.15	.28	.07
1.5	.10	.04	.15	.05	.30	.08
2.0	.13	.05	.20	.07	.40	.10
2.5	.17	.06	.25	.08	.50	.13
3.0	.20	.08	.30	.10	.60	.15
3.5	.23	.09	.35	.12	.70	.18
4.0	.27	.10	.40	.13	.80	.20
4.5	.30	.11	.45	.15	.90	.23
5	.33	.13	.50	.17	1.00	.25
6	.40	.15	.60	.20	1.20	.30
7	.47	.18	.70	.23	1.40	.35
8	.53	.20	.80	.27	1.60	.40
9	.60	.23	.90	.30	1.80	.45
10	.67	.25	1.00	.33	2.00	.50
12	.80	.30	1.20	.40	2.40	.60
15	1.00	.38	1.50	.50	3.00	.75
20	1.33	.50	2.00	.67	4.00	1.00
25	1.67	.63	2.50	.83	5.00	1.25

C. Remove and replace large tool, min per operation

Cutting Time	Soft Steel		Medium Steel, Cast Iron		Stainless Steel	
	HSS	Carbide	HSS	Carbide	HSS	Carbide
.2	.04	.02	.07	.02	.13	.03
.4	.09	.03	.13	.04	.27	.07
.6	.13	.05	.20	.07	.40	.10
.8	.18	.07	.27	.09	.53	.13
1.0	.22	.08	.33	.11	.67	.17
1.2	.27	.10	.40	.13	.80	.20
1.4	.31	.12	.47	.16	.93	.23
1.5	.33	.13	.50	.17	1.00	.25
2.0	.44	.17	.67	.22	1.33	.33
2.5	.56	.21	.83	.28	1.67	.42
3.0	.67	.25	1.00	.33	2.00	.50
3.5	.78	.29	1.17	.39	2.33	.58
4.0	.89	.33	1.33	.44	2.67	.67
4.5	1.00	.38	1.50	.50		.75
5	1.11	.42	1.67	.56		.83
6	1.33	.50	2.00	.67		1.00
7	1.56	.58	2.33	.78		1.17
8	1.78	.67	2.67	.89		1.33

Cutting Time	Soft Steel		Medium Steel, Cast Iron		Stainless Steel	
	HSS	Carbide	HSS	Carbide	HSS	Carbide
9	2.00	.75	3.00	1.00		1.50
10	2.22	.83	3.33	1.11		1.67
12	2.67	1.00		1.33		2.00
15	3.33	1.25		1.67		2.50
20	4.44	1.67		2.22		3.33
25	5.56	2.08		2.78		4.17

3. Drilling and tapping tool life, min per operation

   Remove drill or tap, grind or replace, and install drill or tap

Drill or Tap Dia	Total Drill Cutting Time per Unit	Stainless, Cast Steel, Forgings, Tool Steels	Steel (Medium, Hard)	Steel (Mild), Cast Iron	Aluminum, Brass, Magnesium
¼	.10	.01	.01	.01	.01
	.20	.02	.02	.02	.01
	.50	.06	.05	.05	.03
	1.0	.12	.10	.10	.05
	2.0	.24	.20	.20	.10
	3.0	.36	.30	.30	.15
	4.0	.48	.40	.40	.20
	5.0	.60	.50	.50	.25
¼+ to ½	.10	.01	.01	.01	
	.20	.02	.02	.02	.01
	.50	.04	.04	.04	.02
	1.0	.09	.09	.09	.04
	2.0	.17	.17	.17	.08
	3.0	.26	.26	.26	.12
	4.0	.34	.34	.34	.16
	5.0	.43	.43	.43	.20
½+ to ¾	.10	.01	.01	.01	
	.20	.02	.02	.01	.01
	.50	.04	.04	.03	.02
	1.0	.08	.08	.07	.03
	2.0	.15	.15	.13	.07
	3.0	.23	.23	.20	.10
	4.0	.30	.30	.27	.13
	5.0	.38	.38	.33	.17
¾+ to 1	.10	.01	.01	.01	
	.20	.02	.01	.01	.01
	.50	.04	.03	.03	.01
	1.0	.08	.06	.05	.03
	2.0	.15	.12	.10	.06
	3.0	.23	.18	.15	.09
	4.0	.30	.24	.20	.11
	5.0	.38	.30	.25	.14

Drill or Tap Dia	Total Drill Cutting Time per Unit	Stainless, Cast Steel, Forgings, Tool Steels	Steel (Medium, Hard)	Steel (Mild), Cast Iron	Aluminum, Brass, Magnesium
1+	.10	.01	.01	.01	
	.20	.01	.01	.01	.01
	.50	.03	.03	.03	.01
	1.0	.06	.05	.05	.03
	2.0	.12	.11	.10	.05
	3.0	.18	.16	.15	.08
	4.0	.24	.22	.20	.10
	5.0	.30	.27	.25	.13

BROACHING

12.1 Broaching Machines

DESCRIPTION

Broaching is the process whereby a cutter, called a broach, is used to finish internal or external surfaces such as holes, squares, irregular sections, keyways, teeth of internal gears, splines, and flat surfaces. The broach is an elongated tool having a number of successive teeth of increasing size cutting in a fixed path. A part is often completed in one pass, where the last teeth on the broach conform to finished dimension. The work being broached is usually held in a fixed position and the tool travels through or over the work. The tool, after completing the cut, is returned to the start position, thus completing the cycle. Internal broaches start where the hole has been drilled, reamed or bored, cored, stamped or hot pierced.

A broaching machine consists of the tool, workholding fixture, drive, and frame. The machines may either pull or push the tool through the work. Varieties of models include horizontal surfacing, single-ram vertical, double-ram vertical, large pull down, vertical pull-up electric, and horizontal hydraulic broaching. Even manual small-size presses can be used to push broaches through a hole.

Figure 12.1 is a single-ram broaching machine with variable velocity for surface broaching. This machine has tipdown tables for easier loading. Shuttle tables are optional. Surface velocities range from 5.35 fpm and a return speed of 70 fpm.

Some machines have automatic broach bar engagement while in others, the broach must be retrieved from the inner floor of the machine. In some machines, after disassembly of piece, pushing of a power lever returns the broach bar to the starting position.

FIGURE 12.1 Single-ram vertical broaching machine. (Ex-Cell-O Machine Tool Products)

ESTIMATING DATA DISCUSSION

Setup hours include the standard setup and teardown of the machine, cleaning of fixtures, broaches, guides, gages, etc. used on the job. Distinction is provided for machine and line-up for fixed or loose broaches. The operation elements are for a variety of machines, and the estimator needs to be able to identify machine features.

Pick up, place in position to hold, and remove and place aside may be determined using Element 1 of Table 12.1. Fixtures may be unnecessary, as in the case of a simple nesting die, and various tool designs will have clamping distinctions. If the broach machine is semiautomatic, this element is unnecessary, because the machine has automatic pull or push heads. If additional passes are required, shims of the desired thickness are used, and the time per pass is obtained from Element 4. Inspection and deburring, Elements 6 and 7, may be handled during machining cycle of previous passes, if time is available.

Machining time is found using this relationship:

$$\text{Machine time} = \frac{L \text{ of stroke}}{\text{cutting speed}} + \frac{L \text{ of stroke}}{\text{return speed}}$$

and the estimates of fpm are provided for four material classes. The data are based on a return velocity of 2:1, even though some machines have fixed return velocities. Corrections can be made using the formula if considered important. In duplex broaching machines, it may not be necessary to allow a return stroke, and this time must be deducted from the estimate. Also, the fixture may be loaded during one of the machine cycles, thus the estimate will be machine time only. In the absence of specific knowledge about broach length, assume the length of the broach for one pass as:

Type	Equation
Splines	Broach length = $-5 + 15$ (part Dia) + 8 (part L)
Internal keyways	Broach length = $20 + 40$ (key width) + 85 (key depth)
Round holes	Broach length = $6 + 6$ ($Dia + 6$) part L

For a six-spline having a nominal diameter of 2 in. and a part length of 5 in., the approximate broach length is 62 in.

Speed and ease of application can be increased by avoiding calculation or interpolation if the next higher Table 12.1 values are used.

EXAMPLES

A. Estimate the time to broach a lot of 18,000 automotive front-gear blanks where the gear has 39 internal involute spline teeth. The work will be done on a 50-ton pull-down broaching machine with 42-in. stroke. Two splines must be square with the two faces within 0.003 in. total indicator reading. Before broaching, the faces are ground and the hole is precision-bored with the faces. The part is 4140 material and weighs 7.5 lb. The broach is estimated to be 30-in. long.

Table Number	Process Description	Table Time	Adjustment Factor	Cycle Minutes	Setup Hours
12.1-S	Setup	.5			.5
12.1-1	Handle at 7.5 lb	.17		.17	
12.1-1	Tip over discharge	−.02		−.02	
12.1-2	Location on hole and face in nest	.04		.04	
12.1-4	Engage power lever	.03		.03	
12.1-5	Apply oil	.04		.04	
12.1-8	Machine and retract broach	.58		.58	
	Total			.84	.5
	Total lot estimate	252.5 hr			

B. An airframe part has been previously machined and is shown by Figure 6.4C. A turret lathe was chosen for the first operation, and the hole is predrilled ready for broaching. A quantity of 7075 is required. Find the estimates.

Table Number	Process Description	Table Time	Adjustment Factor	Cycle Minutes	Setup Hours
	Broach square hole				
12.1-S	Setup	0.50			0.50
12.1-1	Handle	0.11		0.11	

Table Number	Process Description	Table Time	Adjustment Factor	Cycle Minutes	Setup Hours
12.1-2	Assm. and disassemble	0.03		0.03	
12.1-8	Machining	0.47		0.47	—
	Broach L = 30 in.				
	Total				
	Lot estimate	72.43 hr		0.61	0.50

TABLE 12.1 BROACHING MACHINES

Setup

    Single-ram machine   .5 hr
    Double-ram machine   .6 hr
    Horizontal machine   .7 hr

Operation elements in estimating minutes

1. Pick up part and put aside
   From tote pan or small carton
   From pallet or skid,

Weight	2–6	7–10	11–15
Min	.11	.17	.25

   Chain hoist   .74
   Less for tilt discharge   .02

2. Assemble and disassemble from fixture
   Locate from 1 hole   .03
   Locate from 2 holes   .03
   Locate from 1 hole and 1 edge   .04
   Locate from 1 hole and 2 edges   .03
       or wing screw or nut   .04
       or palm grip knobs   .08
       or thumbscrew   .06
       or wedge   .10
       or hold-down bar with 2 nuts   .04

3. Assemble part over tail end of broach   .04
   Assemble broach through port to pull head   .02
   Assemble part over shank of bar and assemble broach to pull head   .18
   Disassemble broach from pull head   .08

4. Shim for additional cut, per pass   .05
   Direct oil to part   .02
   Start and stop   .03
   Engage power lever
       No disassemble of broach   .03
       Disassemble broach   .02
       Engage and disengage ram   .05

5. Wire brush broach bar   .15
   Clean fixture with brush   .04
   Apply oil to broach with brush, 1 per 5   .04

6. Inspection
   Go/no-go plug gage   .07
   Spline gage   .05
   Inside micrometer   .10
   Scale rule   .08

7. File burrs off part .07

8. Machine time, min per operation

Broach L	Brass, Bronze, Cast Iron	Mild Steel	Cast Iron, Medium Steel	Hard Steel
6.0	.03	.03	.09	.11
8.8	.04	.05	.13	.16
10.0	.04	.05	.15	.18
11.0	.05	.06	.17	.20
12.1	.05	.06	.18	.22
13.3	.06	.07	.20	.25
14.6	.06	.08	.22	.27
16.1	.07	.08	.24	.30
17.7	.07	.09	.27	.33
19.5	.08	.10	.29	.36
21.4	.09	.11	.32	.39
23.6	.10	.12	.35	.43
25.9	.11	.14	.39	.48
28.5	.12	.15	.43	.53
31.4	.13	.17	.47	.58
34.5	.15	.18	.52	.64
38.0	.16	.20	.57	.70
41.8	.18	.22	.63	.77
45.9	.19	.24	.69	.85
50.5	.21	.27	.76	.93
55.6	.23	.29	.83	1.02
61.2	.26	.32	.92	1.13
67.3	.28	.35	1.01	1.24
74.0	.31	.39	1.11	1.36
fpm return	35.	28.	18.	8.
fpm	70.	56.	36.	16.

GRINDING MACHINES

13.1 Cylindrical Grinding Machines

DESCRIPTION

This machine is used primarily for grinding external cylindrical surfaces, tapered and simple-formed surfaces, and shoulder faces. These data cover axial traverse and plunge grinding. Special single-purpose machines can be rough-estimated, using these data, if machine knowledge is available.

Plain cylindrical grinding machines have swings from 6 to 24 in., and center distances from 18 to 120 in. The power of the wheel motor of regular machines varies from 5 to 25 hp.

Figure 13.1A is a center-type grinding machine with a swing-down internal grinding head. It has an infinitely variable table traverse speed from 2 to 240 ipm and infinite selection for headstock speeds. Pick feed amounts from .0002 to .0016 in. (diameter reduction) in eight steps are available for traverse grinding. Now, consider estimating data for the class of cylindrical grinding machines.

ESTIMATING DATA DISCUSSION

In addition to usual operator, machine, and part instruction chores involved in a setup for cylindrical grinding machines, special considerations can be given to grinding wheel diameter and tolerance requirements. Setup includes the installation of work-holding devices, setting of feed and traverse rates, limit positions for rapid approach, and power infeed and initial truing of installed grinding wheel. The setup values, as given in Table 13.1, are not for semi- or fully automatic operations, since that time may be from one to several hours. The setup data are based upon two factors, wheel diameter and tolerance. The total tolerance, or sum of bilateral tolerance values, are used.

The operational elements are composed of handling, machine operation, grinding, and tolerance control of the part. Handling time is composed of elements to pick up and unload part; and then to mount it in a collet, chuck, face plate, dog, arbor, or mandrel; and if not already secured to machine, then mount the part between centers. Machine operation includes the position of the work for traverse or plunge grinding.

Element 4 provides time for traverse grind, dwell, and sparkout. Entry variables are total stock, material, work velocity, and wheel traverse feed. The relationship used for traverse time is

$$\text{time} = \frac{L \times T_s \times Dia}{(WP)2f_i \pi V}$$

where L = length of part grind, T_s = total stock removed from diameter, Dia = original diameter, W = width of grinding wheel, P = traverse for each work revolution in fraction of wheel width, f_i is the infeed of wheel per pass, and V = workpiece peripheral velocity.

If there are no interfering shoulders with a larger diameter than the diameter to be ground, there is no

FIGURE 13.1A Center-type grinding machine with a swing-down internal grinding head feature. *(Cincinnati Milacron)*

adjustment to part length. An overlap or one-half of the wheel width or ½ in. is adequate. If one shoulder prevents wheel overlap, the wheel grinds to that surface.

The symbol W is wheel width, 1-in., 2-in., . . . , and corresponds to the machine specification being estimated. The symbol P is the traverse for each work revolution in fractions of wheel width, and is given later in the information. The infeed rates, f_i, refer to the penetration of the grinding wheel into the work. The diameter is reduced by twice the amount of wheel advance. Note that some cylindrical grinding machines have cross-slide hand wheels with graduations indicating double the amount of actual wheel advance. This calculation is performed twice, once for rough-grind and once for finish grind, and is summed in the table values. A stock of .002 in. is allowed for each finish grind. The estimating approach assumes a two-infeed and two-traverse feed operation.

In cylindrical traverse grinding, the table values are multiplied by $L \cdot Dia/W$ after the entry values are chosen and determined from the drawing.

For traverse cylindrical grinding and dwell, incremental values for stock removal above .040 in. are given in units of .001 in. Add the additional times to the upper table values and multiply by $L \cdot Dia/W$ or Dia.

Dwell, or tarry, in traverse grinding refers to the controlled delay of the table at the end of the traverse. This delay extends the engagement time of the wheel with the surface at the end position of the traverse. Dwell is used for areas which are not contacted twice during to-and-fro movements of the table, and whenever grinding is close to a shoulder. Entry variables are

Representative Material	*Material Condition*	*V, Velocity fpm*	*f_i = infeed, in. per pass*		*P = traverse for each work revolution in fractions of wheel width*	
			Rough	*Finish*	*Rough*	*Finish*
1. Tool steel	Hardened	50	.001	.0002	¼	⅛
2. Tool steel	Annealed	60	.001	.0004	½	⅙
3. Plain carbon steel,	Hardened	70	.001	.0004	¼	⅛
alloy steel	Hardened	70	.001	.0004	¼	⅛
4. Plain carbon steel,	Annealed	100	.001	.0004	½	⅙
alloy steel,	Annealed	100	.001	.0004	½	⅙
copper alloys	Annealed or					
	cold drawn	100	.001	.0004	⅓	⅙
5. Aluminum alloys	Cold drawn	150	.001	.0005	⅓	⅙
	Solution-treated					

stock removal and material. Dwell time is multiplied by part diameter, *Dia*.

Sparkout overcomes the part or machine deflections during grinding and also improves the finish. Sparkout occurs after the part has attained final size by discontinuing infeed and making passes. A pass is one movement across the face of the wheel in one direction, and three finish grind passes are used. For sparkout time, the entry variable is material or velocity, *V*. Sparkout times are multiplied by the ratio $L \cdot Dia/W$.

For Element 5, plunge grinding, the feed rates per revolution are less than traverse grinding. On the other hand, infeed during plunge grinding is continuous. Infeed rates used by the table are:

	Infeed Per Rev	
Work Material	Rough	Finish
Steel, soft	.0005	.0002
Plain carbon steel, hardened	.0002	.00005
Alloy and tool steel, hardened	.0001	.000025

The resultant reduction of work diameter is twice the amount of cross-slide movement causing wheel infeed.

Entry variables for Element 5 are typical material, diameter, and total stock removed. Incremental additions are possible for increases in diameter and stock removal beyond those shown. A two-feed plunge grind is calculated for the table, an approach allowance constant of .008 in., and 4 rev for sparking out are included.

For the dress wheel, Element 6, truing of the wheel is necessary and the interval depends on performance and wear of the wheel and accuracy-finish requirements of the operation. For a 20-in. wheel, 6500 fpm, and a traverse feed rate of .003 in. per wheel rev, a travel rate of 3.720 in./min axially is used, which is prorated over 20 parts.

In hand-controlled operations, gaging or operator-inspection, Element 7, is necessary. Micrometer, snap gages, air gages, or electronic devices are used. Automatic size control, which concludes the grind operation when a preset size is reached, virtually requires no time, and is not shown.

EXAMPLES

A. An alloy steel shaft turned to 1.512/1.515-in. *Dia* 13-in. long is to be traverse ground to 1.4982/1.4987 in. diameter. There is sufficient material on one end to use a driving dog and give sufficient wheel clearance. Wheel diameter is 24 in. and width is 2 in. A lot of 210 is planned. Find the unit and lot estimate.

Table Number	Process Description	Table Time	Adjustment Factor	Cycle Minutes	Setup Hours
13.1-S	Wheel diameter	.75			.75
13.1-S	Tolerance of .0005 in	.20			.20
13.1-1	Pick up part, aside, 6.5 lb	.07		.07	
13.1-2	Load, unload in dog	.13		.13	
13.1-2	Mount in centers	.10		.10	
13.1-2	Grease centers	.08		.08	
13.1-3	Operate machine, start, stop	.05		.05	
13.1-3	Change speed	.01		.01	
13.1-3	Position wheel to work	.16		.16	
13.1-4	Traverse grind .017, $V = 70$, $P = \frac{1}{4}$ and $\frac{1}{8}$, $L \cdot Dia/W = 13 \times 1.5/2$	.25	$13 \times 1.5/2$	2.44	
13.1-5	Dress wheel	.20		.20	
13.1-6	Inspection, 3 places	.15		.15	
	Total			3.39	.95
	Lot estimate	12.81 hr			

B. The shank of a hardened carbon steel component is plunge ground to a 1.625-in. *Dia* and length of 1¾ in. with a 2-in. wide wheel. The work will be held by a collet. A grinding allowance of .012/.015 in. is expected for a lot of 2000 parts. Find the unit, lot, and shop estimate.

Table Number	Process Description	Table Time	Adjustment Factor	Cycle Minutes	Setup Hours
13.1-S	Setup .50 + .05	.55			.55
13.1-1	Pick up part	.07		.07	

Table Number	Process Description	Table Time	Adjustment Factor	Cycle Minutes	Setup Hours
13.1-1	Load, unload in collet	.09		.09	
13.1-3	Operate machine	.05		.05	
13.1-3	Move wheel to work	.07		.07	
13.1-4	Plunge grind	.39		.39	
13.1-6	Dress wheel	.20		.20	
13.1-7	Electronic check	.06		.06	
	Total			.93	.55
	Lot estimate	31.55 hr			
	Shop estimate	64.5 pc/hr			

C. Estimate the unit, setup, and total lot time for an aircraft part where the material is E4340 aircraft quality bar. Figure 13.1B shows the partially dimensioned assembled part, and the surface to be ground is given by the .4688/.4684-in. dimension. The length of the ground surface is 4.754 in. and traverse grinding with a cylindrial grinder is chosen. A relief next to the hub allows for grinder clearance. Surface requirement is 8 microinch. The lot requirement is 1095 pieces. The machine selected for the grinding operation will have a 20-in. OD and a 2-in. width. The previous operation left a total of .012-in. stock for the finish grind.

Table Number	Process Description	Table Time	Adjustment Factor	Cycle Minutes	Setup Hours
13.1-S	Setup for 20-in. wheel	0.50			0.50
13.1-S	Setup tolerance for .0004-in. total	0.20			0.20
13.1-1	Handle pickup, away	0.05		0.05	
13.1-2	Load and unload between centers	0.10		0.10	
13.1-2	Grease centers every fifth piece	0.08	⅕	0.02	
13.1-3	Start and stop	0.05		0.05	
13.1-3	Wheel to and from work	0.03		0.03	
13.1-3	Guard door open, close	0.05		0.05	
13.1-3	Coolant on and off	0.04		0.04	
13.1-4	Grind, dwell $L = 5.254$, $W = 2$ in. $Dia = .4688$ in., .012-in. stock removal	0.19	1.232	0.23	
13.1-6	Dress wheel	0.20		0.20	
13.1-7	Inspection, electronic check	0.06		0.06	
	Total			0.83	0.70
	Lot estimate	15.85 hr			

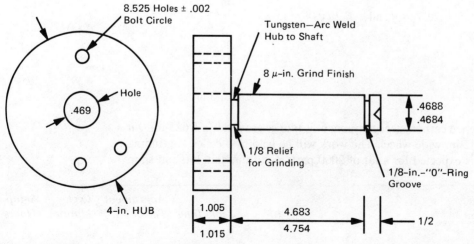

FIGURE 13.1B Part for cylindrical grinding.

TABLE 13.1 CYLINDRICAL GRINDING MACHINES

Setup

Wheel diameter to 12 in.	.25 hr
to 20 in.	.50 hr
to 30 in.	.75 hr
Tolerance .005 +	.05 hr
.004 to .001	.15 hr
.0009 to .00025	.20 hr

Operation elements in estimating minutes

1. Pick up and aside part
    - Small, 0 to 5 lb — .05
    - Medium, 5+ to 13 lb — .07
    - Large, 13+ lb — .09
    - Chain hoist — .74

2. Load and unload part
    - Collet — .09
    - Chuck — .36
    - Face plate — .66
    - Dog — .13
    - Quick-acting dog — .08
    - Expanding arbor — .10
    - Press on-and-off mandrel — .24
        - Between and remove from centers — .10
        - Grease centers — .08

3. Operate machine
    - Start and stop — .05
    - Change speed — .01
    - Move wheel to-and-from work — .07
    - Position wheel to work and feed depth of grind
        - Manual — .16
        - Automatic — .03
    - Against shoulder for plunge, manual — .30
    - Open and close guard cover — .05
    - Turn coolant on and off — .04

4. Traverse grind, dwell, and sparkout

Grind Stock	Representative Material	Time per Work Rev in Fractions of Wheel Width			Dwell	Sparkout
		¼ & ⅛	⅓ & ⅙	½ & ⅙		
.020	Hardened tool steel	.48	.36	.30	.15	.13
.020	Annealed tool steel	.30	.22	.17	.10	.10
.020	Hardened carbon and alloy steel	.25	.19	.15	.09	.09
.020	Annealed steels, copper alloys	.18	.13	.10	.06	.06
.020	Aluminum	.11	.08	.06	.04	.04
.025	Hardened tool steel	.55	.41	.33	.17	.13
.025	Annealed tool steel	.35	.26	.20	.12	.10
.025	Hardened carbon and alloy steel	.30	.22	.17	.10	.09
.025	Annealed steels, copper alloys	.21	.16	.12	.07	.06
.025	Aluminum	.13	.10	.07	.05	.04

Grind Stock	Representative Material	Time per Work Rev in Fractions of Wheel Width			Dwell	Sparkout
		¼ & ⅛	⅓ & ⅙	½ & ⅙		
.030	Hardened tool steel	.61	.45	.37	.20	.13
.030	Annealed tool steel	.40	.30	.23	.14	.10
.030	Hardened carbon and alloy steel	.35	.26	.19	.12	.09
.030	Annealed steels, copper alloys	.24	.18	.14	.09	.06
.030	Aluminum	.15	.11	.08	.06	.04
.040	Hardened tool steel	.74	.55	.43	.25	.13
.040	Annealed tool steel	.51	.38	.28	.19	.10
.040	Hardened carbon and alloy steel	.44	.33	.24	.16	.09
.040	Annealed steels, copper alloys	.31	.23	.17	.11	.06
.040	Aluminum	.20	.15	.11	.07	.04
Add'l .001	Hardened tool steel	.013	.009	.007	.004	0
	Annealed tool steel	.011	.009	.006	.005	0
	Hardened carbon and alloy steel	.010	.007	.005	.004	0
	Annealed steels, copper alloys	.008	.005	.004	.003	0
	Aluminum	.005	.003	.003	.002	0
Multiply table's value by .... L = grind length, Dia = diameter, W = wheel width		$L \cdot Dia/W$	$L \cdot Dia/W$	$L \cdot Dia/W$	Dia	$L \cdot Dia/W$

5. Plunge grind, min

	Aluminum, Annealed Steels, Copper Alloys				Hardened Carbon and Alloy Steels				Tool Steels			
	Total Stock Removed				Total Stock Removed				Total Stock Removed			
Dia	.010	.020	.030	Add'l .001	.010	.020	.030	Add'l .001	.010	.020	.030	Add'l .001
1	.04	.07	.09	.003	.16	.26	.35	.01	.37	.58	.80	.02
1.5	.06	.10	.14	.005	.25	.39	.53	.01	.55	.88	1.20	.04
2	.08	.14	.19	.007	.33	.52	.70	.02	.73	1.17	1.61	.04
3	.13	.20	.28	.008	.49	.77	1.05	.03	1.10	1.75	2.41	.06
4	.17	.27	.38	.010	.66	1.03	1.41	.03	1.47	2.34	3.21	.09
5	.21	.34	.47	.013	.82	1.29	1.76	.04	1.83	2.92	4.01	.11
6	.25	.41	.57	.016	.99	1.55	2.11	.06	2.20	3.51	4.82	.13
Add'l	.04	.06	.09		.16	.26	.35		.37	.58	.81	

6. Dress wheel
    1-in. wheel, 1.93 min; per part   .10
    2-in. wheel, 3.86 min; per part   .20
    Add'l in., 1.93 min; per part   .10

7. Inspection
    Micrometer 3 places   .15
    Micrometer 2 places   .10
    Electronic check   .06
    Other inspection   See Table 22.1-3

13.2 Centerless Grinding Machines

DESCRIPTION

Centerless grinders are designed to support and feed the stock by using two wheels and a workrest. The wheels, turning in the same direction, are the grinding and regulating wheels. The regulating wheel is smaller and provides pressure. Its tilted angular position forces stock to move atop a workrest while the part is being ground. The rate at which parts move through the machine is governed by the speed and angle of the regulator wheel. The width of the grinding wheel is from 4 in. to 20 in. for larger size machines. Work diameters can be from ¼ in. or less to 6 in. or 10 in., depending on the machine. The two types of centerless grinding estimating data considered here are the through-feed method for straight cylindrical surfaces without interfering shoulders, and the infeed method for parts that have a shoulder. With the through-feed method, the work traverses through the machine. The infeed method is similar to plunge grinding with the cylindrical-type grinder.

The process does not require center drilling of ends, and loading is simple. Both short and large quantity runs are common.

One centerless grinding machine is shown in Figure 13.2. This machine has a 30-hp spindle drive motor and can accommodate wheels up to 10-in. wide and 24-in. in diameter. The hydraulic truing unit is mounted to the bed. A *V*-bar grinding attachment allows the OD grinding of bar stock up to 1.5-in. OD by 100-in. long.

ESTIMATING DATA DISCUSSION

The data given for setup are considered average as changes from through-feed to infeed or vice versa and installation of chutes affect the climate. A complete wheel change requires about 15 min if an extra spindle is available, while 45 min is required using the same wheel.

Operation elements are provided for infeed grinding. Handling includes loading and unloading part to work blade. If the work is ejected by wheel retraction, .01 min is deducted. Although stock affects time to grind, the estimate is related to diameter, that being chosen for ease of application reasons as the principal time driver. The time is for one-plunge. The threshold

FIGURE 13.2 Centerless grinding machine. *(Cincinnati Milacron)*

time is for a diameter of .4 in. or less; for diameters in excess of 1½ in., add .28 min/in. For instance, a 2-in. part would be allowed .38 min. Tolerance checking is based upon a frequency of 50 pc between micrometer checkings. For other frequencies, the micrometer check-time can be divided by typical shop practice. Wheel dress time is related to 400 grinds, but this depends upon part shape, material, machine, grind stock available, and other factors.

For through-feed grinding, the handling time is compared to grind time. If handling time is smaller, usual practice is to overlook it in the estimate. If the load time is greater, the difference is added to the estimate. Through-feed grind time is for one pass. While practices vary, some companies establish a manufacturing specification of roughing out of .008 to .010 in. and a finishing or sparkout of .0005 in. to .002 in. The grind time is related to length of grind diameter for one pass. If the grind time is sufficient, other elements can be assumed to take place during this period.

EXAMPLES

A. Free-machining steel parts, 6-in. long, have a 1.125-in. long shoulder, leaving 4.75 in. of .547-in. *Dia* stock. Stock removal to finish dimension is .010 in. to be removed in two grinds, .008 in., and a .002-in. cleanup to final tolerance of ± .001. A moderate lot quantity of 350 is required and the infeed method of centerless grinding is planned. Part weight is .48 lb. Estimate total lot time for two grinds combined as one operation.

Table Number	Process Description	Table Time	Adjustment Factor	Cycle Minutes	Setup Hours
13.2-S	Setup	1.2			1.2
13.2-1	Load and unload part	.03	2	.06	
13.2-1	Deduct for chute ejection	−.01	2	−.02	
13.2-2	Wheel to work	.04	2	.08	
13.2-3	Two infeed grinds	.10	2	.20	
13.2-4	Move wheel back from work	.05	2	.10	
13.2-8	Check tolerance for rough	0			
13.2-8	Check tolerance for finish	.005		.005	
13.2-9	Dress wheel for rough	0			
13.2-9	Dress wheel for finish	.01		.01	
	Total			.435	1.2
	Lot estimate	3.74 hr			

B. Estimate the time to centerless grind the shaft shown by Figure 13.1B. This grinding operation precedes the assembly and the finish grind operation using the cylindrical grinder. This present operation will start with the raw stock which is cut to length and has been grooved, start drilled, etc. There are 1095 units in the lot, and the requirement is to rough and finish grind .015-in. stock to a dimension of .481 in. Two passes are assumed.

Table Number	Process Description	Table Time	Adjustment Factor	Cycle Minutes	Setup Hours
13.2-S	Setup	1.20			1.20
13.2-5	Through-feed handle	0.04		0.04	
13.2-6	Through-feed 6.27-in. long shaft Dia = .496-in. std. dimension; rough pass of .008 in.	0.10		0.10	
13.2-6	Finished dim = .481 in. for a grind removal of .015-in. total; second pass	0.10		0.10	
	Total			0.24	1.20
	Lot estimate	5.58 hr			

C. Stainless steel parts, AISI 303, 8.318-in. long by 1.316 ± .002-in. *Dia* are ground using through-feed, and two passes are required to bring the stock to finish specification. Nearly continuous production is anticipated. Estimate the unit time for one pass.

Table Number	Process Description	Table Time	Adjustment Factor	Cycle Minutes	Setup Hours
13.2-5	Load part via trough	.01		.01	
13.2-5	Remove part from machine	0		0	
13.2-6	Through-feed one pass and use 10.7 in.	.14		.14	
Note:	Handling time can be done during machine time				
13.2-7	Gather parts	.005		.005	
13.2-8	Check tolerances	.005		.005	
13.2-11	Dress wheel	.005		.005	
	Total			.155	

TABLE 13.2 CENTERLESS GRINDING MACHINES

Setup 1.2 hr

Operation elements in estimating minutes

1. Handle part to grinder, infeed grinding

Weight	3	5	12
Min	.03	.04	.05

   Deduct for chute ejection .01

2. Bring wheel to work, infeed grinder .04

3. Infeed grind, one plunge

Dia	Min
.4	.09
.6	.10
.7	.12
.9	.14
1.1	.16
1.2	.18
1.3	.21
1.5	.24
Add'l in.	.28

4. Move wheel back from work, infeed grinding .05

5. Through-feed grinding
   Load part from tote pan to grinder .04
   Fill gravity trough, .19/occurrence
   Remove part from machine 0

6. Through-feed grind, one pass

L	Min
.5	.02
1.1	.03
2.0	.04
3.2	.05
4.9	.07
7.3	.10
10.7	.14
15.4	.20
Add'l	.012

7. Gather parts, place in tote pan	.005
8. Check tolerances, .18/occurrence, prorated	.005
Other inspection	See Table 22.1-3
9. Dress wheel, 4.00/occurrence, prorated	.005

13.3 Honing Machines

DESCRIPTION

Honing is a low-velocity abrading process. Since material removal is accomplished at lower cutting speeds than in grinding, heating and pressure are less, resulting in size control. Abrasives are aluminum oxide, silicon carbide, or diamond. The grits, bonded together, are formed into sticks. Honing, depending on the machine, may be done on internal or external surfaces. There are manual honing units and fixtured honing tools. Size control is managed by gaging the tool or bore. Tools for honing may be mounted on a variety of machines—drill press, electric drill, or any tool that will rotate and reciprocate and allow a floating abrading action. Honing is possible for a wide variety of metallics, ceramics, fiberglass, tool steels, etc. A machine for power stroking is shown in Figure 13.3. Bore diameters from 0.060 in. to 3.75 in. are possible. Other machines are capable up to 12 in. diameter.

ESTIMATING DATA DISCUSSION

Some machines have automatic cycle start and shutoff. Spindle speed selection, stroking rates, universal fixtures, gages, and diamond dressing affect the operation elements. The honing Element 7 is for a variety of materials, ranging from soft steel to hard tool steel using aluminum oxide, diamond, or silicon carbide abrasives. The range of stock removal is for final sizing, cleanup, or rework of bore diameters. Thus, Element 7 is a general purpose honing element. The entry is based upon diameter and .001 in. of stock removal (meaning final bore diameter minus initial bore diameter).

Element 7 also provides for honing cylinders where larger-stock removal is emphasized for mild steel. The distinction between the entries depends on whether the abrasive is either diamond or aluminum oxide. A bore of 2⅝-in. OD × 11⅛-in. length having .030-in. stock removal would find the factor as .25 for

FIGURE 13.3 Power-stroked honing machine. *(Sunnen Products Company)*

diamond-imbedded grinding stones (going to the next larger diameter to avoid tabular interpolation), and a time of 2.78 min (= .25 × 11.125).

For larger diameter or long-length bores, the operator may be working on a different machine simultaneously because of the opportunity to do other work. This "sharing of time" between two or more operations allows a reduction of time opportunity by the estimator.

EXAMPLES

A. Estimate the time to hone a bushing, D3 steel, 58 Rockwell C, 2.16-in. long, .6225/.6250-in. bore. The previous operation was reaming. Mean stock removal from the diameter is .005 in. A lot of 1400 is required.

Table Number	Process Description	Table Time	Adjustment Factor	Cycle Minutes	Setup Hours
13.3-S	Setup	.30			.30
13.3-1	Handle	.09		.09	
13.3-2	Start and stop	.04		.04	
13.3-3	Set honing dial	.06		.06	
13.3-5	Install fixture	.25		.25	
13.3-7A	Hone	.68	2.16	1.47	
	Hone life, 1400 × .66 sq in. = 928 sq in.				
	Total			1.91	.30
	Lot estimate	44.87 hr			

TABLE 13.3 HONING MACHINES

Setup .3 hr

Operation elements in estimating minutes

1. Load and unload part on tray
   - Light — .09
   - 25 lb + — .16

2. Start and stop — .04

3. Set honing dial — .06

4. Air-clean part — .13
   - Cloth clean part — .17
   - Wipe part with hand — .05
   - Wipe off excess oil or grease — .49

5. Place part on mandrel and remove, manual honing — .08
   - 25 lb + — .19
   - Install fixture to hold part for power honing and remove — .25

6. Micrometer check ± .0001 in. — .48
   - Go/no-go gage — .08
   - Other inspection — See Table 22.1-3

7. Honing

   A. Honing (per in. of bore *L*), maintenance or cleanup

Bore Dia	Stock Removal				
	.001	.002	.003	.004	.005
.32	.08	.15	.30	.45	.60
.5	.08	.19	.34	.49	.64
.64	.08	.23	.38	.53	.68
.95	.17	.32	.47	.62	.77
1.27	.25	.40	.55	.70	.85
2.	.43	.58	.73	.88	1.03
3.	.68	.83	.90	1.13	1.28
4.	.95	1.09	1.24	1.39	1.54

Additional: sq in. of surface area		.08
Per ea mil of stock removal		.15
Hone life: 3800 sq in. of surface area		
Replace set of stones		1.00

B. Honing (per in. of bore L), mild steel with diamond-imbedded stones

	Stock Removal					
Bore Dia	.010	.020	.030	.040	.050	.100
2	.16	.19	.21	.24	.26	.39
3	.17	.21	.25	.28	.32	.51
4	.18	.23	.28	.33	.38	.64
5	.20	.26	.32	.38	.45	.76
6	.21	.28	.36	.43	.51	.88

C. Honing (per in. of bore L), mild steel with aluminum oxide stones

	Stock Removal					
Bore Dia	.010	.020	.030	.040	.050	.100
2	.17	.22	.27	.32	.37	.63
3	.19	.27	.34	.42	.50	.89
4	.22	.32	.42	.52	.62	1.14
5	.24	.37	.50	.62	.75	1.39
6	.27	.42	.57	.72	.88	1.64

8. Hand file edge of bore

	Bore Dia				Add'l per in. of
Material	.32	.64	.95	1.27	Circumference
Soft	.05	.07	.09	.11	.018
Rockwell C 20–35	.06	.08	.10	.12	.021
Rockwell C 35+	.07	.10	.12	.14	.024

13.4 Surface Grinding Machines

DESCRIPTION

The grinding of flat or plane surfaces is defined as surface grinding. The machines considered in this estimating guide are the planer with a reciprocating table and a rotating table. Each type has the potential variation of the grinding-wheel spindle in either a horizontal or vertical position. Magnetic chucks are a feature of this machine. Heavy or light stock removal, finish, and tolerance are the usual advantages claimed for these machines. Surfaces need not be regular (i.e., unbroken areas) to accomplish surface grinding. Straight or recessed wheels grinding on the outside face or circumference and segmental or ring-shaped wheels are the wheel designs. In rotary-surface grinding, a vertical spindle can be tilted to reduce the wheel area in contact with the work to allow greater depths, less heating, and higher utilization or available horsepower. For finish grinding, the spindle is returned to a perpendicular position which presents a flat grinding surface to the workpiece.

Horizontal-spindle, reciprocating-table surface grinders include small grinders, die-block grinders, and large grinders. The machines are rated by width × height adjustment × length, and range from 12 × 16 × 36 in. to 48 × 36 × 240 in. Table feeds are hydraulic, with speeds up to 125 fpm. Wheel head is power raising and lowering on larger models and manual on the lighter and smaller machines. Wheel crossfeed is hydraulic transverse, with feeds variable up to the width of the wheel. Electronic reversing controls are available. On the smaller machines the grinding wheels may be 20 in. in diameter with a 3-in face or smaller. On larger machines, typical diameters may be 20, 28, 32, or 36 in. with a 6-in. face. A 12-in. wide table horizontal grinder is shown in Figure 13.4A.

FIGURE 13.4A Horizontal reciprocating table grinder. (*Mattison Grinder*)

Vertical-spindle, rotary-table surface grinding machines include models having table diameters ranging from less than 30 in. to 168 in. Table speed ranges may be fixed or have variable rpm within a range. Spindle horsepower, depending on the model, can range up to 300 hp. Wheel head feeds are continuous or rachet, depending on the manufacturer, and are hand- or automatic-fed, variable from 0.005 to .165 in./min. Feed wheels may be graduated in steps of .001 in. Tiltable spindles for rough grinding allow for grain penetration. Automatic electronic gaging gives continuous measurement of the work during the cutting cycle. Figure 13.4B is a 300-hp machine with a 72-in. magnetic worktable.

ESTIMATING DATA DISCUSSION

The setup is listed as .6 hr and includes ordinary machine chores. It does not include a wheel change, which is an optional part of the run time. If estimating practice is to include a wheel change, Element 12 is added to .6 hr for a total of .9 hr.

The initial listed element is for load and unload, where the entry variable is part weight. Two conditions are given: from table for small and medium sized parts, and from the floor for larger parts. Loading may also refer to fixtures; holding bars, for nonmagnetic workpieces; for bracing; or leveling. The time is found in Element 3.

In rotary grinding, the chuck is positioned under the wheel. For planer-type grinders, the table is positioned for grinding, then Element 4 is used. The number of pieces in the fixture or on the chuck is the entry variable for Element 4 and is determined from knowledge of the part surface area, packing on the table, and the developed area of a fixture, if used. Specific machine table areas must be known.

Grinding times are calculated using Elements 5 (traverse grind), 6 (plunge grinding), and 7 (rotary table grinding). Entry variables for traverse and plunge grinding for planer-type grinding are stock width, depth of grind, and machine grind-wheel width. For plunge grinding they are depth of grind and grinding velocity. For rotary-table grinding, entry variables are depth of grind and material grinding rate.

Stock width is maximum dimension required for grinding. The part width dimension is used directly, as an overlap of one-wheel face width is already included in the calculations. If a stock width other than those shown is required, the "Add'l 1" can be added or subtracted from table values and an approximation will result. Naturally, if stock width is less than wheel width, the plunge grinding data are used. Otherwise, for stock widths greater than wheel widths, traverse grinding is employed. Obviously, the estimator will have in mind a particular machine where wheel width is known.

Depth of grind is the deliberate overstock remaining from a previous operation. Stock allowances for a grinding approach and material irregularities are included in the calculations, though they are not specifically shown as depth of grind. For instance, a .005-in. grind stock assumes a .005-in. stock allowance for approach and irregularities, and thus there is a .010 total amount for grind removal. Also, for .020-in, stock, there is a stock approach of .008 in. which provides a total of .028-in. stock. The minimum and maximum stock allowance for approach and irregularities are .005 and .008 in. The maximum .008 in., is provided once .020-in. stock is reached. Additional or lesser

FIGURE 13.4B Heavy rotary table, vertical spindle grinder. (*Mattison Grinder*)

Surface Grinding Machines 299

stock allowances can be calculated for depth of grind or for approach irregularities using the "Add'l .001 in." number.

The table time to traverse grind is for 1 in. of length per a surface velocity of 1 fpm. Once entry values of stock width, stock depth, and wheel width are known and a value selected, the tabular factor is multiplied by the ratio L/V. Determination of length (L) starts with knowledge of maximum part grind and then an approach and overtravel length is added. This approach-overtravel length may be 1 wheel dia. The length dimension is in in.

The time values are based upon 90% of total stock being rough ground, leaving 10% for finish grind in a two-phase operation. A sparkout of four passes over all area is included in the calculations of Element 5.

Downfeed and crossfeeds assumed for each wheel width within the calculations are:

Wheel Width	1½	3	6	Plunge
Crossfeed, rough, in.	¾	1½	3	
Crossfeed, finish, in.	⅜	¾	1½	
Downfeed, rough, in.	.0009	.001	.001	.0008
Downfeed, finish, in.	.00045	.0005	.0005	.0004

Selection of the entry value of surface velocity for plunge grinding is necessary. Adjustments to table values are possible in the "Add'l 1" surface velocity rpm, if improved estimating accuracy is thought necessary. Suggested grinding velocities (V) are given as:

Material	Cast Iron	Soft Steels	Hard Steels
Traverse velocity, fpm	80	80	90
Plunge velocity, fpm	60	70	80

In the 5 to 12 hp planer-type grinder machines, the common rate of stock removal is ¼ to 2 cu in./min; in 3½ to 5 hp it is 1 cu in./min in cast iron, and 10 hp removes 1 cu in./min in steel. Deflection or rigidity, broken or continuous surface area, and wet or dry grinding can affect these values.

For traverse and plunge grinding, the tabular value is multiplied by L/V and L respectively. L is the length in in. of the part plus a distance for approach and overtravel, perhaps approximated by a wheel dia. V is selected from the above tabular values.

Rotary table grinding time, Element 7, is calculated differently from planer-type grinding. The material removal rate is related to grinding-head hp available, type of material, and the nature of the surface. The hp required per cu in. of material per min is given as:

Material Grinding Rate, hp/cu in./min

Area, sq. in.	300	600	1000	2000
Cast Iron				
Continuous	5	6	7	8
Broken up	4	5	6	6
Steel				
Continuous	7	8	9	10
Broken up	6	7	8	9

Thus, for cast iron, if the developed area is about 550 sq in., and is unrelieved and continuous, the required hp/cu in./min = 6; if that area is broken up the requirements are less, or 5 hp/cu in./min, because chip clearance is increased. The aluminum rate is 4 hp/cu in./min. Having a depth of grind, and a material grinding rate, the tabular value from Element 6 is multiplied by the ratio of $Area/hp_m$, where the hp_m is the machine-grinding head horsepower. Area is that which is seen by the segmented wheel and may be larger than the calculated and apparent area.

Element 8, dwell or sparkout, is for high-tolerance grinding. Elements 5 and 6 include four sparkout passes and are for commercial tolerances. Tolerances for tenths or rotary-table grinding may require sparkout.

If the work is hidden, i.e., prevented from being checked, such as for rotary surface grinding, it is necessary to expose the work for dimension checking and Element 9 is applied.

The rotary-table segmented grinding is self-dressing, but the planer-type horizontal wheels are dressed via a diamond point. Element 12 accounts for both dressing and wheel changing. Wheel wear is contingent upon several factors, and one rule is .015 in. of grinding wheel reduction per .125 in. removed for cast iron, and .050 in. per .125 in. for steel. Thus, it is possible to estimate the proportion of wheel change time.

EXAMPLES

A. Estimate the time to surface grind four sides of AISI 6150 steel index dogs having irregular dimensions, but an overall dimension of ¾ × 2⅝ × ⅝ in. The dogs are to be loaded 384 at a time on a 35 hp planer-type surface grinder with a 20- × 6-in. grinding wheel where table width is 30 in. Stock removal is .010 in. per side. Determine unit time for a large quantity.

Table Number	Process Description	Table Time	Adjustment Factor	Cycle Minutes	Setup Hours
13-4-1	Load and unload from table, 1 unit Load 32 across for a 20-in. stock width, and 12 along the traverse length for 31.5-in. of metal	.07		.07	
13.4-5	Traverse grind 1 side, 20 × 31.5. Overtravel and approach = 10 in. for a length of 41.5 in. Velocity = 90 fpm.	23.7	$\frac{41.5}{90 \times 384}$	.03	
13.4-10	Check with gage	.17	1/384	0	
13.4-11	Wash grinding grits, .69 + .004(384 − 69)/384	.005		.005	
13.4-12	Dress wheel	2.0	1/384	.005	
Repeat	Duplicate above elements for 3 other sides	.33		.33	
	Total			.44	
	Hr/100 units	0.733			

B. Leaded steel plates having dimensions of 11¼-in. long by 2½-in. wide and 1-in. thick are rough ground seven at a time to remove .016 in. per side using a planer-type surface grinder. The operation is plunge grinding using a 3-in. wheel. Determine lot time for 120 pc.

Table Number	Process Description	Table Time	Adjustment Factor	Cycle Minutes	Setup Hours
13.4-S	Setup	0.6			.6
13.4-1	Load and unload, 2.9 lb	0.10		.10	
13.4-6	Plunge grind at 80 fpm Approach and overtravel distance of 14 in. is assumed for a total length of 92.75 in.	.044	92.75/.7	.59	
13.4-10	Check sample with depth gage	.60	½	.09	
13.4-11	Wash grits off parts	.44	½	.06	
13.4-11	Clean table	.37	½	.05	
13.4-12	Dress wheel	.63	½	.09	
Repeat	Duplicate above elements for second side	.98		.98	
	Total			1.96	.6
	Lot estimate	4.52 hr			

C. Estimate a rotary-grinding job of a 60 × 60 × 3-in. 1020 steel plate, where stock removal is ⅛ in. from each side. Stock removal is 450 cu in. per side. Assume a 200 hp spindle motor machine is available for the job. The area is 3600 sq. in. per side. Find the unit time in min and hr/1 unit.

Table Number	Process Description	Table Time	Adjustment Factor	Cycle Minutes	Setup Hours
13.4-1	Load and unload, crane	5.00		5.00	
13.4-4	Position chuck, 1 pc	.31		.31	
13.4-7	Use 10 hp/cu in./min. and factor of 1.25 is read, multiply (1.25) × (3600)/200	1.25	3600/200	22.50	
13.4-8	Sparkout, 1.93 + (3600 − 1148).0015	5.61		5.61	
13.4-9	Expose work	.13		.13	
13.4-10	Check size thickness	.60		.60	
13.4-11	Wash part	.37		.37	
Repeat	Part is turned over and above elements are repeated for second side	34.52		34.52	
	Total			69.04	
	Hr/1 unit	1.51			

TABLE 13.4   SURFACE GRINDING MACHINES

Setup 0.6 hr

Operation elements in estimating minutes

1. Load and unload part or fixture onto magnetic chuck, lb

From Table	From Floor	Min	From Table	From Floor	Min
3.6		.07	36.0	19.3	.21
5.3		.08	41.0	24.0	.24
7.3		.09	46.5	29.1	.26
9.4		.10	52.5	34.7	.28
11.7		.11		40.9	.31
14.3		.12		47.7	.34
17.1	1.7	.13		55.2	.38
				Add'l 1	.004
20.2	4.6	.15		Add'l 1	.005
23.6	7.7	.16	Electric hoist		1.12
27.4	11.2	.18	Hand double-chain hoist		4
31.5	15.1	.19	Bay crane		5–20

2. Tighten part in fixture
   - Nut, washer or strap .55
   - Loosen and tighten nut only .27
   - Index chuck while tightening .04

3. Position and remove holding bars on magnetic chuck

No. of Bars	Min	No. of Bars	Min
2	.06	11	.78
3	.14	12	.86
4	.22	13	.94
5	.30	14	1.02
6	.38	15	1.10
7	.46	16	1.18
8	.54	17	1.26
9	.62	18	1.33
10	.70	Add'l	.08

4. Position chuck under grinding wheel

No. Pc	Min
3	.31
6	.32
9	.33
12	.34
15	.35
18	.37
21	.38
24	.39
27	.40
30	.41
Add'l	.004

5. Traverse grind[1]

Stock Width	Depth of Grind	Min to Grind Width for 1 in. of Length per 1 fpm		
		Wheel Width		
		1½	3	6
3	.005	11.9	8.3	
	.010	16.5	11.4	
	.015	21.1	14.6	
	.020	25.7	17.7	
	.025	29.5	20.3	
	.030	33.3	22.8	
	Add'l .001	.75	.50	
6	.005	20.8	12.8	8.3
	.010	28.8	17.2	11.4
	.015	36.9	21.8	14.6
	.020	45.0	26.5	17.7
	.025	51.7	30.3	20.3
	.030	58.3	34.3	22.8
	Add'l .001	1.42	.75	.50
10	.010	45.3	24.8	15.3
	.015	58.0	31.6	19.4
	.020	70.7	38.3	23.6
	.025	91.8	49.5	30.4
	.030	112.8	60.7	37.3
	Add'l .001	2.17	1.17	.67
20	.010	86.6	43.9	23.7
	.020	134.9	67.7	38.3
	.030	175.1	87.5	49.5
	.040	215.3	107.3	60.7
	.050	255.6	127.2	71.9
	Add'l .001	4.00	2.00	1.08
Add'l 1		6.83		
30	.010		60.2	34.3
	.020		91.4	53.0
	.030		125.6	68.5
	.040		154.0	84.0
	.050		182.4	99.5
	Add'l .001		2.92	1.58
Add'l 1			4.00	
40	.010			43.9
	.020			67.7
	.030			87.5
	.040			107.3
	.050			127.2
	Add'l .001			2.00
Add'l 1				1.58

[1] Multiply table time by L/V to obtain min.
L = in., V = fpm

6. Traverse plunge grind[2]

Depth of Grind	Min to Grind 1 in. L					
	Grinding Velocity, fpm					
	30	50	60	70	80	Add'l 1
.005	.059	.035	.029	.025	.022	−.0003
.006	.063	.038	.032	.028	.024	−.0003
.007	.068	.041	.034	.029	.026	−.0003
.008	.073	.043	.037	.031	.027	−.0003
.009	.077	.046	.038	.033	.029	−.0003
.010	.086	.052	.043	.037	.033	−.0004
.011	.091	.054	.045	.039	.034	−.0004
.012	.095	.058	.048	.041	.036	−.0004
.013	.100	.060	.050	.043	.038	−.0004
.014	.104	.063	.053	.045	.039	−.0005
.015	.113	.068	.057	.048	.043	−.0005
.016	.118	.071	.059	.051	.044	−.0005
.017	.123	.073	.062	.053	.046	−.0006
.018	.128	.077	.063	.054	.048	−.0006
.019	.132	.079	.066	.057	.049	−.0006
.020	.142	.084	.071	.061	.053	−.0007
.022	.150	.090	.075	.064	.057	−.0007
.024	.159	.096	.080	.068	.060	−.0008
.025	.163	.098	.082	.070	.062	−.0008
.028	.178	.107	.088	.076	.067	−.0008
.030	.187	.112	.093	.080	.070	−.0008
Add'l .001	.004	.003	.003	.002	.001	

[2] Multiply table time by length + approach and overtravel, or L, to obtain min, L is in.

7. Rotary table grind[3]

Depth of Grind	Material Grinding Rate, hp/cu in./min								
	4	5	6	7	8	9	10	20	Add'l 1
.010	.04	.05	.06	.07	.08	.09	.10	.20	.01
.015	.06	.08	.09	.10	.12	.13	.15	.30	.015
.020	.08	.10	.12	.14	.16	.18	.20	.40	.02
.025	.10	.13	.15	.17	.20	.22	.25	.50	.025
.030	.12	.15	.18	.21	.24	.27	.30	.60	.03
.040	.16	.20	.24	.28	.32	.36	.40	.80	.04
.050	.20	.25	.30	.35	.40	.45	.50	1.00	.05
.060	.24	.30	.36	.42	.48	.54	.60	1.20	.06
.070	.28	.35	.42	.49	.56	.63	.70	1.40	.07
.080	.32	.40	.48	.56	.64	.72	.80	1.60	.08
.090	.36	.45	.54	.63	.72	.81	.90	1.80	.09
.100	.40	.50	.60	.70	.80	.90	1.00	2.00	.10
.125	.50	.63	.75	.87	1.00	1.12	1.25	2.50	.13
.150	.60	.75	.90	1.05	1.20	1.35	1.50	3.00	.15
.175	.70	.88	1.05	1.27	1.40	1.57	1.75	3.50	.18
.200	.80	1.00	1.20	1.40	1.60	1.80	2.00	4.00	.20
.250	1.00	1.25	1.50	1.75	2.00	2.25	2.50	2.75	.25
Add'l .001	.004	.005	.006	.007	.008	.009	.01	.02	

[1] Multiply table time by A/hp_m to obtain min, A = sq. in.

8. Dwell or sparkout

Area	Min
20	.21
53	.26
96	.32
149	.40
215	.51
298	.63
402	.79
531	.99
692	1.23
894	1.54
1148	1.93
Add'l 1	.0015

9. Expose work for dimension checking

No. Pc on Chuck	Min
7	.13
14	.14
21	.15
28	.16
35	.17
42	.19
49	.20
Add'l 1	.002

10. Check with profile gage or scale .17
    Check sample with depth micrometer .60
    Other inspection See Table 22.1-3

11. Wash grinding grit off parts with coolant

No. Pc on Chuck	Min
3	.41
9	.44
15	.46
21	.49
27	.51
33	.54
39	.56
45	.59
51	.62
57	.64
63	.67
69	.69
Add'l 1	.004

Clean chuck or table or single large part .37

Surface Grinding Machines 305   **13.4**

12. Change wheel or dress wheel
    Change wheel 16.00
    Dress wheel (traverse and plunge)

Wheel Width	1½	3	6
Rough Grind	.47	.63	1.00
Finish Grind	.92	1.28	2.00

    (Prorate over no. of pc.)

13.5 Internal Grinding Machines

DESCRIPTION

Internal grinding machines operate with rapidly rotating grinding wheels, and in the majority of machines, the workpiece is also rotated. Additionally, a traverse movement will bring the wheel into the work area and retract the wheel at the end of the operation. Wheel retraction may also be needed for wheel dressing, gaging, etc. A transverse movement has the dual role of the wheel approach to the surface of the work, which in some cases may be in a recessed location, and secondly, a feed movement during stock removal. The feed may be at different rates for roughing and finishing.

Internal grinders range from machines having a minimum capacity of about 1 mm to about 32+ in. maximum hole diameter. Hole lengths vary up to 24 in. Features of some manufacturers' models include diamond and electronic sizing, automatic dressing with counter for frequency of dressing, and automatic compensation. A typical grinder is shown in Figure 13.5.

Planetary internal grinders and grinders where the workpiece is rotated by rolls are not considered in these data. Now, consider estimating data for the general class of internal grinding machines.

ESTIMATING DATA DISCUSSION

A value of 0.7 hr is considered average for small- and large-size internal grinders. Operational elements are similar to external cylindrical grinding.

Internal grinding time calculations are similar to those of external machines. The basic formula for traverse grinding is used, but velocities and feeds are reduced. A distinction is made for large and small (smaller than 1 in.) bores or slender holes where length to diameter ratios require less grinding wheel pressure. Hardened alloy and tool steel materials, if velocity requirements are considered similar, can be estimated as small holes.

For internal grinding, the following parameters were adopted:

	Traverse Regular Bores	Traverse Small Holes	Plunge
Velocity, fpm	60	30	40
Infeed, rough	.0006	.0006	.0004
Infeed, finish	.0002	.0002	.0001

Table values are multiplied by the ratio $L \cdot Dia/W$, where L = in.-length of grind, accounting for shoulders or hole bottom; Dia = in.-internal diameter; and W = in.-width of grind wheel. If wheel width is wider than hole stroke, the internal grind is considered a plunge grind, even though some machines impart a slight oscillation to the grinding stone. For longer holes, having lengths which are a multiple of the wheel width, the general practices of traverse grinding are observed. Sparkout assumes four dead passes to obtain final finish and tolerance. A stock allowance for runout and approach, initially at .004 for the .010-in. total stock, is gradually increased to .008 in. for .024-in. and higher stock allowances. The runout and approach values are included in the calculations and need not be added to the stock allowance, which is determined by subtraction of initial and final diameters. In traverse and plunge grinding, the practice of allotting 10% of the total stock for a finish grind stock allowance is observed by the data.

Inspection times are indicated as per piece, following the description, and then are prorated over 10 units.

FIGURE 13.5 Internal chucking grinder. *(Hermes Machines Tool Company, Inc.)*

EXAMPLES

A. A workpiece is to be ground internally and final bore dimensions are 7 in., 2.5-in. length, and the material is fully heat treated to give an R_c value between 59–62. The wheel chosen for the job is 4.5-in. diameter and 1.25-in. wide. Grinding allowance is .030 in. Estimate the floor to floor time.

Table Number	Process Description	Table Time	Adjustment Factor	Cycle Minutes	Setup Hours
13.5-1	Chain hoist	.74		.74	
13.5-2	Load and unload part	.14		.14	
13.5-3	Start and stop work	.03		.03	
13.5-3	Guard cover	.05		.05	
13.5-3	Start and stop grind wheel	.02		.02	
13.5-4	Traverse grind	.67	(2.5 × 7)/1.25	9.52	
13.5-4	Sparkout	.06	(2.5 × 7)/1.25	.84	
13.5-6	Dress wheel	.07	2	.14	
13.3-7	Inspection, electronic	.01		.01	
	Total			11.49	

B. Estimate the total lot time for grinding an .867-in. ID by 1-in. long spacer. Material is soft steel, and the amount of stock is .010 in. Lot quantity is 861, wheel diameter is ⅝-in., and the length of the wheel is 2 in. Assume dry grinding conditions.

Table Number	Process Description	Table Time	Adjustment Factor	Cycle Minutes	Setup Hours
13.5-S	Setup	.6			.6
13.5-1	Pick up small part	.05		.05	
13.5-2	Load and unload	.09		.09	
13.5-3	Operate machine	.03		.03	
13.5-5	Plunge grind	.19		.19	
13.5-6	Dress wheel	.07		.07	
13.5-7	Air gaging	.01		.01	
	Total			.44	.6
	Lot estimate	6.91 hr			

TABLE 13.5 INTERNAL GRINDING MACHINES

Setup .6 hr

Operation elements in estimating minutes

1. Pickup and aside part
   - Small, 0–5 lb .05
   - Medium, 5+–13 lb .07
   - Large, 13+ lb .09
   - Chain hoist .74

2. Load and unload part
   - Collet .09
   - In chuck jaws .07
   - Align part in fixture .14

3. Operate machine
   - Start and stop workhead .03
   - Change workhead speed .002
   - Start and stop grind wheel .02
   - Engage and disengage table speed lever .04
   - Open and close workhead guard cover .05
   - Move truing lever to-and-from position .04
   - Operate runout treadle .01

4. Traverse grind—internal

Total Stock	Regular Holes and Bores	Small and Slender Holes
.005	.12	.23
.010	.24	.49
.012	.28	.56
.014	.33	.66
.016	.39	.77
.018	.42	.84
.020	.48	.95
.022	.51	1.01
.024	.56	1.13
.026	.60	1.20
.028	.64	1.26
.030	.67	1.34

Total Stock	Regular Holes and Bores	Small and Slender Holes
.032	.71	1.41
.034	.74	1.49
Add'l .001	.02	.05
Sparkout	.06	.13
Multiply table values	$\dfrac{L \cdot Dia}{W}$	$\dfrac{L \cdot Dia}{W}$

5. Plunge grind—internal

Dia	Total Stock Removed					Add'l .001
	.005	.010	.015	.020	.030	
1	.11	.19	.26	.34	.47	.01
1.5	.18	.28	.40	.50	.71	.01
2	.24	.38	.54	.67	.93	.02
3	.35	.57	.80	1.01	1.40	.04
4	.48	.75	1.07	1.34	1.87	.06
5	.59	.95	1.33	1.69	2.33	.07
6	.71	1.13	1.61	2.03	2.80	.08
Add'l 1.0	1.3	.19	.26	.33	.47	

6. Dress wheel, prorated .07

7. Inspection
    Plug gaging, manual, .20 min, prorated .02
    Inside micrometer, .46 min, prorated .016
    Mechanical dial indicators, .15 min, prorated .015
    Rear-inserted power-operated plug, .04 min, prorated .004
    Air or electronic indicators, .10 min, prorated .01
    Other inspection See Table 22.1-3

13.6 Free Abrasive Grinding Machines

DESCRIPTION

Free abrasive grinding machines produce almost geometrically true surfaces on parts, correct minor surface imperfections, and improve dimensional accuracy to provide a very close fit between two contacting surfaces. Although free abrasive grinding is a material-removing operation, it is not always economical for that purpose. The method is used on flat, cylindrical, or specially formed surfaces. The part surfaces are in contact with an abrasive in such a way that fresh abrasive contacts are being made. Consider the estimating data for the general case of semiautomatic free abrasive grinding machines of external flat surfaces.

Machines are available which provide up to four rings. Rings may range in internal diameter (ID) from 4½ in. to 36 in. Figure 13.6 is a 4-ring machine. As the rotating table turns, it carries the abrasive grains across the surface of each part; the pressure plate is above the ring.

ESTIMATING DATA DISCUSSION

Handling is for one-hand or two-hand held loading of parts. It is possible to load on a separate table during grinding time, in which case only the first rings are considered for the estimate. Subsequent retaining rings are loaded during the lapping time. For elements performed during machine lapping, the estimated

FIGURE 13.6 Free abrasive grinding machine. *(Speedfam)*

A small part is easily held with one hand, while a large part requires two hands. Turnover may be during lapping time. For some machines which may have automatic handling provisions, turnover is mechanical, that is, free of operator involvement. However, this automatic turnover element does interrupt machine operation and 0.75 min per ring is required.

If work is done during the abrasive grinding element, additional time is zero. Parts can be loaded into spare rings and subsequently eased onto the lapping table. Some machines, however, may be stopped during turnover.

A start and stop element is required. For lapping machines which have a free weight, a time is provided to clean, place, and remove weight for each ring. On some other machines, the backing plate is lowered under machine control. A release of pressure plate per ring is also available.

The unit estimate depends upon the number of parts loaded into each retaining ring. The abrasive action is performed simultaneously on several or many parts. The number of parts per ring is given below.

Stock removal rates are expressed as min/in. For example, if a medium hardness steel (40–45 Rc) requires an 8-surface roughness requirement, and 0.002 in. metal removal is necessary, the time for each side is $.002 \times 12{,}000 = 24$ min. This time is divided by the number of rings and the number of parts loaded into each ring.

time is zero, except whenever off-loading elements exceed the abrasive time.

No. parts per ring

Ring Dia, in.	Part Dia, in.								
	.2	.4	.8	1	2	4	6	9	10
6¼	950	220	55	32	7	1	1		
9		380	100	60	15	3	1	1	
18			400	260	66	25	5	1	1
24			750	480	115	28	11	3	4
32				875	210	50	21	9	8

EXAMPLES

A. A flat cylindrical surface must have free abrasively ground sides in a vertical machine. A ring fixture with 12 openings is loaded. The part material is alloy steel and has been previously rough ground, but 0.003 in. per side stock removal is required. Find the lot time for 8,000 units. The machine has four rings. Ring loading is not available during lapping.

Table Number	Process Description	Table Time	Adjustment Factor	Cycle Minutes	Setup Hours
13.6-S	Setup	.5			.5
13.6-1	Load and unload	.12		.12	
13.6-1	Turn over for second side	.05		.05	
13.6-2	Start and stop	.05	½₁₂ × 4	.001	
13.6-2	Clean rings twice	.30	²⁄₁₂	.05	
13.6-3	Free abrasive grind, 8 surface finish	12,000	.003/4 × 12	.75	

Table Number	Process Description	Table Time	Adjustment Factor	Cycle Minutes	Setup Hours
13.6-3	Free abrasive grind, second side	.75		.75	—
	Total			1.721	.5
	Hr/100 units	2.869			
	Lot estimate	230 hr			

B. A 1-in. OD ½-in. ID soft steel washer must have .003 in. per side removed for a 16 RMS finish on both sides. A free abrasive grinding machine with 4 rings is available. Loading is unavailable during grinding time. Machine control of pressure plates is an available machine feature. A 9-in. ring size machine is assumed. Find the unit estimate.

Table Number	Process Description	Table Time	Adjustment Factor	Cycle Minutes	Setup Hours
13.6-S	Setup	.5			.5
13.6-1	Load and unload small part	.03		.03	
13.6-1	Turn over for second side	.01		.01	
13.6-2	Start and stop	.05	2/240	0	
13.6-3	Lap first side: 5000 (.003 ÷ 240)	5000	.003/240	.063	
13.6-3	Lap second side			.063	
13.6-2	Lower backup wheels, release	0		0	—
	Total			.166	.5

TABLE 13.6 FREE ABRASIVE GRINDING MACHINES

Setup   .5 hr

Operation elements in estimating minutes

1. Handling
   - Load and unload small part   .03
   - Load and unload medium part   .07
   - Load and unload large part   .12
   - Turn over for other side,
       - small   .01
       - medium   .03
       - large
   - Turn over ring and all parts automatically   .75
   - Preload parts into spare retaining rings   0
   - Slide ring with parts onto table   .01
   - Remove retaining ring   .08
   - Sandwich-turn parts   .15

2. Machine operation
   - Start and stop   .05
   - Clean, place and remove weight per lapping ring   .38/no. pcs
   - Lower backup wheel, per ring   .03
   - Release pressure plate, per ring   .02
   - Clean ring, each occurrence   .30

3. Free abrasive grinding
   Stock removal rates,[1] min/in.

Material	Surface Finish				
	4	6	8	16	25
Aluminum				1,500	600
Cast iron		12,000	7,500	5,000	3,300
Glass				600	400
Nickel		30,000			6,000
Sintered iron				6,000	1,875
Steel (soft)			10,000	5,000	3,000
Steel (medium)	20,000		12,000	7,500	5,000
Steel (hard)	30,000		15,000	7,500	

[1] Multiply by stock removal in thousandths thickness and divide by no. parts in ring.

4. Miscellaneous elements, per occurrence
   Insert in envelope for protection .09
   Inspect thickness or finish .14
   Wipe small part .03
   Inspection See Table 22.1-3

13.7 Disk Grinding Machines

DESCRIPTION

Disk grinding consists of grinding a side of a workpiece on the flat side or face of an abrasive grinding disk. The grinding disk is mounted on a steel wheel which is attached to a motorized spindle assembly.

To remove stock or material from the parallel faces of a part, a double disk grinding machine is used. The machine has two power driven spindles, and each supports an abrasive disk. The spindles are mounted so that the abrasive disks are opposite each other, (see Figure 13.7A). Thus, as the work is carried between the opposed abrasive disks during the grinding operation, the disks can grind opposite and parallel sides or faces.

Flat surfaces offer possibilities for disk grinding, especially parallel flat surfaces that permit grinding the faces simultaneously.

The advantage in double disk grinding is with a part with two faces that are approximately equal in area. It is more economical to finish two surfaces or faces in one operation. This combines high production while generating a size tolerance, flat and parallel surfaces, and a required surface finish.

Figure 13.7B is a double-disk grinding machine. Machines can be equipped with a variety of fixtures to meet special needs or greater production. An oscillating arm fixture contains an opening where individual parts are loaded and unloaded. Rotary-type fixtures

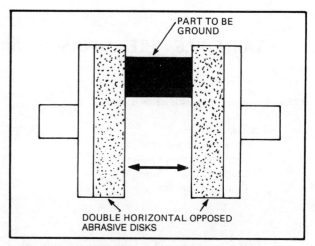

FIGURE 13.7A The operation of abrasive disks during the grinding process.

are for medium or small sized parts where high production and accuracy are required. A through-feed type fixture is also possible. Feeding methods range from manual to automatic. Parts feeder and conveyor unloading are also possible.

Some machines may have only single-sided disk grinding.

FIGURE 13.7B Double disk grinding machine. *(Bendix-Besly)*

ESTIMATING DATA DISCUSSION

Setup conditions relate closely to production quantity. For long runs there is the necessity to clean the machine, perhaps at the beginning of each shift. For short runs where there are changes in fixtures, materials, and grinding requirements, other conditions are provided.

Element 1 deals with disk grinding. The data analysis averaged a variety of feeding conditions ranging from individual loading to automatic feeding. If the loading is completely automatic and a tender is unnecessary, then direct labor time is zero. These estimates then indicate the production rate.

The estimating "driver" is the outside area multiplied by the stock removal. Outside area is the maximum x-y part envelope. Stock removal is for both sides and per pass. If only one disk grinder wheel is available, the amount of stock removal is also less.

When the material is aluminum, we use the part area, stock removal, and width of the part. The lookup procedure for the table needs three pieces of information. The area of the part uses the major x-y dimensions as if the part is a rectangle. If the width is over 5 in., or if the width is over 3 in. and has a total indicator reading for flatness of ±.005, use the column designated "W > 5 in." Otherwise use the column "W < 5 in." Table lookup rules may be used to facilitate speed. For example, if the area and total stock removal are not listed, the estimator may want to jump up to higher value. For ceramic parts, the stock removal is .005 in., and represents high-volume experiences.

$$\text{No. of passes} = 1 + \frac{\text{Total mat'l to be removed} - \text{Finish cut}}{\text{Mat'l removed per pass (see table below)}}$$

Size of Part	Rough Cut	Finish Cut	Material
Round 0–1 *Dia*	.010	.010	Bronze, aluminum, brass
Round 0–1 *Dia*	.005	.002–.003	Stainless steel, hardened steel
Round 1–1½ in. *Dia*	.003	.0015–.002	Stainless steel, hardened steel
	.010	.005	Bronze, aluminum, brass

Irregular shapes can be estimated by figuring on the longest dimension as compared to the diameters, as figured above.

EXAMPLES

A. A SAE 5130 Rockwell 45 pinion gear is to be disk ground using a double disk grinding machine. Stock removal is .009 in. and a flatness of .0005 in. and parallelism of .0005 in. is required. A finish of 20 RMS is necessary. The pinion gear has a 1.323-in. pitch diameter and a width of .75 in. Find the unit estimate. Assume a very long run.

Table Number	Process Description	Table Time	Adjustment Factor	Cycle Minutes	Setup Hours
13.7-1	Area of face ≈ 1.368 sq. in.; stock removal = .009 in.; 1.368 × .009 = .012	.038		.038	
	Total			.038	

B. A double horizontal wet disk grinder with a hydraulically driven oscillating fixture is to be estimated for a gray cast iron valve body. A total of .093-in. stock removal is necessary. The area of the value body faces is irregular, but the maximum x-y dimensions are 2½ × 3⁵⁄₁₆ in. A quantity of 5500 units is planned.

Table Number	Process Description	Table Time	Adjustment Factor	Cycle Minutes	Setup Hours
13.7-S	Setup	3.0			3.0
13.7-1	Outside area = 8.3 sq in.; stock removal = .093 in.; 8.3 × .093 = .77	.25		.25	—
	Total			.25	3.0
	Hr/100 units	.417			
	Lot estimate	25.94 hr			

C. An aluminum part having major dimensions in the *x-y* plane is 3.7 × 5.9 in. Stock removal is .015 in. Each blank is later sheared into two pieces. A quantity of 80 blanks is planned. Find the cycle time, unit estimate, hr/100 units, and lot time.

Table Number	Process Description	Table Time	Adjustment Factor	Cycle Minutes	Setup Hours
13.7-S	Setup basic	.75			.75
13.7-1	Area = 21.83 sq in.	.51	½	.255	—
	Total			.255	.75
	Cycle estimate	.51			
	Hr/100 units	.425			
	Lot estimate	1.43 hr			

D. A ceramic ring has an order of 200,000 units. The tolerances are ±.0005 in. and the OD is 1.5 in. Find the time for the lot.

Table Number	Process Description	Table Time	Adjustment Factor	Cycle Minutes	Setup Hours
13.7-S	Change wheels	4.0			4.0
13.7-S	Adjust wheels	.2			.2
13.7-S	Dress wheels	3.4			3.4
	Total setup				7.9
13.7-1	Disc grind ceramic part	.015		.015	—
	Total			.015	7.9
	Hr/10,000 units	2.500			
	Lot estimate	57.9 hr			

TABLE 13.7 DISK GRINDING MACHINES

Setup

Dress wheel and reset machine	.75 hr
Change abrasive wheels (each material change)	4.0 hr
Dress wheel (every 3 hours of operation)	.2 hr
Remove, replace feed wheels	1.0–6.0 hr
Change bushings of feedwheel	2.0 hr

Operation elements in estimating minutes

1. Disk Grind

    A. Steel:

Outside Area × Stock Removal	.001	.003	.005	.01	.02	.03
Min	.035	.036	.036	.037	.038	.040

Outside Area × Stock Removal	.04	.05	.06	.07	.08	.09
Min	.042	.043	.045	.047	.048	.050

Outside Area × Stock Removal	.1	.2	.3	.4	.5	Add'l .1
Min	.052	.069	.085	.10	.12	.016

B. Cast iron:
    Piston rings, 2–6 in. *Dia*, .0057-in. max
    Stock removal   .0035 ea

Outside Area × Stock Removal	.5	1.0	2.0	3.0	5.0	10
Min	.12	.25	.50	.75	1.25	2.50

C. Aluminum:

Stock Removal, in.

	.006		.012		.015		.018		.024	
Area	W < 5	W > 5	W < 5	W > 5	W < 5	W > 5	W < 5	W > 5	W < 5	W > 5
3	.09		.09		.20		.20		.21	
9	.10		.10		.22		.23		.24	
15	.13	.27	.14	.28	.27	.41	.28	.42	.31	.45
21	.19	.34	.20	.35	.34	.49	.36	.51	.39	.54
27	.19	.34	.21	.36	.36	.51	.39	.54	.43	.58
33	.19	.34	.21	.36	.38	.53	.43	.56	.46	.61
39	.20	.35	.22	.37	.40	.55	.44	.59	.49	.64
47	.20	.35	.23	.38	.42	.57	.47	.62	.54	.69
57		.37		.40		.62		.67		.75
69		.38		.43		.68		.73		.84
84		.40		.45		.73		.79		.92
100		.42		.48		.79		.86		1.01
120		.44		.52		.86		.95		1.12

D. Ceramic:

Dia	.5	.8	1	1.1	1.3	1.5	1.8
Min	.007	.008	.009	.010	.011	.015	.024

2. Wheel dressing
    Automatic from microprocessor control   0
    Manual, prorated   .002/unit

3. Inspection   See Table 22.1-3

13.8 Vertical Internal Grinding Machines

DESCRIPTION

The grinding of internal bores, cams, and other contoured surfaces can be done on a vertical internal grinding machine. Parts are loaded on a horizontal rotating table and chucking is possible. Some vertical grinding wheel spindles can rotate up to 60,000 rpm while the worktable revolves at speeds between 30 and 100 rpm. To permit grinding of contoured surfaces, the worktable is attached to a master cam, which is an enlarged concentric version of the ID to be ground. This master cam rotates against two fixed rolls spaced 90° apart on the cam housing.

An axial traverse motion brings the grinding wheel into the work area, which also allows for vertical traverse grinding. A crossfeed motion brings the grinding wheel into contact with the workpiece and then continues to feed into the piece until the necessary stock has been removed. Maximum stock removal is about .035 in. on the diameter. Feed rates differ between rough and finish grinding.

Due to the size of the machine considered, the outside diameter of the workpiece does not exceed 6 in. Part height is also limited.

ESTIMATING DATA DISCUSSION

The general setup includes adjustments for feed rates, wheel approach, diamond wheel dressing, sparkout, and test cycles. The time needed to change a machine cam is also included under setup. This element is not required if the same cam ID is required in subsequent part lots. Setup values are listed in hours.

The operation elements are composed of part handling, machine start and stop, rough grinding with sparkout, wheel dressing, finish grinding with sparkout, and grinding wheel changes. Part handling consists of loading and unloading the machine, with variations allowed if the machine work fixture is removed for loading. Since part size is limited to objects with 6-in. Dia or less, this factor is considered constant in the estimating table. Machine start and stop includes the time to push the start button, as well as the automatic machine operations of opening and closing the door, and starting and stopping the coolant flow.

The grinding cycles are more complicated due to the number of variables. These include workspeed variations, part diameter, part length, spindle diameter and length, grinding wheel and part material, revolutions for sparkout, and the amount of stock removed. For these estimations, it is assumed that 90% of the stock is removed during rough grinding. The rough sparkout varies with the required finish. For parts with tight tolerances, it is not uncommon to have the sparkout last for over a half minute. One minute is the maximum. After the wheel is dressed the final 10% of stock is removed in the finish grind cycle, with application of the finish sparkout as above. For ideal grinding, the grinding wheel diameter should roughly equal .75 Dia, where Dia = equivalent bore diameter of the cam. Also, for these estimates, the infeed is .0003 in. for rough grinding and .0001 in. for finish grinding.

In internal grinding there are two possibilities. If the spindle is shorter than the length of the part, a traverse grinding operation is necessary. For rough grinding, the axial wheel advances 75% of the grinding wheel width per workpiece revolution. One quarter of the nominal width is also allowed for overlap. For finish grinding, the axial wheel advances only 25%. The values given in the table must be multiplied by the quantity

$$\frac{L \times Dia}{W}$$

where L = length of part, in.;
Dia = equivalent bore diameter (adjusted from cam ID); and
W = width of grinding wheel, in.

This computed value will be the estimated time in minutes. If the grinding wheel width is wider than the part length, a plunge grinding operation is necessary. These values can be taken from the element table for plunge grinding. To find work surface speed from rpm, use the following equation:

$$V = \pi DR/12$$

Due to grinding wheel wear, the wheel will have to be changed or dressed periodically. Since available grinding wheel abrasives differ as widely as material hardnesses, this value is prorated for a quantity, specifically 40 parts per grinding wheel. Finally, operator inspection of parts is done internal to the grinding element, therefore no additional time is allotted for this activity.

EXAMPLES

A. A lot of 200 pump liners is to be ground. The equivalent cam diameter is .80 in., and the height is 1.20 in. The part is loaded directly into the machine so the fixture need not be removed. The grinding wheel is 2-in. long and revolves at 60,000 rpm. The worktable rotates at 40 rpm. The total stock to be removed is .015 in. on the radius. Find unit and lot estimates.

Table Number	Process Description	Table Time	Adjustment Factor	Cycle Minutes	Setup Hours
13.8-S	General setup	1.3			1.3
13.8-1	Load and unload	.46		.46	
13.8-2	Start and stop	.13		.13	
13.8-4	Rough grind (plunge) .014 in. stock	1.17		1.17	
13.8-5	Rough sparkout	.50		.50	
13.8-6	Dress wheel	.06		.00	
13.8-4	Finish grind (plunge) .001 in. stock	.25		.25	
13.8-5	Finish sparkout	.58		.58	
13.8-8	Change grinding wheel	.15		.15	
	Total			3.30	1.3
	Lot estimate	12.30 hr			
	Hr/100 units	5.50			

B. A lot of 30 parts with internal cam IDs is to be ground. The equivalent cam diameter is 3 in., and the height of the part is also 3 in. Work rotates at 30 rpm and the spindle width is 2 in. A total of .013 in. of stock is to be removed. Find unit and lot estimates.

Table Number	Process Description	Table Time	Adjustment Factor	Cycle Minutes	Setup Hours
13.8-S	Setup	1.3			1.3
13.8-1	Load and unload fixture	1.66		1.66	
13.8-2	Start and stop	.13		.13	
13.8-3	Rough grind (traverse) .012 in. stock	.47	3 × ¾	2.11	
13.8-5	Rough sparkout	.50		.50	
13.8-6	Dress wheel	.06		.06	
13.8-3	Finish grind (traverse) .0010 in. stock	.35	3 × ¾	1.57	
13.8-5	Finish sparkout	.58		.58	
	Total			6.61	1.3
	Lot estimate	4.60 hr			
	Hr/100 units	11.01			

C. An internal cam ID is to be ground with an equivalent diameter of .9 in., and height of 3.25. The master cam must be changed since this cam differs in shape from that of the last lot run. Spindle width is 2 in., and the worktable rotates at 45 rpm. Stock removal is .015 in. Lot size is 80. Find unit and lot estimates for the total grinding process.

Table Number	Process Description	Table Time	Adjustment Factor	Cycle Minutes	Setup Hours
13.8-S1	Setup, 2 times	2.6			2.6
13.8-S2	Master cam change, 2 times	1.9	2		3.8
13.8-1	Load and unload, 2 times	.92		.92	
13.8-2	Start/stop, 2 times	.13	2	.26	
13.8-3	Rough grind (traverse) .014 in stock	1.63	9 × 3.25/2	2.38	
13.8-3	Rough grind (traverse) .010 in. stock	1.16	9 × 3.25/2	1.70	
13.8-5	Rough sparkout, 2 times	.50	2	1.00	

Vertical Internal Grinding Machines

Table Number	Process Description	Table Time	Adjustment Factor	Cycle Minutes	Setup Hours
13.8-6	Dress wheel, 2 times	.06	2	.12	
13.8-3	Finish grind (traverse) .001 in. stock	1.05	.9 × 3.25/2	1.54	
13.8-3	Finish grind (traverse) .001 in. stock	1.05	.9 × 3.25/2	1.54	
13.8-8	Change grinding wheel	.15		.15	
	Total			9.61	6.40
	Hr/100 units	16.35			
	Lot time for 80 units	16.01 hr			

TABLE 13.8 VERTICAL INTERNAL GRINDING MACHINES

Setup

Basic	1.3 hr
Master cam change	1.9 hr

Operation elements in estimating minutes

1. Handle part

Removal fixture	1.66
Machine fixture	.46

2. Start and stop .13

3. Traverse grinding[1]

Total Stock Removed	Work Surface Velocity (fpm)					
	10	20	30	40	50	100
Rough						
.005	.59	.29	.19	.15	.12	.07
.010	1.16	.59	.39	.29	.23	.12
.012	1.40	.69	.47	.35	.28	.13
.014	1.63	.81	.55	.41	.32	.16
.016	1.87	.93	.61	.47	.39	.19
.018	2.09	1.04	.69	.52	.41	.21
.020	2.32	1.16	.77	.57	.47	.23
Finish						
.0010	1.05	.52	.35	.26	.21	.10
.0012	1.26	.63	.42	.31	.25	.13
.0015	1.57	.78	.52	.39	.31	.16
.0020	2.09	1.05	.70	.52	.42	.21

[1] Multiply by $L(Dia)/W$, L = length of part, Dia = bore diameter, W = wheel width.

4. Plunge grinding

Total Stock Removed	Work Table rpm					
	30	40	50	60	70	100
Rough						
.005	.55	.42	.33	.28	.24	.17
.010	1.11	.83	.67	.55	.48	.33
.012	1.33	1.00	.80	.67	.57	.40
.014	1.55	1.17	.93	.78	.67	.47
.016	1.78	1.33	1.07	.89	.76	.53
.018	2.00	1.50	1.20	1.00	.80	.60

Total Stock Removed	Work Table rpm					
	30	40	50	60	70	100
Finish						
.0010	.33	.25	.20	.17	.14	.10
.0012	.40	.30	.24	.20	.17	.12
.0015	.50	.37	.30	.25	.21	.15
.0020	.67	.50	.40	.33	.28	.20

5. Sparkout
   Rough grind   .50
   Finish grind   .58

6. Diamond dress wheel, prorated   .06

7. Inspection   See Table 22.1-3

8. Grinding wheel change = 6 min/40 parts, prorated   .15

13.9 Numerically Controlled Internal Grinding Machines

DESCRIPTION

Numerically controlled internal grinding machines operate in much the same way as manual internal grinding machines. With numerically controlled machines the process is either programmed into the console which regulates the operation or punched onto punch cards and then read into the console. These cards will tell the machine exactly what needs to be done. Media can also include punched paper tape or computer data. This includes surfaces which need to be ground, how much material should be taken off with each rough grind and finish grind, and what needs to be done during wheel dressing. The machine also keeps track of when a new grind wheel is needed. Another feature of the numerically controlled machine is that any of the data can be changed at any time during the run cycle.

The actual internal grinding machine operates in the following fashion. It has a rapidly rotating grinding wheel, and in most machines the workpiece will also be rapidly rotated. A separate horizontal movement will position the grind wheel on the grind surface and at the conclusion of the operation will withdraw the wheel to the initial position. The horizontal movements of the wheel also will carry the wheel to the diamond dresser for automatic wheel dressing. In some machines the machine decides the wheel power at which grinding is to be done according to what surfaces are being ground and the wheel and material used.

Numerically controlled internal grinders have a wide range of capacities. The minimum bore diameter is approximately .1 in. while the maximum outside diameter of the workpiece is about 12 in. The maximum length of the hole is roughly 9 in. Features of some models include automatic compensation for wheel wear and can sometimes grind up to 10 surfaces. A typical numerically controlled internal grinder is shown by Figure 13.9.

Manual internal grinders are not considered with these data. Consider estimating data for general class of numerically controlled internal grinding machines.

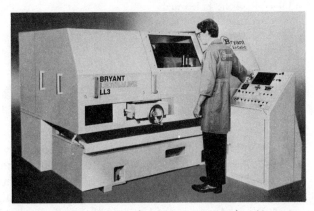

FIGURE 13.9 Bryant Lectraline LL3 CWC Grinder. *(Courtesy Bryant Grinder Corp.)*

ESTIMATING DATA DISCUSSION

A value of 3.5 hours is average for setup of numerically controlled internal grinders. This value will vary to some extent, however, due to increased number of surfaces to be ground. Setup for the numerically controlled machine includes programming of steps to be made so the machine will know what to do. Setup will also include dressing the grind wheel for the first time and running trial parts. Setup will not, however, include changing a worn wheel during cycle time. If cycle time is to include a wheel change, Element 5 should be consulted. Finding gaging materials and setting them to the correct measurements will also be included in setup time.

Data in Element 1 will vary somewhat for various types of fixtures because of different loading features. This element should be used once for each unit. Data for starting machine and opening and closing machine door should also be used once for each unit. Data in Element 2 should be used whenever needed. Element 5 should also be used only once for each unit.

In traverse grinding, Element 5, the values are to be multiplied by $L \cdot Dia/W$ where L = Length of grinding dimensions. Dia is grind dimension prior to material removal and W = grind wheel contact length. Slender holes and bores may use 20 or 30 sfpm while normal grind operations will use 50 sfpm. These data use .0003 in. and .0001 in. for infeed rates for rough and finish grinding. The estimator may want to use a policy of 90% of the total stock for finish grind allowance, leaving 10% of the total stock for finish allowance. Sparkout is a case where there is no infeed but only rotation to achieve finish or final dimension. Total stock removal for these data imply diametral reduction.

Plunge grinding, Element 5, uses the diameter of the bore and total stock removal as entry values. Velocity of 40 spfm, rough infeed of .0004 in, and finish infeed of .0001 in. are the parameters adopted for this table. Stock removal has been allocated as 90% for rough and 10% for finish.

A variety of inspection selections are provided by Element 7. If the estimator believes that additional inspection time is necessary a tight tolerance adjustment is available.

Element 7 includes a time to grind the wheel. If it is necessary to replace a worn wheel, a time value is provided. However, the adjustment factor is necessary to prorate that time to individual units.

EXAMPLES

A. Estimate the time to grind an internal bore of .5-in. *Dia* and 1-in. deep. Total stock removal is .010 in. The number of surfaces is 1. Lot quantity is 78 units. Find the unit cycle, setup, and the lot estimate.

Table Number	Process Description	Table Time	Adjustment Factor	Cycle Minutes	Setup Hours
13.9-S	Setup for one surface	3			3
13.9-1	Load in fixture	.18		.18	
13.9-2	Start machine	.018		.018	
13.9-2	Open, close door	.033		.033	
13.9-1	Blow chips from fixture and ground piece	.069	2	.138	
13.9-3	Move wheel to surface	.20		.20	
13.9-4	Grind hole rough Use 30 sfpm, .009 in.; ½-in. wheel = W	.39	1 × .5/5	.39	
13.9-4	Grind surface finish	.35	1 × .5/5	.35	
13.9-6	Inspection air gage, every part	.10		.10	
	Total			1.209	3
	Hr/100	2.015			
	Lot estimate	4.57 hr			

B. Find the unit and lot time for 100 units for a jaw crusher bearing raceway part where three surfaces are to be ground. Stock removal from the face is .015 in. Stock removed from the gear ID and small ID is .020 in. and .018 in. The face is 5-in. OD, and L of surface grind is 2.5. Gear ID is 4.0 and L = 2.9. Small ID = 2.5 and L = 1.2 in. Allow 10% for finish grind. Wheel *Dia* is 2.0 in. Use rule of using higher table when time driver falls in between tabular values.

Table Number	Process Description	Table Time	Adjustment Factor	Cycle Minutes	Setup Hours
13.9-S	Setup for three surfaces	4			4
13.9-1	Load and unload part	.18		.18	
13.9-2	Start machine	.018		.018	
13.9-2	Open and close door	.033		.033	
13.9-3	Position for first surface	.20		.20	
13.9-4	Traverse grind face; use 50 spfm, rough; stock removal .014 in.	.32	2.5 × ⅝	2.00	
13.9-4	Finish grind .001	.21	2.5 × ⅝	1.31	
13.9-3	Position for second surface	.15		.15	
13.9-4	Gear ID .018 in., rough	.41	2.9 × ⅝	2.38	
13.9-4	Grind ID .002 in., finish	.42	2.9 × ⅝	2.44	
13.9-4	Sparkout	.06	2.9 × ⅝	.35	
13.9-5	Plunge grind 2.5 in.; L = 1.2 in., stock removal = .018 in.			1.35	
		1.01		1.01	
13.9-6	Inspection inside micrometer, 3 dim	.045	3	.135	
13.9-7	Dress wheel	.07	2	.14	
	Total			10.346	4
	Hr/100	17.243			
	Lot estimate	21.24 hr			

C. Plunge grind one dimension of the jaw crusher bearing raceway. The stock removal is .027 in. max and the Dia = 5.255 in. before stock removal. There are 100 units in the lot. Find the total cycle minutes, setup hour, and the lot hours.

Table Number	Process Description	Table Time	Adjustment Factor	Cycle Minutes	Setup Hours
13.9-S	Setup for one surface	3			3
13.9-1	Load and unload chuck jaws	.40		.40	
13.9-2	Start machine	.018		.018	
13.9-2	Open and close door	.033		.033	
13.9-3	Position wheel	.20		.20	
13.9-5	Plunge grind	2.80		2.80	
13.9-6	Inspection	.045		.045	
13.9-7	Dress wheel	.07		.07	
	Total			3.57	3
	Hr/100	5.94			
	Lot estimate	8.94 hr			

TABLE 13.9 NUMERICALLY CONTROLLED INTERNAL GRINDING MACHINES

Setup

1 or 2 surfaces	3 hr
3 or 4 surfaces	4 hr
over 4 surfaces	5 hr

Operation elements in estimating minutes

1. Load and unload part

Fixture	.18
Collet	.44
Chuck jaws	.40
Blow chips from fixture or part	.069

2. Operate machine

Start machine	.018
Open and close door	.033
Change data in machine (prorate over lot quantity)	1.19

3. Position grind wheel before first surface grind .20
   Position grind wheel before secondary grind surfaces .15

4. Traverse grinding internal

Total Stock Removed	Work Surface Velocity (sfpm)					
	10	20	30	40	50	100
Rough						
.005	.59	.29	.19	.15	.12	.07
.010	1.16	.59	.39	.29	.23	.12
.012	1.40	.69	.47	.35	.28	.13
.014	1.63	.81	.55	.41	.32	.16
.016	1.87	.93	.61	.47	.39	.19
.018	2.09	1.04	.69	.52	.41	.21
.020	2.32	1.16	.77	.57	.47	.23
Add'l .001					.02	
Sparkout					.06	
Finish						
.0010	1.05	.52	.35	.26	.21	.10
.0012	1.26	.63	.42	.31	.25	.13
.0015	1.57	.78	.52	.39	.31	.16
.0020	2.09	1.05	.70	.52	.42	.21
Sparkout					.06	

[1] Multiply table value by $L(Dia)/W$ where L = length of grind dimension and any overtravel, in.; Dia = bore diameter, in.; and W = wheel width, in.

5. Plunge grind internal

Dia	Total Stock Removed					Add'l .001
	.005	.010	.015	.020	.030	
1	.11	.19	.26	.34	.47	.01
1.5	.18	.28	.40	.50	.71	.01
2	.24	.38	.54	.67	.93	.02
3	.35	.57	.80	1.01	1.40	.04
4	.48	.75	1.07	1.34	1.87	.06
5	.59	.95	1.33	1.69	2.33	.07
6	.71	1.13	1.61	2.03	2.80	.08
Add'l 1.0	.13	.19	.26	.33	.47	

6. Inspection, .005-in. standard commercial tolerance
   - Plug gaging, manual, .20 min every 10 parts .02
   - Inside micrometer, .46 min every 10 parts .005 in tolerance .046
   - Mechanical dial indicators, .15 min every 10 parts .015
   - Air or electronic indicators, .10 min .01
   - .0005 in. tight tolerance adjustment .15
   - .0002 in. tight tolerance adjustment .25

7. Dress wheel, prorated .07
   Change wheel, prorate over quantity 5.13

GEAR CUTTING

14.1 Gear Shaper Machines

DESCRIPTION

The cutter gear generating process for cutting involute gears uses the principle that any two involute gears of the same pitch will mesh together. If one gear behaves as a cutter and is given a reciprocating motion, as in a gear shaper machine, it will be capable of generating conjugate tooth forms in a gear blank. A gear shaper cutter that provides this action is shown, mounted in a gear shaper, in Figure 14.1. In operation, the cutter and the blank rotate at the same pitch line velocity. In addition, a reciprocating motion is given to the cutter. A rotary feed mechanism is arranged to have the cutter automatically feed to the desired depth while both cutter and work are rotating. Internal and external teeth can be shaped.

Figure 14.1 is a machine that cuts spur, helical or herringbone gears, internal or external, and other non-involute shapes up to a 40-in. pitch diameter and a 6-in. face width. Now consider estimating data for the general class of gear shaper machines.

ESTIMATING DATA DISCUSSION

Element 1, handling, includes both loading and unloading to the shaper table. Entry variable is weight. Table clamps, fixture, or chuck conditions are provided. Chip removal, part cleaning, and clamping is included in the handling element.

Element 2 includes machine start, position, and

FIGURE 14.1 A gear shaper machine capable of shaping internal and external gears up to 36 pitch diameter (*PD*). *(Fellows Corporation)*

engage. The constant time of .25 min is always provided. In addition to the constant time, machine time includes infeed time and rotary feed time for rough and finish cut.

The materials, their Brinell hardness number and cutting velocity, follow:

Material	Bhn	fpm
Low-carbon steel, free machining	170	93
Low-carbon steel	150	81
Medium-carbon steel	215	47
Stainless steels	180	55
Steel castings	230	43

Material	Bhn	fpm
Cast iron	soft	67
Cast iron	medium	51
Brass		100

The steps for estimating machining time involve finding the strokes per minute, infeed time, and rotary feed time. The equation for spm is

$$N = \frac{12 \times \text{fpm}}{2 \times \text{stroke length}}$$

Strokes Per Minute (spm) For Straight Spur Gears

Material	Face Width							
	½	1	1½	2	2½	3	3½	4
Low-carbon steel, free-machining	992	510	336	251	203	168	144	127
Steel, low-carbon	864	444	293	219	177	147	125	110
Steel, medium	501	258	170	127	103	85	73	64
Stainless steels	586	301	199	149	120	100	85	75
Steel casting	458	236	155	116	94	78	67	59
Cast iron, Soft	714	367	242	181	146	121	104	91
Medium	544	280	184	138	111	92	79	69
Brass	1066	548	361	270	218	181	155	136
Helix angle, degrees				15	30	45	60	
Factor				.97	.87	.71	.50	

Multiply spm for helical gears by factor.

Diametral Pitch	No. Cuts	Roughing	Semifinishing	Finishing
24–64	1	.014	.012	.004
	2	.012/.020	.012/.020	.006
	3		.012/.020/.020	.008
10–24	1	.030	0.20	
	2	.015/.030	.030/.045	.025/.040
	3		.030/.030/.045	.030/.030
7–10	1	.020		
	2		.030/.040	.020/.035
	3		.030/.030/.040	.025/.025/.035
4–7	1	.015		
	2		.020/.030	.020/.025
	3		.020/.020/.030	.020/.020/.025

Stroke length is related to face width plus approach and overtravel. For a helix gear, multiply N by the factor. Infeed time has the equation $\frac{D + .030}{.002 \times N}$, where D is tooth depth and .002 in./stroke is the feed rate. Rotary feed time is $\frac{PD \times \pi}{.020 \times N}$, where PD is the pitch diameter.

The rotary feed per stroke will vary with material, number of cuts, and finish required. The value .020 is adopted. Other values may be chosen by the estimator and by ratio; the tabular value can be adjusted. For instance, if .030 is preferred, the table value is multiplied by ⅔. The number of cuts may be 1, 2, or 3. This depends upon diameter pitch (*PD*) and its relationship to rotary feed per stroke. But .020 in. per stroke is an average value and can be used if other information is unavailable.

More than one machine may be tended by the operator, so the unit time is divided by the number of machines to approximate the direct-labor effort.

After finding the strokes per minute (spm) Element 3 is used and the nearest entry values are adopted.

EXAMPLES

A. A steel forging, 7 pitch, 80 teeth, .257-in. depth of tooth, and a length of face of 3⅝ in. is to be estimated with gear shaping. Weight is 32 lb. Two cuts are anticipated. Find the lot time for 272 units. Two machines are tended by one operator.

Table Number	Process Description	Table Time	Adjustment Factor	Cycle Minutes	Setup Hours
14.1-S	Setup for spur gear	.6			.6
14.1-S	Fixture	.3			.3
14.1-1	Handle	1.11	½	.56	
14.1-2	Machine time constant Strokes per minute, 59	.25	½	.13	
14.1-3	Using $N = 61$ and $DP = 8$, infeed	2.48	½	1.24	
14.1-4	Rough rotary feed, $N = 61$, $DP = 8$	20.8	½	10.4	
22.1-3	Check gear over pins	.20	½	.10	
14.1-4	Finish rotary feed, $N = 61$, $DP = 8$	20.8	½ × .020/.030	6.93	
22.1-3	Check gear over pins	.20	½	.10	
11.4-2	Tool wear and replacement for 35 min	2.34	½	1.67	
	Total			21.12	.9
	Floor-to-floor time	42.23 min			
	Lot estimate	96.66 hr			

TABLE 14.1 GEAR SHAPER MACHINES

Setup	.6 hr
Helical gears	1.0
For .001 total composite error or less, add	.2
For fixture, add	.3

Operation elements in estimating minutes

1. Handling

   Manual load and unload blank

Weight	1	5	25	50
To table clamps	.68	1.14	1.97	3.14
To fixture	.40	1.03	.78	1.11
To chuck	.52	1.06	.98	2.54

Jib load	4.00
Turnover	65% of basic handling

2. Start, position, engage   .25

3. Infeed shape, min

spm	Diametral Pitch and Whole Depth									
	2 1.08	3 .72	4 .54	8 .27	12 .18	18 .12	24 .09	30 .07	36 .06	40 .05
50	11.09	7.49	5.69	3.00	2.10	1.50	1.20	1.02	.90	.84
55	10.08	6.81	5.18	2.72	1.91	1.36	1.09	.93	.82	.76
61	9.16	6.19	4.70	2.48	1.73	1.24	.99	.84	.74	.69
67	8.33	5.63	4.28	2.25	1.58	1.13	.90	.77	.68	.63
73	7.57	5.12	3.89	2.05	1.43	1.02	.82	.70	.61	.57
81	6.88	4.65	3.53	1.86	1.30	.93	.74	.63	.56	.52
89	6.26	4.23	3.21	1.69	1.18	.85	.68	.58	.51	.47
97	5.69	3.84	2.92	1.54	1.08	.77	.62	.52	.46	.43
107	5.17	3.49	2.66	1.40	.98	.70	.56	.48	.42	.39
118	4.70	3.18	2.41	1.27	.89	.64	.51	.43	.38	.36
130	4.27	2.89	2.19	1.16	.81	.58	.46	.39	.35	.32
143	3.89	2.63	2.00	1.05	.74	.53	.42	.36	.32	.29
157	3.53	2.39	1.81	.95	.67	.48	.38	.32	.29	.27
173	3.21	2.17	1.65	.87	.61	.43	.35	.30	.26	.24
190	2.92	1.97	1.50	.79	.55	.39	.32	.27	.24	.22
209	2.65	1.79	1.36	.72	.50	.36	.29	.24	.22	.20
230	2.41	1.63	1.24	.65	.46	.33	.26	.22	.20	.18
253	2.19	1.48	1.13	.59	.41	.30	.24	.20	.18	.17
278	1.99	1.35	1.02	.54	.38	.27	.22	.18	.16	
306	1.81	1.22	.93	.499	.34	.24	.20	.17		
336	1.65	1.11	.85	.45	.31	.22	.18	.15		
370	1.50	1.01	.77	.40	.28	.20	.16	.14		
407	1.36	.92	.70	.37	.26	.18	.15	.13		
448	1.24	.84	.64	.33	.23	.17	.13	.11		
492	1.13	.76	.58	.30	.21	.15	.12			
542	1.02	.69	.53	.28	.19	.14	.11			
596	.93	.63	.48	.25	.18	.13	.10			

4. Rotary feed shape, min

spm	Pitch Diameter and Circumference									
	1 3.14	1½ 4.7	2 6.3	2½ 7.9	3 9.4	3½ 11.0	4 12.6	5 15.7	8 25.1	10 31.4	15 47.1
50	3.14	4.71	6.28	7.85	9.4	11.0	12.6	15.7	25.1	31.4	47.1
55	2.86	4.28	5.71	7.14	8.6	10.0	11.4	14.3	22.8	28.6	42.8
61	2.60	3.89	5.19	6.49	7.8	9.1	10.4	13.0	20.8	26.0	38.9
67	2.36	3.54	4.72	5.90	7.1	8.3	9.4	11.8	18.9	23.6	35.4
73	2.15	3.22	4.29	5.36	6.4	7.5	8.6	10.7	17.2	21.5	32.2
81	1.96	2.93	3.90	4.88	5.9	6.8	7.8	9.8	15.6	19.5	29.3
89	1.77	2.66	3.55	4.43	5.3	6.2	7.1	8.9	14.2	17.7	26.6
97	1.61	2.42	3.22	4.03	4.8	5.6	6.4	8.1	12.9	16.1	24.2
107	1.47	2.20	2.93	3.66	4.4	5.1	5.9	7.3	11.7	14.7	22.0
118	1.33	2.00	2.66	3.33	4.0	4.7	5.3	6.7	10.7	13.3	20.0
130	1.21	1.82	2.42	3.03	3.6	4.2	4.8	6.1	9.7	12.1	18.2
143	1.10	1.65	2.20	2.75	3.3	3.9	4.4	5.5	8.8	11.0	16.5

spm	Pitch Diameter and Circumference										
	1 3.14	1½ 4.7	2 6.3	2½ 7.9	3 9.4	3½ 11.0	4 12.6	5 15.7	8 25.1	10 31.4	15 47.1
157	1.00	1.50	2.00	2.50	3.0	3.5	4.0	5.0	8.0	10.0	15.0
173	.91	1.37	1.82	2.28	2.7	3.2	3.6	4.6	7.3	9.1	13.7
190	.83	1.24	1.65	2.07	2.5	2.9	3.3	4.1	6.6	8.3	12.4
209	.75	1.13	1.50	1.88	2.3	2.6	3.0	3.8	6.0	7.5	11.3
230	.68	1.03	1.37	1.71	2.1	2.4	2.7	3.4	5.5	6.8	10.3
253	.62	.93	1.24	1.55	1.9	2.2	2.5	3.1	5.0	6.2	9.3
278	.57	.85	1.13	1.41	1.7	2.0	2.3	2.8	4.5	5.7	8.5
306	.51	.77	1.03	1.28	1.5	1.8	2.1	2.6	4.1	5.1	7.7
336	.47	.70	.93	1.17	1.4	1.6	1.9	2.3	3.7	4.7	7.0
370	.42	.64	.85	1.06	1.3	1.5	1.7	2.1	3.4	4.2	6.4
407	.39	.58	.77	.96	1.2	1.4	1.5	1.9	3.1	3.9	5.8
448	.35	.53	.70	.88	1.1	1.2	1.4	1.8	2.8	3.5	5.3
492	.32	.48	.64	.80	1.0	1.1	1.3	1.6	2.6	3.2	4.8
542	.29	.43	.58	.72	.9	1.0	1.2	1.4	2.3	2.9	4.3
596	.26	.40	.53	.66	.8	.9	1.1	1.3	2.1	2.6	4.0

5. Tool wear and replacement   See Table 11.4-2

6. Inspection   See Table 22.1-3
   Indicate surface and true up gear blank   1.20

14.2 Hobbing Machines

DESCRIPTION

Hobbing is defined as a generating process consisting of a rotating and advancing fluted steel worm cutter which passes a revolving blank. In this process all motions are rotary, with no reciprocating or indexing movements. In the actual process of cutting, the gear and the hob rotate together as in mesh.

Figure 14.2A is a hobbing machine with radial feed, axial feed, automatic double cut, and automatic double cut with crown. Also included are variable speeds and feeds, differential, and automatic hob shift.

ESTIMATING DATA DISCUSSION

Element 1 deals with handling elements. The estimator will know whether one part or several are simultaneously hobbed. If a gear blank is arbor loaded, tightening and loosening of nuts is also appropriate.

The hob to use for spur gears depends upon the pitch of the teeth and number of teeth; but for helical gear, the hob is selected with consideration for tooth profile of lead angle, normal diametrical pitch, and cutter diameter. The entry variable is diametral pitch and hob diameter, and the formula for approach and overrun is

$\{WD(HD - WD)\}^{1/2} + 1/8$, where WD
$\qquad$ = whole depth of gear tooth

HD = hob diameter, and the ⅛ in. accounts for safety stock. This distance can be used for spur gears or splines or where hob swivel is less than 5°. This table is shown on the next page, and it is used to adjust the length of hob machining.

The hob has a high angle setting for gears having a helix angle. The entry factors are hob diameter, work diameter, swivel angle of the hob, and diametral pitch. The formula depends upon these four variables. This is shown on the following page.

Overrun distance for gears having a helix angle is also shown. The top value is for the 14½° involute system. (This is shown on the following page.)

For Element 3, the entry variables are material, hob diameter, and swivel. The factor is multiplied by the number of teeth and length of cut which includes the distance between the face of the gear and the ap-

FIGURE 14.2A Hobbing machine. *(Barber Colman)*

proach. A swivel angle is used for helical gears. The "0" is selected for straight spur gears. If a multiple thread hob is used, divide by the number of threads. Normally the thread number is one. The hob diameters are roughly matched to diametral pitch. For hob diameters not shown, interpolate values. If two arbors for one machine are available, the loading of arbors is done during machine time. Some companies may have two or more machines tended by a single operator, in which case the estimated unit time is divided by the number of machines tended by the operator. Spur gears and spline hobbing require an approach and overrun distance to be added to the drawing length of the part.

Approach and overrun for spur gear and spline hobbing, in.

Diametral Pitch	Hob Diameter								
	$1\frac{7}{8}$	$2\frac{1}{2}$	3	$3\frac{1}{2}$	4	$4\frac{1}{2}$	5	6	7
1									3.36
$1\frac{1}{4}$									3.14
$1\frac{1}{2}$									2.95
$1\frac{3}{4}$									2.55	2.69
2								2.43	2.65
$2\frac{1}{4}$								2.32	2.53
$2\frac{1}{2}$								2.01	2.23	2.43
$2\frac{3}{4}$								1.94	2.15	2.33
3						1.66	1.77	1.88	2.07	2.25
4			1.28	1.39	1.49	1.59	1.68	1.84	1.99

Diametral Pitch	Hob Diameter								
	1⅞	2½	3	3½	4	4½	5	6	7
5			1.18	1.28	1.37	1.45	1.53	1.67	1.81
6			1.10	1.19	1.27	1.34	1.42	1.55	1.67
7			1.04	1.12	1.19	1.26	1.33	1.45	1.56
8			.98	1.06	1.13	1.19	1.26	1.37	1.47
9			.94	1.01	1.07	1.14	1.19	1.30	1.40
10			.90	.97	1.03	1.09	1.14	1.24	1.34
12			.87	.93	.99	1.04	1.10	1.19	1.28
14			.84	.90	.95	1.01	1.06	1.15	1.23
16	.64	.73	.79	.84	.89	.94	.99	1.07	1.15
18	.61	.69	.74	.80	.84	.89	.93	1.01	1.08
20	.58	.66	.71	.76	.81	.85	.89	.97	1.03
24	.56	.63	.68	.73	.77	.81	.85	.92	.99
28	.53	.59	.64	.68	.72	.76	.79	.85	.91
32	.50	.56							
36	.47	.53							
40	.44	.49							

Finishing only .13

Approach where hob is swiveled, in.

Hob Dia	2½						3½					
Work Dia	2			3			3			4½			6		
Swivel Angle	15°	30°	45°	15°	30°	45°	15°	30°	45°	15°	30°	45°	15°	30°	45°
Diametral Pitch 4	1.5	1.7	2.1	1.5	1.7	2.2	1.8	2.0	2.5	1.8	2.1	2.7	1.5	1.9	2.7
6	1.5	1.6	1.9	1.5	1.7	2.0	1.8	2.0	2.3	1.8	2.0	2.5	1.5	1.8	2.4
10	1.5	1.6	1.8	1.5	1.6	1.9	1.8	1.9	2.2	1.8	2.0	2.2	1.6	1.7	2.1
14	1.5	1.6	1.7	1.6	1.6	1.8	1.8	1.9	2.1	1.9	1.9	2.1	1.6	1.7	2.0
18	1.6	1.6	1.7	1.6	1.6	1.7	1.9	1.9	2.0	1.9	1.9	2.1	1.6	1.7	1.9
24	1.6	1.6	1.7	1.6	1.6	1.7	1.9	1.9	2.0	1.9	1.9	2.0	1.6	1.6	1.8

Overrun where hob is swiveled, in.

		Helix Angle								
		15°			30°			45°		
Swivel Angle		15°	30°	45°	15°	30°	45°	15°	30°	45°
Diametral Pitch	4	.3	.6	1.1	1.2	.5	.9	.2	.4	.8
		.2	.4	.7	.8	.4	.7	.1	.3	.5
	6	.2	.4	.7	1.1	.3	.6	.1	.3	.5
		.1	.3	.5	.8	.2	.4	.1	.2	.3
	10	.1	.3	.4	1.0	.2	.4	.1	.2	.3
		.1	.2	.3	.7	.2	.3	.1	.1	.2
	14	.1	.2	.3	1.0	.2	.3	.1	.1	.2
		.1	.1	.2	.7	.1	.2		.1	.2
	18	.1	.1	.2	1.0	.1	.2		.1	.2
			.1	.2	.7	.1	.1		.1	.1
	24	.1	.1	.2	1.0	.1	.2		.1	.1
		.1	.1	.1	.7	.1			.1	.1

(Top value = 14½° involute.)
(Lower value = 20° involute system.)

Hobbing Machines

14.2

EXAMPLES

A. Nine gear blanks are loaded on an arbor. The gears are 45-tooth, 16-pitch mild steel spur gear having a face width of ½ in. A 16-pitch gear will use a 2½-in. hob. Use one arbor. Find unit time.

Table Number	Process Description	Table Time	Adjustment Factor	Cycle Minutes	Setup Hours
14.2-S	Setup	.54			.54
14.2-1	Load and unload nine gears	.11		.11	
14.2-1	Set nut and tighten, release	.29	⅑	.032	
14.2-2	Advance carriage	.14	⅑	.016	
14.2-2	Lower or raise work spindle	.14	⅑	.016	
14.2-2	Start machine	.05	⅑	.006	
14.2-3	Machine nine gears. Approach is .73 in. Length of cut is .73 + 9 × .5 = 5.2. Machine time is 5.2 × 45 × .07	.07	⅑ × 5.2 × 45	1.82	
14.2-2	Clear carriage	.15	⅑	.017	
14.2-2	Check work	0		0	
11.4-2	Tool wear allowance, 1.31 min/unit, large tool	.31	⅑	.034	
	Total			2.05	.54
	Cycle estimate for nine gear blanks	18.45 min			

B. A spur gear made of medium-carbon alloy steel and three gears are loaded on an arbor for cutting. The gear is 24 *DP,* 51 teeth, 20° full depth involute and a face width of 0.438 in. Find the lot time for 18 units. A 3-in. hob is chosen.

Table Number	Process Description	Table Time	Adjustment Factor	Cycle Minutes	Setup Hours
14.2-S	Setup	.54			.54
14.2-1	Pick up part and install	.11		.11	
14.2-1	Set nut, tighten, loosen	.29	⅓	.097	
14.2-2	Advance steady rest and tighten	.23	⅓	.077	
14.2-2	Start machine	.05	⅓	.017	
14.2-2	Advance carriage	.14	⅓	.047	
14.2-2	Engage feed	.02	⅓	.007	
14.2-3	Machine time. Approach, .68 in., 3 × .438 = 1.31, machine distance = 1.99. Machine time = 1.99 × 51 × .13	.13	⅓ × 1.99 × 51	4.40	
	Total			4.75	.54
	Cycle estimate	14.25 min			
	Lot estimate	1.97 hr			

C. Estimate the external spline shown on Figure 14.2B. The part is called a centrifugal breather. Material is forged 2½-solid barstock. While the part requires over 25 separate operations, this estimate is concerned with the hobbing of the 51-tooth 1.59-diametral pitch having a spine OD of 1.65 in. The length of the spline is .6 in., and the approach and overtravel and safety stock = .55 in. It is necessary to compute the approach and overtravel using the formula, as the table does not provide this information. The hob data uses a medium carbon steel material with standard HSS hob material. For a quantity of 4100 a special grade of a coated tungsten carbide cutter might be adopted. Assume that one operator operates one machine.

Table Number	Process Description	Table Time	Adjustment Factor	Cycle Minutes	Setup Hours
14.2-S	Setup for spline	0.54			0.54
14.2-S	Setup for tolerance	0.24			0.24
14.2-1	Handling for spline shaft	0.60		0.60	

Table Number	Process Description	Table Time	Adjustment Factor	Cycle Minutes	Setup Hours
14.2-2	Start and stop	0.05		0.05	
14.2-2	Engage feed on	0.02		0.02	
14.2-2	Clear carriage	0.15		0.15	
14.2-3	Hobbing using 3 in. HSS hob, 0 degrees swivel angle. Spline 51 teeth, used med. carbon steel. Length of cut = .55 for approach + .65 in. for spline length = 1.15 in.	0.13	51 × 1.15	7.62	
	Total			8.44	0.78
	Lot estimate	577.51 hr			

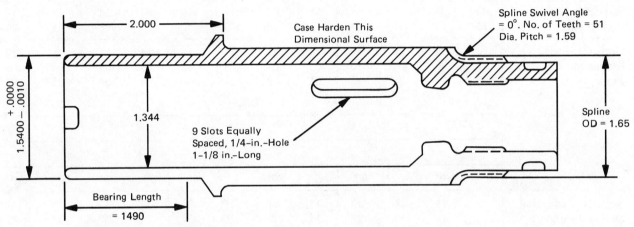

FIGURE 14.2B Example for hob estimate.

TABLE 14.2 HOBBING MACHINES

Setup

Spur gears, splines	.54 hr
Helical gears	1.04
For .001 total composite error or less, add	.24

Operation elements in estimating minutes

1. Handling
   - Chuckgear blanks
     - Small, .09; large,   .15
   - Chuck shaft and lay aside   .37
   - Load and unload gear on arbor   .11
   - Set nut on work arbor and tighten, release   .20
   - Mark gears   .30
   - Load spline shaft or pinion   .60

2. Machine operation
   - Start machine   .05
   - Advance carriage manually   .14
   - Engage feed   .02
   - Lower or raise work spindle   .14
   - Clear carriage   .15
   - Advance steady rest and tighten in position   .23
   - Indicate part for alignment   .40

3. Hobbing. min = no. teeth × L × table value

Material	Hob Diameter											
	1⅞				2¼				3			
	Swivel Angle											
	0°	30°	45°	60°	0°	30°	45°	60°	0°	30°	45°	60°
Steel, free-machining	.04	.05	.06	.08	.07	.08	.10	.14	.07	.08	.10	.14
Steel, low-carbon	.05	.06	.07	.10	.07	.08	.10	.14	.08	.09	.11	.16
Steel, medium-carbon	.08	.09	.12	.16	.11	.13	.15	.22	.13	.16	.16	.27
Stainless, 300 series	.07	.08	.10	.14	.09	.11	.13	.19	.11	.13	.16	.23
Stainless, 400 series	.06	.07	.08	.12	.08	.09	.11	.16	.09	.10	.13	.18
Carbon steel castings	.12	.14	.17	.24	.16	.19	.23	.32	.25	.28	.35	.49
Gray iron	.20	.23	.28	.40	.26	.31	.37	.53	.12	.13	.16	.23
Aluminum	.05	.06	.07	.10	.06	.07	.09	.13	.03	.03	.04	.06
Brass	.06	.07	.08	.12	.08	.09	.11	.16	.03	.04	.05	.06

Material	Hob Diameter											
	4				5				6			
	Swivel Angle											
	0°	30°	45°	60°	0°	30°	45°	60°	0°	30°	45°	60°
Steel, free-machining	.12	.13	.16	.23	.15	.17	.21	.29	.23	.27	.33	.47
Steel, low-carbon	.12	.14	.17	.23	.15	.17	.21	.29	.23	.27	.33	.47
Steel, medium-carbon	.19	.22	.27	.39	.24	.28	.34	.48	.39	.45	.55	.78
Stainless, 300 series	.19	.21	.26	.37	.23	.27	.33	.46	.37	.43	.53	.74
Stainless, 400 series	.14	.16	.20	.28	.18	.20	.25	.35	.28	.33	.40	.57
Carbon steel castings	.37	.43	.52	.74	.46	.53	.65	.92	.74	.85	1.04	1.48
Gray iron	.15	.17	.21	.30	.19	.22	.27	.38	.30	.35	.43	.60
Aluminum	.03	.04	.05	.07	.04	.05	.06	.08	.07	.08	.09	.13
Brass	.04	.04	.05	.08	.05	.06	.07	.10	.08	.09	.11	.16

THREAD CUTTING and FORM ROLLING

15.1 Thread Cutting and Form Rolling Machines

DESCRIPTION

Rotating spindle with stationary work held by vise grips is the type of thread cutting described here. A universal threading machine can be furnished with one or more spindles. Self-opening die heads are assembled tools incorporating inserted cutting components called chasers, the cutting edges not being integral with the tool body. Figure 15.1 is an example of a two-spindle threading machine.

Threads can be rolled in any material sufficiently plastic to withstand the forces of cold working. Factors that influence a material's resistance to deformation are material hardness, internal friction, yield point, and work hardening. Rolling methods are infeed, automatic continuous, and forced-through feed. Roll burnishing can be done on form rolling machines. These machines can be equipped with an automatic parts feeder.

FIGURE 15.1 Double-spindle thread cutting machine. *(Teledyne Landis Machine)*

ESTIMATING DATA DISCUSSION

Small handheld parts are covered by these data. Handling as described for Element 1 visualizes a one-spindle machine. On two or more spindles, the handling may be done during a machining cycle.

Entry variables for Element 1 are material and threads per in. To obtain minutes, multiply the factor by the diameter × length of thread ($Dia \times L$). Surface velocities for the materials are:

Material	Cutting Velocity
Steel, mild	25–50
Steel, hard	10–35
Malleable iron	30–60
Brass	45–70
Aluminum	50–80

The lower values of the cutting velocity range are assigned to the larger thread pitches while the higher end of the range is appropriate to the finer threads per inch.

In form rolling, the manual operation includes labor loading and rolling, and one machine is assumed. For automatic feed, an operator will watch one or more machines. One unit time may be divided by the number of machines being tended by the operator to get an estimate of direct labor time.

EXAMPLES

A. An instrument case having a 6¼-in. 18-pitch, ⁷⁄₁₆-in. long external thread and a mating case cover with 6¼-in., 18-pitch, ½-in., long internal thread is made of cold rolled steel. A double-spindle thread cutting machine is selected. Find the sets per hour (sets/hr).

Table Number	Process Description	Table Time	Adjustment Factor	Cycle Minutes	Setup Hours
15.1-1	Pick up external case and load	.08		.08	
15.1-1	Advance work to die	.04		.04	
15.1-2	Thread, .12 × ⁷⁄₁₆ × 6¼	.12	⁷⁄₁₆ × 6¼	.33	
15.1-1	Release and aside of part	.09		.09	
	For 2nd spindle, some work can be done during machining cycle of other part				
Repeat	Handling from above	.21		.21	
15.1-2	Thread, .12 × ½ × 6¼	.12	½ × 6¼	.38	
	Effective time per set = machine time			.71	
	and 20% interference time	.14		.14	
	Unit estimate			.85	
	Sets/hr	70			

B. An Acme thread stem, 1⅜-in. diameter, 4-pitch, 3G Acme, 4½-in. long is made of cold rolled steel. A single spindle machine is used. Find the unit time.

Table Number	Process Description	Table Time	Adjustment Factor	Cycle Minutes	Setup Hours
15.1-1	Pick up piece	.08		.08	
15.1-1	Advance	.04		.04	
15.1-2	Thread	.05	4.5 × 1.375	.31	
15.1-1	Release and aside	.09		.09	
	Total			.52	

C. A 15-in. long automobile upper control suspension control arm has two ends automatically rolled with 1-in. long ⅝-in. diameter 18-pitch threads. The parts are fed automatically and deposited onto a conveyor belt. Find the cycle time and pc/hr.

Table Number	Process Description	Table Time	Adjustment Factor	Cycle Minutes	Setup Hours
15.1-3B	Total thread length for ⅝-in. diameter	0.182		.0182	
	Total			.0182	
	Shop estimate	3297 pc/hr			

TABLE 15.1 THREAD CUTTING AND FORM ROLLING MACHINES

Setup .5 hr

Operation elements in estimating minutes

1. Thread cutting
   Pick up part and load in vise .08

Advance work to die	.04
Clear work from die head	.02
Release part from vise	.04
Aside of part	.05
Bulk load container per quantity	2.00/no.
Bulk unload container per quantity	1.00/no.

2. Thread cutting, min = table value $\times L \times Dia$

	Threads per Inch													
Material	5	6	7	8	9	10	12	14	16	18	20	24	28	32
Steel (mild) free machining	.05	.05	.06	.07	.08	.09	.10	.10	.12	.12	.12	.14	.15	.17
Steel (hard), stainless	.13	.13	.15	.17	.16	.17	.17	.18	.14	.13	.15	.18	.21	.24
Malleable cast iron	.04	.05	.06	.07	.06	.07	.08	.08	.09	.09	.09	.10	.12	.14
Brass	.03	.03	.04	.05	.05	.06	.06	.07	.08	.09	.09	.10	.10	.12
Aluminum, plastic	.03	.03	.04	.04	.04	.05	.05	.06	.06	.07	.07	.08	.09	.10

3. Form rolling

   A. Manual feed, min/ea

	Diameter		
Length	5/8	3/4	1
.5	.0176	.0216	.0296
.75	.0178	.0218	.0298
1	.0180	.0220	.0300
1.5	.0184	.0224	.0303
2	.0188	.0228	.0307
3	.0196	.0236	.0315

   B. Automatic feed, min./ea

	Diameter					
Length	1/8	3/16	1/4	3/8	1/2	5/8
.5	.0064	.0078	.0091	.0019	.0146	.0173
.75	.0066	.0079	.0093	.0120	.0147	.0175
1	.0067	.0081	.0094	.0122	.0149	.0176
1.5	.0070	.0084	.0125	.0125	.0152	.0179
2			.0100	.0128	.0155	.0182
3					.0161	.0188

4. Tool replacement   See Table 11.4

WELDING and JOINING

16.1 Shielded Metal-Arc, Flux-Cored Arc, and Submerged Arc Welding Processes

DESCRIPTION

Shielded metal-arc welding (SMAW) is effected by melting with the heat of an arc between a coated metal electrode and the base metal. Flux-cored arc welding (FCAW) produces coalescence by heating with an arc between a continuous, consumable electrode and the work. Submerged arc welding (SAW) is a process wherein coalescence is produced by heating an arc between a bare metal electrode and the work. The arc is shielded by a blanket of loose granular fusible material deposited on the work in advance of the arc. Pressure is not used in any of these methods and filler is obtained from the electrode. While these three methods are dissimilar, many work features are common; thus, permitting their grouping.

SMAW, frequently called "stick electrode," is a common method of welding though its prominence in production is challenged by other methods. In this process, the arc is stuck between the electrically ground work and a 9- to 18-in. length of covered metal rod clamped to a holder. These estimating data cover manual SMAW. As the covered rod becomes shorter, the welder stops the process to replace the stub with a new electrode. The versatility of SMAW and the low cost and simplicity of equipment are important advantages to some production people.

FCAW involves an electrode with self-shielding characteristics in coil form and feeding it mechanically to the arc. These wires contain the flux in the core. With the press of a trigger, the operator feeds the electrode to the arc. The operator uses a gun instead of an electrode holder. Manual FCAW means operator manipulation of the gun while automatic FCAW is more mechanized, especially in gun control.

The SAW process differs from other arc welding processes in that a blanket of fusible, granular material (or flux) shields the arc and molten metal. The arc is struck between the workpiece and a bare wire electrode tip is submerged in the flux. As the arc is completely covered by the flux, it is not visible and the weld is run without the flash, spatter, and sparks that characterize the open arc processes.

Now consider the estimating data for the general class of arc welding processes.

ESTIMATING DATA DISCUSSION

These estimating data cover sheet metal, plate, or other weldment materials, notably carbon and low-alloy steels. Those data are for commercial welding. Code welding, implying various specifications and certified welds, are not covered, unless the estimator extends the information by judgmental processes.

These data are developed with a general approach to the elements, especially the handling and jigging. If a more convenient approach to estimating is desired, the handling data of Resistance Spot Welding (RSW) machines can be used especially for sheet metal work. For lighter or medium weight weldments, the handling information and approach used by Gas Metal-Arc Welding (GMAW) can be substituted. If the handling

FIGURE 16.1A Self-shielded flux-cored electrode process for welding snowplow blades. *(The Lincoln Electric Company)*

elemental substitution with RSW or GMAW for estimating is used, the estimator should consistently follow that plan. Note that heavy welding is only covered in Table 16.1, and the process elements are from the table.

FIGURE 16.1B Typical gun for semiautomatic submerged-arc welding with the gun designed for fluidized flux feeding. *(The Lincoln Electric Company)*

Selection of the elements of the operation begins with handling the weldment base which is positioned for welding. Another part is fitted, clamped or jigged, if necessary, and tack welded or welded in place. This procedure may be repeated until the welding is completed. Or, the part may be progressively assembled and tacked until the weldment is fully assembled, and then the assembly finished welded. There are variations to this element pattern.

Entry variables for the pickup elements are length and width or weight. For sheet metal gage work, use the maximum X and Y coordinate dimension. As parts are joined, these entry variables increase, and as required, the entry variables become larger when using the table. The estimates are one-worker. If a welder and helper are used, double the appropriate elemental time.

The pickup, or get time, for sheet metal is used in a number of contexts. If weldment weight exceeds plant safety requirements, various hoist times are available and are listed. Depending upon hoist facilities, it may be necessary to add for C-clamps or pressure clamps, etc.

Element 2 deals with position materials. Simple position to a line or line with one or more closed ends are provided.

Element 3 deals with clamps of various kinds and includes the get, open, close, remove, and aside time. Fitup time depends upon previous operation, tolerances, and condition of the material.

Element 4 is for part weldment motion before and during welding. A variety of degrees and positionable devices are provided.

Element 5 is for work supporting arc time. The

change electrode includes pushing helmet up, flip electrode butt aside, assemble new electrode to holder, and flip helmet down. These motions are identified and listed separately.

Element 6 covers tack time. Welding times are covered by Elements 7 through 9. Any interruptible element that may occur is not included. While entry variables for welding could be size of rod, fillet size, amperage, joint design, or ac vs. dc, we have based the entry variable upon method of welding, thickness, and type of joint as the estimator is more assured of this information at the time of the estimate.

The material is assumed to be carbon or low-alloy steel. Welding is for one side, one fillet, and no gap between butt joints unless exception is given in the remarks column. The welding time may involve multiple passes to fill a joint, but the welding time is simply weld length multiplied by minutes per inch ($L \times$ min/in.). The estimator does not multiply this time by the passes. Joint preparation is not covered here unless gouging by the welder is assumed, and judgment using welding effort for gouging is required.

In these joints, the flat condition is used unless exceptions are noted in the remarks column. Vertical time would be more as overhead welding, which is not common in industrial welding. Penetration, of course, varies, but this information is averaged for the data.

For the SAW process, a backing strip between a gapped joint may be used, and this backing bar may be welded to the assembly. Whether a backing strip is used depends upon the weld design.

In manual welding for SMAW, FCAW, and SAW, the tack time includes raise and lower helmet, strike and break or stickout, and tack.

The welding time is the weld length multiplied by the factor expressed as min/in. It is arc time only. Judgment for materials other than carbon and low-alloy steel is necessary.

As we assume no shielding gas for FCAW, weld cleaning is necessary to remove the flux and splatter for the three methods. Element 10 covers the possibilities.

EXAMPLES

A. Determine direct labor time to fabricate a section of a hopper car using a manual and automatic SAW. Six ¾-in. plates (three 40 × 80 in., two 60 × 80 in., and one 18 × 64 in.) are installed over frames. The welding schedule follows:

   a. Weld three 40 × 80 in. SAW automatic, butt welds, two fillets
   b. Weld two 60 × 80 in. SAW manual, tee weld vertical
   c. Weld one 18 × 64 in. SAW manual, butt weld, two fillets

All plate welds have no gap, and weld length is full perimeter. Assume power hoist hand-traverse material handling. Two fillets implies one fillet per side. As each hopper car requires a setup, the unit time is the setup and cycle time total. Only the material is taken to the hopper car, and only when the hopper car is completed is it disposed.

Table Number	Process Description	Table Time	Adjustment Factor	Cycle Minutes	Setup Hours
16.1-S	Setup manual SAW	.5			.5
16.1-S	Setup automatic SAW	2.0			2.0
16.1-1	Pick up six plates, 6 × .54	.54	6	3.24	
16.1-1	Add for 12 C-clamps, 12 × .47	.47	12	5.64	
16.1-2C	Position material, jiggle hoist, 24 times for fitup	.25	24	6.00	
16.1-3	Open and close toggle clamps, four per plate, 4 × 6 × .08	.08	4 × 6	1.92	
16.1-3	Raise and lower helmet, two sides, 48 edges (2 × 6 × 4), 48 × .02	.02	48	.96	
16.1-9D	Weld 3 40 × 80-in. plates, automatic 3 × 240 × .10	.10	3 × 240	72.00	
16.1-9D	Weld 2 60 × 80-in. plates, manual 2 × 280 × .23	.23	2 × 280	128.80	
16.1-9D	Weld 1 18 × 64-in. manual butt, 1 × 164 × .10	.10	164	16.40	
16.1-10	Pick up air hammer, assume two times per edge, and per side, or 2 × 6 × 4 × 2 (2 times) (6 plates) (4 edges) (2 sides) × .04	.04	96	3.84	
16.1-10	Air-hammer chip flux on top of weld for linear perimeter distance of weld, or				
	a. 1440 × .01	.01	1440	14.40	
	b. 560 × .01	.01	560	5.60	
	c. 164 × .01	.01	164	1.64	

Table Number	Process Description	Table Time	Adjustment Factor	Cycle Minutes	Setup Hours
16.1-10	Wire brush weld surface for linear perimeter distance of weld, or				
	a. 1440 × .01	.01	1440	14.40	
	b. 560 × .01	.01	560	5.60	
	c. 164 × .01	.01	164	1.64	
16.1-10	Push broom clean of welding area, and hopper area where welding conducted, or 120 sq ft or .01 × 120	.01	120	1.20	
	Total			283.28	2.5
	Total direct labor for one hopper car	7.22 hr			

B. Estimate the time to weld six ½-in. gussets to a hopper car. The gussets are 8 lb/ea and are located on 3-ft. centers. The gusset is clamp-held in position and tack welded in six locations, followed by 14 in. of weld. Manual FCAW is used. Include the setup as a part of the unit time, assuming the setup is required for each hopper car.

Table Number	Process Description	Table Time	Adjustment Factor	Cycle Minutes	Setup Hours
16.1-S	Setup	.4			.4
16.1-1	Pick up six gussets	.08	6	.48	
16.1-2D	Position one part to line, two ends are closed, 6 × .05	.05	6	.30	
16.1-3	Apply and remove two C-clamps, 6 × 2 × .23	.23	6 × 2	2.76	
16.1-5	Ground cables, 6 × .19	.19	6	1.14	
16.1-5	Pick up, aside electrode gun, 2 × 6 × .08	.08	2 × 6	.96	
16.1-5	Put on, take off gloves, 1 × .14	.14	1	.14	
16.1-6	Tack weld, 6 × 6 × .09	.09	6 × 6	3.24	
16.1-8B	Butt weld, 6 × 14 × .28	.28	6 × 14	23.52	
Judgment	Operator position between gussets, 6 × 1.00	1.00	6	6.00	
16.1-10	Pick up and aside wire brush, 6 × .04	.04	6	.24	
16.1-10	Brush weld, 6 × 14 × .01	.01	6 × 14	.84	
	Total			39.62	.4
	Total direct labor for 1 hopper car	1.06 hr			

C. A box beam is composed of four rectangular ³⁄₁₆-in. plates internally stiffened by cross bracing at open ends. Each end has a platform welded to the beam. SMAW welding is used with 9-in. rods. A locating fixture is designed to weld two plates, then three plates, and finally four, and will rotate easily. The fixture, built at bench height, is also designed for welding the internal end stiffeners and platforms. The schedule of parts and welds is as follows:

    a. 4³⁄₁₆-in. plates, 16 × 36 in., corner or butt welded
    b. 4¼-in. stiffeners, 4 × 15½ in., butt welded
    c. 2½-in. floor plates, 20 × 20 in., butt welded

The clean steel plates have been oxygen-fuel cut and kerf edge is fair. Lot size is 180. Find setup and hr/100 units. The units are for commercial service and weld quality requirements are good.

Table Number	Process Description	Table Time	Adjustment Factor	Cycle Minutes	Setup Hours
16.1-S	Setup for manual SMAW	.3			.3
16.1-1	Pick up two 16 × 36 × ³⁄₁₆-in plates, 30.6 lb., 2 × .09	.09	2	.18	
16.1-2E	Position part to nest fixture, wrench two times, 2 × .07	.07	2	.14	

Table Number	Process Description	Table Time	Adjustment Factor	Cycle Minutes	Setup Hours
16.1-3	Open and close three toggle clamps, 3 × .06	.06	3	.18	
16.1-3	Use hammer to force position, two times	.33	2	.60	
16.1-6	Manual tack first corner internally, 3 × .09	.09	3	.27	
16.1-7B	Weld 36-in. of length corner internally, 36 × .07	.07	36	2.52	
16.1-5	Electrode requirement is 0.102 lb/ft, using E7024 rods, ⅛ in. × 14-in. long, so rod changes are about 3 per ft. Change electrodes, 9 × .09	.09	9	.81	
16.1-4B	Rotate fixture easily by hand	.03		.03	
Repeat	Use appropriate elements above to tack and corner weld third plate	4.67		4.67	
16.1-1	Pick up fourth ⅜₆-in. plate	.09		.09	
16.1-2E	Position part to nest fixture, wrench two times, 2 × .07	.07	2	.14	
16.1-3	Open and close three toggle clamps 3 × .06	.06	3	.18	
16.1-3	Hammer to force position	.66		.66	
16.1-6	Manual tack fourth corner external	.27		.27	
16.1-7B	Weld 36 in. of length, butt external, 36 × .12	.12	36	4.32	
16.1-5	Change electrodes, 9 × .09	.09	9	.81	
16.1-1	Load two stiffeners, 4.39 lb/ea. 2 × .07	.07	2	.14	
16.1-2B	Position parts to edge, four edges, 4 × .04	.04	4	.16	
16.1-2E	Apply square, two times	.21	2	.42	
Judgment	Clamp and unclamp four special clamps (use C-clamps), 4 × .23	.23	4	.92	
16.1-7B	Weld each end of stiffener, use butt time, 4 × 17 × .04 ends	.04	4 × 17	2.72	
	Replace electrodes five times	.09	5	.45	
Repeat	Use above 6 elements for second end	4.74		4.74	
16.1-4B	Rotate part 90°	.03		.03	
16.1-1	Pick up floor plate, chain hoist	.90		.90	
16.1-2C	Position material to corner, jiggle hoist	.25		.25	
16.1-2E	Apply and remove square	.21		.21	
16.1-3	Open and close toggle clamps, 2 × .06	.06	2	.12	
16.1-6	Tack each edge twice, four edges, 2 × 4 × .09	.09	2 × 4	.72	
16.1-4B	Rotate part 90°, 4 times	.03	4	.12	
16.1-5	Replace rods twice	.09	2	.18	
16.1-7B	Weld corner external, ½-in. plate, 4 × 16 × .10	.10	4 × 16	4.80	
16.1-5	Rod replacement estimated 4 per ft, 4 × 4 × .09	.09	4 × 4	1.44	
Repeat	Use above 10 elements for second floor plate	8.76		8.76	
16.1-11	Weldment aside, bridge crane	1.01		1.01	
16.1-5	Pick up and aside electrode holder. Estimate 1 per tack, 2 per weld, each part, each reposition during welding, start and stop, etc. approximately 75 times, 75 × .08	.08	75	6.00	—
	Total			50.02	.3
	Shop estimate	1.2 pc/hr			
	Hr/100 units	83.367			
	Lot estimate	150.36 hr			

TABLE 16.1 SHIELDED METAL-ARC, FLUX-CORED ARC, AND SUBMERGED ARC WELDING PROCESSES

Setup

Shielded metal arc	.3 hr
Flux-cored arc, manual	.4 hr
Flux-cored arc, automatic	2–4.0 hr
Submerged arc, manual	.5 hr
Submerged arc, automatic	2–4.0 hr

Operation elements in estimating minutes

1. Pick up material

Width	Length								
	3	6	9	12	24	36	48	72	96
3	.03	.03	.04	.04	.06	.07	.08	.11	.13
6		.04	.05	.05	.06	.08	.09	.13	.15
12				.06	.08	.09	.12	.15	.17
24					.09	.11	.12	.16	.18
36						.13	.15	.18	.20
48							.18	.22	.24

Pick up sheet metal, also reverse or turnover, or aside, each piece
  Pick up part, less than 7 lb   .07
  Pick up part, between 7–13 lb   .08
  Pick up part, greater than 13 lb   .09
Chain hoist   .89
Power hoist, hand traverse   .54
Power bridge crane   1.01
  Add'l for C-clamp   .47/ea
  Add'l for pressure clamp   .06/ea

2. Position material

A. Position sheet metal in fixture or jig, also relocate, 1 piece

Width	Length								
	3	6	9	12	24	36	48	72	96
3	.03	.03	.03	.04	.04	.05	.06	.08	.10
6		.03	.03	.04	.05	.06	.07	.09	.11
12				.04	.06	.06	.07	.09	.12
24					.06	.07	.08	.13	.15
36						.08	.10	.15	.17
48							.12	.17	.19

B. Position one part to line or edge (no mechanical aids)

Weight	0–7	7–13	13+	Hoist
Min	.04	.05	.06	.08

C. Position one part to line (1 end closed)

Weight	0–7	7–13	13+	Jiggle Hoist
Min	.03	.04	.04	.25

D. Position one part to line (2 ends closed)

Weight	0–7	7–13	13+	Jiggle Hoist
Min	.03	.05	.06	.25

E. Position one part to corner

Weight	0–7	7–13	13+	Jiggle Hoist
Min	.03	.04	.05	.25

      Position one part to nest fixture, using wrench   .07
      Apply and remove square or rule   .21

3. Clamp and unclamp; lock and unlock parts in fixture

	Small	Large
Open and close one toggle clamp	.06	.08
Apply and remove one vise grip or one quick-action clamp	.08	.13
Apply and remove one C-clamp	.23	.44
Apply and remove bar clamp	.19	.24
Apply and remove collar and wing nut	.10	.15
Use hammer to force position	.12	.33

4. Reposition weldment or fixture

  A. For sheet metal, use Element 1
     Apply and remove spacer   .14
     Turn part 90°

Weight	0–7	7–13	13+
Min	.03	.04	.08

  B. Chain hoist
     Power hoist, hand traverse   .58
     Power bridge crane   .41
     Add'l for C-clamp, pressure clamp   .79

     Turn part 180° or over

Weight	0–7	7–13	13+
Min	.06	.07	.13

     Chain hoist
     Power hoist, hand traverse   .88
     Power bridge crane   .58
     Add'l for C-clamps, pressure clamp   1.08

  C. Mechanically rotated fixture by operator
     Wheel or roller type: 90°, .03; 180°, .06

Degrees	45°	90°	180°
Min	.05	.08	.13

     Crank-type fixture

Degrees	15°	30°	45°	60°	75°	90°
Min	.19	.33	.47	.62	.76	.92

5. Equipment operation
      Attach and remove ground cables   .19
      Pick up and lay aside electrode holder   .08
      Change electrodes   .09
      Flip electrode aside   .02
      Pick up and lay aside face shield, helmet   .10
      Open and close goggles   .06
      Raise and lower helmet   .02
      Put on, take off gloves   .14
      Stamp   .12

6. Post-welding
      Manual tack   .09/ea

Preheat Use welding specifications
Postheat Use welding specifications

7. Shielded metal-arc welding, manual, min/in.

   A.

Joint	Gage				
	18	16	14	12	10
Butt	.05	.04	.04	.05	.06
Corner	.07	.07	.06	.05	.06
Tee			.08	.07	.07
Lap	.05	.05	.05	.06	.07
Edge	.02	.03	.03	.03	.03

   B.

Joint	Plate							
	3/16	1/4	5/16	3/8	1/2	5/8	3/4	1
Butt	.12	.17	.22	.27	.35	.44	.53	.70
Corner	.07	.07	.07	.08	.10	.16	.23	.34
Tee	.07	.08	.08	.08	.10	.21	.30	.47

8. Flux-cored arc welding, manual, min/in.

   A.

Joint	Gage		Remarks
	12	10	
Butt		.05	Gap, 30° sloping sides
Tee	.04	.05	Flat
Tee	.04	.04	Vertical
Lap	.01	.02	Flat

   B.

Joint	Plate								Remarks
	3/16	1/4	5/16	3/8	1/2	5/8	3/4	1	
Butt	.07	.09	.14	.16	.28	.38	.51	.77	Gap, 30°, sloping sides
Butt				.34	.48				J-grooved, gap
Butt				.99	1.27	2.04	2.25	3.50	Gap, overhead
Tee	.95	.97		.15	.22	.34			Flat
Tee	.04	.06							Vertical
Lap	.03	.05	.07	.10					Flat

   C. Flux cored arc welding, automatic, min/in.

Joint	Gage						Remarks
	18	16	14	12	10	3/16	
Butt		.005	.008	.009	.01	.01	No gap
Butt	.005	.006	.008	.01	.01	.02	Gap
Corner		.007	.009	.01	.01	.01	
Lap		.006	.008	.01	.01	.01	

D.

Joint	Plate 1/4	1	Remarks
Butt	.19	.23	Gap, vertical with shoe

9. Submerged arc, manual, min/in.

A.

Joint	Gage 14	12	10	Remarks
Butt	.02	.03	.03	Gap
Butt		.05	.05	No gap, two fillets
Tee	.03	.02	.02	
Lap	.02	.02	.02	

B.

Joint	Plate 3/16	1/4	5/16	3/8	1/2	5/8	3/4	1	Remarks
Butt	.04	.07	.08	.10					Gap
Butt	.05	.06	.07	.08	.11	.14	.19	.35	Two fillets
Tee	.04	.04	.05	.06	.08	.14	.23		
Tee	.04	.05	.06	.07	.10				Two fillets
Lap	03	.04	.05	.07					

C. Submerged arc welding, automatic, min/in.

Joint	Gage 16	14	12	10	Remarks
Butt	.01	.01	.02	.02	Gap, steel backing
Tee		.01	.01	.01	
Tee				.02	Two fillets
Corner	.007	.01	.01	.01	Backing plate
Edge	.007	.01	.01	.01	

D.

Joint	Plate 3/16	1/4	5/16	3/8	1/2	5/8	3/4	Remarks
Butt	.02	.03	.04	.05	.07			Gap, steel backing
Butt		.03	.04	.05	.07	.10	.10	Two fillets
Tee	.02	.02	.03	.04	.07	.10	.14	
Tee	.03	.04	.05					Two fillets
Lap	.02	.03	.04	.05				

10. Cleaning

Pickup and aside chipping hammer, air hammer, hose or brush	.04
Manually chip weld	.02/in.
Air-hammer chip weld	.01/in.
Blow off weld, in.	.006/in.
Wire brush, in.	.01/in.
Push broom area	.01/sq ft

11. Weldment aside

    For sheet metal, Use Element 1

Weight	0–7	7–13	13+
Min	.07	.08	.09

Chain hoist	.89
Power hoist, hand traverse	.54
Power bridge crane	1.01
Add'l for C-clamp, pressure clamp	

16.2 Gas Metal-Arc and Gas Tungsten-Arc Welding Processes

DESCRIPTION

Gas Metal-Arc Welding (GMAW) produces coalescence by heating with an arc between a continuous and consumable filler metal electrode and the work. Shielding is obtained entirely from an externally-supplied gas. Sometimes this method is called MIG (metal inert gas).

Gas Tungsten-Arc Welding (GTAW) produces coalescence by heating with an arc between a non-consumable tungsten electrode and the work. Shielding is obtained from a gas. A filler may or may not be used. This process is frequently called TIG (tungsten inert gas). Welding is possible in almost all positions. Slag removal is unnecessary. These processes are used with the commercially-important metals. Continuous joint or tack welding methods also are possible.

Figure 16.2A is a GMAW process. Here, 11-gauge, 316-stainless steel is jointed to ⅜-in. carbon steel using an average travel speed of 16 ipm and an argon CO_2 gas mixture. Figure 16.2B is a GTAW process, where 3-in. aluminum pipes are being joined with a filler material. GMAW is used for spotwelding two materials from one side. (See Figure 16.2C.)

The estimating data is now considered for the general class of GMAW and GTAW processes.

ESTIMATING DATA DISCUSSION

Four possibilities exist for an approach to handling. A simple consolidated approach uses the inches

FIGURE 16.2A Gas metal-arc welding (GMAW) process joining 316 stainless steel to ⅜ in. carbon steel. *(Union Carbide Corporation)*

FIGURE 16.2B Gas tungsten-arc welding (GTAW) using filler metal to weld aluminum piping. *(Union Carbide Corporation)*

FIGURE 16.2C Gas metal-arc welding (GMAW) used in spot welding two materials. *(Union Carbine Corporation)*

of weld plus the number of tacks as the entry variable to determine handling time. On the basis of time-study observations, the number of joined components, fixturing, jigging, clamps, etc., is related to this driver. Either the welded inch length or the number of tack welds may be zero ("0") or a positive (+) number. This element is complete, as it includes pickup materials, clamp, unclamp, fit, and aside welded components. The data are based on aluminum structural materials which, after assembly, is too heavy for operator handling.

The second approach to handing is for sheet metal materials and is a get table, move one part to the welding table. The same element would be used for subsequent parts. This table is also used for relocate and aside.

A girth approach ($= L + W + H$) for sheet metal is given in Resistance Spot Welding (RSW). For heavy welding, the data as given by Table 16.1 must be used.

Element 3 is welding time, expressed in min/in. for GMAW and GTAW. The time is total, as multiple passes are averaged in the data. This is especially true for thicker materials where multiple passes are required. The weld time is pure arc weld. Interruptions are itemized as separate elements. Tackweld includes: pickup of gun, tack 1 in., gun aside, and tilt helmet.

Entry variables for welding time are either GMAW or GTAW, material, joint, and thickness. Materials are considered clean, and welded quality is commercial. Code quality characteristics are not considered. A spotweld value for carbon steel involves position between GMAW spots and spot.

While slag chipping and clean up are not an element, blow off, clean, or wipe includes pickup of air hose, etc.

EXAMPLES

A. Estimate the unit and lot time to weld 76 in. of ¼-in. aluminum using the GMAW process. Blow off and a stamp to certify the weld are necessary. Lot size is 200 parts.

Table Number	Process Description	Table Time	Adjustment Factor	Cycle Minutes	Setup Hours
16.2-S	Setup	.4			.4
16.2-1A	Handle components for 76 in. of weld	4.23		4.23	
16.2-2	Attach cable	.19		.19	
16.2-2	Pick up, lay aside gun, estimate 8 times for 4 components	.08	8	.64	

Gas Metal-Arc and Gas Tungsten-Arc Welding Processes

Table Number	Process Description	Table Time	Adjustment Factor	Cycle Minutes	Setup Hours
16.2-2	Gloves	.14		.14	
16.2-2	Stamp	.12		.12	
16.2-4A	Butt weld ¼-in. aluminum plate, 76 in.	.10	76	7.60	
16.2-6	Blow off, .28 + (76–48) .0047	.41		.41	
	Total			13.33	.4
	Lot estimate	44.833 hr			
	Shop estimate	4.5 pc/hr			

B. Estimate the direct-labor time to weld 22 in. and tack six times with GMAW on a ⅜-in. aluminum plate. The lot quantity for this minor assembly is 150.

Table Number	Process Description	Table Time	Adjustment Factor	Cycle Minutes	Setup Hours
16.2-S	Setup	.44			.44
16.2-1A	Weld in. + no. of tacks, 22 + 6 = 28	1.97		1.97	
16.2-4A	Fillet weld	.05	22	1.10	
16.2-3	Tack weld, six times	.09	6	.54	
16.2-2	Attach cable	.19		.19	
16.2-2	Pick up gun, four times	.08	4	.32	
16.2-6	Blow off	.23		.23	
	Total			4.35	.44
	Lot estimate	11.275 hr			

TABLE 16.2 GAS METAL-ARC GAS TUNGSTEN-ARC WELDING PROCESSES

Set up .4 hr

Operation elements in estimating minutes

1. Handling components

   A.

Weld In. Plus No. of Tacks	Min	Weld In. Plus No. of Tacks	Min
1.0	.76	39.4	2.39
2.8	.84	45.0	2.62
4.8	.92	51.2	2.82
6.9	1.01	58.0	3.17
9.3	1.11	65.5	3.49
12.0	1.22	73.8	3.84
14.8	1.35	82.8	4.23
18.0	1.48	92.8	4.65
21.5	1.63	103.8	5.11
25.4	1.79	115.8	5.62
29.6	1.97	129.1	6.19
34.3	2.17	Add'l	.042

B. Get one sheet and load on table, also for relocate, dispose

Width	Length							
	3	6	9	12	18	36	60	84
3	.02	.03	.04	.04	.05	.07	.09	.12
6		.04	.05	.05	.06	.08	.10	.13
12				.05	.07	.09	.14	.16
18					.07	.09	.14	.17
36						.12	.16	.20

2. Equipment operation
   - Attach, remove ground cable   .19
   - Pick up, lay aside gun   .08
   - Raise and lower helmet   .02
   - Put on, take off gloves   .14
   - Stamp   .12

3. Welding
   - Manual tack   .09/ea
   - Preheat   Use welding specifications for time
   - Postheat   Use welding specifications for time

4. Gas metal-arc welding, min/in.

   A. Aluminum

Joint	Thickness								Remarks
	1/8	3/16	1/4	3/8	1/2	5/8	3/4	1	
Butt	.04	.09	.10	.13	.15	.19	.19		No. gap, flat
Butt					.32		.49	.77	Gap, 70°V, two sides
Butt	.09	.09	.10						Gap, horizontal two sides
Butt				.17	.21		.31	.64	Gap, 70°V, horizontal, two sides
Fillet			.04	.05	.19				Horizontal

   B. Copper

Thk	1/8	1/4	3/8	1/2
Butt	.04	.12	.16	.21

   C. Stainless Steel

Joint	Thickness					Remarks
	1/16	1/8	1/4	3/8	1/2	
Butt		.06	.12	.14	.29	60°V, backing plate
Tee, lap	.06	.08				

   D. Carbon Steel

Thk	1/8	3/16	Remarks
Butt	.06	.10	Spray transfer

   Spot weld and position   .17/spot

5. Gas tungsten-arc welding, min/in.

   A. Aluminum

Thk	1/32	.050	1/8	1/4	1/2	3/4	Remarks
Butt	.02	.03	.04	.05	.05	.17	No filler rod

   B.

Thk	1/16	1/8	3/16	1/4
Butt, corner	.10	.11	.12	.13
Fillet, lap	.12	.13	.14	.16

6. Blow off.

Weld In. + No. of Tacks	Min	Weld In. + No. of Tacks	Min
3	.07	26	.18
7	.09	35	.23
12	.12	48	.28
18	.15	Add'l	.0047

16.3 Resistance Spot Welding Machines

DESCRIPTION

Resistance Spot Welding (RSW) has two or more sheets of metal held between metal electrodes. A welding cycle is started with the electrodes contacting the metal under pressure before the current is applied for a period known as squeeze time. A low voltage current of sufficient amperage is passed between the electrodes causing the metal in contact to be rapidly raised to welding temperatures. As soon as the temperature is reached, the pressure between the electrodes squeezes the metal together and completes the weld. A pressure dwell completes the cycle. A nugget cools and forms the weld.

A picture of a stationary single-spot welding machine is shown in Figure 16.3A. But these estimating data are intended for the general class of RSW machines.

ESTIMATING DATA DISCUSSION

Element 1 includes loading the major part from a skid, box, bear cage, etc. to a table. Following the final weld, it also includes remove subassembly aside. The time includes, both for replacement and removal, some occasional clamp and fixturing time. The independent variable is girth, $L + W + H$, and girth is the smallest-sized box of $L + W + H$ in. dimensions

FIGURE 16.3A Single-spot welding machine joining unalloyed annealed titanium. (Sciaky Brothers, Inc.)

which will enclose any shape. In the case of a sphere, girth is the sum of three diameters. Girth is an easy calculation, and spot welded parts are reasonably correlated to girth. Element 2 is for loading component parts from skid, cage, tote box, etc. It includes moving major parts where occasional clamps or fixtures are necessary for positioning. The time variable also depends upon girth. The estimator, in order to expedite application, may want to use the higher table value rather than interpolate values. For instance, if $L + W + H = 26$, use .43 min.

Repositioning of the welded part may be required during spot welding. A 90°-reposition is a corner turn; 180° implies a roll over of the narrow dimension, while end-for-end is a flip of the longer dimension. Spot welding is covered for aluminum and steel, and several gages were a part of the observations. For an assembly exceeding $L + W + H = 80$, the table includes two operators. Spot-weld time must be doubled for two operators. The distance between spots and metal thickness are average for Element 4. Occasional clean tips is also included.

EXAMPLES

A. Two sheets of .010-in. Titanium 4901 material are to be spot welded. The major part, an unusual configuration, has a minimum girth of 69 in. The secondary part has a girth of 27.3 in. and 23 in-line spots are required. A holding fixture with level clamps positions both parts relative to each other. Commercial tolerances are required. The practice of using the next higher table value is followed. A lot of eight units is required.

Table Number	Process Description	Table Time	Adjustment Factor	Cycle Minutes	Setup Hours
16.2-S	Setup	.35			.35
16.3-1	Handle major part, $L + W + H = 69$	.61		.61	
16.3-2	Handle secondary part, $L + W + H = 27.3$	.55		.55	
16.3-4	Spotweld 23 times	1.68		1.68	—
	Total			2.84	.35
	Lot estimate	.73 hr			
	Shop estimate	21 pc/hr			

B. A partially dimensioned sketch is given by Figure 16.3B. Other operations have preceded the spot welding work. In this operation two formed angles are to be spot welded to the formed 5052-H32 aluminum hatch cover and 32 total spots are required. Outside approximate dimensions are 9 × 12 in. There are 716 units in the lot. Find the cycle, setup, and lot estimates.

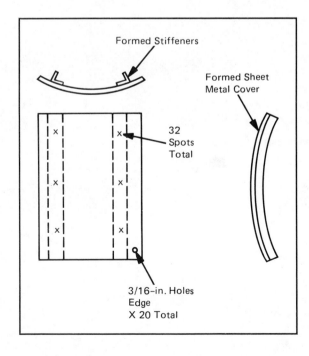

FIGURE 16.3B Part for spotwelding estimate.

Table Number	Process Description	Table Time	Adjustment Factor	Cycle Minutes	Setup Hours
16.3-S	Setup for three parts where $L + W < 30$ in.	0.30			0.30
16.3-1	Load and unload where $L + W + H < 23$ in.	0.43		0.43	
16.3-2	Load two components where $L + W + H < 15$	0.31	2	0.62	
16.3-3	Repositioning	0.08		0.08	
16.3-4	Spot, clean tip 32 spots	2.34		2.34	
	Total			3.47	0.30

TABLE 16.3 RESISTANCE SPOT WELDING MACHINES

Setup hr

Components	1, 2, or 3	4 or 5	6 and up
$L + W + H < 60$	.30	.35	.40
$L + W + H \geq 60$	.35	.40	.50

Operation elements in estimating minutes

1. Load major part and unload welded assembly

$L + W + H$	Min	$L + W + H$	Min
3.0	.34	100.6	.70
7.5	.36	110.0	.74
12.3	.37	119.8	.78
17.3	.39	130.2	.82
22.5	.41	141.1	.86
28.0	.43	152.5	.90
33.8	.45	164.5	.94
39.8	.48	177.1	.99
46.2	.50	190.3	1.04
52.9	.53	204.2	1.09
59.9	.55	218.8	1.15
67.2	.58	234.1	1.21
75.0	.61	250.2	1.27
83.1	.64	267.0	1.33
91.6	.67	Add'l	.004

2. Load each component part, position with fixture, clamp to major part

$L + W + H$	Min	$L + W + H$	Min
3.0	.07	20.0	.38
3.4	.08	22.2	.41
3.9	.09	24.5	.45
4.4	.10	27.1	.50
4.9	.11	29.9	.55
5.6	.12	33.0	.60
6.2	.13	36.4	.67
7.0	.14	40.2	.73
7.8	.16	44.3	.80
8.7	.18	48.9	.89

L + W + H	Min	L + W + H	Min
9.7	.19	53.9	.97
10.8	.21	59.4	1.07
12.0	.23	65.4	1.18
13.3	.26	72.1	1.30
14.7	.28	79.4	1.43
16.3	.31	87.5	1.57
18.1	.34	Add'l	.016

3. Repositioning part during spotwelding

L + W + H	90°	180°	End-for-End
15	.07	.08	.08
35	.08	.09	.11
55	.09	.10	.14
75	.10	.12	.16
95	.12	.13	.19
115	.13	.14	.21
135	.14	.15	.24
155	.16	.17	.27
Add'l	.001	.001	.001

4. Spotweld, reposition for next spotweld, and occasional clean tip

No. Spots	Aluminum	Steel	No. Spots	Aluminum	Steel
1	.09		15	1.02	.95
2	.16		16	1.08	1.02
3	.22	.07	17	1.15	1.09
4	.29	.14	18	1.21	1.17
5	.36	.22	19	1.28	1.24
6	.42	.29	20	1.35	1.31
7	.49	.36	22	1.48	1.46
8	.55	.44	25	1.68	1.68
9	.62	.51	28	1.87	1.90
10	.69	.58	30	2.01	2.04
11	.75	.66	35	2.34	2.41
12	.82	.73	40	2.67	2.77
13	.88	.80	45	3.00	3.14
14	.95	.88	Add'l	.06	.07

16.4 Torch, Dip, or Furnace Brazing Processes

DESCRIPTION

Brazing and soldering are processes which unite metals. This is accomplished by using a third joining metal, which is introduced into the joint in a liquid state before it solidifies. Brazing and soldering have a filler metal with a melting point higher and lower than 800°F respectively. The filler metal is drawn into the joint by capillary attraction. In these applications, heat is applied to the parts to be joined. It is not applied directly to the filler material alone. Various heat sources can be used as flames, molten salt, or radiant energy. If gas flames are used, it is called torch brazing. Similarly, dip brazing would involve molten salt.

The torch is handheld. The soldering iron is not evaluated here. An oxyacetylene or oxyhydrogen torch applies heat locally, and the filler metal, applied in wire rod, is melted into the joint. Fluxes are applied by pastes or are included as part of a coated filler metal. Some filler metal is prepared in the forms of rings, washers, or special shapes to fit the joint. If a protective atmosphere or environment is provided during the brazing, a flux may not be required.

A pot-type furnace is shown in Figure 16.4. These pots can be fuel-fired or electrically heated. In addition to dip brazing, they can be used for carburizing, hardening, melting, salt-bath drawing, tempering, and tinning. The pot in Figure 16.4 is 16-in. OD and has a 20-in. depth, with a capacity of 1.89 cu ft. Larger units have a capacity to 5.5 cu ft.

ESTIMATING DATA DISCUSSION

The setup differs with respect to whether jigging is available. There are four operation elements. In the first element, a preassembled assembly is available, and there is no assembly by the operator. If the parts can be held together by friction or gravity, the jig is considered "positioning." If clamping or holding of some sort is required, a holding jig time is selected. For sheet metal parts, five or more components, or when close tolerances are specified, an aligning jig value is picked for the estimate.

In torch brazing, Element 4, three conditions are available for selection. Simple linear inches are the time driver for no-jig or jig conditions. If two or more diameters are joined, the diameter, as is, is selected for the time driver. The time driver is based upon total joint or seam length. For instance, for a cap joint, use the total outside seam length. Either a handheld rod or special filler is used.

Depending upon the atmosphere or containment for brazing, a flux may or may not be needed. With a molten-salt pot, for example, it may be unnecessary to require a flux. Most often a jig is unlikely for dip brazing.

For furnace brazing and soldering, the confinement may range from a protective atmosphere oven to a large furnace. However, the handling time to load and unload is covered in Element 5. It does not include process time. The entry variable is squat area. Only one layer of parts on the sled or of the furnace or oven is considered.

In assembly aside, Element 7, the $L + W$ time driver is the two greatest dimensions of the assembly. Packaging is average time for pack and unpack using wrap, bag, pads, paper, etc. Greater detail can be found in the packaging estimating data.

FIGURE 16.4 A pot-type gas-fired furnace that is used for dip brazing. *(Sunbeam Equipment Corporation)*

EXAMPLES

A. A small mechanical assembly, weighing less than 3 lb, is to be torch brazed. Lot quantities are indeterminate. An aligning jig is required because of critical tolerances. The joints are lap. Linear length of the lap joint is 13.25 in. Determine the unit time and hr/100 units.

Table Number	Process Description	Table Time	Adjustment Factor	Cycle Minutes	Setup Hours
16.4-1	Aligning jig	1.64		1.64	
16.4-2	Flux	.36		.36	
16.4-2	Torch braze 13.25 in	4.70		4.70	
16.4-7	Assembly aside	.77		.77	
	Total			7.47	
	hr/100 units	12.450			

B. An assembly is to be dipped brazed. Parts are clean, and a flux is not required. Metal-to-metal sliding fit and gravity keeps the parts together. The four parts, which slip together, are previously bench assembled, and now are to be brazed with a filler metal. The part area is 21.25 sq in. and has two separate seams. Find a lot estimate for 625 units and the shop estimate.

Table Number	Process Description	Table Time	Adjustment Factor	Cycle Minutes	Setup Hours
16.4-S	No jig setup	.15			.15
16.4-1	Preassembled parts	.05		.05	
16.4-3	Load braze wire to 2 seams and area, 21.25 sq in.	.48		.48	
16.4-6	Dip braze	.38		.38	
16.4-7	Assembly aside	.05		.04	
	Total			.96	.15
	Lot estimate	10.15 hr			
	Shop estimate	63 pc/hr			

TABLE 16.4 TORCH, DIP, OR FURNACE BRAZING

Setup

no jig	.15 hr
jig	.20 hr

Operation elements in estimating minutes

1. Load parts to welding bench and position

Preassembled, no jig	.05
Positioning jig for machined parts	.17
Holding jig for four or less components, or tolerances not close or simple assembly	.53
Aligning jig for sheet metal parts, or five or more components, or close tolerances, or complex assembly	1.64

2. Apply flux to seam

Brush L	Syringe L	Min	Brush L	Syringe L	Min
1.0	2.7	.10	7.8	17.4	.24
1.5	3.8	.11	9.0	19.9	.27
2.1	5.0	.13	10.3	22.7	.30
2.7	6.3	.14	11.8	25.8	.33
3.3	7.7	.15	13.3	29.2	.36
4.1	9.3	.17	15.1	32.9	.39
4.9	11.0	.18	17.0	37.0	.43
5.8	12.9	.20	Add'l		.02
6.8	15.0	.22		Add'l	.01

3. Load bare wire to assembly

	No. Separate Seams							
Part Area	1	2	3	4	5	6	7	8
2	.28	.41	.54	.69	.82	.95	1.08	1.21
19	.31	.44	.58	.71	.84	.97	1.10	1.23
45	.34	.48	.61	.75	.87	1.00	1.13	1.27
70	.38	.51	.64	.77	.90	1.03	1.18	1.31

Add'l seam .13
Add'l area .0014

4. Torch braze or solder

No. Jig in.	Dia in.	Jig in.	Min	No. Jig in.	Dia in.	Jig in.	Min
	.26	.7	.36	15.3	1.92	6.1	2.19
	.29	.8	.39	17.2	2.11	6.8	2.41
	.33	.9	.43	19.2	2.33	7.5	2.65
1.1	.37	1.1	.48	21.4	2.57	8.3	2.92
1.5	.41	1.2	.52	23.8	2.84	9.1	3.21
2.0	.46	1.3	.58	26.4	3.13	10.1	3.53
2.4	.51	1.5	.63	29.4	3.45	11.1	3.88
3.0	.57	1.7	.70	32.6	3.80	12.3	4.27
3.6	.63	1.9	.77	36.1	4.18	13.6	4.70
4.2	.70	2.1	.85	40.0	4.61	14.9	5.17
4.9	.78	2.4	.93	44.3	5.07	16.5	5.68
5.7	.86	2.7	1.02			18.2	6.25
6.5	.95	3.0	1.12			20.0	6.88
7.4	1.05	3.3	1.24			22.0	7.57
8.5	1.17	3.7	1.36			24.3	8.32
9.6	1.29	4.1	1.50			26.8	9.16
10.8	1.42	4.5	1.65	Add'l			.13
12.2	1.57	5.0	1.80		Add'l		1.12
13.7	1.74	5.5	1.99			Add'l	.34

5. Furnace braze

Area	Min	Area	Min	Area	Min	Area	Min
1.0	0.3	4.2	.13	7.3	.23	21.3	.66
1.4	.04	4.5	.14	7.7	.24	24.5	.76
1.7	.05	4.9	.15	8.0	.25	28.5	.88
2.1	.06	5.2	.16	9.2	.29	32.4	1.01
2.4	.07	5.6	.17	10.6	.33	37.2	1.16
2.8	.08	5.9	.18	12.2	.33	42.9	1.34
3.1	.09	6.3	.19	14.1	.44	49.3	1.54
3.5	.11	6.6	.20	16.2	.50	56.6	1.77
3.8	.12	7.0	.22	18.6	.58	Add'l	.03

6. Dip braze

Area	Min	Area	Min	Area	Min
3.0	.005	8.6	.11	17.7	.29
3.6	.01	9.2	.12	19.1	.31
4.1	.02	9.8	.13	20.8	.35
4.7	.03	10.5	.15	22.5	.38
5.3	.05	11.2	.16	24.5	.42
5.8	.06	12.1	.18	26.7	.46
6.4	.07	13.0	.20	29.0	.52
7.0	.08	14.0	.22	31.7	.56
7.5	.09	15.1	.24	34.7	.61
8.1	.10	16.3	.26	Add'l	.02

7. Assembly aside
   No jig .05
   Remove from jig
     $L + W < 6$ in. .33
     $L + W \geq 6$ in. .77

8. Package for protection .15

16.5 Ultrasonic Plastic Welding Machines

DESCRIPTION

Ultrasonic welders produce mechanical vibratory power by means of high-frequency sound waves. Everything that makes a sound vibrates, and everything that vibrates makes a sound. However, not all sounds are audible. Ultrasonics refers to sound beyond the audible spectrum, normally considered to be above 18,000 Hertz.

The essential components required to apply ultrasonic energy include a power supply, converter, booster horn, horn, and assembly stand. The power supply, or ultrasonic generator, supplies high-frequency electrical energy to the converter, a component that changes electrical energy into mechanical vibratory energy. Attached to the converter is an amplitude-modifying device known as a booster horn. The booster horn can either increase or decrease the amplitude of vibration supplied to the horn, which is the device that transmits the ultrasonic energy to the part. The mechanical assembly that houses all of the above components is referred to as an assembly stand, or welding press. Its function is to bring the horn face into contact with the plastic workpiece, apply appropriate pressure, and then retract the horn upon completion of the welding cycle.

The use of ultrasonics enables high energy to be imparted to a plastic part at force levels that will not stress, crack, or produce residual deflection of the material being welded. When a thermoplastic is subjected to ultrasonic vibrations, a combination of surface and intermolecular friction produces a temperature rise at the interface area. If the temperature rise is sufficient to melt the resin, a fusion or molecular bond will result and a uniform weld will be produced.

The most common use of ultrasonic energy in plastics assembly is the process of welding plastic parts. The weldability of certain thermoplastics depends on a variety of factors that affect the ultrasonic energy requirements and energy transmission characteristics of the various resins. In ultrasonic welding operations the vibrations are transmitted through the horn into the part and then proceed to the interface of the two parts being welded. Here vibratory energy is converted to heat by means of friction, which melts and fuses the plastic. When the vibrations cease, the plastic solidifies while under pressure, thus producing a weld at the joining surfaces. The shape of the two joining surfaces, referred to as the joint design, is important in achieving optimum welding results.

Insertion is the process of embedding a metal component, known as an insert, in a thermoplastic part. A hole slightly smaller than the insert is molded into the plastic part. This hole provides a certain degree of interference and also serves to guide the insert into place. The insert is usually designed with knurls, flutes, undercuts, or threads in order to resist loads that will be applied on the finished part. In this process ultrasonic vibrations travel to the metal insert and plastic interface. Heat is produced at this interface by means of the insert vibrating against the plastic. This in turn allows the plastic to briefly melt, thereby allowing the insert to be driven into place. The molten plastic flows into the voids between the insert and part, and once solidified locks the insert in place. In most ultrasonic insertion applications the plastic component is fixed and the insert is driven into place by the horn. It is also possible, but not as common, to have the horn contact the plastic part and drive it over the insert.

ESTIMATING DATA DISCUSSION

The setup hours include the setup and teardown of the machine and fixtures, in addition to a tuning procedure which involves adjusting the power supply to the resonant frequency of the converter. Also taken to be part of the setup is the setting of the three energy controlling variables of the machine, namely: weld time, pressure, and amplitude.

The handling element involves picking up the main part from a box, skid, etc., and positioning the part on the fixture. Additional handling time is added if the fixture is complex. The handling element also accounts for the unloading of the finished part from the machine and placing it in a pallet, box, on bench, etc.

Element 2 is for obtaining and positioning the parts that are to be welded to the main part. More than one part may need to be obtained and positioned, as is the case when welding inserts. A repositioning of the part in the fixture is also required when more than one weld exists per part. This is accounted for in Element 3. Additional time is added to the previous element when the repositioning is done in a large or intricate fixture.

The actual machine cycle of the ultrasonic welder is accounted for in Element 4. After the part has completed the welding cycle it is sometimes necessary to perform further operations. Such processes as blowing off the part, removing flash, and weld inspection are considered as miscellaneous.

EXAMPLES

A. A polycarbonate panel having a girth $L + W + H = 26.0$ in. has three metal inserts that are to be welded into the part. After welding, the part is blown off with an air hose. Lot quantity = 300. Calculate the unit and lot estimate.

Table Number	Process Description	Table Time	Adjustment Factor	Cycle Minutes	Setup Hours
16.5-S	Setup	.8			.8
16.5-1	Handle	.10		.10	
16.5-2	Pick up and position three inserts	.13		.13	
16.5-3	Reposition part after each weld	.06		.06	
16.5-4	Weld three inserts	.10		.10	
16.5-5	Blow off part	.06		.06	
	Total			.45	.8
	Lot estimate	3.05 hr			

B. A black Lexan cartridge base with girth $L + W + H = 19.5$ in. has a small rectangular flap that will be ultrasonically welded onto the part. Assume a lot quantity of 1500. Calculate unit and lot estimates.

Table Number	Process Description	Table Time	Adjustment Factor	Cycle Minutes	Setup Hours
16.5-S	Setup	.8			.8
16.5-1	Handle	.10		.10	
16.5-2	Obtain and position second part	.03		.03	
16.5-3	Ultrasonically weld	.06		.06	
	Total			.19	.8
	Lot estimate	5.55 hr			

C. A 10 percent glass-filled polycarbonate medical part will have three plastic plugs welded onto the end of the part. The fixture that holds the main part requires additional operator time. Find the min/unit.

Table Number	Process Description	Table Time	Adjustment Factor	Cycle Minutes	Setup Hours
16.5-S	Setup	.8			.8
16.5-1	Handle	.10		.10	
16.5-1	Add for intricate fixture	.07		.07	
16.5-2	Pick up and position three plugs	.13		.13	
16.5-3	Reposition part on table after weld	.06		.06	
16.5-4	Weld three plugs	.10		.10	
	Total			.46	.8

D. A black polycarbonate medical part weighing .29 lb needs to have six metal inserts ultrasonically welded into it. After each part is finished with the welding process it requires the removal of excess flash. Assume a lot quantity of 2500. Calculate the unit estimate, lot hours, hr/100, and pc/hr.

Table Number	Process Description	Table Time	Adjustment Factor	Cycle Minutes	Setup Hours
16.5-S	Setup	.8			.8
16.5-1	Handle	.10		.10	
16.5-2	Pick up and position six inserts	.29		.29	
16.5-3	Reposition part after each weld	.12		.12	
16.5-4	Weld six inserts	.15		.15	
16.5-5	Remove flash	.20		.20	
	Total			.86	.8
	Lot estimate	36.63 hr			
	Hr/100 units	1.43			
	Pc/hr	69			

TABLE 16.5 ULTRASONIC PLASTIC WELDING MACHINES

Setup .8 hr

Operation elements in estimating minutes

1. Handling .10
   Add for intricate or large fixtures .07

2. Obtain and position additional parts

No. Parts	Min	No. Parts	Min
1	.03	5	.24
2	.08	6	.29
3	.13	7	.34
4	.18	8	.39

3. Reposition part in fixture after each weld

No. Parts	Min	No. Parts	Min
1	0	5	.10
2	.04	6	.12
3	.06	7	.14
4	.08	8	.16

Add for intricate or large fixtures .09

4. Ultrasonically Weld

No. Parts	Min	No. Parts	Min
1	.06	5	.13
3	.08	6	.15
3	.10	7	.16
4	.11	8	.18

5. Miscellaneous
Blow off part, get and aside air hose included .06
Remove flash from part .20
Inspect part .07

FURNACES

17.1 Heat Treat Furnaces

DESCRIPTION

Heat treatment is the operation of heating and cooling a metal part in its solid state to change physical properties. While a variety of hardening, tempering, annealing, and normalizing operations can be identified, these estimating data are concerned with furnace loading and unloading. The specific heat-treating operation is immaterial; though operator variations certainly affect how material is handled, these variations have been averaged by the analysis.

Furnaces considered by these estimating data are heavy duty, gas tight, and a production-section pusher. Often there is an unheated pre-heat section followed by a high temperature section. Both ends are equipped with a manual-lift vertical door with protective flame curtains. The operating temperature may be as high as 1800°F. A typical furnace is shown in Figure 17.1. The tunnel cross section of Figure 17.1 is 12-in. wide × 8-in. high and 7-ft long.

FIGURE 17.1 Manual pusher, atmospheric-tight, electric furnace. *(Pereny Equipment Company, Inc.)*

ESTIMATING DATA DISCUSSION

The setup of .10 hr includes the usual chores, but does not include waiting for furnace warmup or cooldown.

Element 1 is for loading and unloading parts onto a sled, tray, basket, or brick outside of the furnace proper, or on these flat surfaces within the cool chamber. The entry variable is rough-box volume, or cu in., of the part. The volume does not include the sled, tray, etc. For a fixtured part, the volume does include the fixture. For multiple part fixtures, the total volume is prorated over the number of parts. No furnace wait time is included.

Loading and unloading the part onto a fixture for distortion prevention is provided by Element 2. Flat, ring, or conical-shaped parts are the configurations provided. Entry variable is the part width of fixture.

Element 3 deals with putting on, taking off gloves, open and close one furnace door, and load and unload one part (which may weigh up to 30 lb.). It does not include any furnace time.

EXAMPLES

A. A beryllium copper part, previously formed in a sheet metal operation, is to be precipitation hardened in a furnace operation. The part is reasonably flat, even though it is formed. To avoid distortion, the part is loaded into a multiple unit fixture and then heat treated. Find the unit time if the part is $\frac{1}{16} \times 2.025 \times 3.0$ in.

Table Number	Process Description	Table Time	Adjustment Factor	Cycle Minutes	Setup Hours
17.2-2A	Load flat-shaped part onto fixture	.23		.23	
17.1-1	Load and unload fixtured part, net volume prorated over units = 3.1 cu in.	.21		.21	
	Total			.44	

B. An aluminum casting in "O" condition is to be stabilized with a furnace operation. Part volume = 23 cu in. Determine the shop estimate for direct-labor production.

Table Number	Process Description	Table Time	Adjustment Factor	Cycle Minutes	Setup Hours
17.1-1	Load and unload part in furnace	.26 min		.26	
	Total			.26	

C. An annealed part, 4340, is to be hardened to RC 45. The part is fully finished, and the partially dimensioned sketch is given by Figure 6.4C. A quantity of 7075 is planned. Find the estimates.

Table Number	Process Description	Table Time	Adjustment Factor	Cycle Minutes	Setup Hours
	Heat treat to RC 45				
17.1-S	Setup	0.10			0.10
17.1-1	Handle Vol < 20 cu in	0.23		0.23	—
	Total			0.23	0.10
	Lot estimate	27.22 hr			

TABLE 17.1   HEAT TREAT FURNACES

Setup .10 hr

Operation elements in estimating minutes

1. Load and unload part in furnace

Volume	Min	Volume	Min	Volume	Min	Volume	Min
1.0	.20	97	.35	269	.63	866	1.61
7.0	.21	108	.37	288	.67	964	1.77
13.3	.22	119	.39	308	.70	1073	1.95
20.0	.23	131	.41	330	.73	1192	2.14
26.9	.24	143	.43	352	.77	1323	2.36
34.2	.25	156	.45	376	.81	1467	2.59
41.9	.26	170	.47	401	.85	1626	2.85
50.0	.28	185	.50	427	.89	Add'l	.002
58.4	.28	200	.52	454	.94		
67.3	.31	216	.55	621	1.21		
76.6	.32	233	.58	695	1.33		
86	.34	250	.60	776	1.46		

2. Load and unload part on fixture

    A. Flat shape

Part W	Min	Part W	Min	Part W	Min
0.5	.10	1.8	.21	4.7	.44
0.6	.11	2.1	.23	5.2	.49
0.7	.12	2.4	.25	5.8	.54
0.9	.13	2.7	.28	6.5	.59
1.0	.14	3.0	.30	7.2	.65
1.2	.16	3.4	.33	8.0	.71
1.4	.17	3.8	.37	8.8	.78
1.6	.19	4.2	.40	Add'l	.08

   B. Rings or conical shape

Dia	Min	Dia	Min
.5	.23	2.5	.46
.8	.26	3.1	.52
1.1	.30	3.8	.60
1.5	.35	4.5	.69
2.0	.40	Add'l	.12

3. Load and unload one part to furnace .80

4. Put on and remove asbestos gloves .18

DEBURRING

18.1 Drill Press Machine Deburring

DESCRIPTION

These data provide for hole or edge burr removal times using floor or bench-mounted drill press machines. The operation of deburring can be processed many ways—notably by machines, portable tools, hand methods, or tumbling. The data here, however, are limited to drill presses only. A variety of spindle-mounted bits can be assumed, and a few are given.

Figure 18.1 is a 10×14-in. (tilt-table) 15-in. drill press, where spindle drive is variable in the 500–4000 rpm range. Discussion, typical estimates, and data for the general class of drill press deburring machines follow.

ESTIMATING DATA DISCUSSION

Handling is provided for three categories of parts, flat-like, box-like, and machined. The maximum length plus width, $L + W$, is the entry variable for sheet metal. The smallest box dimensions, or girth, $L + W + H$, is the entry for box structures. A qualitative description for ease or difficulty of handling is the entry variable for machined parts. This handling includes both pickup and aside. It also includes the initial position of the drill-press tool. But if special handling requirements are necessary for tweezers, pliers, etc., add 25 percent. If a fixture is used, apply the drill-press data.

It should be noted that some parts may fit any of these descriptions, but consistency for a class is necessary to maintain long-term application of the data for best results.

Deburring, or machine time, covers hole-to-hole positioning of the parts and break edges for metals. For nonmetals, such as plastics, other tables are used. The deburr is done with spindle-mounted tools such

FIGURE 18.1 A 15-in. floor drill press with variable spindle rpm. *(Clausing Corporation Machine Tool Group)*

as drill, countersink, tap, reamer, wire brush, router-bit, sanding disks or sanding drums, grinding burrs, etc.

In Element 3 the hole depths are shallow, thus qualitative judgment of adding 25 percent per time for four-thread depths, is advised in re-tapping.

If the holes are to be deburred on both sides, double the hole count.

EXAMPLES

A. A small machine part, which is difficult to handle, has two side holes to be deburred. The part, bagged, is a lot of 1500. Estimate lot time to deburr using a drill press tooled with a countersink. The holes are on opposite sides.

Table Number	Process Description	Table Time	Adjustment Factor	Cycle Minutes	Setup Hours
18.1-S	Setup	.05			.05
18.1-1C	Very small, difficult to handle, and stack	.05		.05	
18.1-2A	Reposition	.02		.02	
18.1-3A	Deburr two holes	.07		.07	
18.1-4	Pack and unpack	.09	2	.18	—
	Total			.32	.05
	Lot estimate	8.05 hr			

B. A lot of 32 sheet metal parts, which have been braked into a U-shape, have 32 holes accessible to drill press deburring. The holes are to be deburred on both sides. The girth measurements are 40 = 10 + 20 + 10 in. The holes are in three planes, but on both surfaces. Find the unit estimate.

Table Number	Process Description	Table Time	Adjustment Factor	Cycle Minutes	Setup Hours
18.1-1B	Box-like handling	.38		.38	
18.1-2C	Reposition to present five planes for deburring Original plane in Element 1	.15	5	.75	
18.1-3A	32 holes, both sides, 1.07 × 2	1.07	2	2.14	
18.1-4	Cardboard layer for skid, but layer covers 5 parts, 2 × .05/5	.05	⅖	.02	
	Total			3.29	

C. Deburr a magnesium die cast part which has been cast, deflashed, and drill and tapped on an NC drill press. The prior operation to this operation is given as Example C for equipment 9.4. Notice Figure 2.9C for the part. The reader may want to examine that operation to understand this estimate. The quantity is 500 and we are to find the estimates for direct labor time. The work is to be done a sensitive drill press and a bench for some of the manual work required.

Table Number	Process Description	Table Time	Adjustment Factor	Cycle Minutes	Setup Hours
	Drill press deburr holes which are accessible, hand deburr inside difficult holes				
18.1-S	Setup	0.05			0.05
18.1-1B	Box-like parts hdl	0.07		0.07	
18.1-1C	Pick up and drop	0.05		0.05	
18.1-3A	Deburr the big hole	0.03		0.03	
18.1-3A	Deburr four body holes	0.13		0.13	
18.5-1	Handling, repos for hand deburring of difficult to deburr holes	0.07		0.07	
18.5-2	Tool handling	0.03		0.03	
18.5-3A	Deburr holes by hand eight holes under tab and four inside holes	0.59		0.59	
	Total			0.97	0.05
	Lot hours	8.13			

TABLE 18.1 DRILL PRESS MACHINE DEBURRING

Setup .05 hr

Operation elements in estimating minutes

1. Position and aside

    A. Flat or sheet, extrusion, aluminum and steel, .020–.050 in.

L + W	Min	L + W	Min
9.0	.14	64.6	.31
15.4	.16	79.3	.36
22.7	.18	96.2	.42
31.1	.21	115.7	.48
40.7	.24	138.0	.55
51.8	.27	163.7	.63

    B. Box-like parts

L + W + H	Min	L + W + H	Min	L + W + H	Min
15.0	.07	29.5	.22	72.8	.66
16.5	.09	33.9	.27	85.8	.70
18.2	.11	39.1	.32	101.4	.95
20.3	.13	45.4	.38	120.2	1.14
22.8	.15	52.9	.46	Add'l	.01
25.9	.18	61.9	.55	Hoist	1.78

    C. Very small, pickup, difficult to control, drop .03
    Small in pan, or box, pickup and drop .02
    Medium, under 3 lb, easy to handle, drop .03
    Large, from bench, under 8 lb, pickup and drop .06
    Very small, pickup, and stack .05
    Small, pickup, and stack .04
    Medium, pickup, and stack .05
    Large, from bench, pickup, and stack .08

2. Reposition to present new surface, edge, plane

    A. Very small, small, medium .02
    Large .04

    B. Sheet metal

L + W	6	12	24	48	72	96	120
Min	.03	.04	.06	.09	.13	.18	.24

    C. Box-like parts

L + W + H	16	21	35	50	75
Min	.03	.05	.09	.15	.22

    Hoist .58

3. Deburr

    A. Holes with drill, re-team, re-tap, chamfer

Holes	Min	Holes	Min	Holes	Min	Holes	Min
1	.03	11	.37	21	.70	31	1.04
2	.07	12	.40	22	.74	32	1.07
3	.10	13	.44	23	.77	33	1.11
4	.13	14	.47	24	.81	34	1.14
5	.17	15	.50	25	.84	35	1.17
6	.20	16	.54	26	.87	36	1.21
7	.23	17	.57	27	.91	37	1.24
8	.27	18	.60	28	.94	38	1.27
9	.30	19	.64	29	.97	39	1.31
10	.34	20	.67	30	1.01	Add'l	.034

   B. Wire brush—horizontally turning, spindle mounted

L + W	1.1	2.5	4.2	7.2
Min	.13	.29	.49	.80

   C. Wheel buff, spindle mounted

Sq In.	.3	1	3.7	6.3	10
Min	.10	.34	1.26	2.14	3.39

   D. Router tool, spindle-mounted

L	4	9	14	19	24	29
Min	.22	.28	.34	.40	.45	.51

4. Blowing off
   - Pickup, aside air hose .04
   - Blow out bottom hole .06
   - Blow off deburring chips .01/hole

   Pack or unpack
   - Sheet metal cardboard layer .05
   - Envelope, bag, box .09

18.2 Abrasive Belt Machine Deburring

DESCRIPTION

The conveyorized abrasive belt deburring machine has a traveling belt which passes under an abrasive belt. The part is pulled through and is surface sanded. Small machines handle parts up to 2½-in. thick at a conveyor feed of 20 ft/min. Larger machines with abrasive belts to 60-in. wide have a conveyor feed of 20–60 ft/min. Top and bottom abrasive heads permit single-pass deburring. Belt selection with 120–220 grit with oil lubrication will also produce a satin finish. But 80–120-grit belts are required for deburring operations. Figure 18.2 is an example of a conveyorized deburring machine with a height opening of 0–6 in.

Some abrasive belt operations are for box-like structures having parallel planar surfaces to the moving abrasive belt. The box-like part is cross fed at a right angle to the belt. The operator depresses the belt down over the entire area. Or the operator can localize the pressure for spot surface deburring. Spot or surface deburring is on the top only.

The vertical abrasive belt machine has a belt ex-

FIGURE 18.2 Abrasive belt deburring and finishing machine. *(Timesavers, Inc.)*

posed which travels at a relatively low speed. In the smaller sizes, the machines are portable, and they can be mounted on any bench. A table allows resting of flat parts during edge burring.

The following data are for the family of abrasive belt machines.

ESTIMATING DATA DISCUSSION

The setup allows for a belt change along with the customary chore tasks. The sheet metal parts can be gathered and relocated next to the conveyor or belt machine for deburring.

Handling elements vary with the machine and whether handling is already included in the process time or if it is for one or two operators.

Element 3 is for laying on traveling belt or inserting in opening of conveyor. Note that this element does not include the part aside; the entry variable is length plus width ($L + W$).

Box-like parts are loaded on a machine table over which the overhead belt travels. The table, sometimes pulled out away from the belt, is crossfed under the moving abrasive belt. For box-like parts, the time does not include aside.

For machine parts, there are several circumstances. The pickup and drop or stack depend upon ease of handling. The nature of abrasive belt machining excludes heavy parts from using this method. The reposition element is for parts that use the 1-in. or 4-in. belt, which are normally vertical.

In conveyor deburring, Element 5, the part entry is length (L). Small and medium refer to the size of the machine, or 4 in. for small machines and 18 in. and 30 in. for medium machines. This machine sands the entire surface and top periphery. The conveyor pulls the part through the machine. The times are total as nominally a small conveyor uses one loader while a medium conveyor uses a loader and unloader operator. In the small conveyor, the parts continue to stack off to the side of the belt.

The entry variable for Element 5 is length (L) and 12- or 16-gage aluminum. The moving belt is pressed down onto the surface of the part, and the part is sanded. Either local or total area can be sanded and belt widths vary. Spot welds, indentations, and some welds may be sanded. Though deburring is the original purpose, some surface preparation for a paint coat may be achieved. The entry variable is usually in the direction of the belt movement.

Element 5 also is deburr edges, holes, and cutoffs, thus has the entry variable of deburred length. The parts are handheld during the deburring process. Flat parts can be rested upon a table and the edge pressed into the abrasive belt.

Because of the operation similarity to abrasive belt conveyor deburring, a roll edge, Element 6, is used for sheet metal parts after a shear, notch, or pierce operation. Handling is included in the time. The edge is rolled over slightly by mechanical action and no abrasive sanding occurs. The time is for one side only.

EXAMPLES

A. A sheet metal part, previously blanked, pierced and notched from both sides, is to be deburred. The part, 9 × 27.25 in., is clean and a large lot is available. Estimate the unit time.

Table Number	Process Description	Table Time	Adjustment Factor	Cycle Minutes	Setup Hours
18.2-1	Turn machine on, off	0		0	
18.2-2	Go through stack every 10 pieces	.015		.015	
18.2-5	Conveyor sand first side	.17		.17	
18.2-2	Carry stack to front	.010		.010	

Table Number	Process Description	Table Time	Adjustment Factor	Cycle Minutes	Setup Hours
18.2-5	Conveyor sand second side	.17		.17	
	Total			.365	

B. A small easy-to-handle machine part is to have two circumferences deburred. A 1-in. belt, nonbacked, is available. The diameter is 1.305 in., and can be handled by one hand, although part presentation to the belt is two-handed. Estimate the unit time.

Table Number	Process Description	Table Time	Adjustment Factor	Cycle Minutes	Setup Hours
18.2-3	Small part pickup, drop	.03		.03	
18.2-5	Deburr circumference, 4.1 in.	.14		.14	
18.2-4	Turn end-for-end	.02		.02	
18.2-5	Deburr second end	.14		.14	
	Total			.33	

C. Estimate the magnesium die casting given in Figure 2.9C with the abrasive belt operation machine. The part has flashing from the die casting operation. Notice the preoperation given as Example B. There are 500 units to be estimated.

Table Number	Process Description	Table Time	Adjustment Factor	Cycle Minutes	Setup Hours
18.2-S	Setup	0.10			0.10
18.2-1	Start and stop for the entire lot	0.04	1/500	0.00	
18.2-3C	Pick up, position	0.05		0.05	
2.9-9	Degate runner, sprue	0.13		0.13	
18.2-4	Reposition for three edges	0.02	3	0.06	
18.2-5C	Vertical belt for about 13 in.	0.29		0.29	
18.2-5C	Vertical belt for add't 8 in. of flashing	0.02	8	0.14	
18.2-7	Blow off	0.25		0.25	
	Total			0.92	0.10
	Lot estimate	7.80 hr			

TABLE 18.2 ABRASIVE BELT MACHINE DEBURRING

Setup .1 hr

Operation elements in estimating minutes

1. Start and stop machine .04

2. Gather stack
   - Several pieces .15
   - Armful .25
   - Carry stack to front of machine .10

3.A. Lay on traveling belt, flat parts (aside not required)

L + W	Min	L + W	Min
2.2	.05	55.6	.12
11.1	.06	64.5	.13
20.0	.07	73.4	.14
28.9	.08	82.3	.15
37.8	.09	91.2	.16
46.7	.10	Add'l	.001

B. Lay on machine table, box-like part, remove to skid

L + W + H	Min	L + W + H	Min	L + W + H	Min
15.0	.07	29.5	.22	72.8	.66
16.5	.09	33.9	.27	85.8	.79
18.2	.11	39.1	.32	101.4	.95
20.3	.13	45.4	.38	120.4	1.14
22.8	.15	52.9	.46	Add'l	.01
25.9	.18	61.9	.55	Hoist	1.78

C. Pickup, position to belt, part aside
Very small pickup, difficult to handle, drop	.03
Small pickup in pan or box, drop	.02
Medium pickup, under 3 lb, easy to handle, drop	.03
Large pickup from bench, under 8 lb, drop	.06
Very small, pickup and stack	.05
Small, pickup and stack	.04
Medium, pickup and stack	.05
Large, pickup and stack	.08

4. Reposition to present new edge, plane, hole, slot, etc.
Very small, small, medium	.02
Large	.04

5. Abrasive belt deburr

A. Conveyor or traveling belt, flat parts, deburr surface

Small L	Medium L	Min	Medium L	Min
1.0		.02	13.6	.12
2.4	1.3	.03	15.1	.13
3.9	2.8	.04	16.6	.14
5.3	4.3	.05	18.2	.15
6.7	5.9	.06	19.7	.16
8.1	7.4	.07	21.3	.17
9.6	9.0	.08	23.9	.18
11.0	10.5	.09	25.3	.19
12.4	12.0	.11	Add'l	.007

B. Moveable table, horizontal overhead belt, heavy deburring, box-like structures, deburr surface

12-ga Al L	16-ga Al L	Min	12-ga Al L	16-ga Al L	Min
8.7	10.8	.47	16.4	21.9	1.34
9.5	11.9	.55	17.2	23.0	1.43
10.2	13.0	.64	18.0	24.1	1.52
11.0	14.1	.73	20.6	28.0	1.82
11.8	15.2	.82	23.9	32.6	2.18
12.5	16.3	.90	27.7	38.1	2.62
13.3	17.5	.99	32.3	44.8	3.14
14.1	18.6	1.08	Add'l		.11
14.9	19.7	1.17		Add'l	.08
15.6	20.8	1.25	Steel		.024/sq in.

Abrasive Belt Machine Deburring   18.2

C. Vertical abrasive belt, deburr edges, cutoffs, excess material

1-In. Belt, Nonbacked L	Min	4-In. Belt Backed L	Min	4-In. Belt Backed L	Min
.1	.07	.1	.07	14.6	1.91
.9	.08	.5	.50	17.2	2.10
1.8	.10	1.0	.89	20.0	2.31
2.8	.12	2.2	.98	23.0	2.54
4.1	.14	3.5	1.08	26.4	2.79
5.7	.17	4.9	1.18	30.2	3.07
7.6	.20	6.5	1.30	34.3	3.38
9.8	.24	8.3	1.43	38.8	3.71
12.5	.29	10.2	1.58	43.8	4.09
Add'l	.018	12.3	1.73	Add'l	.07

6. Roll edges mechanically

L	Min	L	Min
4.0	.03	24.2	.08
7.5	.04	32.7	.10
11.9	.05	43.4	.12
17.3	.06	56.8	.15

7. Blow off part, once
    Machine part   .10
    Sheet metal   .13
    Box-like part   .25

18.3 Pedestal-Machine Deburring and Finishing

DESCRIPTION

These estimating data are for low- and high-velocity sanding or wire and cloth brushing processes with pedestal machines. The purpose of these machines may be metal preparation or deburring, and oftentimes polishing and buffing. An operator will sit or stand and manipulate the part against the rotating surface.

Removal of weld beads flush to the parent surface plane, dishing-out of spot-weld projections, descaling, blending of dents, deburring holes or surfaces, feathering, and imbedding a special scratch pattern are some of the operations.

The pedestal machine uses a rotary aluminum-oxide grit (or other material) belt grinder over high-speed pulleys. A typical universal belt grinder is shown in Figure 18.3. This machine has a 132-in. long belt with a width of 3 in. Contact area of the belt with the part varies with the diameter of the pulley and width of the belt. The larger diameters have greater flat-surface contact. In the figure, the diameter is 12 in., but contact wheel diameters range from 1 to 16 in. With a 12-in. pulley, the abrasive belt speed is 5500 fpm.

ESTIMATING DATA DISCUSSION

The elements are sensitive to the quality requirements for deburring and finishing. Where fine-deburring and primary metal surface preparation are separated is a moot point, since sometimes these operations are consolidated. Polishing can be achieved on pedestal machines and is also considered here. Most polishing operations are intended to obtain luster. Overall, these processes are used to remove welds, blend corners or edges, deburr, and as a primary surface preparation before painting, anodizing, or as a moderate-lot solution for a polishing operation.

FIGURE 18.3 Universal abrasive belt grinder with a 12-in. head. *(Hammond Machinery Builders)*

The handling element includes the pickup-and-aside. Turnover and rehandling to expose other surfaces or planes are not covered. Should the part entry variable exceed the maximum girth as provided by Element 1, the difference in size multiplied by .001 min and added to .93 min gives the handling time. If the part is flat sheet metal rather than boxlike, add length plus width, as thickness seldom affects girth. For convenience, the estimator may wish to use the next higher table value rather than interpolate.

The rotary-belt sanding Element, 3, uses a fixed-machine location, and the part is positioned against the belt. The belt may range in width, from 1 to 4 in. or more, and has a surface velocity sufficient to categorize the operation as near metal-removal. But in this context, the purpose is the blending of weld beads, or the removal of detrimental-appearing spot weld indentations, or burrs. Parts that can be manipulated would be used with this machine. If the parts are awkward and exposure of the bead to the weld is difficult, a portable tool is possible. With this rotary belt element, the entry variable is length. Remember that the width of a manual pass against the belt is restricted to belt width, for instance 4-in.

Element 4 is for buffing or descaling of discoloration, rust, hard carbon, or metal etching. The first time is for a threshold of 25 sq in. Additional areas above 25 sq in. are indicated as per 1 sq in.

Metal polishing can be described as the process of producing a uniform surface of specific characteristics. In most instances, metal polishing is preceded by machining operations and followed by a buffing operation. The two, buffing and polishing, are sometimes separated, but not in these data. Metal polishing removes metal and corrects surface imperfections, but is not used to generate or retain geometric or dimensional accuracy. The wheels in this application are flexible cloth and vary in hardness, to which the desired abrasive is cemented. Flap-wheels can be considered also. Oftentimes in lot production, metal polishing is termed an art form, but estimating data are provided in Element 5. Area, sq in., is the entry variable.

EXAMPLES

A. A sheet metal chassis having maximum box dimensions of 59.5 in. has 24 in. of gas metal-arc welding bead, and it is necessary to blend the bead to the basic metal surface. The pedestal belt grinder with a suitable grit belt is used to do this job. Find the lot estimate for 20 parts.

Table Number	Process Description	Table Time	Adjustment Factor	Cycle Minutes	Setup Hours
18.3-S	Setup	.1			.1
18.3-1	Start, stop machine	.04		.04	
18.3-2	Handle	.58		.58	
18.3-3	24 in. of weld bead	.72		.72	—
	Total			1.34	.1
	Lot estimate	.55 hr			

B. A forging can be handheld and has a light rust over the surface. The irregular dimension, when boxed for girth dimensions, is 17.1 in. The surface area is 36 in. Find the unit estimate.

Table Number	Process Description	Table Time	Adjustment Factor	Cycle Minutes	Setup Hours
18.2-2	Handle	.10		.10	
18.2-4	Remove light rust, .27 + .01(36 − 25)	.38		.38	
18.2-2	Reposition five times	.05	5	.25	
	Unit estimate			.73	

C. A pedestal machine is to be used to deburr milling burrs. Notice Figure 6.4C. The previous operation milled slots (see Example C, Equipment 9.4). The quantity is 7075 units. Find the estimates.

Table Number	Process Description	Table Time	Adjustment Factor	Cycle Minutes	Setup Hours
	Deburr slots				
18.3-S	Setup	0.10			0.10
18.3-1	Start/stop mach	0.04	/250	0.00	
18.3-2	Handling $L + W + H < 15$ in.	0.07		0.07	
18.3-3	Belt grinding				
	Length of burr 1.5 in., four slots as adjustment factor	0.06	4	0.24	
18.3.2	Handling for four slots	0.05	4	0.20	
	Total			0.51	0.10
	Lot estimate	60.26 hr			

TABLE 18.3 PEDESTAL-MACHINE DEBURRING AND FINISHING

Setup .1 hr

Operation elements in estimating minutes

1. Start, stop machine .04

2. Handling

$L + W + H$	Min	$L + W + H$	Min	$L + W + H$	Min
15.0	.07	25.0	.18	48.0	.41
15.5	.08	26.7	.19	53.0	.46
16.1	.09	28.6	.21	58.7	.52
17.4	.10	30.7	.23	65.1	.58
18.2	.11	33.0	.26	72.3	.65
19.8	.12	35.6	.28	80.3	.74
20.7	.13	38.3	.31	89.4	.83
21.7	.14	41.4	.34	99.6	.93
22.7	.15	44.8	.38	Add'l	.001

Reposition side, edge, turnover, each time .05

3. Belt grinding

L	Min	L	Min	L	Min	L	Min
1.0	.04	5.3	.16	10.8	.30	28.8	.79
1.4	.05	5.8	.17	11.9	.33	31.7	.87
1.9	.06	6.2	.18	13.2	.37	35.0	.95
2.3	.08	6.7	.19	14.5	.40	38.5	1.05
2.7	.09	7.1	.20	16.0	.44	42.4	1.15

L	Min	L	Min	L	Min	L	Min
3.2	.10	7.5	.22	17.7	.49	46.7	1.27
3.6	.11	8.0	.23	19.5	.54	51.5	1.39
4.0	.12	8.4	.24	21.5	.59	56.6	1.53
4.5	.13	8.8	.25	23.7	.65	62.4	1.69
4.9	.15	9.8	.28	26.1	.72	Add'l	.03

4. Wire brush or deburr with wire wheel
    Remove surface discoloration, up to 25 sq in.   .15
      Add'l sq in.   .005
    Remove light rust or corrosion, up to 25 sq in.   .27
      Add'l sq in.   .02
    Remove heavy rust or corrosion, up to 25 sq in.   .44
      Add'l sq in.   .03
    Remove hard carbon, metal etching, up to 25 sq in.   .68
      Add'l sq in.   .03

5. Buff or polish with cloth wheel

Area	Min	Area	Min	Area	Min
1.2	.42	2.6	.91	5.6	1.94
1.3	.47	2.9	1.00	6.1	2.14
1.5	.51	3.1	1.10	6.7	2.35
1.6	.56	3.5	1.21	7.4	2.59
1.8	.62	3.8	1.33	8.1	2.85
1.9	.68	4.2	1.46	9.0	3.13
2.1	.75	4.6	1.61	9.8	3.45
2.4	.82	5.1	1.77	Add'l	.35

18.4 Handheld Portable-Tool Deburring

DESCRIPTION

There is a variety of portable tools on the market for deburring. Powered by air, electricity, cable, and hydraulic motor, their description is familiar. Portable-tool deburring methods are popular because of the operator mobility and the low cost of equipment. Sometimes these tools are bench-located; in other situations, the deburring operator will have access to drill presses, pedestal machines, etc., and the portable tools extend the capacity to deburr a wide range of metal parts.

ESTIMATING DATA DISCUSSION

The handling elements include both the pickup, aside, and occasional reposition or turnover to expose various internal or external surfaces, edges, slots, holes, weld beads, spot-weld indentations, scratches, and burrs. With portable-powered tools, one hand can hold the tool while the other simultaneously can reposition the part. Of course, size and difficulty of work may restrict the reposition opportunity.

If the work is awkward, and a turnover or reposition is necessary to expose a new burr, then a lay-down and pickup of the tool, as found in Element 2, is collaterally required. Because of the mobility of the operator, he or she may position the tool and body in different positions relative to the workpiece.

Element 3 uses the air-motor drill, usually with a countersink or bit. The element includes hole-to-hole time and burr for accessible location. If the location is internal, and a reposition is not planned or very difficult to achieve, the hole-to-hole-and-deburr is increased to .15 min. The hole count is for one side. If the number of holes exceeds that which is given, the estimator may add combinations of holes to estimate the number.

For retap or reream metal, or an air-tap of a sheet metal, Element 4 may be used.

Element 5 uses the air-driven rotary flat sander with an abrasive aluminum oxide grit of 50, and a disk usually of 5-in. diameter. As a high-speed surface sander, the element may be used to reduce the crest of the weld bead, blend spot weld indentations, or round sharp corners. Cutouts or holes which may be burred are another application. The tool is held with both hands. Similar remarks apply for Element 6, where the entry variable is length.

Element 7 uses a dual-action sander, i.e., oscillating motion in both the x-y direction that is air-driven with a 3-in wide × 6 in.-long 60–80 grit aluminum oxide paper. The tool may be gripped by one hand. The process may follow a rough-sanding operation. It scratches in a mixed direction and is usually not intended as a basic stock-removal tool, but it does benefit in deburring edges. The pass length suggests that a zone area is sanded.

Single action sanding, Element 7, is for an air-driven portable tool with the scratch action coincident to the long direction of the tool. The data are for 100-grit aluminum oxide. The primary entry variable is surface area, sq in.

Changing of sanding disks or pads is often required on a part basis, especially if the part is large. Otherwise for weld-bead blending, the ratio of one change to 40 in. is appropriate.

EXAMPLES

A. A sheet metal chassis having minimum girth dimensions of $L + W + H = 59.5$ in., has 24 in. of gas metal-arc welding length. Portable rotary tool sanding is required to blend the weld bead with the sheet metal surface. Lot quantity is 20.

Table Number	Process Description	Table Time	Adjustment Factor	Cycle Minutes	Setup Hours
18.4-S	Setup	.1			.1
18.4-1	Handle	.58		.58	
18.4-5A	24-in. weld length	5.31		5.31	
18.4-2	Replace two rotary sanding disks	.14	2	.28	
	Total			6.17	.1
	Lot estimate	2.157 hr			

B. A sheet metal bracket subassembly having a girth $L + W + H = 59.5$ in., has 30 in. of weld bead to sand down, 38 in. of length for dual-action sanding to blend the area, and 825 sq in. of total surface area sanding. Size of lot is 20.

Table Number	Process Description	Table Time	Adjustment Factor	Cycle Minutes	Setup Hours
18.4-S	Setup	.1			.1
18.4-1	Handle	.58		.58	
18.4-5A	38 in. of weld length	8.30		8.30	
18.4-7A	38 in. of dual-action sanding	.55		.55	
18.4-7B	825 sq in. of surface area = .95 + (825 − 336.8).0019 =	1.88		1.88	
18.4-2	Change three pads	.14	3	.42	
	Total			11.73	.1
	Lot estimate	4.009 hr			

TABLE 18.4 HANDHELD PORTABLE-TOOL DEBURRING

Setup   .1 hr

Operation elements in estimating minutes

1. Handling and occasional reposition

L + W + H	Min	L + W + H	Min	L + W + H	Min
15.0	.07	25.0	.18	48.0	.41
15.5	.08	26.7	.19	53.0	.46
16.1	.09	28.6	.21	58.7	.52
17.4	.10	30.7	.23	65.1	.58
18.2	.11	33.0	.26	72.3	.65
19.8	.12	35.6	.28	80.3	.74
20.7	.13	38.3	.31	89.4	.83
21.7	.14	41.4	.34	99.6	.93
22.7	.15	44.8	.38	Add'l	.001

   Special reposition, turnover, each time   .05

2. Tool handling
   - Pickup and aside tool from bench   .03
   - Pickup and aside tool from hook   .04
   - Pickup and aside tool from suspended spring hanger   .02
   - Replace portable sanding paper   .14
   - Open and close goggles   .06

3. Deburr; easy location, air motor

Holes	Min	Holes	Min	Holes	Min	Holes	Min
1	.04	8	.35	15	.65	22	.95
2	.09	9	.39	16	.69	23	.99
3	.13	10	.43	17	.73	24	1.04
4	.17	11	.47	18	.78	25	1.08
5	.22	12	.52	19	.82	26	1.12
6	.26	13	.56	20	.86	27	1.16
7	.30	14	.60	21	.91	Add'l	.043

   Difficult location, each hole   .15

4. Tap sheet metal, or re-tap, re-ream with air motor

Holes	Min	Holes	Min	Holes	Min	Holes	Min
1	.09	6	.52	11	.96	16	1.40
2	.17	7	.61	12	1.05	17	1.49
3	.26	8	.70	13	1.14	18	1.57
4	.35	9	.79	14	1.22	19	1.66
5	.44	10	.87	15	1.31	Add'l	.087

5.A. Rotary sand, 5, 7-in. disk, air motor

L	Min	L	Min	L	Min
1.1	.15	3.7	.71	15.9	3.40
1.3	.19	4.5	.89	19.7	4.25
1.5	.23	5.5	1.11	24.5	5.31
1.8	.29	6.8	1.39	30.5	6.64
2.1	.36	8.3	1.74	38.0	8.30
2.5	.46	10.3	2.18	47.4	10.37
3.0	.57	12.8	2.72	Add'l	.23

B. Rotary sand, 2, 3-in. disk, air motor

L	Min	L	Min	L	Min
1.0	.77	1.8	1.40	3.2	2.30
1.3	1.00	2.0	1.50	3.3	2.50
1.4	1.10	2.2	1.70	3.7	2.80
1.6	1.20	2.5	1.90	4.1	3.10
1.7	1.30	2.8	2.10	Add'l	.77

6. Grinding wheel, electric motor

L	Min	L	Min	L	Min
1.0	.06	6.0	.16	18.3	.39
1.7	.08	7.6	..19	22.4	.47
2.5	.09	9.6	.23	27.4	.56
3.4	.11	12.0	.27	33.3	.67
4.6	.13	14.9	.32	Add'l	.019

7.A. Dual-action sanding with air sander, or 5-in. drum

Pass Length	Min	Pass Length	Min	Pass Length	Min
2.0	.37	38.9	.55	92.9	.80
10.0	.41	50.5	.60	110.0	.88
18.7	.45	63.3	.66	128.7	.97
28.3	.50	77.4	.73	Add'l	.005

B. Single-action sanding with air sander

Area	Min.	Area	Min	Area	Min
25.0	.37	115.5	.54	248.7	.79
44.6	.40	144.4	.59	290.6	.87
66.1	.45	175.9	.65	336.8	.95
89.8	.49	210.6	.72	Add'l	.002

18.5 Hand Deburring

DESCRIPTION

This kind of deburring consists of manual elements, as there are no motorized, air, electrical, etc., devices that the operator is using. The operator is using countersinks, burrs, scrapers, files and other simple contrivances that are operated manually.

ESTIMATING DATA DISCUSSION

Hand operations usually imply the absence of cranes, hoists, etc., and handling is for pickup and aside, load on bench, from pan, skid, etc. With hand deburring, some repositioning by the operator on the part during deburring is possible. If a part has to be turned over, then the tool is set aside, and the special reposition is included. A tool rehandle may be colaterally included.

Element 3 has two tables for deburring holes of ½-in. diameter and smaller using a countersink bit. Also, holes larger than ½ in. are categorized in Element 3 and may use a larger countersink or special tool.

If the material is aluminum and is an edge, as distinguished from a hole, use Element 4. For steel deburring of edges, use Element 4.

For gear tooth deburring, Element 5, allow a reposition for every five teeth. Hand sandpaper, see Element 8, is for a 1-in. pass width.

Item 18.5 of Section IV can be used to estimate direct-labor cost for hand deburring. The constant operational cost is always included along with one or more of the other costs. A cost can be included for each major rehandling, each hole or gear tooth, and breaking of linear edges.

EXAMPLES

A. A small aluminum assembly, $L + W + H = 17.5$ in., has two square punched holes of 4-in. length each, three elongated holes with burr length of 1.6 in. each, and 14 internal holes less than ½ in.-diameter. For this small lot, determine the unit estimate.

Table Number	Process Description	Table Time	Adjustment Factor	Cycle Minutes	Setup Hours
18.5-1	Handle	.11		.11	
18.5-2	Pickup, aside scraper	.03		.03	
18.5-4A	Scrape square holes	.14	2	.28	
18.5-4A	Scrape elongated holes	.10	2	.20	
18.5-2	Pickup, aside countersink	.03		.03	
18.5-3A	Deburr 14 holes	.69		.69	
	Total			1.34	

B. A steel casting has a machine flange with a burr on the inside, irregular opening and on the bolt-circle of 6½-in. ID holes. The casting can be handled and minimum box dimensions for girth are 31.0 in. The irregular opening has a dimension 10.67 in. in length. Using hand labor, estimate the unit time. The forging is medium hardness.

Table Number	Process Description	Table Time	Adjustment Factor	Cycle Minutes	Setup Hours
18.5-1	Handle	.26		.26	
18.5-2	Tool handling for file	.03		.03	
18.5-4B	File inside burr	.31		.31	
18.5-2	Countersink handle	.03		.03	
18.5-3A	Deburr six holes	.29		.29	
	Unit estimate			.92	

TABLE 18.5 HAND DEBURRING

Setup .05 hr

Operation elements in estimating minutes

1. Handling and occasional reposition

L + W + H	Min	L + W + H	Min	L + W + H	Min
15.0	.07	25.0	.18	48.0	.41
15.5	.08	26.7	.19	53.0	.46
16.1	.09	28.6	.21	58.7	.52
17.4	.10	30.7	.23	65.1	.58
18.2	.11	33.0	.26	72.3	.65
19.8	.12	35.6	.28	80.3	.74
20.7	.13	38.3	.31	89.4	.83
21.7	.14	41.4	.34	99.6	.93
22.7	.15	44.8	.38	Add'l	.001

   Special reposition, turnover, end-for-end .05

2. Tool handling
   Pickup and aside tool from bench .03
   Pickup and aside tool from hook .04
   Pickup and aside tool from suspended spring hanger .02

3. Deburr holes

   A. Holes ½-in. diameter and under with countersink

Holes	Min	Holes	Min	Holes	Min
1	.05	6	.29	11	.54
2	.10	7	.34	12	.59
3	.15	8	.39	13	.64
4	.19	9	.44	14	.69
5	.24	10	.49	Add'l	.049

   B. Holes over ½-in. diameter using large diameter scraper

Holes	Min	Holes	Min	Holes	Min
1	.08	4	.32	7	.57
2	.16	5	.41	8	.65
3	.24	6	.49	Add'l	.08

4. Break edges

   A. Break edge with scraper, file on aluminum

L	Min	L	Min	L	Min
1.0	.08	5.0	.14	19.1	.35
1.5	.09	6.9	.17	24.0	.42
2.1	.10	9.2	.20	29.7	.50
2.8	.11	11.9	.24	36.1	.60
3.5	.12	15.1	.29	Add'l	.015

B. Break edge with scraper, file on steel

Soft L	Medium L	Hard L	Min	Soft L	Medium L	Hard L	Min
1.0	.4		.07	12.0	10.2	8.3	.28
1.4	.8	.2	.08	13.5	11.6	9.4	.31
1.8	1.2	.5	.09	15.2	13.0	10.7	.34
2.3	1.6	.9	.10	17.0	14.6	12.1	.37
2.8	2.1	1.3	.11	19.0	16.4	13.6	.41
3.4	2.6	1.7	.12	21.1	18.4	15.3	.45
4.0	3.1	2.2	.13	23.6	20.5	17.1	.49
4.7	3.8	2.8	.14	26.2	22.8	19.1	.54
5.5	4.4	3.3	.16	29.1	25.4	21.3	.60
6.3	5.2	4.0	.17	32.3	28.3	23.8	.65
7.3	6.0	4.7	.19	35.8	31.4	26.5	.72
8.3	6.9	5.5	.21	Add'l			.019
9.4	7.9	6.3	.23		Add'l		.021
10.7	9.0	7.3	.25			Add'l	.024

5. File gear-teeth edges

Gear Diametral Pitch	11+	6–10	3–5	1, 2
Min per Tooth	.07	.09	.09	.12

6. Blowoff, get and aside air hose included
    Bottom hole   .10
        4-in. part, to 5 lb   .10
        8-in. part, 5–15 lb   .17
        12-in. part, 15–40 lb   .26

7. Chase thread with hand tap   .40/in.

8. Sand using sandpaper

1-in. Pass L	2.0	2.4	2.9	3.5	4.3	6.5	9.8	Add'l
Min	.28	.34	.41	.49	.61	.91	1.37	.14

9. Straighten part
    $L + W$ over 60 in.   4.00
    $L + W$ under 60 in.   2.00

10. Assemble and disassemble to deburr
    Handling per component   .30
    Bolt or screw   .60

18.6 Plastic Material Deburring

DESCRIPTION

These data are grouped for plastic materials that are produced by thermoplastic injection, thermosetting plastic, and bench molding. Plastics usually deburr easier than metals. Although plastics themselves vary greatly as to deburring properties, they are categorized jointly as "plastics."

A variety of deburring elements are used, and their descriptions are given in the next section. No new machines, processes, or manual work descriptions are given beyond those already described in the other deburring of metals sections.

ESTIMATING DATA DISCUSSION

The handling of the molded unit may start with several units to be handled as one. Following separation by degating Element 3, the parts are individually stacked, dropped, or tossed. In a degating operation, the work starts with a pattern of several pieces which are eventually separated. The .07 and .21 min are inadequate in this case; therefore, a "dispose smaller piece" time is available.

Handcutting involves the use of the knife and the cutting action. The time is for each gate cut. A bandsaw can be used for degating, as found in Element 4.

Trimming, Element 5, is with a bandsaw or by hand and depends either upon girth or in. Hole deburring may be with a drill press or punch. In prick-punching, Element 6, the pickup of the hammer and point is included in the first hole.

In Element 7, thread deflashing, a tap or die is used for handchasing. Surface or edge preparation involves a file, sandpaper, or scraper, and time is given.

Sand blasting is within a hood and a directed air-grit stream from a gun can be used to prepare surfaces.

EXAMPLES

A. A molded plastic cluster of 10 parts per shot is to be degated by hand, and 10 holes have a slight flash. Estimate the unit time.

Table Number	Process Description	Table Time	Adjustment Factor	Cycle Minutes	Setup Hours
18.6-1	Handle	.07	1/10	.007	
18.6-1	Reposition 20 times for gates and holes	.023	20/10	.046	
18.6-2	Pick up two tools	.03	2	.06	
18.6-3	Hand degate runner	.11		.11	
18.6-6	Open hole by deburring	.05		.05	
18.6-6	Toss each piece aside	.03		.03	
	Total			.30	

B. A phenolic thermosetting resin is molded as a single unit with two inserts. The exposed part of the inserts is about ½ × ½ × ⅜-in. overall, but is covered by a thin coating. Additionally, two gates need to be removed. Find the time per unit.

Table Number	Process Description	Table Time	Adjustment Factor	Cycle Minutes	Setup Hours
18.6-1	Bench to bench	.07		.07	
18.6-3	Saw off gate	.04		.04	
18.6-8	Scrape area, area about 1 in.	.25		.25	
18.6-8	File edges, about 1.5-in. long	.32	1.5	.48	
18.6-2	Pick up two tools	.03	2	.06	
	Total			.93	

TABLE 18.6 PLASTIC MATERIAL DEBURRING

Setup	.05 hr

Operation elements in estimating minutes

1. Handle (per pc or molded unit)
   - Bench to bench, pickup and aside .... .07
   - Skid to skid, pickup and aside .... .21
   - Dispose small piece .... .03
   - Reposition, turn 45° or more to give new surface .... .023
   - Pack or unpack small part, hand held .... .06

2. Pick up tool .... .03

3. Handcut gate or runner
    ⅜ in. or more .23/ea
    Under ⅜ in. .11/ea

4. Saw off gate or runner .02/ea

5. Trim flash with saw .02/in.
   Trim flash by hand
    $L + W + H$ over 12 in. .56/pc
    $L + W + H$ under 12 in. .11/in.

6. Deburr hole with drill or countersink .06/ea
   Open hole with punch .02/ea
   Prick punch with point and hammer .12/first hole
    Add'l hole .04/ea

7. Deflash threads .11/in.

8. File or sand edges .32/in.
   Scrape area .28/sq in.
   Hand-sand area .17/sq in.

9. Sand blast .02/sq in.

10. Clean
    Wash with cleaning fluid .56/sq in.
    Wipe part .45/sq in.
    Blow off .23/sq in.
    Plug open end .65/end

18.7 Loose-Abrasive Deburring Processes

DESCRIPTION

Loose-abrasive deburring is also called tumbling and is a controlled method to remove burrs, scale, flash, oxides, as well as to improve surface finish. Parts to be deburred are placed in a rotating barrel or vibrating tub with an abrasive media, water or oil, and perhaps some chemical compound. As the barrel or tub rotates or vibrates, sliding motions of the media cause an abrading action.

Abrasives are usually aluminum oxide and silicon carbide and exist in a preform geometry. A preform size and geometry are available for the particular part. Cleaners are usually alkaline or acid. Alkaline cleaners must be chemically formulated to avoid part rusting. The cleaners are formulated to do certain things; remove rust, suspend oils, etc. In self-tumbling situations, media are oftentimes not used.

Figure 18.7 shows a horizontal-barrel finishing process with two compartments and a 13.8 cu ft volume capacity. The tumbling velocity is variable, ranging from 6 to 30 rpm.

ESTIMATING DATA DISCUSSION

The basic setup value is given for no media change. If media changes can be anticipated by the estimator, the setup is related to the approximate volume of the compartment or tub.

The entry value is volume, meaning box volume of the part. This minimizes volume calculations. For example, box volume of a turned part is length × diameter × diameter ($L \times Dia \times Dia$). To minimize

FIGURE 18.7 Horizontal tumbling barrel with two compartments, 30 × 16-in. in size and 13.8 cu ft capacity. (Almco Division, King-Seeley Thermos Co)

cost estimating effort, the estimator may wish to follow the policy of using the next higher tabular entry.

The handle, magnet disposing, and pick disposing columns provide the basic handling time. During actual tumbling time, the operator is doing other work. No tumbling or operator waiting time is covered by any of the elements. If a wash or oil dip or both elements are required, they are added to the basic handling value. Wash and oil dip are a function of material corrosion, sensitivity, and pre- or post-operations.

Parts that stick in the screen or flat parts that are difficult to separate from the stones may be pick-disposed where the part is removed, separated, and stacked. Ferrous parts that do not easily screen may be magnet-separated.

If the parts are tumbled without any stones, or a strainer is used to separate the parts, reduce the basic deburr time by one-fourth. For parts with an impingement-damage problem, double the time. For small parts (terminals, nuts, etc.) that are batch loaded and unloaded, a special time of .0005 min is given.

ESTIMATES

A. A machined part having box dimensions of ⅜ × 1¼ × ½ in. is to be tumbled, washed, and oil dipped. No exceptional problems are foreseen. Determine the unit estimate.

Table Number	Process Description	Table Time	Adjustment Factor	Cycle Minutes	Setup Hours
18.7-S	Setup	.1			.1
18.7-1	Volume = .23, deburr	.008		.008	
18.7-1	Wash	.001		.001	
18.7-1	Oil dip	.002		.002	
	Total			.011	

B. A part having box dimensions of 6½ × 3 × 2.125 in. is to be tumbled. Impingement problems are deburring only.

Table Number	Process Description	Table Time	Adjustment Factor	Cycle Minutes	Setup Hours
18.7-1	Volume = 41.4 cu in, and for damage prevention, increase time by 2	.22	2	.44	
	Unit estimate			.44	

C. Estimate the part sketch given by Figure 3.2B. Assume that previous operations have sheared, blanked, pierced, and formed the OEM part. A degreasing operation will follow, and this operation requires that the part be tumbled to remove any burrs. The data of Table 18.7 requires that the entry variable be box cu in. The overall box dimensions of the part are 1.562 × 1.75 × .75 in. Box volume is not the same as displacement volume, and box volume is 2.05 in.^{3}. The lot has 2500 units.

Table Number	Process Description	Table Time	Adjustment Factor	Cycle Minutes	Setup Hours
18.7-S	Setup	.1			.1
18.7-1	Handle	.08		.08	—
	Total			.08	.1
	Hr/100 units	.133			
	Total lot hours	3.43			

TABLE 18.7 LOOSE-ABRASIVE DEBURRING PROCESSES

Setup

No media change	.1 hr
3 cu ft media change	.3 hr
10 cu ft media change	.4 hr
20 cu ft media change	.5 hr

Operation elements in estimating minutes

1. Loose abrasive deburring

Part cu in.	Handle	Wash	Oil Dip	Handle Magnet-Disposing	Handle Pick-Disposing
.01	.0007	.00003	.0007		.02
.03	.001	.0001	.0008		.02
.05	.002	.0002	.0009		.02
.1	.003	.0003	.001		.03
.2	.006	.0007	.001	.007	.03
.3	.008	.001	.002	.010	.03
.4	.01	.001	.002	.014	.03
.5	.01	.002	.002	.02	.03
1	.03	.003	.004	.04	.04
2	.05	.007	.007	.07	.06
3	.08	.01	.01	.11	.07
7	.13	.02	.02		
22	.19	.03	.03		
54	.22	.05	.04		
104	.24	.05	.04		
171	.28	.06	.05		
238	.32	.07	.06		
305	.35	.07	.06		
405	.41	.09	.07		
709	.58	.12	.10		
920	.69	.14	.12		
1170	.83	.17	.14		
1472	1.00	.20	.17		
1873	1.22	.25	.20		
2376	1.49	.30	.25		
2878	1.77	.36	.29		
3380	2.05	.41	.34		
4133	2.46	.50	.41		
Add'l	.0006	.0001	.0001		

Rivets, pins, terminals	.0005
Printed circuit board	.15

18.8 Abrasive-Media Flow Deburring Machines

DESCRIPTION

Abrasive-media flow deburring machines deburr internal edges or surfaces by a controlled force flow of an abrasive laden semi-solid grinding media. The deburring machines hold the workpiece and tooling in a position relative to the media. Fixtures can hold the workpiece in a position and contain, direct, or restrict the media flow to areas of the workpiece where abrasion is desired. In application, the number of holes or surface area does not necessarily influence the cycle time. Abrasion occurs in areas where media flow is restricted.

Abrasive-media flow machines will deburr a hidden or secondary burr and improve internal surface finish. These machines remove burrs that are inconvenient (or impossible) to reach by manual deburring methods. The deburring also will result in a smoother and brighter finish.

The flow of the media is three dimensional. The media is forced up through a fixture and the workpiece into a top cylinder. After a predetermined volume of media or time has been achieved, the media returns from the top cylinder back through the workpiece and fixture to the bottom cylinder. Different media may be selected for different surface finish and burr removal. The media cylinder diameter and pressure can vary depending on the machine. Also, hydraulic force is used to clamp the workpiece and also to give the desired media pressure. The fixtures, depending on the size and shape of the part, may hold one or several workpieces.

Figure 18.8 is a sketch of an abrasive-media flow deburring machine. Media pressures range from 100 to 3000 psi. Delivery rates of more than 100 gallons per minute are possible.

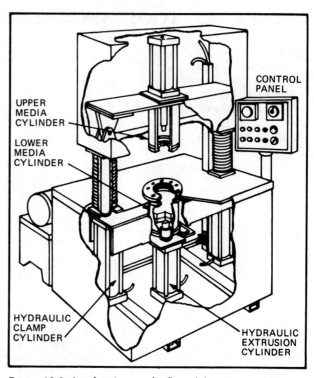

FIGURE 18.8 An abrasive-media flow deburring machine. *(Extrude Hone Corp.)*

ESTIMATING DATA DISCUSSION

The setup allows for the time to obtain fixture and parts and to study the blueprints. The setup time also includes adding media, installing the fixture on the machine, one cycle of the machine for running the first part, and its inspection. If the water cooler needs attention, the setup time is increased.

Element 1 deals with loading and unloading parts in a fixture. Time is determined by knowing the number of parts. Because the fixture guides the abrasive media throughout and around the part, it is usual to load a part into a fixture.

The weight of the fixture and part(s) is used for finding the load and unload time of the fixture and parts onto the machine table. The element includes removing a top plate, removing excess media from the fixture and part(s), loading the fixture on the bottom plate, and loading the fixture on machine.

Brinell hardness (Bhn) is used to give an indication of the deburring time. The relationship shows that harder materials require less time. The deburring of holes, size of burr, or surface area is not a factor in this element. Deburring time is shown in Element 3.

EXAMPLES

A. Internal cavities of a part are inconvenient for manual deburring methods. An abrasive-media flow deburring machine is selected to remove primary and secondary burrs.

A fixture is designed for three parts, and total weight is 40 lb. The part has a hardness of 240 Bhn. Find the cycle and unit estimate.

Table Number	Process Description	Table Time	Adjustment Factor	Cycle Minutes	Setup Hours
18.8-1	Load part in fixture	1.58	⅓	.53	
18.8-2	Load fixture, unload from machine	1.33	⅓	.44	
18.8-3	Deburr, 240 Bhn	4.00	⅓	1.33	
	Total			2.30	
	Floor-to-floor estimate	6.91			

B. A nickel alloy is used for an airfoil surface. The operation is planned to remove traces of the original casting process. Two 8-lb parts are loaded in a 36-lb fixture. A lot of 500 is to be surfaced finished. Find the unit estimate, hr/100 units, and lot estimate.

Table Number	Process Description	Table Time	Adjustment Factor	Cycle Minutes	Setup Hours
18.8-S	Setup	.55			.55
18.8-1	Load and unload part to fixture	1.58	½	.79	
18.8-2	Load and unload 52-lb fixture and parts	1.92	½	.96	
18.8-3	Deburr (Hardness of alloy is unknown, use maximum time.)	4.4	½	2.20	—
	Total			3.95	.55
	Floor-to-floor estimate	7.90			
	Hr/100 units	6.583			
	Lot estimate	33.47			

C. A tool steel extrusion die is produced by electrical discharge machining. The external surfaces of three T-shaped internal slots must be polished. Each die weighs 14 lb and the fixture weighs 26 lb. Quantity is 1600 dies. Four separate setups are planned for the year. Find the unit estimate, hr/100, lot, and annual hourly requirements.

Table Number	Process Description	Table Time	Adjustment Factor	Cycle Minutes	Setup Hours
18.8-S	Setup	.55			.55
18.8-S	Water-cooler	.85			.85
18.8-1	Load and unload one part	.83			.83
18.8-2	Fixture and part, 40 lb	1.33		1.33	
18.8-3	Deburr	2.8		2.80	
	Unit estimate			4.96	
	Hr/100 units	8.267			
	Lot estimate for 1600 parts	34.47 hr			
	Annual time requirement	137.87 hr			

TABLE 18.8 ABRASIVE-MEDIA FLOW DEBURRING MACHINES

Setup
    Basic setup   .55 hr
    Water-cooler attention   .85 hr

Operation elements in estimating minutes

1. Load and unload parts into fixture

No.	Min	No.	Min
1	.83	7	3.08
2	1.20	8	3.45
3	1.58	9	3.83
4	1.95	10	4.20
5	2.33	11	4.57
6	2.70	12	4.95

2. Load and unload parts and fixture onto machine

Parts and Fixtures, Lb	Min	Parts and Fixtures, Lb	Min
14.0	.09	37.2	1.11
18.0	.26	42.2	1.33
22.5	.46	48.2	1.60
25.1	.58	51.7	1.75
28.5	.73	55.4	1.92
30.0	.79	64.0	2.31
33.0	.93	74.4	2.77

3. Abrasive-media flow deburring

Bhn	Min	Bhn	Min
225	4.4	280	3.2
240	4.0	290	3.0
270	3.4	300	2.8

NON-TRADITIONAL MACHINING

19.1 Chemical Machining and Printed Circuit Board Fabrication

DESCRIPTION

Chemical machining, a process for the removal of metal by chemical action, is suited for production of flat, relatively thin parts having difficult configurations, and for designs requiring area, point, or line removal of metal. Metal is removed from desired areas by etching or chemically converting it into a metallic salt which is carried away by the etchant. Chemical machining produces nameplates, blanks, and printed circuit boards. It is a means to selectively remove metal for weight reduction. While the chemical blanking techniques are appropriate for small- and medium-run quantities and thicknesses less than 3/32 in., it is also suitable for hard-to-work metals such as beryllium copper, and thin materials where the finished product cannot have a burr edge as produced by a blanking die. For printed circuit production, chemical machining is an obvious and popular method.

The artwork and negatives used in chemical machining are the "tooling" for the process. The artwork is drawn oversized and reduced in scale for production. When a large number of parts are to be chemically machined, it is necessary to have multiple images on the same film after photographic reduction. Two transparencies, if opposite-side machining is required, are placed together and registered. The process continues with metal preparation, image printing, etching and resist removal, and postfabrication of parts from a multiple-unit panel. Figure 19.1A is a single-chamber etch machine with a variable velocity conveyor. In the etcher, oscillating nozzles spray both sides of the metal. Areas of the metal not protected by a photo resist are dissolved, leaving only the finished part. Quantities of small parts are usually held together by metal tabs incorporated into the original artwork, or by an adhesive or plastic-laminated backing. When etching is complete, photo resist may or may not be removed. The various units of the process can be arranged for separated batchwork, or continuous wet-processing systems are available.

ESTIMATING DATA DISCUSSION

These data deal with electroless etching or subtractive etching using chemicals. The volume of work to be handled depends on whether the shop is producing for its own needs or as a vendor to outside customers. Generally, job shops do not have high-volume runs but handle short to medium quantity runs of many types of work with a few of the larger shops handling high-volume production for commercial or industrial application.

The data are for a noncontinuous process and the estimator selects the appropriate operations to produce the design. Batching is the usual mode. Times are selected from five major parts of Table 19.1: (1) artwork and negative preparation, (2) metal preparation, (3) image processing and masking, (4) etching and post etching processes, and (5) fabrication processes.

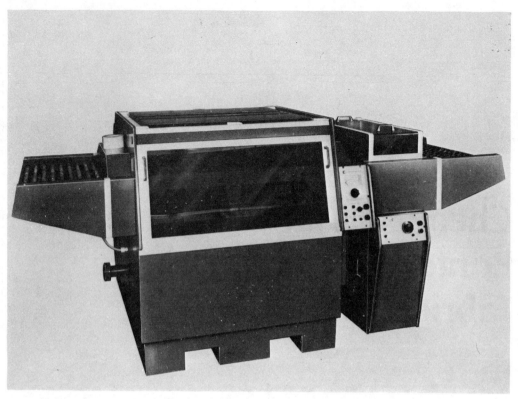

FIGURE 19.1A Conveyorized chemical etching machine. *(Chemcut Corporation)*

The data deal with three types of chemical machining problems: blanking of small thin parts, production of printed circuit boards, and etching of work for weight reduction or other design purposes. Costs resulting from yield and scrap, and chemical supplies, are not provided, and are usually obtained from a historical analysis of the shop's data. As direct-labor time is influenced by many factors, these average data are for a shop using standard and noncontinuous processing equipment. Complexity of design is a factor averaged out in the data. Raw material, while it can vary, is reasonable as to weight, size, girth, etc., and is not heavy. For instance, printed circuit stock material may arrive in 36 × 48 in. sheets, with copper 1.4 mil thick, either one or both sizes, to panels perhaps 12 × 18 in. in size. The data to follow use a 12 × 18-in. panel as typical, although many other sizes are possible. Data are not too sensitive to panel size although it is sensitive to units on the panel. Chemical blanking may start in the sheet, strip, or roll stock. Material for etching and weight reduction may start out as sheet or strip.

A panel is composed of one or many units of the same design. Thus, panel time is ultimately divided by the number of parts placed on the panel.

EXAMPLES

A. Estimate a 2-in. square printed circuit switch having nickel-gold-rhodium fingers, 28 close-tolerance holes, plating on one side, and 1-oz. copper-board two-sided. Raw material is a 12 × 15-in. panel. The desired quantity is 500 switches. Allowing for border and inner-part spacing, there are five columns and eight rows or 40 switches per panel. The number of panels is 13, allowing a 4 percent scrap loss.

Table Number	Process Description	Table Time	Adjustment Factor	Cycle Minutes	Setup Hours
19.1-S1	Artwork preparation	20.00			20.00
19.1-S1	Artwork reduction	1.00			1.00
19.1-S1	Step and repeat	1.00			1.00
19.1-S3	Image processing	0.30			0.30
19.1-S5	Shear strips, blanks	0.20			0.20
19.1-5A	Shear strips	0.14	1/40	0.00	

Table Number	Process Description	Table Time	Adjustment Factor	Cycle Minutes	Setup Hours
19.1-5A	Shear blanks 8/strip	0.16	1/8	0.02	
19.1-S5	Pierce holes in general purpose machine in panel	0.22			0.22
19.1-S5	No. of pierce stat. for blank holes, 2	0.06	2		0.12
19.1-2D	Degrease	0.45	1/40	0.01	
19.1-2C	Liquid hone	1.56	1/40	0.04	
19.1-S5	Liquid hone	0.10			0.10
19.1-S2	Photo resist vat SU	0.10			0.10
19.1-2D	Photo resist vat	0.35	1/40	0.01	
19.1-2D	Load rack, 10 panels	3.00	1/400	0.01	
19.1-S3	Expose	0.30			0.30
19.1-3	Expose panels	2.58	1/40	0.06	
19.1-1	Artwork per panel	8.00	1/40	0.20	
19.1-S3	Develop in spray	0.10			0.10
19.1-3	Develop in spray	0.23	1/40	0.01	
	Nickel and gold plate jig	2.62	1/40	0.07	
4.2-S	Setup silk screen for rhodium protect	0.25			0.25
4.2-1	Manual silk screen	0.56	1/40	0.01	
20.5-4B	Rhodium plate, flat 12 + 15 in.	0.90	1/40	0.02	
19.1-S4	Strip resist setup	0.10			0.10
19.1-4C	Strip resist	0.75	1/40	0.02	
19.1-S4	Etching cupric chlor	0.30			0.30
19.1-4C	Conveyor etch	2.10	1/40	0.05	
19.1-5A	Roll burnish edge one edge, 2-in. long	0.02	2	0.04	
	Total			0.57	24.09
	Lot estimate	28.27 hr			

B. Estimate a 1.75-diameter ring electrical contact, .020-in. thk copper alloy. Raw stock available in 12 × 48-in. strips. Adopting a 12 × 12-in. panel, there are six columns and six rows allowing inner-part and border space. The number of finished parts is 1800 requiring 50 panels and allowing a 4 percent scrap, 52 panels will be made.

Table Number	Process Description	Table Time	Adjustment Factor	Cycle Minutes	Setup Hours
19.1-S1	Artwork preparation	8.00			8.00
19.1-S1	Artwork reduction	1.00			1.00
19.1-S1	Artwork step and repeat	1.00			1.00
19.1-S3	Negative envelope	0.30			0.30
19.1-S5	Shear setup	0.10			0.10
19.1-2A	Shear panel from stock, 36 pc/panel	0.14	1/36	0.00	
19.1-S2	Pierce hanging holes	0.10			0.10
19.1-2D	Pierce hanging holes	0.18	1/36	0.01	
19.1-S2	Metal prepar, clean	0.10			0.10
19.1-2D	Degrease	0.45	1/36	0.01	
19.1-4A	String and takedown	0.51	1/36	0.01	
19.1-S2	Resist	0.10			0.10
19.1-2D	Degrease	0.35	1/36	0.01	
19.1-S2	Drying setup	0.10			0.10
19.1-2D	Drying time 10 panels	3.00	1/360	0.01	
19.1-S3	Image processing expose	0.30			0.30
19.1-3	Image processing expose	2.58	1/36	0.07	
19.1-S3	Image processing developing step	0.10			0.10
19.1-3	Image processing developing step	0.23	1/36	0.01	
19.1-S4	Etching setup	0.30			0.30
19.1.4A	Etching, post-etc	7.61	1/36	0.21	
19.1-S4	Etching strip resist setup	0.10			0.10
19.1-4C	Strip resist	0.75	1/36	0.02	
19.1-S5	Detab setup	0.10			0.10

Table Number	Process Description	Table Time	Adjustment Factor	Cycle Minutes	Setup Hours
19.1-5A	Detab parts from panel	0.36	1/36	0.01	
	Total			0.37	11.70
	Total estimate	23.37 hr			

C. Estimate the time to alkaline-etch an aluminum panel, 3/16 in.-thk, where the panel is 14 × 14 in. Each panel is composed of two parts, and raw stock is 14 × 14 in. The lot size is 20. Artwork and negative preparation are considered as overhead.

Table Number	Process Description	Table Time	Adjustment Factor	Cycle Minutes	Setup Hours
19.1-S2	Metal preparation by piercing holes along top border	0.10			0.10
19.1-2D	Pierce hanging holes two parts per panel	0.18	1/2	0.09	
19.1-S2	Clean setup	0.10			0.10
19.1-2D	Degrease	0.45	1/2	0.23	
19.1-S2	Apply resist setup	0.10			0.10
19.1-2D	Apply resist	0.35	1/2	0.17	
19.1-S3	Image processing expose setup	0.30			0.30
19.1-3	Image processing expose step	2.58	1/2	1.29	
19.1-S3	Image processing developing step	0.10			0.10
19.1-3	Image processing develop step	0.23	1/2	0.12	
19.1-S4	Etching setup	0.30			0.30
19.1-4B	Aluminum panel etch	0.58	1/2	0.29	
19.1-S4	Strip resist setup	0.10			0.10
19.1-4C	Strip resist cycle	1.32	1/2	0.66	
19.1-S5	Shear blanks from panel setup, strips and blank setup	0.10	1/2		0.20
19.1-5A	Shear strip, one hit	0.17	1/2	0.09	
	Shear blank, one hit	0.17	1/2	0.09	
19.1-S5	Roll burnish setup to deburr sharp edge	0.10			0.10
19.1-5A	Roll burnish sharp edge, 42-in. length	0.02	42	0.84	
	Total			3.86	1.40
	Total lot estimate	2.69 hr			

D. Figure 19.1B is a simple and qualitative design for a printed circuit board. A lot quantity is 18. The panel is 12 × 18 × .063-in. copper, both sides. The estimate requires that the board be time estimated for several operations with the exception of stuffing components. The operations to be considered are the following:

Artwork and negative preparation
Metal preparation

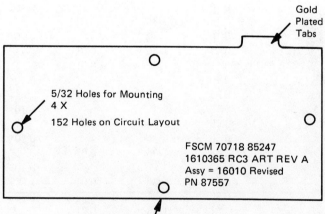

Full Size: 3-1/4 × 7-1/4 in. Finished Overall Outside
Panel Size: 12 × 18 × .063 in. Copper Two Sides

FIGURE 19.1B Printed circuit board.

Image processing and masking
Etching and post etching processes
Fabrication processes
    Shear the finished size from a 12 × 18-in. panel
    Drill the holes
    Solder the circuit
    Gold plate the tab

The following 11 operations use a variety of equipment to demonstrate the estimating approach. The lot size is small, 18 units, and each panel will have six units.

Table Number	Process Description	Table Time	Adjustment Factor	Cycle Minutes	Setup Hours
19.1	Artwork				
19.1-S1	Artwork setup for 12 × 18-in. panel				
	Each panel = 6 units	8.00			8.00
19.1-1	Registration and negative envelope per panel of six	8.00	1/6	1.33	
19.1-S1	Artwork reduction	1.00			1.00
19.1-S1	Artwork step and repeat	1.00			1.00
	Total			1.33	10.00
	Total lot estimate	10.40			
19.1	Shear holes in panel				
19.1-S2	Shear holes for hanging and for p.c. board, four holes per board	0.10			0.10
19.1-2D	Pierce holes panel	0.18	1/6	0.03	
	Total			0.03	0.10
	Total lot estimate	.11			
19.1	Metal preparation				
19.1-S2	Metal prepar, clean	0.10			0.10
19.1-2D	Degrease	0.45	1/6	0.07	
	Total			0.07	0.10
	Total lot estimate	.12			
19.2	Dry film				
19.2-S	Auto lamination of dry resist	0.35		0.35	
19.2-1	Laminate panel, 18-in. dry film application	0.49	1/6	0.08	
19.3-S	Ultraviolet printing	0.05			0.05
19.3-1	Lot preparation for ultraviolet prt	0.20	1/18	0.01	
19.3-2	Board insertion for ultraviolet prt	0.18	1/18	0.01	
19.3-3	Exposure of panel for ultraviolet prt	0.35	1/6	0.06	
19.3-4	Removal of panel for ultraviolet prt	0.27	1/6	0.05	
	Total			0.56	0.05
	Total lot estimate	0.22			
19.4	Develop				
19.4-S	Setup for conveyor developing machine	0.20			0.20
19.4-1	Develop circuit board	0.78	1/6	0.13	
	Total			0.13	0.20
	Total lot estimate	0.24			
19.1	Etch p.c. board				
19.1-S4	Etching setup	0.30			0.30
19.1-4C	Conveyor oper cupric chloride etching	2.10	1/6	0.35	
	Total			0.35	0.30
	Total lot estimate	0.41			
19-1	Strip resist				
19.1-S4	Strip resist setup	0.10			0.10
19.1-4C	Strip resist	0.75	1/6	0.13	
	Total			0.13	0.10
	Total lot estimate	0.14			
23.13	Drill p.c. boards				
23.13-S	Printed circuit bd drilling	0.50			0.50

Table Number	Process Description	Table Time	Adjustment Factor	Cycle Minutes	Setup Hours
23.13-1	Load one board	0.90	1/6	0.15	
23.13-2	Drill .033-in. holes, 152 holes	1.40		1.40	
23.13-3	Unload board	0.70	1/6	0.12	
	Total			1.67	0.50
	Total lot estimate	1.00			
19.1	Shear individual pc				
19.1-S5	Fabrication shear	0.10			0.10
19.1-5A	Fab shear	0.20	1/6	0.03	
19.1-5A	Shear blank	0.03		0.03	
	Total			0.06	0.10
	Total lot estimate	0.12			
23.7	Vertical solder unmounted board				
23.7-S	Setup for vertical solder machine	2.00			2.00
23.7-1	Load board	0.13		0.13	
23.7-2	Machine insert	0.82		0.82	
23.7-3	Flux board	0.21		0.21	
23.7-4	Preheat	0.27		0.27	
23.7-5	Solder	0.18		0.18	
23.7-6	Level	0.27		0.27	
	Total	2.56		1.88	2.00
	Total lot estimate	2.56 hr			
19.5	Gold plate tabs				
19.5-S	Setup for conveyor line	1.00			1.00
19.5-1	Tape press	0.10		0.10	
19.5-2	Start, stop press rollers	0.00		0.00	
19.5-3	Run one edge	0.01		0.01	
19.5-6	First part through	6.21	1/18	0.34	
19.5-7	Overtravel	0.78	1/6	0.13	
	Total			0.59	1.00
	Total lot estimate	1.18 hr			

TABLE 19.1 CHEMICAL MACHINING AND PRINTED CIRCUIT BOARD FABRICATION

Setup operations in estimating hours

S1. Artwork and negative preparation
    Preparation of image of part   8–40 hr
    Artwork reduced to 1/1 photographically by camera   1 hr
    Step and repeat processing of image on one large film for multiple images on blank   1 hr

S2. Metal preparation
    Shear panel from sheet   .2 hr
    Deburr edges of panel by hand-file perimeter   .1 hr
    Clean   .1 hr
    Pierce hanging holes   .1 hr
    Photo-resist vat
        Dip and hang for draining, manual   .1 hr
        Dry film application   See Table 19.2
    Load rack in oven or air dry   .1 hr

S3. Image processing and masking
    Load photosensitive panel in negative envelope, load into vacuum printer, expose, unload, manual   .3 hr
    Develop in solvent spray   .1 hr

	Dye	.1 hr
	Brush touch up, manual	.1 hr
	Silk-screen printing	See Table 4.2
S4.	Etching and post-etching processes	
	Etch, load and unload	.3 hr
	Conveyor operation for 1-oz circuit panel	.3 hr
	Strip resist	.1 hr
S5.	Fabrication processes	
	Pierce holes in printed circuit panel	.22 + .06 (no. stations)
	Pierce holes with pierce die	2.0 hr
	Size holes by drill press	See Table 11.2
	Remove piece from panel by clipping tabs, manual	.1 hr
	Shear strips from panel	.1 hr
	Shear blanks from strip	.1 hr
	Roll burnishing	.1 hr
	Route, engrave, or slot	.1 hr
	Protective coating for intermediate operations	.1 hr

Operation elements in estimating minutes

1. Artwork and negative preparation
   Multiple-part negative and its mirror image are registered for alignment and secured together for a negative envelope to accept sensitized metal panel, per panel — 8–12

2. A. Metal preparation
   Shear panel from sheet
   - 3 × 3 — .07
   - 12 × 18 — .14
   - 18 × 24 — .21

   B. Deburr edges of panel by hand-file perimeter

In.	6	30	Add'l
Min	.16	.59	.02

   C. Clean area

Sq in.	3 × 3	15 × 15	Add'l
Scrub	.12	3.11	.014
Liquid hone	.06	1.56	.007

   D. Degrease — .45/panel
   Pierce hanging holes — .18/panel
   Photo-resist vat
     Dip and hang for draining, manual — .35/panel
     (Note: No drain time included)
     Dry film application — See Table 19.2
   Load rack in oven or air dry
     Load several panels at once, unload — 3.00
       Single load, unload
       (Note: No dry time included) — 1.00

3. Image processing and masking
   Load photosensitive panel in negative envelope, load into vacuum printer, expose, unload, manual — 2.58/panel
   Ultraviolet printing machine — See Table 19.3
   Develop in solvent spray — .23/panel
   Conveyorized developing — See Table 19.4
   Dye — .77/panel

Brush touch up, manual

L + W	1	3	5	8	Add'l
Min	.90	1.04	1.38	.175	.13

Silk-screen printing   See Table 4.2

4. Etching and post-etching processes

    A. Manual, blanking   7.61/panel
        Manual printed circuit   3.33/panel
        String and take down (no wait time)   .51/panel

    B. Alkali etch of aluminum panel

Thickness	Width			
$0+ -\frac{1}{8}$	0+ to 8	8+ to 14		Min/Panel
$\frac{1}{8}+ -\frac{1}{4}$			8+ to 16	
Length	0+ to 16+			.26
		0+ to 16		.40
			0+ to 16	.58

    C. Conveyor operation for 1-oz printed circuit panel   Min/panel
        Cupric chloride   2.10
        Ferric chloride   1.37
        Alkaline   1.40
        Ammonium persulfate   1.63
        Chromic sulfuric   1.75
        High speed cupric   1.30
    Strip resist   .75/panel
    Strip mask for printed circuit tabs   .29/panel
    Strip resist, aluminum etching   1.32/panel
    Plate connector fingers   See Table 19.5

5. Fabrication processes

    A. Pierce holes in printed circuit panel
        Load and unload   .31
        Rotate turret and pierce first hole of each size in part (turret punching machine)   .07 (no. sizes)
        Change punch and die of each hole size   .56 (no. sizes)
            (Note: Single-station punching machine)
        Pierce remaining holes   .017 ea
    Pierce holes with pierce die   .35/panel
    Size holes by drill press   See Table 11.2
    Remove piece from panel by clipping tabs, manual   .36 ea
    Shear strips from panel $(.15 + .02 \times$ no. hits$)$/panel units
    Shear blanks from strip $(.05 + .065 \times$ no. hits$)$/strip units
    Roll burnishing   .02 (linear in.)
    Route, engrave, or slot
        Carbide   .09 (perimeter in.)
        Diamond   .07 (perimeter in.)

    B. Protective coating for intermediate operations

	Printed Circuit Part	Alum. Etching Part, Chem. Blank Part
Apply	.32/panel	.58/unit
Wash Off	.35/unit	.12/unit

19.2 Automatic Lamination Machines (for Printed Circuit Boards)

DESCRIPTION

An automatic laminator is a machine used to apply dry film photoresist on printed circuit boards. Through the use of a microprocessor, this machine automatically measures, cuts, and feeds the dry film. Once the machine is setup, the operation is continuous, including automatic shutoff after the last board has been coated. Two thicknesses of dry film (1.5 and 2.0 mil) are used in the automatic laminating machine. The dry film photoresists consist of photopolymer layers that are sensitive to ultraviolet light. Thus, the laminating machine must be used in an area illuminated by yellow lights. The laminator process begins as the machine strips the dry film of a polyethylene sheet and feeds the film into position above two heated rollers. The dry film is then measured and cut to a specific length set by the operator. Next, the circuit board plate is driven through the hot rollers, and the film is applied. Finally, the laminated plates are transferred by a delivery belt that automatically stacks the plates in a rack. The first responsibility of the laminating operator is to set up the machine. This procedure includes inserting the roll of dry film, setting vacuum controls, and establishing length control for a particular board. After the operation begins, the operator must continuously observe the machine during operation even though the machine is controlled by a microprocessor. If a problem occurs, the machine must be shut down manually.

ESTIMATING DATA DISCUSSION

As an example of data, consider the amount of time required to laminate circuit boards. The total estimate for the process consists of setup time, heatup time, and machine time. The time for loading the film and setting controls (setup time) is independent of the number of boards being coated. The heating setup time is the period required for the two rollers to heat up to a steady temperature of 244°F. This procedure begins when assembly setup is completed and the machine door is closed. When the rollers reach the required temperature of 244°F, the laminating procedure is ready to begin. The period required for the circuit board to move through the coating rollers and into the stacking rack is the lamination time. This time period is dependent on the length of the circuit board. The time values for various lengths are shown in the detailed element table. If the length of the circuit board falls between listed values, use the next higher value.

These data can be combined with Table 19.1, Element 2.

EXAMPLE

A. Estimate the time to dry film laminate a board. Each panel is 12 × 15 in. There are 60 panels. Each panel contains four printed circuit boards. Find the cycle minutes, setup hours, and lot time.

Table Number	Process Description	Table Time	Adjustment Factor	Cycle Minutes	Setup Hours
19.1-S	Set up machine	.35			.35
19.1-S	Heat up machine	.17			.17
19.1-2	Laminate panel	.43	¼	.108	—
	Total			.108	.52
	Hr/100	.179			
	Lot hours	.95			

TABLE 19.2 AUTOMATIC LAMINATION MACHINES (FOR PRINTED CIRCUIT BOARDS)

Setup, basic .35 hr
  Close machine door and heat up rollers .17 hr

Operation elements in estimating minutes

1. Laminate circuit board panels, min per panel

Length	12.0	12.5	13.5	14.5	15.6	17.2	18.4	19.5	20.7	21.9
Min	.43	.44	.45	.46	.47	.48	.49	.50	.51	.52

19.3 Ultraviolet Printing Machines (for Printed Circuit Boards)

DESCRIPTION

An ultraviolet printer is used for exposing dry film photoresist on printed circuit boards. This machine has two vacuum exposure frames. Therefore, when one circuit board is being exposed, the other frame is available to the operator to load another circuit board. Exposure time depends on the thickness (1.5 or 2.0 mil) of the dry film on the circuit board.

The operator first applies an imagine pattern on the laminated circuit boards. This pattern is snapped onto the circuit board. The circuit board is then inserted into the exposure frame and covered. Next, the operator starts the machine and the circuit board is driven into the machine to be exposed. The exposed circuit board is removed when the operator trips a switch on the control panel. As the exposed circuit board is driven out of the machine on one frame, another circuit board is entering on the other frame.

This type of exposure machine contains two ultraviolet light sources. Therefore, both sides of a circuit board can be exposed at the same time. Single-sided boards can also be exposed in the machine. The lights are cooled with a self-contained water cooling system.

ESTIMATING DATA DISCUSSION

Setup time is minor for this machine as several procedures are performed by the operator before and after the machine process. In the cycle the operator covers the laminated circuit boards with an image on both sides of the boards. The time for this procedure does not depend on the size of the board or the thickness of the dry film. This time is referred to as preparation and is listed as Element 1.

Insertion is the next step and consists of the opertor moving the prepared boards onto the machine frame and covering the circuit board with a mylar sheet. Preparation and insertion time is required only for the first board. This work becomes internal to the cycle during exposure.

Exposure time, Element 3, depends on the thickness of dry film applied to the board during the laminating process. The thickness of the dry film is 1.5 or 1.0 mil. After the circuit board has been exposed, it is removed from the machine by the operator and placed in a rack.

EXAMPLES

A. Find the time required to expose 70 circuit boards with a dry film thickness of 1.5 mil. Each exposure is for one board.

Table Number	Process Description	Table Time	Adjustment Factor	Cycle Minutes	Setup Hours
19.3-S	Setup	.05			.05
19.3-1	Preparation	.20	1/70	.003	
19.3-2	Board insertion	.18	1/70	.003	
19.3-3	Exposure	.40		.40	
19.3-4	Removal	.27		.27	
	Total			.67	.05
	Lot estimate	1.71 hr			

Note: Preparation and board insertion are only required for the first board. As one board is exposed, another board is being prepared. Thus, the actual process time for each following board consists of exposure and removal time only.

B. Find the total time required to expose 20 panels with a dry film thickness of 2.0 mil. Each panel has four printed circuit boards.

Table Number	Process Description	Table Time	Adjustment Factor	Cycle Minutes	Setup Hours
19.3-S	Setup	.05			.05
19.3-1	Preparation	.20	1/60	.003	
19.3-2	Board insertion	.18	1/60	.002	
19.3-3	Exposure	.35	1/4	.088	
19.3-4	Removal	.27	1/4	.068	
	Total			.16	.05
	Hr/100	.266			
	Lot estimate	.26 hr			

TABLE 19.3 ULTRAVIOLET PRINTING MACHINES (FOR PRINTED CIRCUIT BOARDS)

Setup	.05 hr

Operation elements in estimating minutes

1. Preparation .20

2. Board Insertion .18

3. Circuit board exposure

Thk	1.5 mil.	2.0 mil.
Min	.40	.35

4. Removal .27

19.4 Conveyorized Developing Machines (for Printed Circuit Boards)

DESCRIPTION

A conveyorized developing machine is used to develop printed circuit boards. The development process includes a wash, rinse, and drying process. The speed of the machine is set according to the thickness of the dry film applied in the laminating process. The actual developing takes place in the first module of the developing system. A conveyor system moves the circuit boards through this module. The solution used for the development process contains potassium carbonate in a water solution. After the circuit boards are moved through the developing solution, they are fed through a wash-rinse-drying cycle. First, the circuit boards are fed through a two-stage water-wash cycle. The conveyor system then transfers the panels to a fresh water rinse module. The circuit boards are finally transferred through a drying module that completes the developing cycle.

ESTIMATING DATA DISCUSSION

These data will provide an estimate for the time required to develop printed circuit boards. Although the development process includes individual treatments (development, wash, rinse, and dry), these different procedures are not separated for estimating.

The setup time is the time required to prepare the machine for operation. The machine must be checked for desired water levels and solution mixtures. This procedure is performed at the start of the shift. The only setup thereafter is the time needed to set controls and desired speeds. The speed of the conveyor system depends on the thickness (1.5 or 2.0 mil) of the dry film on the boards.

The time required for developing circuit boards depends on two variables: the length of the board and the number of boards being developed. The estimator finds the unit estimate for different quantities with a given length. This information is contained in the data table as Element 1. If a length or quantity falls between two values, use the next higher value for the estimate.

EXAMPLES

A. Estimate the time required to develop 40 13.5-in. circuit boards.

Table Number	Process Description	Table Time	Adjustment Factor	Cycle Minutes	Setup Hours
19.4-S	Setup	.2			.2
19.4-5	Control adjustment	.08			.08
19.4-3	Develop	.55		.55	—
	Total			.55	.28
	Lot estimate	.65 hr			

B. Find the unit estimate for a 6-in. circuit board in a batch of 60 boards.

Table Number	Process Description	Table Time	Adjustment Factor	Cycle Minutes	Setup Hours
19.4-S	Setup	.2			.2
19.4-5	Control adjustment	.08			.08
19.4-1	Develop	.31		.31	—
	Total			.31	.28
	Hr/100	.517			
	Lot estimate	.59 hr			

C. Determine the total time required to develop 57 12-in. boards.

Table Number	Process Description	Table Time	Adjustment Factor	Cycle Minutes	Setup Hours
19.4-S	Setup	.2			.2
19.4-S	Control adjustment	.08			.08
19.4-3	Develop	.48		.48	—
	Total			.48	.28
	Lot estimate	.74 hr			

TABLE 19.4 CONVEYORIZED DEVELOPING MACHINES (FOR PRINTED CIRCUIT BOARDS)

Setup, basic	.2 hr
Set controls	.08 hr

Operation elements in estimating minutes

Develop circuit boards, min/ea.

Length (in.)	Number of Boards						
	5	10	20	30	40	50	60
6.0	1.1	.69	.46	.39	.35	.33	.31
6.5	1.17	.70	.48	.40	.36	.34	.33
7.0	1.18	.71	.49	.41	.38	.35	.34
7.5	1.19	.72	.50	.43	.39	.37	.35
8.0	1.20	.74	.52	.44	.41	.38	.37
8.5	1.22	.75	.53	.46	.42	.40	.38
9.1	1.23	.77	.55	.47	.44	.41	.40
9.6	1.24	.78	.56	.49	.45	.43	.41
10.1	1.25	.79	.58	.50	.47	.44	.42
10.7	1.27	.81	.59	.51	.48	.45	.43
11.2	1.28	.82	.60	.52	.49	.46	.45
11.8	1.29	.84	.61	.54	.50	.48	.47
12.3	1.30	.85	.62	.56	.52	.50	.48
12.9	1.32	.87	.64	.57	.54	.51	.49
13.5	1.33	.89	.66	.58	.55	.52	.51
14.0	1.34	.90	.68	.59	.56	.54	.53
14.6	1.36	.91	.70	.61	.58	.56	.54
15.2	1.37	.92	.71	.63	.60	.57	.55
15.8	1.38	.94	.72	.64	.61	.59	.57
16.4	1.40	.95	.73	.65	.63	.61	.58
17.0	1.41	.97	.75	.67	.64	.62	.60
17.6	1.42	.99	.76	.69	.65	.63	.62
18.2	1.44	1.00	.78	.70	.67	.65	.63
18.8	1.45	1.01	.80	.72	.70	.66	.65
19.5	1.47	1.03	.81	.74	.70	.68	.66
20.1	1.48	1.04	.82	.77	.72	.70	.68
20.7	1.50	1.06	.84	.78	.73	.71	.70
21.4	1.51	1.08	.86	.80	.75	.73	.72

19.5 In-Line Plating Processes (for Printed Circuit Boards)

DESCRIPTION

In-line nickel/gold plating machines and others in the same category are used to deposit thin layers of nickel and gold on the connector fingers of printed circuit boards by electrodeposition (electroplating). This coating provides a highly conductive non-corroding interface between the electronic circuit printed on the board and external connectors and cabling. Typical metal deposition thickness range from 50 to 200 millionths and 50 to 300 millionths of an inch for gold and nickel respectively. Figure 19.5 illustrates typical processing equipment.

Areas of the board adjacent to the connector fingers are masked off with protective plastic tape prior to the plating process. The masked edges are pressed between the rollers of a manually-fed tape pressing to seal the tape firmly against the board's surface. A var-

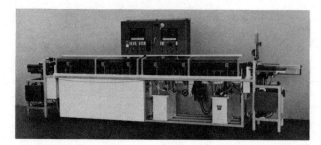

Figure 19.5 Automatic tab plating system. *(Baker Brothers/Systems)*

iable number of boards (from 1 to 50) are then loaded into the feed rack of the plating machine. Two opposed tractor-like rubber belts automatically feed the boards at a constant predetermined speed through a series of acid, rinse, and plating baths. Boards usually travel through the machine edge-to-edge and not one at a time. The connector fingers are stripped of the lead/tin (solder) coating that was originally plated to the surface of the board (and that makes up the circuit wiring), and subsequently replated with nickel and gold. Stripping and plating takes place on only one edge of the board (both front and back) at a time necessitating several complete runs through the machine and possible remasking on boards with connectors along more than one edge. Other variables that can affect total operation time are metal deposition thickness, inches of connector per inch of board, and connector finger width. After plating, the boards are unloaded from the catcher rack and the tape is removed.

ESTIMATING DATA DISCUSSION

In-line electrodeposition machines normally require startup once at the beginning of each day. This can include plugging the machine into an electrical outlet, opening water feed valves, adjusting the control panel, and changing acid and plating solutions if required while allowing time for warm-up. Clear-up at the end of the day (or shift) includes the reverse of setup except for warm-up time. Before starting each lot a test board is usually run through and inspected for correct metal deposition thickness and adhesion strength and the controls (line feed speed and electrode current) adjusted accordingly. After this initial inspection the operator randomly inspects about every 10th board.

The tape pressing operation can be performed on a large number of boards at a time in usually the same room that houses the plating machine. For these reasons a continuous supply of pretaped, prepressed boards is assumed ready on a shelf or rack nearby (10–15 ft) the plating machine. Shelf-to-feed rack carrying time is contained in Element 1, handling. Additional transport time must be added if the pressed boards are initially located in another room or part of the building.

Some circuit boards are solder dipped to make component soldering easier. This thick, fused solder is more resistant to chemical etching than thinner, more uniform plated solder and must be stripped off before the plating process. Stripping can be thought of as plating without the electrode current turned on. The circuit boards travel through the plating machine as though they are being plated but only the acid and rinse baths are active. This effectively doubles the number of plating runs and so also the plating time.

The data cover stripping, plating and tape pressing operations only and not masking operations. The majority of printed circuit boards have single or simple connector configurations. Some boards, however, require connectors on adjacent edges so that remasking during the plating operation is necessary. The remasking operation is usually performed on a large number of boards at a time and not on individuals as they leave the plating machine. For this reason unit time for remasking should be calculated separately. Remember that only one edge can be plated per run through the machine. Average traverse time is a combination of travel time through all solution baths plus the over-travel time necessary for the trailing end of the board to clear the tractor feed mechanism. Average traverse time is calculated using:

$$T_{ave} = (L_m + L_p)/S$$

where T_{ave} = Average traverse time (min)
S = Line feed speed (feet/min)
L_m = Machine length (length of belt feed mechanism, feet)
L_p = PCB plating edge length (feet)

Electroplating is a slow continuous process subject to individual operator efficiency and method of operation and room configuration. Many elements such as board transportation to feed rack, feed rack loading and unloading, metal deposition thickness inspection and tape removal can be accomplished during the long traverse time through the machine. Transportation, loading, inspection and tape removal times are significant only if they are larger than, or the elements are expected to interfere in some way with the average traverse time. A trained operator can usually perform pressing and plating operations simultaneously. Accounting for these elements and the pressing operation is therefore optional especially when the lot size is large or line feed speed slow.

Line feed speed is directly proportional to throughput and is consequently an important operation time driver. It is initially set by the operator and normally held constant throughout the lot. Large plating areas

Plating Machine Line Feed Speed

Connector Finger Width	Nickel Thickness (mils)	Line Fed Speed (ft/min)
Thin	50	1.8
	100	1.6
	150	1.6
	200	1.6
Normal	50	1.6
	100	1.6
	150	1.6
	200	1.6
Thick	50	1.6
	100	1.6
	150	1.4
	200	1.2

(wide connector fingers) and/or unusually thick nickel deposition may require slower line feed speeds. It is suggested that the estimator unfamiliar with plating processes consider 1.6 feet/min as a typical line feed speed value and use for all but unusual circumstances. Electrical current supply adjustments vary metal deposition thickness and compensate for metal salt depletion within the solution baths. These adjustments are left to the discretion of the operator and do not affect the time estimating process.

Element 6 is based on the standard 10-foot machine. Element 7 (overtravel), for machine clearance, (throughput) is the time required for the board to become completely free of the machine once the front edge reaches the end of the drive belt. After waiting for the first board to travel through additional boards emerge at the throughput rate. If no metal thickness are specified, assume 100 millionths of an inch for nickel. Gold thickness can be compensated for through current adjustments by the operator and is relatively unimportant. If Element 7 is to be used again, a return to Element 8 is available.

EXAMPLES

A. A printed circuit board of dimension 12 × 12 in. has a 3-in. connector on one edge. Lot quantities of 300 are expected. Find the unit time to gold plate the connector with 100 millionths of an inch of gold.

Table Number	Process Description	Table Time	Adjustment Factor	Cycle Minutes	Setup Hours
19.5-S	Setup				1.0
19.5-1	Handling/tape press	.10		.10	
19.5-2	Start/stop rollers	.004	$1/300$		
19.5-3	Run one edge through press	.019		.019	
	Tape press subtotal			.117	

Table Number	Process Description	Table Time	Adjustment Factor	Cycle Minutes	Setup Hours
19.5-5	Handling/plating machine	.22		.22	
19.5-6	Wait for first board	8.07	$1/300$	.027	
19.5-7	Overtravel (throughput)	.70		.70	
19.5-9	Check plating thickness	.02		.02	
19.5-10	Check adhesion strength	.01		.01	—
	Total			.977	1.0
	Lot estimate	5.89 hr			

Note: The tape press subtotal is less than throughput so it is neglected.

B. A printed circuit board is made up of two identical boards printed on a single sheet of composite material. The two boards will be separated in a shearing process after plating.

Find the unit time if the board is 17 × 19 inches and there are 6 inches of connector fingers on each of the two short sides. Lot quantity is 2500 and gold thickness is 50 millionths of an inch.

Table Number	*Process Description*	*Table Time*	*Adjustment Factor*	*Cycle Minutes*	*Setup Hours*
19.5-S	Setup				1.0
19.5-1	Handling/tape	.1	½	.03	
19.5-2	Start/stop	.004	1/2500	0	
19.5-3	Run one edge through press	.028		.028	
19.5-4	Flip to new edge	.015	½	.007	
	Tape press subtotal			.078	
19.5-5	Handling/plating machine	.22	½	.11	
19.5-6	Wait for first board	8.07	1/2500	.003	
19.5-7	Over travel (throughput)	1.06	2/2	1.06	
19.5-8	Return to feed rack	.04	½	.02	
19.5-9	Check plating thickness	.02		.02	
19.5-10	Check adhesion strength	.01		.01	
	Subtotal, plating machine			1.77	1.0
	Unit time 1.223				
	Lot estimate	51.96 hr			

Note: The tape press subtotal is less than throughput so it is neglected.

C. Six separate circuits are printed on one sheet of material so that the connectors are located on two opposite edges. Each circuit has a 1-in. connector with thinner than normal fingers. Find the unit estimate if the single sheet dimensions are 10 × 15 in. with connectors along the long edges. Nickel and gold deposition thickness are both 50 millionths of an inch. A total of 24–1000 unit orders are expected. Assume line speed = 1.8 ft/min.

Table Number	*Process Description*	*Table Time*	*Adjustment Factor*	*Cycle Minutes*	*Setup Hours*
19.5-1	Handling/tape press	.10	⅙	.017	
19.5-2	Start/stop rollers	.004	1/1000	0	
19.5-3	Run through press	.019	⅓	.008	
	Tape press subtotal			.025	
19.5-5	Handling/plating machine	.22	⅙	.034	
19.5-6	Wait for first board	6.21	1/1000	.006	
19.5-7	Over travel (throughput)	.78	⅓	.26	
19.5-8	Return to feed rack	.04	⅙	.007	
19.5-9	Check plating thickness	.02		.02	
19.5-10	Check adhesion strength	.01		.01	
	Plating machine subtotal			.337	
	Unit time	.337 min			

Note: The tape press subtotal is less than throughput so it is neglected. Line speed is 1.8 ft/sec.

TABLE 19.5 IN-LINE PLATING MACHINES (FOR PRINTED CIRCUIT BOARDS)

Setup **1.0 hr**

Operation elements in estimating minutes

1.	Handling/tape press	.10
2.	Start/stop press rollers	.004

3. Run one edge through press

Edge Length	6	9	12	15	18	21	24	27	30
Min	.009	.014	.019	.024	.028	.033	.037	.042	.047

4. Flip or rotate to new edge .015
   (Reuse Element 3 for additional edges)

5. Handling/plating machine .22

6. Wait for first PCB to travel through machine (10 feet)

Line Speed	1.0	1.2	1.4	1.6	1.8	2.0
Min	11.30	9.38	8.88	8.07	6.21	5.65

7. Over travel for machine clearance (throughput)

Ft/Min Line Speed	Plating Edge Length								
	6	9	12	15	18	21	24	27	30
1.0	.57	.85	1.13	1.41	1.70	1.98	2.36	2.54	2.82
1.2	.47	.70	.94	1.18	1.41	1.65	1.89	2.12	2.35
1.4	.41	.60	.80	1.00	1.21	1.41	1.61	1.82	2.01
1.6	.35	.53	.70	.88	1.06	1.23	1.41	1.59	1.76
1.8	.32	.47	.52	.78	.83	.94	1.25	1.41	1.57
2.0	.28	.43	.57	.70	.85	.98	1.13	1.27	1.41

8. Return to feed rack for additional edges (Reuse Element 7) .04

9. Carry PCB to bench and check plating thickness .02

10. Tape test gold adhesion strength .01

19.6 Electrical Discharge Machining

DESCRIPTION

The conventional electrical discharge machine (EDM) cuts metals by means of electrical discharge or "spark erosion" between the metal to be cut (negative charge) and an electrode (positive charge). This cutting takes place in a nonconductive fluid known as dielectric.

The process fills the tank with the dielectric fluid and submerges the metal inside connected to a negative charge. An electrode is chosen depending on the shape of the cut needed. It is positioned on the top of the workpiece leaving a small gap between.

After connecting the electrode to a positive charge, spark erosion takes place and it causes a "miniature thunderstorm" between the two metals. Flashes of lightning take place in rapid succession. Each one produces a tiny crater in the surface of the two metals. Metal evaporation occurs where the flash strikes.

Equal amounts of material are not removed from both plates. By an appropriate choice of materials (e.g. electrode of copper, workpiece of steel) and a skillful selection of the opening and closing times of the automatic switch, more material is removed from the steel than from the copper.

During the process the dielectric constantly flows through the tank requiring filtration. Also erosion creates heat, so the dielectric has to be cooled. The capacity of work of the conventional EDM machine is measured by the rate of material removal, cu in./min. The process is described by Figure 19.6.

ESTIMATING DATA DISCUSSION

The setup element is given in hours. The basic setup includes the time to get the first electrode, un-

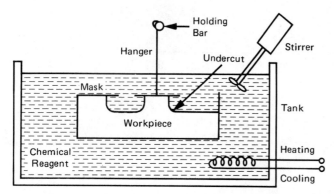

FIGURE 19.6 Electrical discharge machining.

load the old electrode, and load the new one. Also the setup includes the time required to get the fixture (which will hold the workpiece) to the tank, bolt it down, plus the time needed to place and indicate the fixture in two directions. Additionally, it includes the time to install the first workpiece on the fixture and fill the tank with the dielectric fluid.

The operation elements are given in estimating minutes and are arranged into handling and machining categories. The handling consists of loading and unloading the workpieces and changing the electrode. This time is constant for all operations.

The machining time includes filling the tank and the time for the spindle to go up and down. This time is also constant for all operations. The machining time also includes cutting time, which varies from one operation to another depending on the type of material and the volume of cut. Distinction is made between soft material (notably plain carbon steels) and hard materials (or alloys).

EXAMPLES

A. A .215-in. hole with a .375-in. length of cut is to be done in a low carbon steel. A 3/16-cu in. tube electrode is to be used. A six cu mm/min rate of removal is applied in this case. One operator will operate the machine. Find unit estimate.

Table Number	Process Description	Table Time	Adjustment Factor	Cycle Minutes	Setup Hours
19.6-1	Unload and load	.91		.91	
19.6-2	Fill tank to level	.52		.52	
19.6-3	Spindle down and up	.58		.58	
19.6-4A	Cut time, .014 cu in.	9.90		9.90	
19.6-5	Change electrode (every 10 pieces)	1.18	1/10	.12	
	Total			12.02	

B. A lot of 50 pieces of hard carbon steel is to be machined. A hole .197 in. with .235 in. of length is cut. One operator will operate the machine. Find the unit and lot estimates.

Table Number	Process Description	Table Time	Adjustment Factor	Cycle Minutes	Setup Hours
19.6-S	Setup	.30			.30
19.6-1	Unload and load	.91		.91	
19.6-2	Fill tank to level	.52		.52	
19.6-3	Spindle down and up	.58		.58	
19.6-4B	Cut time; .002 in.^{3}	5.2		5.20	
19.6-4	Change electrode (every 15 pieces)	1.18	1/15	.08	
	Total			7.29	.30
	Lot estimate	6.38			

C. A lot of 30 pieces of high carbon alloy steel is to be machined. A volume of .020 cu in. will be removed. Find the unit and lot estimate.

Table Number	Process Description	Table Time	Adjustment Factor	Cycle Minutes	Setup Hours
19.6-S	Setup	.30			.30
19.6-1	Unload and load	.91		.91	
19.6-2	Fill tank to level	.52		.52	
19.6-3	Cut time, .02 cu in.	18.80		18.80	
19.6-5	Change electrode (every 17 pieces)	1.18	1/17	.07	
	Total			20.88	.30
	Lot estimate	10.76 hr			

TABLE 19.6 ELECTRICAL DISCHARGE MACHINING

Setup .30 hr

Operation elements in estimating minutes

1. Unload and load .91

2. Fill tank to level .52

3. Spindle down and up .58

4. Cut time,

   A. Soft materials:

Vol. Cut	.003	.006	.009	.012	.015	.018	.021
Cut Time	3.1	4.8	6.5	8.2	9.9	11.6	13.3

   B. Hard material:

Vol. Cut	.003	.006	.009	.012	.015	.018	.021
Cut Time	5.2	7.5	9.7	12.0	14.2	16.5	18.8

5. Change electrodes 1.18

19.7 Traveling Wire Electrical Discharge Machining

DESCRIPTION

Traveling wire electrical discharge machining (EDM), a metal cutting process that removes metal with an electrical discharge, is suited for production of parts having extraordinary workpiece configurations, close tolerances, the need of high repeatability, and hard to work metals. Wire EDMing produces a variety of parts such as gears, tools, dies, rotors, turbine blades, etc. It is appropriate for small to medium batch quantities. Actual machining times may vary from a half hour to twenty hours. It uses the heat of an electrical spark to vaporize material, thus essentially no cutting forces are involved and parts can be machined with fragile or complex geometries. The sparks are generated one at a time in rapid succession (pulses) between the electrode (wire) and the workpiece. The sparks must have a medium in which to travel, thus a flushing fluid (water) is used to separate the wire and workpiece. Hence the one requirement is that the workpiece must be electrically conductive. A vertically oriented wire is fed into the workpiece continuously traveling from a supply spool to a take-up spool so that it is constantly renewed.

A power supply provides a voltage between wire and workpiece. By means of an adjustable setting one can determine the pulse amplitude and pulse duration—in other words, the on and off times (microseconds). On time refers to metal removal; off time is the period during which the gap is swept clear of removed metal via flushing. So both the intensity of the spark and the time it flows determine the energy expended and consequently the amount of material removed per unit time. Figure 19.7 describes the traveling-wire electrical discharge machining process.

ESTIMATING DATA DISCUSSION

In traveling wire EDM, setup for different parts remain essentially constant. The setup may be reduced when running consecutive batches of parts with similar geometric configurations but with different dimensions. The manual setup elements are constant. The setup values can be found in Table 19.7 and are given in hours. If machining is over 50 hours, add setup time to change wire spool. Similarly, a coolant flush is necessary for each 35 hours of machining.

The operational elements are listed as: handling, first cut constant, additional cut constant, multiple stack piece part run, and machining. Element 1, part handling, is to be allowed once per piece on a constant basis. The element stone and burr is adjusted for parts according to finish requirements and complexity. The value given is base requirement only per piece. Inspection is usually not included because it may be done internally to the long machining cycles, but will be added at least once per batch. Element 2, first cut constant, is allowed once per piece on the first cut made on the piece. Element 3, additional start cut constant, is to be used once for each additional start cut required. Do not use this element if only one cut is being made—i.e., one cut allow 0X, two cuts allow 1X, three cuts allow 2X. When trim cuts are required use only Element 3, one time per trim cut. Surface requirements determine the number of trims needed. The number of start cuts and trim cuts will be determined from the operational sketch. Element 4, multiple stack piece part run constant, is to be used once for each additional part in excess of one on a multiple basis—i.e., two pieces cut at one time allow 1X, three pieces allow 2X, etc.

The actual cutting cycle on traveling wire EDM is computer controlled and has a high degree of variability. Therefore the cutting cycle requires little operator attention time, which allows the operator to run additional machines. If this is the case, when costing the process divide the total cost by the number of machines attended. Wire EDMing is an unconventional machining process which has some peculiar characteristics. The cutting speed is measured as the length of cut times the workpiece thickness per unit time. Also involved is workpiece material and the physical properties of the wire electrode. Surprisingly, the thicker the workpiece the faster will be the cut in square surface area units per unit of time. This may be explained as follows. In EDM not every pulse generated by the electrode produces a spark; but, the longer length of wire electrode in a thicker workpiece provides more opportunities for the spark to occur, i.e., more sparks jump from electrode to workpiece. Hence, this makes the progress more efficient in thicker workpieces.

The machine time is to be calculated as length of cut per feed rate, where feed is a direct function of workpiece thickness, material, and electrode wire. Element 5 is used by determining the required wire electrode diameter and material being cut, as well as part dimensions, length, and depth of cut, from the design. Using the appropriate data, determine machine time in minutes per 1 inch of cut. Once the time per 1 inch has been determined, multiply by the length of cut and the total machine time is estimated. The values of Element 5 are of first rough-cut passes. If trim passes are required use a feed of ½ inch per minute, then divide the length of cut by this feed. This gives trim time and should be added to first cut time to obtain the total machine time. The estimator should note that the feeds used in determining these times are for maximum accuracy and not maximum speed.

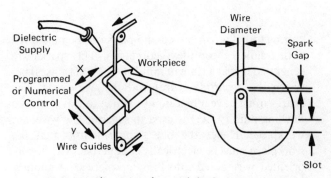

FIGURE 19.7 Traveling wire electrical discharge machining.

EXAMPLES

A. Estimate the time required to EDM a small die with a cut length of 6.3 in. The part has a nominal thickness of 1.8 in., and its material is cold work tool steel (D2). Surface finish requires only one rough cut. Inspection will be included. Wire used is 0.010-in. brass. Find unit estimate.

Table Number	Process Description	Table Time	Adjustment Factor	Cycle Minutes	Setup Hours
19.7-1	Supply piece, wash and blow dry	.84		.84	
19.7-1	Stone and burr piece	.84		.84	
19.7-1	Load and unload	1.00		1.00	
19.7-2	Punch data, wait for head to move to start position	.91		.91	
19.7-2	Thread, tie wire, close shield	.96		.96	
19.7-2	Punch data, wait for head to move down, start cycle	.42		.42	
19.7-5A	Machine time	28.14	6.3	177.28	
19.7-2	Raise shield, punch data, break wire	.56		.56	
19.7-1	After unloading, inspect	2.30		2.30	
	Total			185.11	

B. Estimate the time required to produce five progressive dies made of D2 tool steel. Three holes of ⅜-in. diameter are required. The thickness of the dies is 1.12 in. No finish cuts are needed. Wire use will be .010-in. brass. Setup and inspection are required. Length of cut = (⅜ in.) × 3 × 3.14 = 3.53 in.

Table Number	Process Description	Table Time	Adjustment Factor	Cycle Minutes	Setup Hours
19.7-S	Setup	1.40			1.40
19.7-1	Part handling	4.98		4.98	
19.7-2	First cut	2.85		2.85	
19.7-3	Additional start cut (2×)	2.36	2	4.72	
19.7-5A	Machine				
	Rough (1×)	15.75	3.53	55.60	
	Total hrs = (55.6)5/60 = 4.63 < 38	.00		0	___
	No need to change filters or wire spool				
	Total			68.15	1.40
	Hr/1 unit	1.14			
	Lot estimate	7.08 hr			

C. A lot of 110 aviation pump rotors are to be machined. Material is D2 tool steel quality, length of cut and depth of cut are 2.27 in. and .600 in. respectively, as determined from the design. Two slots with a surface finish requiring two trim cuts per slot are being machined. The slots have equal dimensions. Inspection is required and so is setup for lot. The wire will be .010-in. brass. Find the unit and lot estimates.

Table Number	Process Description	Table Time	Adjustment Factor	Cycle Minutes	Setup Hours
19.7-S	Setup	1.40			1.40
19.7-S	Change wire spool, 122.36/47 = 2.6, (3×)	.60			.60
19.7-S	Change coolant filters, 122.36/38 = 3.22, (3×)	1.64			1.64
19.7-1	Part handling	4.98		4.98	
19.7-2	First cut	2.85		2.85	
19.7-3	Additional start cut (1×)	2.36		2.36	

Traveling Wire Electrical Discharge Machining

Table Number	Process Description	Table Time	Adjustment Factor	Cycle Minutes	Setup Hours
19.7-3	Trim cut (4×)	.42	4	1.68	
19.7-5A	Machine				
	Rough cut (2×)				
	$T = 2.27$ in. $\times$ 10.70 = 24.29	24.29	2	48.58	
	Trim cut (4×)				
	$T = 2.27$ in./.5 in./min = 4.54	4.54	4	18.16	—
	Total hr = (48.58 + 18.16) × 110/60 = 122.36 hr > 50				
	Total			78.61	3.64
	Hr/100 units	131.28			
	Lot estimate	148.05 hr			

D. Estimate the time needed to produce 1000 carbide pump liners requiring four holes of .252-in. diameter and one slot with a length of .84 in. Thickness of each liner is .0160 in. They will be run on a multiple basis of 100. Wire will be .010-in. brass. Setup is required. Carbide is a hard material.

Table Number	Process Description	Table Time	Adjustment Factor	Cycle Minutes	Setup Hours
19.7-S	Setup	1.40			1.40
19.7-1	Handling (no inspection)	2.68	1/100	.0268	
19.7-2	First cut	2.85	1/100	.0285	
19.7-3	Additional start cut	4.72	2/100	.094	
19.7-5B	Machine, rough,	45.46	4.01/100	1.823	
	$T = 4.01 (45.46) = 182.30$, Total time = $(182.3)\ ^{10}\!/_{60} = 30.4$ hr				
	≈ 30 hr (No need to change filter)				
19.7-4	Multiple stack piece,				
	$T = .84 \times 100 = 84.0$	84.0	1/100	.84	—
	Total			2.812	1.40
	Floor-to-floor for 100 units	281.27 min			
	Unit estimate	2.81 min			
	Hr/100 units	4.693			
	Lot Estimate	48.33 hr			

TABLE 19.7 TRAVELING WIRE ELECTRICAL DISCHARGE MACHINING

Setup   1.40 hr
    Change wire spool   .30 hr
    Change coolant filters   .55 hr

Operation elements in estimating minutes

1. Part handling   .84
   Supply piece wash and blow dry   .84
   Stone and burr piece as required
       Rating factor: fine finish (×2.0)   1.00
   Load and unload to and from fixture   2.30
   Inspection   2.85

2. First cut start constant
   Punch data to move head to start position, wait for head to reposition   .91
   Thread, tie wire, close shield   .96
   Punch data for head to move down, wait, start cycle   .42
   Raise shield, punch data to move head, break wire   .56

Relocate part, unclamp and clamp		.47
Thread, tie wire, change weight, close shield		.91
Puch data for head to move down, wait, start cycle		.42
Note: Use for trim cut only.		

4. Multiple stack piece part run constant   1.68
    Supply piece wash and blow dry   .84
    Stone and burr piece   .84

5. Machining, min/in. (where in. is traverse length)

A.

Thickness In.	Material — Tool Steel		
	.008-in. B	.010-in. B	.012-in. B
.083	7.08	6.39	7.73
.125	7.39	6.60	7.90
.25	8.42	7.27	8.42
.50	10.91	8.82	9.57
.75	14.13	10.70	10.88
1.00	18.31	12.98	12.37
1.25	23.73	15.75	14.07
1.50	30.76	19.11	15.99
1.75	—	23.19	18.18
2.00	—	28.14	20.67
2.25	—	34.14	23.50
2.50	—	41.43	26.72
2.75	—	—	30.38
3.00	—	—	34.53

B.

Thickness In.	Material		
	Carbide .010-in. B	Graphite .010-in. B	Copper .010-in. B
.083	—	—	4.03
.125	19.75	7.69	4.30
.25	21.05	8.76	5.23
.50	23.94	11.35	7.75
.75	27.21	14.71	11.48
1.00	30.94	19.06	17.01
1.25	35.17	24.70	25.20
1.50	39.98	—	—
1.75	45.46	—	—
2.00	51.68	—	—
2.25	58.75	—	—
2.50	66.80	—	—
2.75	75.94	—	—
3.00	86.33	—	—

FINISHING

20.1 Mask and Unmask Bench

DISCUSSION

Masking is used to protect surfaces against overspray, for lining, and for cutting-in different paints or colors. Large areas can be protected by paper secured with self-adhesive tape or with a gummed strip. For cutting-in or lining with spray-paint operations, thin cellulose tape can be positioned over marking lines, against edges or other parts to reduce the tendency to build up a thick edge. Trimming fixtures are sometimes attached to the part, and a tape is added to reduce runs onto the part.

Two operations are involved as masking precedes painting while unmasking is done following painting. The masking considered in these estimating data is "bench" work, although masking can be by spraying or dipping processes as well.

ESTIMATING DATA DISCUSSION

Adhesive masking tape and brown Kraft paper are assumed to be available near the masking station, and thus the setup is for typical lot chores.

The position and aside is for flat or boxlike parts. Because there are repositions during the taping time, consideration for reposition is unnecessary. The process of taping causes the part to be repositioned. The $L + W$ or $L + W + H$ are part dimensions.

Element 2 is for the application of paper and tape for flatlike parts. The entry variables are length and width of the masked area, not the part, and the time is per side.

If the part is boxlike, and paper and tape are substantially over three or more of the six surfaces, Element 3 is used. The entry variable is $2L + 2W$, which is the periphery and the height. Thus, for one unit of $2L + 2W$, .15 min can be added to, or subtracted from, table values. Similarly, for one unit length of height, .06 min is added to, or subtracted from, table values.

If tape is exclusively used, Element 4 is chosen. The width across the top is tape width and lengths are under each dimension. This element is for taping an area, and tolerance requirements are not critical.

When a trimming fixture is used to reduce taping costs, it may be necessary to use Element 5. The entry variable of length is the laid-in tape distance. It secures the part to the fixture.

For dimensions that are closely held, and trim work with a knife or razor blade is anticipated, and the tape is worked-in along the edge, Element 6 is employed. The entry variable is length, meaning tape length.

Spot cover or plugs can be estimated using Element 7.

Masking material, if it is to be brushed, is estimated using Element 8. The area is the brushed-masked area, not part area. The length is the mask-edge length. For instance, if a circular section surrounded by another color is to be masked, then the edge length is the circumference. If any of the mask is against a corner, another part, or braked-edge, then the edge length does not include that distance, because operator-care with brush strokes is not as critical.

For the unmasking operation, Elements 1, 9, 10, or 11 would be selected as needed. The entry variable for 9 is length plus width for the masked material.

EXAMPLES

A. A flat part 7.75 × 21.2-in. overall has a 6 × 11.5-in. section masked. Estimate the unit time.

Table Number	Process Description	Table Time	Adjustment Factor	Cycle Minutes	Setup Hours
20.1-1A	Position and aside	.24		.24	
20.1-2	Apply paper and tape, $1.96 - (11.5 - 9).19 - (9 - 6).11$	1.16		1.16	
	Total			1.40	

B. A boxlike part is to be wrapped for overspray protection. Two plugs in taped holes are necessary. The handling dimensions are $L = 6.5$ in., $W = 7.2$ in., and $H = 2$ in. One surface, the top, is to be sprayed. Find the pc/hr.

Table Number	Process Description	Table Time	Adjustment Factor	Cycle Minutes	Setup Hours
20.1-1B	Handle, $L + W + H = 16.2$	.09		.09	
20.1-3	$2L + 2W = 27.4, H = 2$	4.28		4.28	
20.1-7B	Two plugs	.14	2	.26	
	Total			4.63	
	Shop estimate	12.9 pc/hr			

C. A critical machine part has several dimensions that are taped, and an internal area that is liquid masked. Use dimensions from Example B above. The tape is 2 and 4 in. × 7.5-in. long, and one tape is aligned, trimmed, and placed along a circular feature. Estimate the unit time.

Table Number	Process Description	Table Time	Adjustment Factor	Cycle Minutes	Setup Hours
20.1-1B	Handle	.09		.09	
20.1-4	Tape area 2-in. tape × 7.5-in. long	.20		.20	
20.1-6	Length of exact edge is about 15 in., $.57 + (15 - 13.6).05$	.64		.64	
20.1-4	Tape area 4 in. × 7.5-in. long	.39		.39	
20.1-8	Internal area of 17 in. and an edge length of 5.4 in., $4.35 + (5.4 - 4).11 + (17 - 11).28$	6.18		6.18	
				7.50	

Total

D. Estimate the unit time to unmask the part described in Example B above.

Table Number	Process Description	Table Time	Adjustment Factor	Cycle Minutes	Setup Hours
20.1-1B	Handle	.09		.09	
20.1-10B	Remove plugs	.38		.38	
20.1-9	Remove tape, approximately $30 = L + W$	1.01		1.01	
	Total			1.48	

TABLE 20.1 MASK AND UNMASK BENCH

Setup .05 hr

Operation elements in estimating minutes

1. Position and aside, occasional reposition

    A. Flat parts

L + W	9.0	22.7	40.7	64.6	96.2	138	Add'l
Min	.14	.18	.24	.31	.42	.55	.003

    B. Box parts

L + W + H	16.5	20.3	25.9	33.9	45.4	61.9	85.8	120.2	Add'l
Min	.09	.13	.18	.27	.38	.55	.79	1.14	.010

2. Apply paper and tape to flat parts.

Length (L)	Width (W) 9	12	18	24	36
9	1.96				
12	2.49	2.85			
18	3.58	3.95	4.69		
24	4.66	5.03	5.77	6.51	
36	6.83	7.19	7.93	7.93	10.15
48	8.98	9.35	10.10	10.84	12.32
72	13.31	13.68	14.42	15.16	16.64

    Add'l length   .19
    Add'l width   .11

3. Apply paper and tape to box parts

2L + 2W	Height (H) 5	10	15	25
10	1.27	1.56		
20	2.78	3.06	3.34	
30	4.28	4.56	4.86	5.42
40	5.78	6.08	6.36	6.94

    Add'l 2L + 2W   .15
    Add'l H   .06

4. Tape area

Length (L)	Width (W) 2	4	6	8	Min
	6.0				.15
	6.4	4.7			.16
	6.9	4.9			.18
	7.4	5.2	4.2		.20
	8.0	5.5	4.4		.22
	8.6	5.8	4.6		.24

	Width (W)				
	2	4	6	8	Min
Length (L)	9.3	6.1	4.9		.26
	10.1	6.5	5.1	4.9	.29
	10.9	6.9	5.4	5.1	.32
	11.9	7.4	5.7	5.3	.35
	12.9	7.9	6.9	5.6	.39
	14.0	8.4	6.4	5.8	.43
	15.3	9.0	6.8	6.1	.47
	16.6	9.7	7.3	6.5	.52
	19.6	11.2	8.2	7.2	.62
	23.2	12.9	9.4	8.1	.74
	27.5	15.0	10.8	9.1	.89
	32.7	17.6	12.5	10.4	1.07
	38.9	20.6	14.5	11.9	1.28
	46.3	24.3	17.0	13.7	1.54
	55.2	28.7	19.9	15.3	1.85
	Add'l				.03
		Add'l			.07
			Add'l		.11
				Add'l	.14

5. Tape and trim edges for trimming fixture

L	1.0	5.6	10.6	16.2	22.3	29.0	36.3	44.5	Add'l
Min	.65	.72	.79	.87	.96	1.05	1.16	1.27	.015

6. Align tape along edge, trim

L	4.0	4.5	5.0	5.7	6.5	7.4	8.6	10.0	11.6	13.6	Add'l
Min	.11	.13	.16	1.9	.23	.28	.33	.40	4.8	.57	.05

7. Apply spots or plugs

A.
Spots	Min	Plugs	Min
1	.14	1	.32
2	.26	2	.37
3	.38	3	.41
4	.50	4	.46
5	.62	5	.51
6	.74	6	.56
7	.86	7	.60

B.
Spots	Min	Plugs	Min
8	.98	8	.65
9	1.10	9	.70
10	1.22	10	.75
11	1.34	11	.79
12	1.46	12	.84
13	1.58	13	.89
Add'l	.13	Add'l	.05

8. Handbrushing of liquid mask or adherent coat

Length	Area						
	1	2	3	4	6	8	11
0	1.13	1.40	1.69	1.96	2.51	3.07	3.90
.5	1.18	1.46	1.74	2.01	2.57	3.13	3.96
1.5	1.29	1.57	1.85	2.13	2.68	3.24	4.07
3.5	1.40	1.79	2.07	2.35	2.90	3.46	4.29
4	1.57	1.85	2.13	2.41	2.96	3.52	4.35
6.5	1.85	2.13	2.41	2.68	3.24	3.80	4.63
12	2.47	2.74	3.02	3.30	3.85	4.41	5.24
20	3.36	3.64	3.51	4.19	4.75	5.30	6.13
28	4.25	4.53	4.81	5.08	5.64	6.19	4.25
35	5.03	5.31	5.59	5.87	6.42	6.98	7.81

Add'l length .11
Add'l area .28

9. Remove paper and tape or tape from area

L + W	2.0	2.8	3.8	5.0	6.5	8.2	10.3	12.8	15.8	19.4	23.7
Min	.11	.14	.16	.20	.23	.28	.34	.41	.49	.58	.70
L + W	28.9	35.1	42.5	51.4	62.2	75.0	90.5	109.0	131.3	157.9	Add'l
Min	.84	1.01	1.21	1.45	1.74	2.09	2.51	3.02	3.62	4.34	.03

10. Remove spots or plugs

A.
Spots	1	2	3	4	5	6	7	8	9	10	Add'l
Min	.09	.18	.27	.36	.45	.54	.63	.72	.81	.90	.10

B.
Plugs	1	2	3	4	5	6	7	8	9	10	Add'l
Min	.19	.38	.57	.76	.95	1.14	1.33	1.52	1.71	1.90	.21

11. Strip liquid or inherent coating mask .02/sq in

20.2 Booth, Conveyor and Dip Painting, and Conversion Processes

DESCRIPTION

For industrial painting operations, spray painting is a commonly used method. In spray painting, paint is forced into the spray gun by pressure or suction and then atomized at the discharge end or nozzle of the gun. A spray-painting system may include a compressed air supply, pistol-type spray gun, air and fluid lines, material container, stirrers or agitators, pressure regulators, air filters, and a spray booth or conveyor—and clean-air exhaust. Baking in an oven is more common than air drying.

One system is shown in Figure 20.2A, which shows two operators spraying a conveyorized panel.

Electrostatic spraying processes use the attraction of paint droplets by an electrode in an electrostatic

FIGURE 20.2A Industrial spray operation. *(DeVilbiss)*

field. The material can be applied using a conveyor, with and without manual application, or by dipping. The airless-type spraying processes coat articles that are carried about a loop in a conveyor, at the center of which is located a disk-type atomizer. Paint is pumped to the disk and centrifugally released from the periphery as a spray of fine particles that carry the same charge as the disk, but are attracted to the parts. The disk also has reciprocal motion in order to spray large surfaces. Figure 20.2B illustrates this disk.

Dip coating is a means to cover all areas, and is normally confined to smaller parts.

The application of paint may be fully manual, semi-, or fully automatic, although some kind of part loading and unloading is usually found. Brush painting is not considered in this guide.

Several types of finishing materials are provided. Provision is given for conversion coatings or prime coat. Wrinkle, enamel, and plastic resins are evaluated. Additionally, other materials can be used with a conveyor, booth, or dipping methods. Liquid masks which can be sprayed are another finish considered.

ESTIMATING DATA DISCUSSION

Several kinds of operations have been grouped within these data. It depends upon the local conditions that the estimator is assuming. Conversion and painting can be processed by conveyor, dip, booth, or bench methods. One to several operators may be necessary, and the work may be operator-controlled, semiautomatic, or fully automatic.

For handling, a variety of circumstances dictate which element is preferred. For the ordinary load and

FIGURE 20.2B Electrostatic coating disk. *(Ransburg Corporation)*

unload, a part to be finished, say on a turntable or stand, or in a booth, Element 1 with the entry variable of girth is chosen. The booth may have one or two doors and be a walk-in, or in the smaller situation, a bench booth is possible. Power hoist, dolly, or fork truck are other means of handling when a load and unload of a single part is implied.

Rack or tray parts handling, Element 2, is provided for a variety of purposes. The data are for a typical rack or tray 15 × 35 in. in size. Flat parts (on one surface at least) are laid on the tray spaced with minimum spacing. Racks may have hanging hooks and may vary from 1 to 150 hooks. The entry variable is the number of pieces placed on the tray or rack. Tray loading is for parts that are automatically or operator sprayed. Racks with hooks can be used for dipping as well as operator spraying. Automatic painting of rack-held parts is another possibility. The five conditions for Element 2 are described by the column headings.

Element 3 deals with basket parts, where a typical basket is 18 × 14 × 18 in. in size and is perforated. The basket can be hand-dipped, or be dipped by automatic means. The element includes no painting—only handling. While the volume of this basket is 2.5 cu ft., conversions to other sizes can be on the basis of parts per basket, as the only variable is number of parts per basket and the type of loading and unloading. Stacking in a basket is vertical while stacking out of the basket can be horizontal.

The use of single-strand, endless-chain conveyors if found frequently in moderate- to high-production volume shops. Even job shops use this method if the part family is consistent in size. The part is often hooked to the chain by operators, then sprayed or dipped and thru-oven baked (or air dried) and unhooked. Roller coating processes can use these estimating data as they are conveyor controlled. The table entry variables for Element 4 are part spacing center-to-center, expressed in in., and conveyor velocity, expresed in feet per minute (fpm). Element 4 time is for one operator. The selected time must be multiplied by the number of operators to man the conveyor for the part being estimated no matter what the lot size. Consider the example where the line velocity is 8 fpm and the chain has hook eyes every 2 ft. A flat part 20 × 30 in. (with holes in the four edges) can be hooked with the 30-in. dimension horizontally and requires two eyes for two hooks and a free chain-eye in between parts for spray control. In this case the part spacing would be 6 ft. The part could also be hung with the 20-in. dimension parallel to the chain axis, but two chain-eyes are still required for the hooks. Because the long 30-in. dimension is the vertical dimension, open distance between parts is adequate, and the part-spacing center distance is 4 ft.

At 8 fpm and a spacing of 6 ft, the time value is .75 min, but the crew necessary to support this part at this velocity, for example, are four spray painting operators, two hangers, and one relief or swing operator. Thus, the .75 min is multiplied by 7 for an estimate giving 5.25 min/each or .088 hr/unit. Furthermore, if the conveyor is used for part dipping, spraying operators would be unnecessary. Local shop circumstances dictate how Element 4 is to be used. The times in Element 4 can be added or subtracted to give different conveyor velocities.

In spray-booth painting, Element 5, the equipment essentials are the air supply, spray gun, spraying technique, booth, and the paint. Element 5 involves manual cold spraying with a single gun and the booth may be on a bench or in a room with either a dry back or water wash for ventilation. The data are for priming, wrinkle, enamel, and spray mask. Auxiliary work such as wipe clean, say with a dustless rag, or blow off the article are also given because this work is sometimes done prior or subsequent to spraying. The estimator may wish to make a qualitative extension of the data if the booth involves automatic spraying and the operator is not involved or if the operator is a tender in the case of semiautomatic spraying. Entry data are total area to be sprayed (which means the sum of inside and outside areas if both sides are covered). This work is for one piece at a time.

Parts can be rack-hung, dipped, and suspended for baking. After drying, the parts are packed or dumped into containers. A typical hand rack is 12-in. long, has five hooks 3-in. apart for a maximum of 5 parts per rack and 15 per cycle, and the operator is sometimes able to manage three racks at once (note Element 6).

Using parts that have been racked and placed on the drying truck, Element 7 is for painting and replacing the rack on the truck, which is followed by air or oven dry and eventually a pack for protection. Air-pressure or electrostatic methods of spraying are possible, but operators are assumed.

For Element 7, rack spraying using a drying truck with oven or an air dry, part handling was done in Element 2, and Element 7 considers operator spraying where the entry variable is the number of pieces loaded on the rack. Three conditions are specified.

If the rack is conveyorized to the oven, Element 8 is used. Handling to load and unload the rack is given in Element 2.

Element 9 deals with the filling of engraving which involves cleaning, painting, removing excess paint, and touchup. The work is done on a bench, and the entry variable for the element is the length of area, where area is a 2-in. wide path. The area encompasses the engraved characters. In the case of a dial, it equals the mean circumference of the engraved characters times width. A single isolated letter or a small word requires the threshold time for 1 in. Handling, such as Element 1, must be added to Element 9.

Element 10 involves the unpacking and packing of small items, and is abbreviated from Table 24.1.

EXAMPLES

A. A part is painted white on a 2-ft separated-hook chain-eye conveyor. Two parts are hooked on each eye, the second part being hooked to a lower hole in the upper part. Conveyor velocity is 6 fpm to facilitate oven drying. Five operators will man the line. Estimate the unit time.

Table Number	Process Description	Table Time	Adjustment Factor	Cycle Minutes	Setup Hours
20.2-4	Line load, spray, unload, at line speed of 6 fpm and part spacing of 2 ft	.33	½ × 5	.83	
	Total			.83	

B. A plexiglass object, 2 × 3 in. in size, is to have the letters *A* and *B* and the word *MODE,* and a 2-in. line filled. The letters are in different spots and are isolated. The plexiglass has a protective strip on both sides, and one strip must be removed. Find the unit time. There are 183 units in the lot.

Table Number	Process Description	Table Time	Adjustment Factor	Cycle Minutes	Setup Hours
20.2-S	Setup for fill engraving operation	.05			.05
20.2-1A	Load and unload part	.08		.08	
20.2-10	Strip masking from top surface	.14		.14	
20.2-9	Fill letter *A*	.30		.30	
20.2-9	Fill letter *B*	.30		.30	
20.2-9	Fill word *MODE*, ½-in. length	.30		.30	
20.2-9	Fill 2-in. hairline	.47		.47	
	Total			1.59	.05
	Lot estimate	4.90 hr			

C. A top chassis cover is to receive automatic electrostatic spray. Hooks are 18-in. apart. The part designed hook supports two parts. Conveyor velocity is 6 fpm, and two operators tend the conveyor line. Determine the unit estimate and hr/100 units.

Table Number	Process Description	Table Time	Adjustment Factor	Cycle Minutes	Setup Hours
20.2-4	Load and unload part, two operators at conveyor velocity 6 fpm, 18-in. spacing, 2(.17 + .01 × 6)	.46		.46	
20.2-4	Spray	0		0	
	Total			.46	
	Hr/100 units	.77			

D. As a preliminary to painting for protection from corrosion and to aid the bond between metal and the paint coating, a zinc phosphate coating is planned. The application is to be applied by immersion methods. The parts are to be racked, dipped manually, and unloaded. The operator, with racks, can handle 10 parts. Find the unit estimate.

Table Number	Process Description	Table Time	Adjustment Factor	Cycle Minutes	Setup Hours
20.2-S	Setup racks for existing dip-line	.20			.20
20.2-6	Rack parts, 10 units, and dip	.09		.09	
20.2-6	Unload and pack (or stack)	.09		.09	
	Total			.18	

E. A welded steel base is to be zinc chromate sprayed in a booth. The part is to be loaded with a fork truck and placed upon a pedestal for turning. Following the spray work, it is loaded into an oven and subsequently hand sprayed for enamel. Provide the unit time for these two operations. Spray area total = 1721 sq in.

Table Number	Process Description	Table Time	Adjustment Factor	Cycle Minutes	Setup Hours
20.2-S	Setup racks for existing dip-line	.2			.2
20.2-S	Setup second operation	.2			.2
20.2-1A	Load and unload weldment in and out of booth from general area and finally to oven, two operations	3.00	2	6.00	
20.2-5	Wipe clean, 1721 sq in., two operations	.47	2	.94	
20.2-5	Blow clean, two operations	.17	2	.34	
20.2-5	Conversion coating	.82		.82	
20.2-5	Enamel	3.65		3.65	
	Total			11.75	.4

F. A part is to be painted to a military specification color. Notice Figure 6.4C, which is complete except for painting and protection. A conveyor is to be used for 7075 parts and a two-man team is necessary. The adjustment factor provides for the second operator since the data tables are for one operator. Find the estimates.

Table Number	Process Description	Table Time	Adjustment Factor	Cycle Minutes	Setup Hours
20.2-S	Spray paint on conveyor; setup	0.00			0.00
20.2-4	Conveyor; two-man conveyor load, unload, spray 8 fpm speed 12-in. spacing	0.13	2	0.26	
	Total			0.26	0.00
	Lot estimate	30.66 hr			

TABLE 20.2 BOOTH, CONVEYOR AND DIP PAINTING, AND CONVERSION PROCESSES

Setup

Conveyor = 8 ft/min	0 hr
Dip	.20 hr
Booth	.20 hr
Bench	.05 hr
Fill engraving	.05 hr

Operation elements in estimating minutes

1. Handle part to spray booth or bench

    A.

L + W + H	Min	L + W + H	Min	L + W + H	Min
15.8	.08	25.0	.17	33.1	.37
16.6	.09	26.8	.19	35.6	.41
17.5	.10	28.7	.21	38.4	.45
18.5	.11	30.8	.23	41.5	.50
19.5	.12	33.1	.25	44.9	.54
20.7	.13	35.6	.28	48.6	.60
22.0	.14	38.4	.31	52.7	.66
23.5	.16	41.5	.34	Add'l	.01

    B. Load and unload part to spray booth or bench (con't)

Power hoist, hand traverse	1.08
Dolly	.58
Fork truck	3.00

2. Load rack or tray parts

No. Pc	Rack, Place Rack on Dolly	Rack, Place Rack on Conveyor	Turnover	Unload	Unload, Paper Wrap, Pack
1	.30	.27	.07		.60
2	.15	.14	.06		.39
3	.13	.11	.06		.30
4	.11	.10	.05		.27
5	.10	.09	.04	.06	.23
8	.08	.07	.04	.05	.18

No. Pc	Rack, Place Rack on Dolly	Rack, Place Rack on Conveyor	Turnover	Unload	Unload, Paper Wrap, Pack
10	.07	.06	.04	.05	.17
12	.06	.05	.03	.05	.15
15	.05	.05	.02	.05	.13
20	.05	.04	.02	.04	.11
25	.04	.04	.01	.04	.10
30	.04	.03	.01	.03	
50	.04	.03	.01	.02	
75	.03	.03	.01	.01	
100	.03	.03	.01	.01	
150	.03	.02	.01	.01	
200	.03	.02	.01	.01	

3. Load basket parts

No. Pc	Drop Parts In and Out of Basket	Stack Parts In Basket, Dump Out	Stack Parts In Basket, Stack Out
2		.72	.72
5		.27	.27
8		.15	.16
10		.14	.15
12		.12	.14
15		.09	.12
20		.08	.11
25		.07	.11
30	.05	.07	.10
40	.04	.06	.10
50	.03	.05	.09
75	.02	.05	.07
100	.02	.04	.06
150	.02	.04	.05
200	.02	.03	
250	.01	.03	
300	.01		
400	.01		

SPLITTING   "SPLIT"

4. Conveyor-line loading, spraying, unloading, min/pc for 1 operator[1]

| Conveyor Speed, fpm | Part Spacing Centers, In. | | | | | | | | | Add'l |
	3	5	7	12	24	36	48	72	120	
2	.13	.21	.29	.50	1.00	1.50	2.00	3.00	5.00	.04
3	.08	.14	.19	.33	.67	1.00	1.33	2.00	3.33	.03
4	.06	.10	.15	.25	.50	.75	1.00	1.50	2.50	.02
5	.05	.08	.12	.20	.40	.60	.80	1.20	2.00	.02
6	.04	.07	.10	.16	.33	.50	.67	1.00	1.67	.01
7	.04	.06	.08	.14	.29	.43	.57	.86	1.43	.01
8	.03	.05	.07	.13	(.25)	.38	.50	.75	1.25	.01
9	.03	.05	.06	.11	.22	.33	.44	.67	1.11	.009
10	.03	.04	.06	.10	.20	.30	.40	.60	1.00	.009

Conveyor Speed, fpm	Part Spacing Centers, In.									
	3	5	7	12	24	36	48	72	120	Add'l
11	.02	.04	.05	.09	.18	.27	.36	.55	.91	.008
12	.02	.03	.05	.08	.17	.25	.33	.50	.83	.007
13	.02	.03	.04	.08	.15	.23	.31	.46	.77	.007
14	.018	.03	.04	.07	.14	.21	.29	.43	.71	.006
15	.017	.03	.04	.07	.13	.20	.27	.40	.67	.006
16	.016	.03	.04	.06	.13	.19	.25	.38	.63	.005
Add'l	.001	.002	.002	.004	.007	.01	.01	.02	.04	

[1] Multiply by number of operators on the line.

5. Prime, paint, or finish in booth

Area, Sq In.	Prime or Conversion Coating	Wrinkle	Enamel	Epoxide Resin	Mask	Wipe Clean	Blow Clean
24	.12	.34	.15	.53	.04		
32	.14	.37	.20	.55	.05		
50	.16	.40	.25	.57	.07		
58	.17	.41	.28	.58	.08	.13	
69	.18	.44	.32	.59	.10	.14	
83	.20	.47	.36	.61	.12	.14	
100	.22	.51	.42	.63	.14	.14	
115	.23	.55	.46	65	.17	.14	
132	.25	.59	.51	.67	.19	.15	.04
152	.26	.63	.57	.70	.22	.15	.05
175	.28	.69	.63	.73	.25	.16	.05
201	.30	.75	.70	.76	.29	.16	.05
231	.32	.82	.77	.80	.33	.17	.05
266	.34	.90	.86	.84	.38	.17	.05
306	.36	.99	.95	.89	.44	.18	.06
352	.38	1.10	1.06	.95	.50	.19	.06
405	.41	1.22	1.17	1.02	.58	.20	.06
465	.44	1.36	1.30	1.10	.67	.21	.07
535	.46	1.53	1.44	1.19	.77	.22	.07
615	.49	1.71	1.60	1.29	.88	.24	.08
708	.53	1.93	1.77	1.41	1.01	.25	.09
814	.56	2.18	1.97	1.54	1.17	.27	.09
937	.60	2.46	2.18	1.70		.29	.10
1078	.64	2.79	2.42	1.88		.32	.11
1240	.68	3.17	2.69			.35	.12
1428	.72	3.61	2.98			.38	.14
1640	.77	4.10	3.30			.42	.15
1800	.82	4.66	3.65			.47	.17
2170	.87	5.34	4.06			.52	.19
2500	.93	6.10	4.51			.58	.21
Add'l	.0002	.0023	.0017	.0013	.0014	.0002	.0001

6. Rack and hand dip, manual

No. Pc	Rack and Dip	Unload and Pack	Dump
1	.54	.36	
2	.28	.20	
3	.21	.12	.02
5	.12	.12	.02
6	.12	.09	.02
10	.09	.09	.02
15	.08	.09	.02

7. Rack spraying with drying dolly, oven or air dry

No. Pc	Spray One Side Flat Parts	Turn Part and Spray Other Side	Spray Inside, Turn Over, Spray Outside
1	.48	.90	1.08
2	.26	.48	.57
3	.18	.30	.36
4	.11	.25	.27
5	.08	.22	.25
8	.06	.18	.20
10	.05	.15	.17
12	.05	.13	.16
15	.05	.10	.14
20	.04	.07	.11
25	.03	.06	.09
30	.03	.05	.08
50	.02		
75	.01		
100	.01		
150	.01		
200	.01		

8. Rack spraying with rack loaded to conveyor and oven

No. Pc	Spray One Side Flat Parts	Turn Part and Spray Other Side	Spray Inside, Turn Over, Spray Outside
1	.42	.84	.96
2	.23	.45	.51
3	.14	.30	.34
4	.10	.21	.30
5	.07	.18	.27
8	.05	.13	.19
10	.05	.11	.16
12	.04	.09	.13
15	.04	.08	.11
20	.03	.06	.09
25	.03	.05	.07
30	.03	.05	.06

No. Pc	Spray One Side Flat Parts	Turn Part and Spray Other Side	Spray Inside, Turn Over, Spray Outside
50	.02	.04	.05
75	.02		
100	.01		
150	.01		
200	.01		

9. Fill engraving

L of Area	Min	L of Area	Min	L of Area	Min
Word or					
1.0	.30	3.2	1.16	7.1	3.47
2.0	.47	3.6	1.40	8.3	4.17
2.2	.56	4.1	1.67	9.7	5.00
2.3	.67	4.6	2.01	11.4	6.00
2.6	.81	5.3	2.41	13.4	7.20
2.9	.97	6.1	2.89	Add'l	.59

10. Pack and unpack
    Box ..... .12
    Envelope ..... .18
    Brown paper ..... .18
    Strip 1 pc of mask ..... .14

20.3 Metal Chemical Cleaning Processes

DESCRIPTION

Chemical cleaning depends upon a solvent and/or chemical action between the basis material and the dirt, oil, grease, etc. The cleaning solvents considered here are the chlorinated solvents or the vapor degreasers. The cleaning media may be trichlorethylene, perchlorethylene, and methylene chloride. The chemical actions are the bright dip and the chromatic dips. In the bright dip for copper and copper alloys, the medium is sulphuric nitric acid. For aluminum materials, the medium is chromate. While not a chemical cleaning method, estimates for steam cleaning are provided.

The means of production can be manual or automatic, in tanks, submerged or above floor, or in small containers. The parts are lowered into a tank in which the solvent has been heated to its boiling point causing the solvent to vaporize. As the hot vapors meet the cold parts, the vapors condense and dissolve the dirt. The basic degreasing methods of vapor, liquid vapor, spray vapor, and ultrasonic liquid vapor can be utilized with a manually-generated or conveyor system degreaser. Figure 20.3 shows a manually-operated model.

ESTIMATING DATA DISCUSSION

The data are consolidated, that is, each type of chemical cleaning has the required handling, tank-to-tank movement, and delay for the cleaning. Entry variables are girth. For flat parts, girth is usually taken to mean length plus width ($L + W$). For cylindrical parts, girth is defined to the maximum diameter times two plus length (max diameter $\times 2 + L$). Box-type parts are computed as the maximum sum of the dimensions in the X-Y-Z planes.

Degreasing elements are separated on the basis of single or multiple-item degreasing. For bright and chromate dip, the part configuration, whether it is

FIGURE 20.3 Gas-heated 8 × 8 × 8-ft vapor degreaser. *(Phillips Manufacturing Company)*

more flat-like, box-like, or cylindrical-like, is another governing entry variable.

Element 5 deals with the process time for steam cleaning using a direct nozzle against a rack, or individual part or basket of parts. A typical rack would have six hooks. The quantity of parts loaded into the basket must be determined by the estimator. Other loading estimates can be found in other tables.

Basket, barrel, tank or pan distinctions are not provided for these data.

EXAMPLES

A. A part has a 1.5-in. dia and is 4-in. long. A chromatic dip for this aluminum material is planned for appearance and protection sake. Find the unit estimate.

Table Number	Process Description	Table Time	Adjustment Factor	Cycle Minutes	Setup Hours
20.3-4	$L + W + H = 4 + 2 \times 1.5 = 7$	.16		.16	
	Total			.16	

B. Estimate the process time to organically coat a magnesium die cast product. A figure of the die cast product is given as Figure 2.9C. There are 500 parts in the lot. The data supplied in this section does not entirely meet the magnesium specification, but are adopted because they are the nearest set of information.

Table Number	Process Description	Table Time	Adjustment Factor	Cycle Minutes	Setup Hours
20.3-S	Setup	0.00			0.00
20.3-3	Bright dip as box of $L + W + H = 12.97$ in. Bright dip is approx	0.39		0.39	
	Total			0.39	0.00
	Lot estimate	3.25 hr			

C. Estimate Figure 3.2B for a degreasing operation. This sheet metal part is finished except for cleaning. The lot of 2500 arrives jumbled, and will be degreased in a basket. The time driver is given as $L + W + H = 4.062$ from the blank size of $1.562 \times 1.75 \times .750$ in.

Table Number	Process Description	Table Time	Adjustment Factor	Cycle Minutes	Setup Hours
18.7-S	Setup for no media change	0.10			0.10
18.7-1	Abrasive deburr for handle only as degrease follows	0.08		0.08	—
	Total			0.08	0.10
	Total lot estimate	3.43 hr			

D. A part, box-shaped, with $L + W + H = 19.3$ in., and can be multiple degreased. Find the shop estimate.

Table Number	Process Description	Table Time	Adjustment Factor	Cycle Minutes	Setup Hours
20.3-2	For $L + W + H = 19.3$	.055		.055	
	Total			.055	
	Shop estimate	1090 pc/hr			

TABLE 20.3 METAL CHEMICAL CLEANING PROCESSES

Setup   0 hr

Operation elements in estimating minutes

1. Degrease one part

$L + W + H$	Min	$L + W + H$	Min	$L + W + H$	Min
.2	.01	11	.14	53	.37
.5	.02	13	.15	59	.39
.8	.03	16	.17	63	.41
1.5	.04	18	.19	66	.42
2.5	.05	21	.20	70	.44
3.1	.06	22	.21	75	.46
3.7	.07	25	.23	80	.48
4.4	.08	30	.25	85	.49
5.5	.09	31	.26	95	.53
6.7	.10	37	.29	100	.55
9.5	.12	40	.31	115	.60

2. Degrease many parts simultaneously

$L + W + H$	Min	$L + W + H$	Min	$L + W + H$	Min
.2	.0003	11	.03	53	.15
.5	.0009	13	.035	59	.17
.8	.0015	16	.04	63	.18
1.5	.003	18	.05	66	.19
2.5	.005	21	.055	70	.21
3.1	.007	22	.06	75	.22
3.7	.008	25	.07	80	.24
4.4	.01	30	.08	85	.26
5.5	.015	31	.085	95	.29
6.7	.02	37	.10	100	.31
9.5	.025	40	.11	115	.34

3. Bright dip

Flat	L + W + H		Min	Flat	L + W + H		Min
	Box	Cylindrical			Box	Cylindrical	
6.5	2.3	4.0	.006	16	6.5	9.2	.16
6.7	2.4	4.1	.008	18	7.4	10.5	.19
6.8	2.5	4.2	.010	26	11	15	.33
7.0			.014	30.5	13	17.5	.39
7.4	2.6	4.5	.020	35	15	20	.47
7.5	2.8	4.6	.022	41.5	17.5	23	.57
8.0	3.0	4.8	.029	48.5	21	27	.68
		5.0	.035	57	24.5	32	.82
9.1	3.5		.047	67	29	37.5	.98
10.1	3.9	6.0	.063	79.5	34	44	1.18
	4.7	7.0	.092	94	41	52	1.41
33	5.2	7.6	.11	111	48	62	1.69

4. Chromatic dip

Flat	L + W + H		Min	Flat	L + W + H		Min
	Box	Cylindrical			Box	Cylindrical	
7.6	2.2	3.7	.002	13	5.7	8.2	.20
7.7	2.3	3.8	.007		6.4	9.1	.24
7.8		3.9	.011	15	7.3	10	.28
		4.0	.014		8.3	11	.34
8.0			.016	18	9.5	13	.41
	2.5		.018	20	11	15	.49
8.5		4.5	.037	23	13	17	.59
	3.0		.049	26	15	20	.70
9.0		5.0	.056	30	18	23	.84
9.5	3.5		.075	34	21	27	1.01
10	4.0	6.0	.099	39	24	32	1.21
10.5			.11	46	29	37	1.46
11	4.6	6.8	.14	53	34	34	1.75
12	5.1	7.4	.16	63	40	52	2.10

5. Steam clean parts

Rack of parts	3.84
Medium part	1.03
Large part	2.68
Basket of parts to 8 × 18 × 60 in.	2.80
Basket of parts to 2 × 4 × 4 ft	3.80

20.4 Blast-Cleaning Machines

DESCRIPTION

Blast cleaning in one of its various forms cleans by the tumbling action of castings, weldments, etc., on one another as the mill rotates. The machine consists of a cleaning barrel, which is formed by an endless apron conveyor. The work is tumbled beneath a blasting unit located above the load, and metallic shot is blasted onto the objects. A tumbler blaster is shown in Figure 20.4.

A variety of blast-cleaning machine configurations exists. Compartmentalized tables which allow sepa-

FIGURE 20.4 Tumbler blaster using metallic shot. Vibratory conveyor unloads parts after cleaning into hopper. *(Wheelabrator-Frye, Inc.)*

rated handling during cleaning of other compartments is available. Swing-out doors which have rotating tables and allow for handling during cleaning, are also available. Double-door arrangements allow loading during cleaning. Conveyorized methods permit loading and unloading on hooks.

ESTIMATING DATA DISCUSSION

The selection of the handling element depends upon the machine type that is assumed and the size of the article that is being blast-cleaned. For the ordinary load and unload, an operator will handle the part from the floor or a table and the entry variable is weight. Hoists, crane, fork truck, or rail insertion are modifications of the operator load and unload. The loading may be to a turn table, ring, or arbor. The larger parts are obviously set upon a flat surface. A turnover operation, if required, assumes 75 percent of the basic load and unload time.

Bulk load assumes a dump into the hopper and undumping into a tote box. The entry variable is the expected number of units to be dumped. If automatic unloading is expected, remove 25 percent. In some cases, a rack is used having hooks and the part is hooked. The rack has multiple hooks and for smaller parts, the entry variable is the number of hooks on the rack.

If semiautomatic operation is used, the start and stop element is required. For swing-out door blasting units, times are given.

The time for shot blasting and tumbling is given by Element 3. It depends upon the machine configu-

Blast-Cleaning Machines 429

ration and nature of object blasted. In the tumbling blasting method, the entry for gray iron castings is batch number, which relates to the capacity of the tumbling unit. Similarly, for a spinner-hanger, the cleaning time is related to the number of units connected to the eyehook within a compartment. If a monorail is used, the time per unit depends on the hook spacing as the conveyor travel speed is 3 ft/min.

EXAMPLES

A. A gray sand casting weighing 36 lb is received in boxes from the shake-out table. Some sand inclusions remain. A spin-hanger cabinet is used, and three parts are suspended on the hanger. The parts are loaded, and then the unit is revolved, allowing the parts to be blasted during loading. Determine the unit time for a lot of 200 castings.

Table Number	Process Description	Table Time	Adjustment Factor	Cycle Minutes	Setup Hours
20.4-S	Setup	.2			.2
20.4-1A	Load three parts from floor	.31		.31	
20.4-3A	Blast clean using spinner-hanger cabinet	2.00	⅓	.66	
Remark	Use max of blast cleaning time (blast cleaning requires more time than loading unit)			—	—
	Total			.97	.2

B. Steel castings, weighing 14 tons, are loaded on a revolving table, which is guided by tracks for blasting in a car-type room. Two castings are stacked on the table. A turnover is required. Find the unit time.

Table Number	Process Description	Table Time	Adjustment Factor	Cycle Minutes	Setup Hours
20.4-1	Bay crane loading on table, ea	15		15.00	
20.4-3A	Blast steel casting, 14/30 × 20	20	14/30	9.3	
20.4-1A	Turnover	15	3/4	11.25	
20.4-3A	Blast steel casting	9.3		9.3	
	Total			44.85	

C. Cores and rims of industrial casterwheels are shot blasted prior to the bonding of rubber to surfaces. Ten units are loaded on a rod. Simultaneously, other wheels are being cleaned on other quadrants of the multi-table compartment machine. Find the hours per 1000 units (hr/1000 units) to clean off welding flux and splatter.

Table Number	Process Description	Table Time	Adjustment Factor	Cycle Minutes	Setup Hours
20.4-1A	2-lb units loaded on spindle	.07		.07	
20.4-3A	Clean 50 in. of area	.004	50	.20	
Remark	Cleaning exceeds loading				
	Total			.20	
	Hr/1000 units	3.333			

TABLE 20.4 BLAST-CLEANING MACHINES

Setup
    Batch or lot of parts   .2 hr
    Conveyor, automatic   0 hr

Operation elements in estimating minutes

1. Handling

    A. Load and unload one part, lb

From Table	From Floor	Min
3.6		.07
5.3		.08
7.3		.09
9.4		.10
11.7		.11
14.3		.12
17.1	1.7	.13
20.2	4.6	.15
23.6	7.7	.16
27.4	11.2	.18
31.5	15.1	.19
36.0	19.3	.21
41.0	24.0	.24

From Table	From Floor	Min
46.5	19.3	.21
52.5	24.0	.24
	29.1	.26
	34.7	.28
	40.9	.31
	47.7	.34
	55.2	.38
Add'l		.004
	Add'l	.005
Electric hoist		1.12
Hand double chain hoist		4
Bay crane		5–20
Fork truck		2–10
Rail		10–30

    B. Bulk load and unload into machine

No. Pc	Min	No. Pc	Min
30	.05	75	.02
40	.04	250	.01
50	.03	500	.005

    Automatic load and unload   no time

    C. Place parts on rack, unrack

No. Pc	Min	No. Pc	Min
1	.33	10	.11
2	.20	12	.10
3	.17	20	.08
4	.16	30	.07
5	.15	50	.06
8	.12	75	.04

2. Machine operation
    - Start and stop   .10
    - Close and open doors
        - Large   1.50
        - Small   .75

3. Blast cleaning

    A. Tumbling-blasting
    - Furnace-hardening steel scale   .003–.01/lb
    - Steel castings   .005–.02/lb
    - Gray iron castings (sand inclusions)   15/batch
    - Compression molded plastic (defashing)   4/batch

    Swing-out, rotating-table cabinet
    - Weldments (rust, mill scale, and welding flux, small splatter)   4/side
    - (Prorate by number of units loaded on table)

    Multi-table
    - Welded pieces (mill scale, welding fluxes, spatter)   .004/sq in.
    - (Prorate by number of units loaded)
    - Drums   .3/ea

Spinner-hanger cabinet
    Gray iron castings   2/hook
       (Prorate by number of parts on hook)
Car-type room
    Bulk steel castings to 30 tons   5–20/ea

B. Monorail, overhead

Hook Spacing	3	5	7	12	24	36	48
Min (for One Operator)	.08	.14	.19	.33	.67	1.00	1.33

Roller, horizontal
    Steel beams, angle plate, trusses, girders   4/ea

20.5 Metal Electroplating and Oxide Coating Processes

DESCRIPTION

Electroplating is a means of applying decorative and protective coatings to metals. In commercial plating, the object to be plated is placed in a tank containing a suitable electrolyte. The anode consists of a plate of pure metal while the part is the cathode. The tank contains a solution of salts of the metal to be applied. A dc current is required. When the current is flowing, metal from the anode replenishes the electrolyte solution while ions of the dissolved metal are deposited on the workpiece in a solid state. Oxide coating, i.e., anodizing for aluminum, involves an electrolyte of sulphuric, oxalic, or chromic acid with the part to be anodized as the anode. Since the coating is produced by oxidation, it is permanent and an integral part of the basis material. The coating is porous, which enables organic coatings and dyes to be applied to the surface of aluminum. Electroforming is not considered in these data. Conversion coatings are covered in Table 20.3, while other metal preparation and cleaning is given in Table 20.4.

As it affects direct labor time and cost, the tanks, barrels, baskets, hoists, racks, and timers or other controls are the important time drivers. For instance, open-surface or hooded tanks affect the time and style of loading. The racks are critical for loading, as the smallest tank in the line sets the dimension and ultimately the number handled per load. Rack configurations are numerous—rings, box, quad-point, and various designs can be easily constructed to suit special parts. Work-holder contacts such as pressure-type, hook and loose-hole, wire to strip, and nut and screw are possibilities.

Plating can be manual, semi-, and fully automatic. Bath time, however, is restricted to specification, which in turn depends upon dc density, part packing, etc.

Figure 20.5 is an example of a polypropylene barrel and hoist assembly installed in a row programmed-hoist barrel electroplating line for zinc and chromate coating of fasteners.

ESTIMATING DATA DISCUSSION

These estimating data may be for one or several operations and are combined. The estimator may wish to organize the estimate collectively or separately. For instance, the operation string-up/takedown may be by operators other than electroplaters.

Setup is separated for pre-plating operations such as string-up/take-down, rack, and barrel, or steel plate.

The string-up/take-down element, while it can be separated into two different operations, is combined here. The string-up is 70 percent of the total time. In oxide coating processes it may be necessary to use aluminum strip for anodizing instead of wire, and the entry variable is panel $L + W$ size, and hanging length. The variable hanging length means the vertical distance the part will hang.

Cleaning elements are given in number 2. Blow off seams using manual methods have the length of seam as the entry variable. Liquid hone starts with the small parts in the booth, start machine, blast part, stop, and spray to rinse. The small parts could be in a basket, and the load would be blast. Medium parts are those which are individually worked, and the entry variable is $L + W + H$.

Electroplating is divided into manually-controlled and automatic. In manually-controlled, the entry var-

FIGURE 20.5 Horizontal oscillating barrel and hoist assembly for programmed hoist barrel electroplating line. *(Udylite Equipment Systems Division of Oxy Metal Industries Corporation)*

iable for electroplating estimating data, Element 4, is based upon part configuration, girth $L + W + H$, and the type of plate. The three part-type geometries are box-like, cylinder-like, and flat-like. This selection is due to part packing within the container or rack. The girth is the minimum X-Y-Z box container that will fit the three configurations. For cylinders, it is $L + 2D$, where D is the part maximum diameters. Usually, thickness of flat parts does not affect the entry value chosen.

There are many specifications in the field of electrodeposition. In silver plating, for instance, the tank density, basis material, and thickness are only a few of the specifications which affect the estimate. Even so, the many specifications are provided in a composite estimate. Element 4 is for multiple parts for small- and moderate-lot production where racks and baskets are common. Typical baskets are hexagonal with 17-in. flats and 36-in. flat-to-flat; circular with 10-in. *Dia* and 8-in. *H*; and rectangular style 48 × 28 in. and 28-in. *H*. Racks have about 150 hooks on a 36 × 52-in. base to a single-hook with manual hoist control. Tanks, of course, are suitably sized to match these requirements, but tank sizes do differ.

The estimate is affected by the number to be plated. If barrel plating is to be used, multiply the table value by .75. To find the manual time for a quantity of one, the data are extended this way. The time for 40, 40, and 50 table entry values found in Element 4 are multiplied by a factor.

Process	Factor
Cadmium plate steel	4
Copper plate copper	2
Gold plate aluminum	7
Gold plate copper	4
Nickel plate copper	3
Nickel-rhodium plate copper	4*
Passivate stainless steel	4
Silver plate aluminum	5
Silver plate copper	3
Tin plate aluminum	7
Tin plate copper	2

*Use nickel value.

Thus, the one-unit time considers the process of metal cleaning, rinse, acid or alkaline dip, rinse, metal plate, rinse, etc. according to various commercial and federal specifications. On the basis of manual batch time, other estimate values can be constructed for differing batch size, or rack or basket loading.

Elements 4 and 5 are helpful if ease is desired. Many parts can be smaller than the minimum entry values listed. In this case, the estimator will use the threshold value, as the typical estimates point out.

The above discussion is for manually dipped parts, i.e., by hand or hoist, and requires that the part be wired, clipped, or somehow loaded to the rack or basket.

Whenever automatic plating is used and manual racking and unracking are required, Element 6 is used. Entry variables are maximum dimension and weight. While there could be a contradiction as the two may not agree for a part, the better choice would be to adopt the larger time value. The racks have 25 hooks, and the operator presents the part to hook in a way to hang it through a port hole.

It is possible that parts will be unracked or disposed to a package operation directly after the electroplating. The estimator may want to use Table 24.1 for these operations.

EXAMPLES

A. A steel stamping, similar to a scissor half, is to be cadmium plated using still-plating methods. Part dimensions are 4.5 in. (L), 1.2 in. (W), and ⅛ in. (max thk). The finger hole is easily available for hooking. Lot size is 1750. Find the unit estimate for several operations.

Table Number	Process Description	Table Time	Adjustment Factor	Cycle Minutes	Setup Hours
20.5-S	Racking	.05			.05
20.5-S	Still plating	.10			.10
20.5-1A	Hang part in its hole	.28		.28	
20.5-4A	Cadmium plate, 5.8-in. flat part, and using the threshold value	.018		.018	—
	Total			.298	.15

B. Estimate A above if the object is to be barrel plated.

Table Number	Process Description	Table Time	Adjustment Factor	Cycle Minutes	Setup Hours
20.5-S	Setup the barrel plate	.20			.20
20.5-4A	Cadmium barrel plate	.018	.75	.014	—
	Total			.014	.20

TABLE 20.5 METAL ELECTROPLATING AND OXIDE COATING PROCESSES

Setup

String-up/take-down, racking	.05 hr
Still plating	.10
Barrel plating	.20
Automatic-conveyor line	1–8 hr

Operation elements in estimating minutes

1. String-up, take down

    A. Wire

Hanging L	.5	1.0	1.5	2.0	2.5	3.0	3.5	5
Min	.14	.16	.17	1.9	.20	.22	.23	.28

Hanging L	6	7	8	9	10	13	15
Min	.31	.34	.37	.40	.43	.52	.58

B. Aluminum strip

	L + W		
L	0–4 in.	4+–12 in.	12+
.2	.16		
.6	.17		
1.2	.18	.29	
1.9	.19	.32	
2.5	.20	.35	
3.0	.21	.37	
3.5		.40	
4.0		.42	
4.5		.44	
5.5		.49	
10		.70	.63
15			.82
20			1.00

C. Alligator clip .08
   Hang part in hole on rack, and unhang
      Maximum single dimension to 3 in. .06
      Maximum single dimension, 3+ to 9 in. .09
      Maximum single dimension, 9+ to 20 in. .16
      Longest dimension, 20+ in. .35

2. Clean

A. Blow off seams

L	.5	1	2	4	Add'l
Min	.50	.55	.63	.81	.09

B. Liquid hone

L + W + H	1	2	4	8	16	32	Add'l
Min	.52	.59	.73	1.01	1.34	2.32	.07

3. Strip liquid mask

L + W + H	1	2	4	6	Add'l
Min	.01	.02	.08	.13	.03

Install robber wire .48
Remove robber wire .27

4. Electroplating, min/pc

A.

L + W + H			Bright Alloy	Cadmium	Copper	Gold
Box	Cylinder	Flat				
2	3	6			.028	
		7	.004	.018	.068	.012
		8	.020	.044	.11	.022
	4	9	.037	.069	.15	.057
3	5	10	.062	.11	.21	.10
	6	12	.099	.17	.30	.16
4	7	14	.14	.22	.40	.22
5	8	16	.18	.30	.51	.30
6	9	18	.23	.37	.62	.38
7	10	20	.27	.44	.74	.45
8	12	23	.34	.54	.90	.57
9	14	26	.40	.65	1.07	.68
11	16	30	.49	.79	1.29	.83
16	20	35	.67	1.06	1.73	1.13
18	23		.76	1.20	1.94	1.28
20	26		.84	1.33	2.16	1.42
24	30	40	.98	1.54	2.49	1.65
40	40	50	1.46	2.28	3.68	2.46
Add'l			.075	.078	.12	.085
	Add'l		.061	.062	.10	.068
		Add'l	.053	.036	.057	.038

B.

L + W + H			Nickel	Rhodium	Silver	Tin
Box	Cylinder	Flat				
2	3	6		.014		.014
		7	.015	.029	.023	.049
		8	.044	.057	.057	.085
	4	9	.072	.086	.092	.12
3	5	10	.12	.13	.15	.18
	6	12	.18	.20	.22	.26
4	7	14	.24	.26	.30	.34
5	8	16	.32	.34	.40	.44
6	9	18	.40	.42	.49	.54
7	10	20	.48	.50	.59	.64
8	12	23	.60	.62	.73	.78
9	82	26	.71	.74	.87	.93
11	16	30	.87	.90	1.06	1.13
16	20	35	1.18	1.21	1.43	1.52
18	23		1.32	1.36	1.62	1.71
20	26		1.48	1.52	1.80	1.90
24	30	40	1.71	1.75	2.08	2.19
40	40	50	2.54	2.60	3.09	3.24
Add'l			.088	.089	.11	.11
	Add'l		.070	.072	.085	.088
		Add'l	.039	.040	.048	.050

5. Oxide coating and other processes, min/unit

L + W + H			Anodize	Blacken	Chemical Polish	Passivate
Box	Cylinder	Flat				
2	3	6	.018			
		7	.043	.004	.010	
		8	.069	.020	.030	
	4	9	.095	.037	.050	.005
3	5	10	.13	.062	.081	.028
	6	12	.19	.099	.13	.063
4	7	14	.25	.14	.17	.10
5	8	16	.32	.18	.23	.14
6	9	18	.40	.23	.28	.18
7	10	20	.47	.27	.34	.22
8	12	23	.57	.34	.42	.29
9	14	26	.68	.40	.50	.35
11	16	30	.82	.49	.61	.43
16	20	35	1.10	.67	.82	.59
18	23		1.23	.76	.93	.67
20	26		1.37	.84	1.03	.75
24	30	40	1.58	.98	1.20	.87
40	40	50	2.34	1.46	1.78	1.31
Add'l			.080	.051	.061	.046
	Add'l		.064	.040	.049	.036
		Add'l	.036	.023	.027	.022

6. Rack and unrack part on hook

Max Dim	Weight	No. Pc/Rack									
		1	3	5	8	10	12	15	18	20	25
0 < Dim ≤ 3 in.	0 < Wt ≤ .2 lb							.07	.07	.06	.06
3 < Dim ≤ 9	.2 < Wt ≤ 3					.11	.09	.08	.07	.06	.06
9 < Dim ≤ 20	3 < Wt ≤ 8		.33	.21	.12	.12	.10	.09			
20 < Dim ≤ 36	8 < Wt ≤ 12	.90	.35	.24	.13	.13					

ASSEMBLY

21.1 Bench Assembly

DESCRIPTION

The estimating data in this section are for general bench assembly. They are listed with the purpose of the work element to be done, and not on the basis of fundamental motions. No machines are involved in bench assembly, although a range of hand or powered-hand tools is included. The data are also not associated with a reach or move distance, as this information may be unavailable at the time of estimating.

The type of work done on the bench is too vast to attempt to provide extensive description. Instead, we have chosen to identify key elements that are versatile and broadly usable.

ESTIMATING DATA DISCUSSION

No setup is allowed for bench assembly. Throughout the data, reference is made to part size. While ambiguity is difficult to prevent, the part size is identified as very small, small, medium, large, and very large. These descriptions are not classified relative to a specific machine but to manual handling characteristics. The parts classification is included in the following table.

Part Size	Remarks
Very small	When handled individually, the parts are difficult to control, though they can be handled easily in handfuls. Tweezers are sometimes used.
Small	Small parts are easy to manage by handfuls or with fingers. No dimension exceeds 3 in. and weight does not exceed ¼ lb.
Medium	Medium-sized parts are sometimes handled by double handfuls. Their maximum dimension and weight are 9 in. and 3 lb.
Large	Each large part is handled separately, and a large part requires the use of two hands. The maximum dimension and weight are 20 in. and 8 lb.
Very large	A very large part has a maximum dimension and weight of 36 in. and 12 lb.

It should be understood that "very large" is in the context of bench assembly. This description is unacceptable for the other processes and machines described in this estimating guide.

The data are expressed in min/occurrence and are three-place decimals under .10 min. As the data are multiplied by the frequency of occurrence, rounding off of these elements leads to errors.

Elements 3 and 4 are concerned with the same part. For each different part, Element 3 is first used. For each subsequent part, Element 4 is applied. Other elements are self-explanatory.

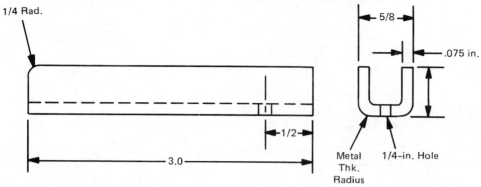

FIGURE 21.1 Part assembled to major part.

EXAMPLE

A. A frame has several components attached using screws, nuts, and washers. Stack bins and ordinary hand tools are laid out for the job. Estimate the job, as described below.

Table Number	Process Description	Table Time	Adjustment Factor	Cycle Minutes	Setup Hours
21.1-1	Get and aside basic part	.045		.045	
21.1-3	Get small component, place on restricted frame, .027 + .015	.042		.042	
21.1-7	Get screw and washer, place, 7 times	.099	7	.69	
21.1-7	Get nut, start, seven times	.083	7	.58	
21.1-7	Spin tight six threads, seven times, 7 × .045 + 7 × .015 × 5	.84		.84	
21.1-3	Get very small component	.03		.03	
21.1-4	Get and place three more components into three very small holes	.045	3	.14	
21.1-7	Get screw, nut, and washer, 8 times, and place and start, 8 × .083 + 8 × .024	.86		.86	
21.1-7	Spin tight six threads, eight times, 8 × .045 + 8 × .015 × 5	.96		.96	
21.1-9	Air clean, medium	.063		.063	
21.1-9	Rubber stamp	.13		.13	
	Total			4.38	

B. Four sheet metal parts, 50531, are to be assembled to a heavier part, 50532, which becomes 50530. A simplified sketch of the sheet metal part is given as Figure 21.1. For reference, the sketch for Figure 6.4C is PN 50532. The quantity to be assembled is 7075 units. Bench assembly is proposed. The adjustment factor allows for multiple screws, for example. Find the estimates.

Table Number	Process Description	Table Time	Adjustment Factor	Cycle Minutes	Setup Hours
	Bench assembly				
21.1-S	Setup	0.00			0.00
21.1-1	Handle part no. 50532, large part	0.09		0.09	
21.1-1	Handle four minor parts 50531	0.04	4	0.17	
21.1-3	Place, position large part	0.07		0.07	
21.1-3	Place, position	0.03	4	0.12	
21.1-4	Get, place add'l four screws	0.04	4	0.16	
21.1-6	Handle tool	0.05	1	0.05	
21.1-7	Fasteners	0.10	4	0.40	
21.1-7	Fasteners	0.02	4	0.10	
	Total			1.14	0.00
	Lot estimate	134.43 hr			

TABLE 21.1 BENCH ASSEMBLY

Setup 0

Operation elements in estimating minutes

1. Handle basic part

Part Size	Type of Handling			
	Toss	Place	Stack	Conveyor
Very small	.027	.039	.042	
Small	.021	.036	.039	
Medium	.030	.042	.045	.042
Large	.078	.084	.090	.054
Very large		.13	.14	.090

2. Tumble or turn parts

Part Size	0°–90°	90°–180°
Very small	.009	.018
Small	.012	.024
Medium	.015	.030
Large	.030	.060
Very large	.039	.078

		Part Size	On Open Bench	Into Angle or Vee	Into Formed Nest	Over 1 Pin or into Hole	Over 2 Pins or 1 Pin Nested	Add for Close Tol or Care	Add for Other Part Aligned
3.	Place and position part	Very small	.018	.024	.033	.040	.045	.018	.018
		Small	.021	.027	.030	.027	.039	.015	.012
		Medium	.024	.030	.036	.033	.051	.015	.018
		Large	.051	.060	.066	.063	.084	.021	.030
		Very large	.063	.075	.084	.078	.11	.030	.036
4.	Get and place add'l part	Very small	.036	.042	.048	.045	.060	.018	.018
		Small	.027	.036	.042	.039	.054	.015	.012
		Medium	.042	.048	.057	.054	.069	.015	.018
		Large	.078	.087	.093	.090	.11	.021	.030
		Very large	.11	.12	.13	.12	.14	.030	.040
5.	Get and place multiple part (multiple get)	Very small	.024	.033	.042	.036	.048	.018	.018
		Small	.024	.033	.045	.036	.048	.015	.012
		Medium	.033	.045	.051	.048	.057	.015	.018

6. Handle tool
   - Simple grasp   .036
   - Difficult grasp   .048

7. Fasteners
   - Get screw or nut and place (multiple get)   .024
   - Get screw and washer, place   .099
   - Get screw, two washers, place   .15
   - Get screw or nut, start 1½ threads   .083
   - Get screw, washer, start   .13
   - Get screw, place on split or magnetic screwdriver, and start or get nut, place, start with spin tight   .11
   - Get screw, washer, place on driver and start   .15
   - Get screw, two washers, place on driver, start   .19
   - Add for starting sheet metal or self tapping screws   .096

    Run screw or nut down
        Hand, each thread .015
        Hand tool, 1 standard thread .045
            Ea add'l .015
    Power driver, 1st 10 threads .024
        Ea add'l 10 threads .009
    Open end wrench, ea thread .060

8. Fixture handling and clamping
    Close and open toggle clamp .042
    Close and open cam acting lever clamp .051
    Close and open swing or slide clamp .030
    Tighten and loosen thumb or wing nut .084
    Tighten and loosen nut or bolt with wrench .24
    Place and remove holding pin .06
    Place and remove locating arbor or nest .14
    Place and remove clamping plate .054
    Tap part in and out of fixture .072
    Pry part out of fixture .092
    Operate air clamp .015
    Index fixture one station .042
    Place clamp or spacer on part .060
    Remove clamp or spacer from part .030

9. Miscellaneous elements
    Drill and pin with roll pin, 1st hole 1.40
        Add'l hole .68
    Chase one piece, 1st hole .16
        Add'l hole .07
    Install helicoil spring .52
    Rubber stamp once .13
    Air clean
        Small .048
        Medium .063
        Large .093
        Very large .13

21.2 Riveting and Assembly Machines

DESCRIPTION

In riveting, a one-piece fastener consisting of a head and body is passed through an aligned hole of two or more pieces. It is then clinched, or a second head is formed on the body end. While the rivets have a variety of materials, sizes, and shapes for various purposes, our concerns are for the machines which do this operation.

Riveting machines may be floor mounted, bench mounted, or even hand held. Whether they are manual or automatic, the rivet is fed from the hopper to a track or manually loaded that drops it, shank down, in the center of the upper jaws. A driver action pushes the rivet onto a spring-mounted plunger on the lower arm of the machine. A lower die clinches the rivet. The power to drive the rivet may be pneumatic or electric.

Hardware can mean a number of things. For instance, standoffs, terminals, pemnuts, press nuts, springs, clips, eyelets, bushing, tubes, or anything that is pressed-in and is multiple-carried by one hand is considered hardware.

Assembly machines are specially designed, and one is shown in Figure 21.2. The direct-labor costs vary with the machine design requirements.

FIGURE 21.2 A specially designed machine for automatic syringe assembly. *(Hill Rockford)*

ESTIMATING DATA DISCUSSION

A setup depends on whether the operation is the first light-mechanical assembly or not. Following setups require less time.

The handling element deals only with parts. It involves the handling of any rivet, screw, etc. The entry variable is location (from a skid or a bench). Obviously, the larger part is on the skid. Additional parts location can be from either a skid or bench, as well. In light-mechanical assembly, a hoist is unlikely, and the weight of the part is within operator safety limits. The times for one are total.

Element 2 includes get hardware, position hardware, position part, and cycle (or press). Element 3, "rivet," represents a rivet, stake, roll, flare, or spin type of a machine element. It includes position part and machine cycle. The entry variable is the number of cycles.

Several special fixture handling and clamping elements in 4 are given. These are associated with light-mechanical assembly. Similarly, Element 5 lists some miscellaneous machine times. Specially designed assembly machines depend upon cycle time designed for the unit. They may or may not require operator attendance for work such as filling tubs with parts and watching.

EXAMPLES

A. A major part, located in a skid, has 2 large skid-located parts, and 8 minor tote-box parts to be assembled by riveting. For hardware-mounting there are 27 holes into which rivets are inserted and pressed. Once the hardware is attached, 15 rivets are rolled. Determine the unit time and the hr/100 units.

Table Number	Process Description	Table Time	Adjustment Factor	Cycle Minutes	Setup Hours
21.2-S	Setup, first operation	.25			.25
21.2-1	Get basic parts and two skid-mounted parts	.50		.50	
21.2-1	Get eight minor parts	.31		.31	
21.2-2	Hardware mounting, 27 cycles	1.90		1.90	
21.2-3	Rivet, 15 cycles	.70		.70	
21.2-1	Assembly aside to skid	.13		.13	
	Total			3.54	.25
	Hr/100 units	5.90			

TABLE 21.2 RIVETING AND ASSEMBLY MACHINES

Setup

    First operation   .25 hr
    Additional operations   .10 hr

Operation elements in estimating minutes

1. Handling

Major and Minor Parts		From Skid	From Bench
Get basic part		.13	.03
Get minor part no.	1	31	.07
	2	.50	.10
	3	.68	.14
	4	.86	.17
	5	1.05	.21
	6	1.23	.24
	7	1.42	.28
	8	1.60	.31
	9	1.79	.35
	10	1.97	.38
Add'l.		.18	.035

Aside to skid   .13
Aside to bench   .03

2. Press in hardware

No. Cycles	Min	No. Cycles	Min	No. Cycles	Min	No. Cycles	Min
1	.06	11	.77	21	1.48	31	2.19
2	.13	12	.84	22	1.55	32	2.26
3	.20	13	.91	23	1.62	33	2.33
4	.27	14	.98	24	1.69	34	2.40
5	.34	15	1.05	25	1.76	35	2.47
6	.41	16	1.12	26	1.83	36	2.54
7	.48	17	1.19	27	1.90	37	2.62
8	.55	18	1.27	28	1.98	37	2.69
9	.63	19	1.34	29	2.05	39	2.67
10	.70	20	1.41	30	2.12	Add'l	.071

3. Rivet

No. Cycles	Min	No. Cycles	Min	No. Cycles	Min	No. Cycles	Min
1	.06	11	.52	21	.97	31	1.43
2	.11	12	.56	22	1.02	32	1.48
3	.15	13	.61	23	1.07	33	1.52
4	.20	14	.65	24	1.11	34	1.57
5	.24	15	.70	25	1.16	35	1.61
6	.29	16	.75	26	1.20	36	1.66
7	.33	17	.79	27	1.25	37	1.71
8	.38	18	.84	28	1.29	38	1.75
9	.43	19	.88	29	1.34	39	1.80
10	.47	20	.93	30	1.38	Add'l	.046

4. Special fixture handling and clamping
   - Place and remove locating pin or arbor .09
   - Open and close toggle clamp .14
   - Open and close cam acting lever .04
   - Open and close swing or slide clamp .05
   - Tighten and loosen thumb screw or wing nut .03
   - Tighten and loosen nut or bolt with wrench .08
   - Operate air clamp .24
   - Pry part out of fixture .02
   - Pry part into fixture .10
   - Tap part in and out of fixture .07

5. Miscellaneous machine times
   - Arbor press with lever .03
   - Arbor press with wheel or small hydraulic press .04
   - Hopper feed screw setters .03
   - Small bench welder .06

6. Miscellaneous light-mechanical assembly
   - Centerpunch .07
   - Helicoil insert in predrilled hole .48

21.3 Robot Machines

DESCRIPTION

Industrial robots are machines that load, process, transfer, position, and unload. These data deal with all these elements, with the exception of processing. (Consult other tables for processing.)

Industrial robots consist of the manipulator (or mechanical unit), which actually performs the functions; controller, which stores data and directs the movements of the manipulator; and the power supply. The manipulator is a series of mechanical linkages and joints that allows movement in various directions. These mechanisms are driven by actuators, which may be pneumatic or hydraulic cylinders, hydraulic rotary actuators, or electric motors. The actuators may be coupled directly to the mechanical links or joints or, they may drive indirectly through gears, chains, or ball screws. Feedback devices are installed to sense the positions of the various links and joints and to transmit this information to the controller. The feedback devices may be limit switches or position measuring devices such as encoders, potentiometers or resolvers, or tachometers.

Figure 21.3A is a robot handling two die cast machines. The die cast machines are 800-ton units and have 21-ft 6-in centers.

ESTIMATING DATA DISCUSSION

Most robot applications use more than two independent axes. To estimate the traverse time required for each axis, which may involve the working-tip at-

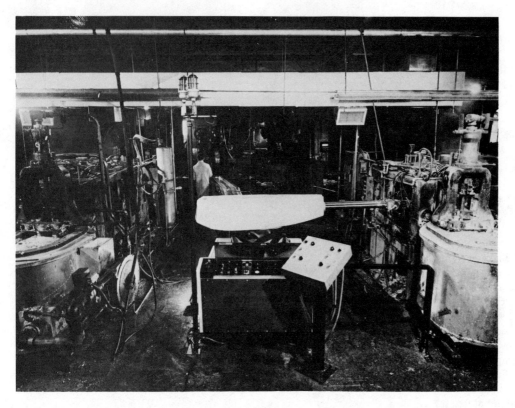

FIGURE 21.3A A specially designed machine for automatic syringe assembly. *(Hill Rockford)*

titude and path, and then calculating the time for the longest motion, would be unnecessarily laborious, as velocities (wrist movements, etc.) are nominally expressed in degrees/second (deg/sec). Or, in-out reach (radial travel) is expressed by robot manufacturers in terms of in./sec.

Element 1 deals with moving from a rest position to the proximity of the first area of processing. Similarly, it offers an estimate for retracting from processing to the rest position. The robot tool can be any choice, but its weight should be known. The extension distance is also necessary to know.

After positioning to the work area, adjusting the robot tool attitude to be normal to the metal is given by Element 2. Element 2 is used with any minor zone change where a robot tool attitude change is required. "Simple" implies no interference.

Element 3 is used for a group of sequential skips in which the distance is small and no significant change in the attitude of the gun is required.

If the tool spins, to dry or centrifuge, etc., a spin estimate value is given.

Table 21.3 provides estimates for robot tool movement. (Use other tables for processing values.)

EXAMPLES

A. A body side and roof rail are spot welded using a robot machine. The side and rail have 29 and 28 spots. A spot-welding gun is 35 lb, and a maximum reach gives 750 in.-lb. Spot weld separation is 2 in. for wheel well and 1.5 in. for the roof rail. Obstructions exist between the weld zones, which require complex convolute paths for the robot gun tip. Clamps are necessary within the welding zones, and inner movement is necessary. Movements for the gun from rest to the workpiece are complex. Find the unit estimate and the hr/10,000 units. It is pointed out that if the robot is unattended, direct labor does not exist, and a direct labor estimate is unnecessary. The cost of the robot is then "overhead".

Table Number	Process Description	Table Time	Adjustment Factor	Cycle Minutes	Setup Hours
21.3-1	Approach to weld zone, 750 in.-lb., complex movement	.081		.081	
21.3-2	Position to work area	.025		.025	
	Weld 29 times, 29 × .01, steel material	.01	29	.29	
21.3-3	Point-to-point tool movement, 2 in.	.009	29	.261	

Table Number	Process Description	Table Time	Adjustment Factor	Cycle Minutes	Setup Hours
21.3-2	Position tool to roof rail, 750 in.-lb.	.025		.025	
	Weld 28 times	.01	28	.28	
21.3-3	Point-to-point tool movement, 1.5	.008	28	.224	
21.3-1	Retract tool to rest position, 750 in-lb	.081		.081	
	Total			1.267	
	Hr/10,000 units	211.167			

B. An auto assembly line is anticipating the use of a robot to install a roof garnish on one of its car lines. Notice Figure 21.3B. The work will be watched by direct labor and an estimate is required to determine the cost of the work and the line time for this station. Two robots are assigned this work for this module line station. The first robot loads the garnish from the waiting side conveyor station and it will position it tightly in and over the prepared receptacle. A second robot will use self-drilling taps and fix the garnish to the roof. There are 12 self-drilling screws. The purpose of the estimate is to find the station time for the two robots such that the operator tender can be estimated for labor cost contribution. While the quantity is many thousands, the estimate for one unit is required. Any startup efforts to put into place the tooling and robots is estimated with the total tooling cost.

Table Number	Process Description	Table Time	Adjustment Factor	Cycle Minutes	Setup Hours
21.3-S	Setup for line robot	0.00			0.00
21.3-1	Approach-retract 2000 in. lb to conveyor stack for waiting garnish. Both approach and retract to wait for next robot finish	0.10	2	0.21	
21.3-1	Grip garnish, estim.	0.15		0.15	
21.3-2	Positioning garnish to top of car frame and reposition and vision delay, five times	0.03	5	0.14	
21.3-3	Point-to-point 2.5 and control loading two ends and middle for three times on rooftop. Robot loader swings away.	0.01	3	0.03	
	Second Robot Tool:				
21.3-S	Setup for second robot which has self tapping head and threads, 12 times	0.00			0.00
21.3-1	Approach-retract from rest to active work and to rest pos	0.06	2	0.12	
21.3-2	Positioning for 12 drill and tap holes	0.02	12	0.23	
11.2-8	Tapping with self drilling screw, 12 times, depth .25 in. .15 min/in.	0.04	12	0.45	
	Total			1.32	0.00

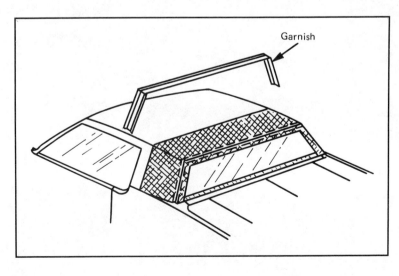

FIGURE 21.3B Robot assembly of garnish.

TABLE 21.3 ROBOT MACHINES

Setup

Operation elements in estimating minutes

1. Approach or retract robot tool to work

In.-lb	250	500	750	1000	2000	3000
Simple	.049	.052	.056	.059	.073	.087
Complex	.072	.077	.081	.086	.104	.122

2. Positioning of robot tool to work area, ea area

In.-lb	250	500	750	1000	2000	3000
Simple	.013	.015	.017	.019	.027	
Complex	.023	.024	.025	.026	.027	.032

3. Point-to-point movement of robot tool without processing, each time

Distance	.5	1.0	1.5	2.0	2.5	3.0
Min	.006	.007	.008	.009	.011	.012

4. Spin   .01/rev

5. Processing, spot welding, drill, mill, paint, die cast, cut, mold, etc.   See other tables

INSPECTION

22.1 Machine, Process, and Bench Inspection

DESCRIPTION

These data are for production workers checking their work. It is unlike the data labeled "Inspection Tables and Machines," which is for receiving, tool, source, or final inspection customarily done by inspectors.

A variety of measuring instruments are used, such as direct-reading, angular, and plane instruments. Principles for the readings may be mechanical, optical, electronic/electrical, and pneumatic. The gages can be read as direct, digital, analog, comparative, or projecting. Furthermore, gages can be classified into master, inspection, and manufacturing, which deal with their truth of measurement.

It is the manufacturing gages, located on or about the machine, process, or bench, that are considered here. There are many types too numerous to mention, but they are listed in Table 22.1. Other gages, notably master and inspection grade, are not process or machine based, and they are found in Table 22.2.

ESTIMATING DATA DISCUSSION

There is no setup for Table 22.1 since it is provided by other machine, process, or bench data.

Two categories are evaluated. In the first, the workpiece is checked following completion. The inspection occurs during an automatic cycle of a following part and time required for checking is less than the automatic element.

In the second category, the work is mounted on the machine or process and inspection occurs during an interruption of the cycle, or during the process without elemental interruption. This is called in-processing inspection.

The key to the calculation of inspection time is whether or not the interruption increases direct-labor time for the unit estimate. If there is an interruption, the possibility exists for an inspection time to be added to the unit estimate, given the time for the inspection and the frequency of inspection to the number of parts is significant when compared to the unit cycle estimate. The work is done by production workers, or by inspectors at the work stations, and cycle time is interrupted.

The times include pickup of the measuring device, adjust to dimension, read dimension, and lay aside. The times do not include part handling. If that is necessary, then the machine or process data are used.

Element 3 lists the measuring devices and the times for various total dimensional tolerances.

The following table, gaging frequency, relates total dimensional tolerance to gaging frequency. As expected, the frequency is inversely proportional to the functional dimensional tolerance. For example, a .005-in. tolerance would require two checks per one part while a .010-in. tolerance would require one check per 25 parts. These ratios are judgmental, although they are comparative to typical company practices. These frequencies are used as Adjustment Factors.

Adjustment Factor for Inspection

Total Dimensional Tolerance, In.	Gaging Frequency per No. of Parts or Occurrences
–.0002	3/1
.0003–.0005	2/1
.0006–.001	1/1
.002 –.005	1/5
.006 –.010	1/25
.011 –.030	1/50
Fraction	1/50
Feature	1/100

These ratios are used as adjustment factors whenever the production operator will be doing part or assembly inspection during the production cycle. In many cases throughout the *AM Cost Estimator*, the tables also provide specific inspection elements that may be appropriate for that machine or process or bench work. Table 22.1-3 values may be used to supplement those values, or they can be used whenever the list given by the processing table is insufficient. The usual plan is to know the tolerance that will require inspection and then establish the method that will be used. The tolerance establishes the gaging frequency which in turn becomes the adjustment factor. You will notice the tighter that the tolerance, the greater the time required. A tolerance of .0003–.0005 in. requires two times the time given by Table 22.1-3 while a .002–.005-in. range requires one-fifth of the time of Table 22.1-3.

EXAMPLES

A. A steel casting weighing 700 lb is drilled, counterbored, and tapped on an NC machining center. Although there are 37 holes, only several are critical, and they are listed in the schedule below. Tape control is stopped pending inspection for specified dimensions using the ratio identified in Element 4. Find the unit estimate of inspection time that would be added to the direct-labor NC machine-center production estimate.

Table Number	Process Description	Table Time	Adjustment Factor	Cycle Minutes	Setup Hours
22.1-3	Vernier caliper .005-in. ID of bore, .13/5, where five is gaging frequency	.13	1/5	.026	
22.1-3	Vernier caliper fractional step	.10	1/50	.002	
22.1-3	Plug gage three bores with go/no-go, each .002 in. total tolerance	.08	3/5	.048	
22.1-3	Machinist micrometer check of five .001-in. ID	.16	5/1	.80	
22.1-3	Depth check step with depth micrometer for .005-in. total tolerance	.29	1/5	.058	
	Total			.934	

B. A radial drill machines a forging having two center-to-center locating dimension, and a five-internal thread is also checked. Work is done subsequent to production and during the boring of two holes. What time will the unit estimate require? The schedule of inspection is given below. (Note that gaging frequency is shown in the table in the Estimating Data Discussion above.)

Table Number	Process Description	Table Time	Adjustment Factor	Cycle Minutes	Setup Hours
22.1-3	Plug gage two bores. .005-in. tolerance	.15	2/5	.060	
22.1-3	Make two depths, .010-in. tolerance	.29	2/25	.023	
22.1-3	Go/no-go gage check of two bores simultaneously, .002 in.	.22	1/5	.044	
	Total			.13	

C. A part is drilled, reamed, and tapped in a box jig. Inspection is done following production of the part. Evaluate the unit estimate. As the work is handled by a sensitive drill press, the time is added to the estimate.

Table Number	Process Description	Table Time	Adjustment Factor	Cycle Minutes	Setup Hours
22.1-3	Plug gages, two at .002-in. tolerance	.16	2/5	.064	
22.1-3	Snap thread gage, one occurrence at .010 pitch tolerance	.17	1/25	.007	

TABLE 22.1 MACHINE, PROCESS, AND BENCH INSPECTION

Setup   0

Operation elements in estimating minutes

1. Handling   See other machine, process, or bench tables.
2. Machining, processing, or benchwork   See other machine, process, or bench tables.
3. Inspection

Measuring Device	Total Dimensional Tolerance					Feature
	Fraction	.005	.001	.0005	.0002	
Scale	.11					
Scale-square	.10					
Vernier caliper	.10	.13				
Plug gage, end	.06	.07	.08	.09		
Go/no-go, 2 end	.11	.15	.16	.17	.20	
Machinist micrometer, inside		.16	.16			
Machinist micrometer, outside		.14	.16	.29		
Dial indicator and surface reference		.10	.11	.24		
Snap gages		.07	.09	.11		
Thread snap gages			.17			
Thread gage, plug or ring			.30			
Depth micrometer			.29			
Dial bore gage			.17	.18		
Visual						.05
Pneumatic bore gage		.18	.19	.20		
Transfer caliper, ID	.25					
Transfer caliper, OD	.14					
Pin and surface inspection gage		.5	.5	.6		
Go/no-go templet	.22	.22				
Machine-mounted electronic gages, internal and readout		.06	.06	.06	.06	
Machine-mounted electronic gages, external and readout		.05	.05	.05	.05	
Radius gage	.06					.06
Thickness gage	.10					.10
Protractor	.10	.13				
Dial indicator, plug member		.5	.5	.6		
Gears over pins	.15	.20	.4	.6		

Values are adjusted by gaging frequency.

22.2 Inspection Table and Machines

DESCRIPTION

Inspection is either a direct- or indirect-labor operation. Where it is classed as indirect, the usual procedure is to charge appropriate expenses and eventually determine inspection labor costs as overhead. For indirect inspection labor these estimating data are not usually considered. When inspection is classified as direct, and planning operations stipulate the nature of work, these estimating data can be used.

Inspection operations are found in receiving and final areas, as well as patrol, departmental, or conveyor-line functions. The inspection may correspond to a sampling plan, 100 percent of all parts, proof-test for a tool, or as an approval of a part from a setup, thus allowing the machine operator to continue on with production.

These data are for an inspection table, booth, or room where the equipment is available. Roving inspection for occasional part inspection to determine changing dimensions or attribute quality is not evaluated.

A coordinate measuring machine is useful where production quantity is low to medium, and usually in one setup manages to check hole locations on different faces from plane-to-plane dimensions. The part to be measured is placed on the work table and aligned. A probe is moved to a reference point and then moved to various points to be checked. Electronic readout or computer printout options are available. Figure 22.2 is typical.

ESTIMATING DATA DISCUSSION

All parts, whether sampled or 100 percent fully inspected, are handled, and are allotted time for a primary handle. Contrariwise, subsequent repositioning or tumbling, or turning over, are specifically excluded,

FIGURE 22.2 Coordinate measuring machine. (*Boice Division, Mechanical Technology Inc.*)

as that time is excluded in the application of the inspection of tools and equipment while the part is being inspected.

In random selection of a part, a time of .1 is allowed which permits several styles for random-part picking.

In many of the inspection elements, the entry variables are the number of similar dimensions and sample size. As the data are sensitive to two variables, the procedure is to first select the row specifying the sample size or next lower value and to move horizontally across to the number of dimensions. Values for additional dimensions or sample units are provided. Because the data are two-variable linear, some tables do not show values for the lower ranges of sample size and dimensions, as these values would be negative because of the negative intercept. For those tabular cells which show no time value, the preferred procedure is to choose a threshold value. A reverse procedure will lead to a different threshold or minimum value.

The minimum value is for handling the instrument, gage, etc., and the part simultaneously to measurement. Many tables show additional time for dimensions and sample unit. An overall average time for inspection is also listed, and is called "expected."

Element 3 is for thimble and barrel micrometers, including inside, outside, blade, telescope, depth, etc. If eight different mikes are used, then the entry variable is "8." Plug gaging is for go/no-go and includes checking of both go and no-go. Element 3, visual check, is for feature appearance and surface defects. In Element 13, either a device for teeth counting or manual counting is permitted. Elements 15 through 18 allow for a variety of setups to mechanically verify dimensions.

Table 22.2 is for inspection only. It does not consider the time to correct defective material. Gaging frequency, implying the smaller the tolerance the more frequent dimensional checks, is handled by specifying sample size per sampling plan.

Machine inspection, as provided by Element 19, is able to check hole locations on different faces, plane-to-plane dimensions, concentricities, etc., and checks out a part in much the same way as an NC machine makes it. This elemental time is per dimension checked.

EXAMPLES

A. A lot of 60 turned parts is randomly sampled for three units. Eight dimensions are critically checked, two features are visually inspected, and eight dimensions are minor but necessitate scale or vernier caliper confirmation. Determine unit time for inspected parts, prorated time per unit, and find lot time. The schedule of the inspected dimensions are below. All work is for table inspection.

Table Number	Process Description	Table Time	Adjustment Factor	Cycle Minutes	Setup Hours
22.2-S	Bench setup	.05			.05
22.2-1	Load, unload part, very large turning	.2	3/60	.010	
22.2-2	Random selection	.1	3/60	.005	
22.2-3	Five miked dimensions, three external and two internal	4.6	1/60	.077	
22.2-4	One dimension that is plug-checked	.6	1/60	.010	
22.2-10	Thread gaging	1.4	1/60	.023	
22.2-16	Bench center, runout check	4.7	1/60	.078	
22.2-8	Visual check of two features	.2	1/60	.003	
22.2-9	Vernier check of eight dimensions	2.8	1/60	.047	
	Total			.253	.05
	Floor-to-floor time	15.2 min			
	Lot time	.31 hr			

B. First-part or proof inspection is required of a stamped part to assure die performance. Inspection is mechanical table, and the optical comparator is used for difficult dimensions. All print dimensions are verified. Find the time for first-part inspection.

Table Number	Process Description	Table Time	Adjustment Factor	Cycle Minutes	Setup Hours
22.2-S	Table setup	.05			.05
22.2-4	Plug gage, four holes	.3		.3	
22.2-6	Radius check, two corners	.1		.1	
22.2-9	Caliper check length and width, three notches for five dimensions	1.5		1.5	

Table Number	Process Description	Table Time	Adjustment Factor	Cycle Minutes	Setup Hours
22.2-12	Compare four center and three hole-to-hole lengths for 7 dimension	7.6		7.6	
22.1-8	Visually check three features	.1		.1	
22.2-3	Mike thickness of sheet stock	1.5		1.5	
	Total			11.1	.05
	Proof inspection	.24 hr			

C. A casting classed as critical is 100 percent inspected for major dimensions. Other dimensions and features can be sampled following military procedures. These include minor, not-mating tolerances, and surface roughness. The lot is 40 units, and sampling tables indicate the sample size is seven. The feature and dimension schedule are given in the description below. All work is bench. Part size is medium.

Table Number	Process Description	Table Time	Adjustment Factor	Cycle Minutes	Setup Hours
22.2-S	Table inspection. Sample-work considered first.	.05			.05
22.2-1	Load and unload part	.1	7/40	.018	
22.2-2	Random selection	.1	7/40	.018	
22.2-3	Micrometer check, blade, OD, ID, and depth types, nine dimensions, $11.0 + 2 \times .15 = 12.0$, 7 units	12.0	7/40	.30	
22.2-4	Two dimensions are plug checked, $2.5 - 3 \times .5 = 1.0$, 7 units	1.0	7/40	.025	
22.2-5	Scale four pads, eight dimensions, $3.1 + 3 \times .8 + 2 \times .3 = 6.1$, 7 units	6.1	7/40	.153	
22.2-7	Counterbore is thickness checked, $1.5 + 2 \times .4 = 2.3$, 7 units	2.3	7/40	.058	
22.2-9	Caliper check of five dimensions, $2.2 + 2 \times .2 = 2.6$, 7 units	2.6	7/40	.065	
	Estimate surface roughness, two areas	.4	7/40	.010	
	Sample unit inspection subtotal for 7 units, prorated			.645	
	100 percent unit inspection. All parts are reloaded and unloaded				
22.1-1	Load and unload part	.1		.1	
22.1-3	ID check of bore, $13.3 + 15(.5) = 20.8$, 40 units	20.8	7/40	.5	
22.2-4	Plug check of two-line bored holes $10 + 15(.5) = 17.5$	17.5	7/40	.438	
22.2-17	Bolt circle of four tapped holes, $14.7 + 15(.5) = 22.20$, 40 units	22.20	7/40	.56	
22.2-15	Flat plate, dial indicator for parallel surface, $6.8 + 15(2.1) = 38.3$, 40 units	38.3	7/40	.96	
22.2-17	Dial indicate to plug on flat plate, $14.7 + 15(.5) = 22.5$, 40 units	22.5	7/40	.56	
	Total				
	100 percent sampled-part unit estimate			3.12	.05
	100 percent lot inspection estimate, unit	3.76 min			
	100 percent lot inspection and sample inspection	2.51			

D. A magnesium die cast part is to be inspected by an inspector as regular operation. Note Figure 2.9C which has been finished by several operations. Inspection is to be concluded to meet a military specification. The quantity is 500 units. The adjustment factor is applied for tolerance difficulty or for quantity effects. The lot of 500 is sample inspected. Determine the estimates.

Table Number	Process Description	Table Time	Adjustment Factor	Cycle Minutes	Setup Hours
	Inspection per mil std 6021 class C casting with sample plans, not 100 percent checked				
22.2-S	Setup	0.05			0.05
22.2-1	Handle	0.10		0.10	
22.2-3	Micrometer 2.85 hole every two parts	0.50	.5	0.25	

Table Number	Process Description	Table Time	Adjustment Factor	Cycle Minutes	Setup Hours
22.2-2	Miscellaneous	0.10	1/500	0.00	
22.2-10	Thread gaging for 25 sample size but prorated over lot of 500 parts	22.00	1/500	0.04	
	Total			0.30	0.05
	Lot estimate	3.34 hr			

TABLE 22.2 INSPECTION TABLE AND MACHINES

Setup
    Table   .05 hr
    Machine   .25 hr

Operation elements in estimating minutes

1. Handle one at a time; medium, large   .1
    One at a time very large   .2
    Bulk, very small to small   .0005–.05

2. Miscellaneous elements
    Random part selections   .1
    Stamp part with rubber stamp,
        1st part   .1
        Each three add'l parts   .1
    Air clean   .1
    Wipe with shop cloth,
        1st surface   .1
        Five add'l surfaces   .1
    Remove part from envelope, return   .1
    Wrap, unwrap with Kraft paper   .2
    Tagging lot or part   .1
    Write signature, minor statements   .1

3. Micrometer gaging

Sample Size	No. of Miked Dimensions/Part									
	1	2	3	4	5	6	7	8	9	10
1					1.5	3.1	4.7	6.3	7.9	9.5
2				.7	2.3	3.9	5.5	7.1	8.7	10.3
3				1.4	4.6	5.4	6.2	7.8	9.4	11.0
5			1.3	2.9	4.5	6.1	7.7	9.4	11.0	12.6
10	1.9	3.5	5.1	6.7	8.3	9.9	11.5	13.1	14.7	16.4
15	5.7	7.3	8.9	10.5	12.1	13.7	15.3	16.9	18.5	20.1
25	13.3	14.9	16.5	18.1	19.7	21.3	22.9	24.5	26.1	27.7

Add'l sample unit   .5
Expected time to make 1 dimension/unit   .2

4. Plug gaging

Sample Size	No. of Dimensions/Part									
	1	2	3	4	5	6	7	8	9	10
1					.3	1.1	1.9	2.7	3.4	4.2
2				.1	.8	1.6	2.4	3.2	3.9	4.7
3				.6	1.3	2.1	2.9	3.7	4.4	5.2
5			.8	1.6	2.3	3.1	3.9	4.7	5.4	6.2
10	1.7	2.5	3.3	4.1	4.8	5.6	6.4	7.1	7.9	8.7
15	4.2	5.0	5.8	6.6	7.3	8.1	8.9	9.6	10.4	11.2
25	9.2	10.0	10.8	11.5	12.3	13.1	13.9	14.6	15.4	16.2

Add'l dimension .8
Add'l sample unit .5
Expected time to plug 1 dimension/unit .1

5. Scale and snap gage

Sample Size	No. of Like Dimensions/Part				
	1	2	3	4	5
1			.3	1.1	1.9
2			.5	1.4	2.2
3			.8	1.7	2.5
5		.6	1.4	2.2	3.1
10	1.2	2.0	2.8	3.7	4.5
15	2.6	3.4	4.2	5.1	5.9
25	5.4	6.2	7.1	7.9	8.7

Add'l dimension .8
Add'l sample unit .3
Expected time to measure 1 dimension/unit .1

6. Radius gage

Sample Size	No. of Like Dimensions/Part				
	1	2	3	4	5
1			.1	.3	.5
2		.1	.2	.4	.6
3		.2	.3	.5	.7
5	.2	.4	.6	.7	.9
10	.8	.9	1.1	1.3	1.5

Add'l dimension .2
Add'l sample unit .1
Expected time to measure 1 radius/unit .04

7. Thickness gage

Sample Size	No. of Like Dimensions/Part				
	1	2	3	4	5
1		.3	.8	1.4	1.9
2	.1	.7	1.3	1.8	2.4
3	.6	1.1	1.7	2.3	2.8
5	1.5	2.1	2.6	3.2	3.7
10	3.7	4.3	4.9	5.4	6.0

Add'l dimension .6
Add'l sample unit .4
Expected time to measure 1 dimension/unit .18

8. Visual check

Sample Size	No. of Like Dimensions/Part				
	1	2	3	4	5
1			.1	.3	.5
2			.2	.4	.6
3		.2	.4	.5	.7
5	.2	.4	.6	.8	1.0
10	.9	1.1	1.3	1.5	1.7

Add'l dimension .2
Add'l sample unit .1
Expected time to visually check 1 dimension/unit .04

9. Vernier caliper and protractor

Sample Size	No. of Like Dimensions/Part									
	1	2	3	4	5	6	7	8	9	10
1	.2	.6	.9	1.2	1.5	1.8	2.1	2.5	2.8	3.1
2	.4	.7	1.1	1.4	1.7	2.0	2.3	2.6	2.9	3.3
3	.6	.9	1.2	1.6	1.9	2.2	2.5	2.8	3.1	3.4
5	1.0	1.3	1.6	1.9	2.2	2.5	2.9	3.2	3.5	3.8
10	1.9	2.2	2.5	2.8	3.1	3.4	3.8	4.1	4.4	4.7
15	2.8	3.1	3.4	3.7	4.0	4.4	4.7	5.0	5.3	5.6

Add'l dimension .3
Add'l sample unit .2
Expected time to measure 1 dimension/unit .08

10. Plug and ring thread gaging

Sample Size	No. of Like Dimensions/Part									
	1	2	3	4	5	6	7	8	9	10
1					1.0	3.1	5.3	7.4	9.5	11.6
2				.1	2.2	4.4	6.5	8.6	10.7	12.8
3				1.4	3.5	5.6	7.7	9.8	12.0	14.1
5			1.7	3.8	5.9	8.1	10.2	12.3	14.4	16.5
10	3.6	5.7	7.8	10.0	12.1	14.2	16.3	18.4	20.6	22.7
15	9.7	11.9	14.0	16.1	18.2	20.3	22.5	24.6	26.7	28.8
25	22.0	24.2	26.3	28.4	30.5	32.6	34.8	36.9	39.0	41.1

Add'l dimension 2.1
Add'l sample unit 1.2
Expected time to thread gage 1 dimension/unit .26

11. Ring gage

Sample Size	No. of Like Dimensions/Part									
	1	2	3	4	5	6	7	8	9	10
1				.2	.7	1.1	1.6	2.0	2.5	2.9
2				.4	.9	1.3	1.8	2.2	2.7	3.1
3			.2	.6	1.1	1.5	2.0	2.4	2.8	3.3

| | No. of Like Dimensions/Part | | | | | | | | | |
Sample Size	1	2	3	4	5	6	7	8	9	10
5		.1	.5	1.0	1.4	1.9	2.3	2.8	3.2	3.7
10	.6	1.0	1.5	1.9	2.4	2.8	3.3	3.7	4.1	4.6
15	1.5	2.0	2.4	2.8	3.3	3.7	4.2	4.6	5.1	5.5

Add'l dimension .4
Add'l sample unit .2
Expected time to measure 1 dimension/unit .06

12. Optical comparator

| | No. of Like Dimensions/Part | | | | | | | | | |
Sample Size	1	2	3	4	5	6	7	8	9	10
1				.9	3.1	5.4	7.6	9.9	12.1	14.4
2		.6	2.8	5.1	7.3	9.6	11.8	14.1	16.3	18.6
3	2.5	4.8	7.0	9.3	11.5	13.8	16.0	18.3	20.5	22.8
5	11.0	13.2	15.5	17.7	20.0	22.2	24.5	26.7	29.0	31.2
10	32.0	34.3	36.5	38.8	41.0	43.3	45.5	47.8	50.0	52.3
15	53.1	55.3	57.6	59.8	62.1	64.3	66.6	68.8	71.1	73.3
25	95.2	97.4	99.7	102	104	106	109	111	113	115

Add'l dimension 2.2
Add'l sample unit 4.2
Expected time to measure 1 dimension/unit .77

13. Counting teeth

| | No. of Teeth/Part | | | | | | | | | |
Sample Size	10	20	40	60	80	100	120	140	160	200
1				.3	.9	1.6	2.2	2.8	3.5	4.7
2			.5	1.1	1.7	2.4	3.0	3.6	4.2	5.5
3	.3	.6	1.3	1.9	2.5	3.1	3.8	4.4	5.0	6.3
5	1.9	2.2	2.8	3.4	4.1	4.7	5.3	6.0	6.6	7.8

Add'l teeth .03
Add'l sample unit .78

14. Mike gears over wires

Sample Size	1	2	3	5	10	15	25	Add'l
Min	2.3	3.0	3.7	5.1	8.6	12.2	19.3	.7

15. Bench micrometer, flat plate

Sample Size	1	2	3	5	10	15	25	Add'l
Min	2.3	2.5	2.7	3.0	4.0	4.9	6.8	2.1

16. Bench center and dial indicator

Sample Size	1	2	3	5	10	15	25	Add'l
Min	3.9	4.3	4.7	5.5	7.4	9.3	13.3	3.6

17. Bench centers, dial indicator, plug member

Sample Size	1	2	3	5	10	15	25	Add'l
Min	3.1	3.6	3.8	4.8	7.3	9.7	14.7	.5

18. V-Block, dial indicator, and flat plate

Sample Size	1	2	3	5	10	15	25	Add'l
Min	2.3	2.7	3.0	3.8	5.7	7.6	11.4	.4

19. Coordinate measuring machine .75/dimension

ELECTRONIC FABRICATION

23.1 Component Sequencing Machines

DESCRIPTION

The electrical component sequencer is a computer-controlled machine which selects electrical components, i.e., resistors, capacitors, diodes, etc., and places them on a tape in a specific order. The tape is used by other automatic machines (see Tables 23.2 and 23.3 for examples) which place these components onto circuit boards at high rates of speed. The sequencer has the capability of sequencing any series of standard components. The number of components in a certain sequence may be from ten components to over a hundred components. The lot or run may consist of any desired amount of these sequences. A "sequence" is the terminology used to mean one printed circuit board. The operator initiates the computer program to operate the machine in the desired manner according to instructions. The operation of the sequencer is fully automatic and only requires periodic adjustments and corrections (replacing missing components, adjusting alignment, clearing scrap, etc.). The number of input reels which have only one type of electrical component vary according to need. Activation of operating mechanisms, such as the heads which align and cut off component leads, is powered pneumatically. A machine may have the ability to stop feed automatically when a component is missing from a sequence when indicated by an optical sensor. The computer also indicates which part is missing. Figure 23.1 is an example of an axial lead component sequencer with loading rates to 25,000 components per hour.

ESTIMATING DATA DISCUSSION

The term "head change" means loading a particular component part reel and placing the tape of identical components in the head, which will cut off and place the component onto the conveyor. There are a number of these heads available, which allows for a sequence of different kinds of ordered components. The setup is performed by two operators and is reflected in the time. The head is either removed from the mounting by loosening a nut or the head is supplied with a quick release and snap-on mechanism. The number of head changes depends on the setup of the previous run and the needs of the sequence to be performed. Each head change is multiplied by the value. A printed circuit board may require ten components, which may be the same as the previous run. The number of head changes presents a confusing choice if the estimator is unaware of how the boards are to be scheduled. For purpose of our examples, we have adopted the rule that 50% of the heads need to be changed. The estimator will need to consider his or her experience and be guided by that.

Element 1, "sequence," represents the actual start and run of the machine. The feed is constant but is interrupted by various adjustments which must be made to ensure the quality of the output sequence. These interruptions are included within the data.

The estimating method selects the time needed to produce one sequence (which will be components on one circuit board) and use the number of sequences as

FIGURE 23.1 Universal Instruments Model 2596 Axial Lead Component Sequencer.

the lot. The unit estimate is the complete sequence rather than a single electronic component. Total time for the run is the number of sequences desired multiplied by the unit estimate.

Element 1 is an extensive listing of a constant time. It was prepared as a listing to aid estimating convenience. If the number of components falls between two table values, use the higher value as little error is introduced. Note that there is only one table, as the handling and other irregular work elements have been consolidated into this one element.

EXAMPLES

A. A printed circuit board will use 91 components. The 91 components are arranged in sequence, and 650 sequences are required for the lot. Assume that 32 head changes are necessary. Find the cycle, hr/100 units, and the lot estimate.

Table Number	Process Description	Table Time	Adjustment Factor	Cycle Minutes	Setup Hours
23.1-S	Head changes	.05	32		1.60
23.1-1	91 components in sequence	1.00		1.00	—
	Total			1.00	1.60
	hr/100 units	1.667			
	Lot estimate	12.44 hr			

B. A printed circuit card has a bill of material listing of 182 components. Two hundred and fifty sequences are needed. We anticipated that 18 head changes are required. Find the cycle, hr/100 units, lot estimates, and the shop value (pc/hr).

Table Number	Process Description	Table Time	Adjustment Factor	Cycle Minutes	Setup Hours
23.1-S	Head changes	.05	18		.90
23.1-1	182 components in sequence	2.00		2.00	—
	Total			2.00	.90
	Hr/100 units	3.34			
	Lot estimate	9.25 hr			
	Pc/hr	30			

C. Use Table 23.1 to estimate a card type of 36 components. There are 55 cards necessary. Assume that five head changes are necessary. Find the unit, hr/100 units, and lot estimates.

Table Number	Process Description	Table Time	Adjustment Factor	Cycle Minutes	Setup Hours
23.1-S	Head changes	.05	5		.25
23.1-1	36 components	.40		.40	—
	Total			.40	.25
	hr/100 units	.668			
	Lot estimate	.62 hr			

TABLE 23.1 COMPONENT SEQUENCING MACHINES

Setup, each head change .05 hr

Operation elements in estimating minutes

1. Sequence

Components	Min	Components	Min	Components	Min
10	.11	42	.45	132	1.40
11	.12	47	.50	141	1.50
12	.13	52	.55	150	1.60
13	.14	56	.60	160	1.70
14	.15	61	.65	169	1.80
15	.16	66	.70	179	1.90
16	.17	71	.75	188	2.0
17	.18	75	.80	282	3.0
18	.19	80	.85	376	4.0
19	.20	85	.90	470	5.0
20	.21	89	.95	564	6.0
21	.22	94	1.00	650	7.0
22	.23	99	1.05	752	8.0
23	.24	103	1.10	847	9.0
24	.25	108	1.15	941	10.0
28	.30	113	1.20	1035	11.0
33	.35	118	1.25	1129	12.0
38	.40	122	1.30	Add'l	.011

23.2 Component Insertion Machines

DESCRIPTION

The following estimating data are for component inserting machines used in the circuit board industry. The machine is used for inserting micro-chips that have been packaged in a certain size case (usually .3-in. or .6-in. wide). The packages can have any number of leads but usually have 8, 16, or 32. The machine can be powered mechanically or hydraulically. Figure 23.2 is an example of a dual inline package inserting machine.

Small clamps hold a board to a table. The table moves in an x-y plane, positioning the board for insertion of the package. Next, a head lowers in the z-direction inserting the correct package into the printed circuit (PC) holes while cutting the leads underneath

FIGURE 23.2 Universal Instruments Model 6772A Multi-Module Dip Inserter.

and crimping either with an outside or inside crimp at varying angles. The head can place any size package according to its size. Packages are chosen above the machine, where they are held in tubes. The tables can hold from one to several boards depending on the sizes of the table and designs.

ESTIMATING DATA DISCUSSION

Setup of the machine requires that a computer program be loaded into memory. The computer controls the movement of the table in the x-y plane, movement of the package loading head in the z direction, and the correct choice of package corresponding to its position on the circuit board. The operator loads the tubes and containers which are within easy reach

With setup completed, the next step is to begin the process. Elements are relatively simple. Element 1 includes picking up an empty board, placing it on the machine, removing the completed board, and placing it in a stack. If the table is holding two or more boards it is often possible for the operator to perform this task while the machine works on other boards.

The entry for Element 2 is the number of components. It may be convenient to use a higher value on the table if the exact number is not shown.

The estimator may wish to find the productive hour cost for this operation. This PHC reflects a base area wage that is national U.S. norm data.

EXAMPLES

A. A board must have 10 chips mounted. Table loading will accommodate three boards, but loading of two boards is internal to the machine automatic cycle of inserting components. The job consists of 50 boards. Find the unit estimate to process one board.

Table Number	Process Description	Table Time	Adjustment Factor	Cycle Minutes	Setup Hours
23.2-S	Basic setup	.1			.1
23.2-S	Fill containers, small	.1			.1
23.2-S	Load program, manual	.3			.3
	Setup subtotal				.5

Table Number	Process Description	Table Time	Adjustment Factor	Cycle Minutes	Setup Hours
23.2-1	Load one board	.24		.24	
	Load two boards, internal to automatic cycle	0.0		0.0	
	Insert 30 chips	.48		.48	
	Total cycle time			.72	
	Cycle estimate	.24 min			
	hr/100	.400			
	Pc/hr	250			

B. A board must be manufactured one-by-one for a quantity of 1000. Each board will have 90 chips. Program loading to the machine is with satellite link. Find the setup and unit estimate and hr/100 units for the job.

Table Number	Process Description	Table Time	Adjustment Factor	Cycle Minutes	Setup Hours
23.2-S	Basic setup	.1			.1
23.2-S	Fill containers	.2			.2
23.2-S	Load program, satellite	.2			.2
	Setup subtotal				.5
23.2-1	Load one board	.24		.24	
	Insert 90 chips	1.44		1.44	
	Cycle estimate			1.68	
	hr/100 units	2.800			

C. Five hundred boards having 55 chips each are to be made on a machine table which allows side-by-side processing of two boards. The machine stops to load each board; however, this machine action is really no different than single-board processing. The program is centrally loaded. The inserting head is also scheduled for a change. Find the lot time.

Table Number	Process Description	Table Time	Adjustment Factor	Cycle Minutes	Setup Hours
23.2-S	Basic setup	.1			.1
23.2-S	Fill containers	.2	2		.4
23.2-S	Load computer	.1			.1
	Total setup				.6
23.2-1	Handle	.24		.24	
23.2-2	Insert components, 55	.88		.88	
	Cycle estimate			1.12	
	hr/100 units	1.867			
	Lot estimate	9.93 hr			

TABLE 23.2   COMPONENT INSERTION MACHINES

Setup
    Basic   .1 hr
        Fill containers with material:
            small job   .1 hr
            large job   .2 hr
        Change head for different width of package   .1 hr

Operation elements in estimating minutes

1.   Handling, combined   .24
      Pick up board   .04
      Place on table and secure   .10
      Remove from table   .05
      Place in stack   .05

2. Insert components

No. of Components	5	8	10	15	18	20	25
Min	.08	.13	.16	.24	.29	.32	.40
No. of Components	28	30	35	40	45	50	55
Min	.45	.48	.56	.64	.72	.80	.88
No. of Components	60	65	70	75	80	90	100
Min	.96	1.04	1.12	1.20	1.28	1.44	1.60

23.3 Axial-Lead Component Insertion Machines

DESCRIPTION

Axial-lead component insertion machines automatically insert axial-shaped electrical components into printed circuit board (PC) holes. Raw materials for the machine are a printed circuit board and a reel or tape. The components are previously sequenced and placed on the tape. The components are axial-shaped, such as resistors, and have two wires or leads extending from the ends of the resistor. The wire diameters are about .015 to .037 in. depending upon the machine or design. The body length of the resistor may vary. The variable distance that is programmed is from center-to-center of the form leads. The rolls will have many sequences. One sequence may have sufficient numbers for the axial-shape components as required by the PC board. The machines considered here are numerically controlled, which directs table positioning and machine head motion. Figure 23.3 is an example of a single-head variable center distance axial-component insertion machine.

The components have been previously sequenced into a roll via a sequence machine which is considered by Table 23.1. The roll of ordered components is mounted on the insertion machine and the components are fed onto the upper machine head.

Printed circuit boards are loaded into a fixture mounted upon a movable table. The boards are located in the fixture using pins and are clamped. A variety of fixture holding devices are possible with one or more boards or x-y fixtures.

The machine process is as follows. The table moves to a designated position over the PC board as commanded by the NC control. The upper-machine head cuts the component from the roll of tape and inserts it into the two holes. A lower-machine head raises and cuts the leads to length and bends them to either 90° or 45°. The heads retract and the table moves to a new position. The process of insertion continues.

The operator can override an automatic stop to insert more than one board without restarting the machine.

ESTIMATING DATA DISCUSSION

The basic setup time, given in hours, includes getting the computer tape and instruction manual, reading the tape into the computer, getting the lot of boards to be inserted and the sequenced roll of components, examining the manual, and adjusting the fixture. Time may be added to the basic time for a fixture change.

The operation elements include the loading and unloading of a board, start, and the insertion of components into one board. If more than one board can be loaded into a fixture and there is a delay in the cycle time, additional handling may be considered. Usually, the unload and load time will be used once since loading of multiple boards occurs internal to the machine cycle. If an x-y fixture is used, the insertion time should only be used once since the times are for a completely inserted board. If the number of components is less than 35, the handling time should be added for as many boards per fixture since the operator does not have enough time to remove them internal to the automatic cycle of inserting components.

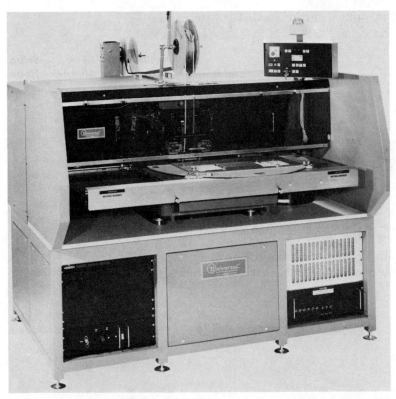

FIGURE 23.3 Universal Instruments Model 6287A Variable Center Distance Insertion Machine.

EXAMPLES

A. Insert 380 components into a 36 × 24-in. printed circuit board. Order size is 50 boards. The fixture holds one board and the operation requires a special fixture. Determine setup and the unit estimate.

Table Number	Process Description	Table Time	Adjustment Factor	Cycle Minutes	Setup Hours
23.3-S	Setup, basic	.45			.45
23.3-S	Fixture	.05			.05
	Total setup				.50
23.3-1	Unload, load, and start machine	.31		.31	
23.3-2	Insert components	2.40		2.40	
	Total			2.71	

B. Insert 210 components into a 12 × 8-in. printed circuit board. The order size is 100 boards. The fixture holds two boards. Determine the hr/100 units.

Table Number	Process Description	Table Time	Adjustment Factor	Cycle Minutes	Setup Hours
23.3-S	Setup	.45			.45
23.3-1	Unload, load, start machine	.31	½	.15	
23.3-2	Insert components	1.38	½	.69	
23.3-2	Insert second board	1.38	½	.69	
	Total			1.54	.45
	Unit estimate	1.54 min			
	hr/100 units	2.567			

Axial-Lead Component Insertion Machines

C. Insert 180 components into 12 × 8-in. printed circuit board. The fixture is an *x-y* fixture.

Table Number	Process Description	Table Time	Adjustment Factor	Cycle Minutes	Setup Hours
23.3-S	Setup	.45			.45
23.3-S	Fixture	.05			.05
	Total setup				.50
23.3-1	Unload, load, start machine	.31		.31	
23.3-2	Insertion of components	1.20		1.20	
	Total			1.51	
	hr/100 units	2.517			
	Lot estimate	1.00 hr			

D. Insert 25 components into a 12 × 6-in. printed circuit board. The order size is 220 boards. The fixture holds three boards. No new fixture is needed. Determine the unit estimate.

Table Number	Process Description	Table Time	Adjustment Factor	Cycle Minutes	Setup Hours
23.3-S	Setup	.45			.45
23.3-1	Unload, load, start machine,	.31	⅓	.103	
	Handling of 2nd board	.31	⅓	.103	
23.3-1	Handling of 3rd board	.31	⅓	.103	
23.3-2	Insert components	.26	⅓	.087	
23.3-2	Insert 2nd board	.26	⅓	.087	
23.3-2	Insert 3rd board	.26	⅓	.087	
	Total			.57	.45
	Unit estimate	.57 min			

E. Insert 30 components into a 12 × 8-in. printed circuit board. The fixture holds two boards. Determine the cycle time and unit estimate.

Table Number	Process Description	Table Time	Adjustment Factor	Cycle Minutes	Setup Hours
23.3-S	Setup	.45			.45
23.3-1	Unload, load, start machine	.31	½	.155	
23.3-1	Handling 2nd board	.31	½	.155	
23.3-2	Insert components	.26	½	.13	
23.3-2	Insert 2nd board	.26	½	.13	
	Total			.57	.45
	Unit estimate	57			

TABLE 23.3 AXIAL-LEAD COMPONENT INSERTION MACHINES

Setup
    Basic   .45 hr
    Fixture additional   .05 hr

Operation elements in estimating minutes

1. Load, unload, start machine   .31

2. Insert components for one board

No. of Components	Min
30	.26
36	.29
43	.34
53	.39
62	.45
73	.52
86	.60
102	.67
118	.79
139	.91
162	1.04
188	1.20
218	1.38
252	1.58
292	1.82
337	2.09
390	2.40
450	2.76
519	3.18
Add'l	.0006

23.4 Light-Directed Component-Selection Insertion Machines

DESCRIPTION

Automatic printed circuit board stuffing machines are designed to aid the operator in the insertion of electronic components into a printed circuit board. Parts to be inserted are kept in a set of trays next to the operator. By placing a light over the surface of the board, the machine directs the operator where to insert the component. The machine will indicate the direction of the polarized parts, crimp the leads down, and cut off any excess. The machine controls the order of parts to be inserted by moving the trays up or down or rotating them. This allows the operator access to the tray containing the desired component.

Numerous machines are available to aid in loading printed circuit boards. These data are confined to machines that show the part to be loaded, position of the part, and its preparation for soldering. Although the machine does not insert the part, time is eliminated for searching for the part in the bins and finding its location on the board.

Figure 23.4 shows a typical circuit board assembly

FIGURE 23.4 Light-directed component-selection insertion machine.

machine. The circuit board is moved under the illumination head and a rectangle of light indicates the part's location and orientation. The parts bin is automatically cycled to expose the correct part for the operator. Parts of the DIP type (dual in line package) are stored in the DIP part dispenser rack. A light below each slot indicates which part is to be loaded. After the part is inserted, the operator holds the part in position while the leads of the component are cut to size and crimped down. Cycling of the machine is computer controlled and is cued by the operator with a foot pedal.

ESTIMATING DATA DISCUSSION

Setup provides time to fill the bins of the part trays. This involves a review of the schematic for the circuit board, acquiring parts from the plant's part store, and filling the specified bins with each component. It should be noted that the time to program the machine to sequence is not included due to the high variability in debugging. Multiple setups may be necessary depending on the lot quantity since the bins may not have sufficient volume to hold the components.

The task of installing a single component on the circuit board has been broken down to separated elements to facilitate an accurate estimate. A total of four separate elements were chosen: reach, install insulator, insert, and cut crimp advance.

The time from the advance of the machine over the next component position to the time the operator fetches the next part out of the bin and positions it over the circuit board is defined as the reach, Element 2. The time to complete this task depends on the ability to untangle a part and on the reach distance; both are constant for a work area. For consecutive parts from the same bin, only one reach may be required. In some cases, it is necessary to install a small insulator, Element 3, or heat sink, on the part after it has been fetched. The part and its insulator are stored in the same bin disengaged from one another.

The time to get the terminals of a part into their respective holes is defined as the insert time, Element 4. Time increases with the number of terminals or leads because all terminals must be partially lined up with their respective holes.

After the part is inserted the machine will either cut crimp and advance for two terminal devices or merely advance for components with more than two terminals. The time for each operation is constant for a single component.

The tables give an estimate of assembly time per circuit board. Insertion time is given in numbers of parts with the same number of terminals, and number of terminals per part to a maximum of 16 terminals. Devices with more than 16 terminals do not require more insertion time than the maximums given. Parts that require sockets or other types of minor assembly before insertion should be estimated using data for installing insulators.

EXAMPLES

A. Determine both the unit and lot estimate for 20 boards if each needs 100 parts inserted (all two-terminal) The parts do not need crimping.

Table Number	Process Description	Table Time	Adjustment Factor	Cycle Minutes	Setup Hours
23.4-S	Setup	0.5			.5
23.4-1	Load board	1.08		1.08	
23.4-2	Reach (100 parts)	2.95		2.95	
23.4-4	Insert (100 2-terminal)	12.12		12.12	
23.4-5	Advance (100 parts)	2.59		2.59	
23.4-6	Inspect board	1.00		1.00	
	Total			19.74	.5
	Lot estimate	7.08 hr			

B. Determine the unit estimate for a board which will have 50 2-terminal, 10 3-terminal, 5 8-terminal and 3 16-terminal devices inserted. The parts are not crimped down.

Table Number	Process Description	Table Time	Adjustment Factor	Cycle Minutes	Setup Hours
23.4-S	Setup	0.5			.5
23.4-1	Load board	1.08		1.08	
23.4-2	Reach (68 parts)	2.34		2.34	
23.4-4	Insert (50 2-terminal)	6.32		6.32	

Table Number	Process Description	Table Time	Adjustment Factor	Cycle Minutes	Setup Hours
23.4-4	Insert (10 3-terminal)	1.65		1.65	
23.4-4	Insert (5 8-terminal)	2.29		2.29	
23.4-4	Insert (3 16-terminal)	4.37		4.37	
23.4-5B	Advance (68 parts)	1.89		1.89	
	Inspect board	1.00		1.00	
	Total			20.94	.5

C. Determine the lot estimate for a board which will have the same specifications as in the previous example with the addition of terminal insulators on the 3- and 8-terminal parts.

Table Number	Process Description	Table Time	Adjustment Factor	Cycle Minutes	Setup Hours
23.4-S	Setup	.5			.5
23.4-1	Load board	1.08		1.08	
23.4-2	Reach (68 parts)	2.34		2.34	
23.4-3	Install insulator (10 3-terminal)	1.47		1.47	
23.4-3	Install insulator (5 8-terminal)	.74		.74	
23.4-4	Insert (50 2-terminal)	6.32		6.32	
23.4-4	Insert 10 3-terminal)	1.65		1.65	
23.4-4	Insert (5 8-terminal)	2.29		2.29	
23.4-4	Insert (3 16-terminal)	4.37		4.37	
23.4-7	Advance (68 parts)	1.89		1.89	
23.4-9	Inspect board	1.00		1.00	
	Total			23.15	.5
	hr/100	38.593			
	Lot estimate	31.37 hr			

D. Determine both the unit and lot estimates for a board with a total of 90 parts. Forty of the parts will be 2-terminal and will need to be crimped. There will also be 10 each of 3, 6, 8, 14, and 16-terminal devices. These will not require crimping. The 6 and 8-terminal devices will need terminal insulators and there will also be six board insulators. Fifteen boards are needed.

Table Number	Process Description	Table Time	Adjustment Factor	Cycle Minutes	Setup Hours
23.4-S	Setup	.5			.5
23.4-1	Load board	1.08		1.08	
23.4-2	Reach (90 parts)	2.74		2.74	
23.4-3	Install insulator (10 6-terminal)	1.47		1.47	
23.4-3	Install insulator (10 8-terminal)	1.47		1.47	
23.4-4	Insert part (40 2-terminal)	5.22		5.22	
23.4-4	Insert part (10 3-terminal)	1.65		1.65	
23.4-4	Insert part 10 6-terminal)	2.95		2.95	
23.4-4	Insert part (10 8-terminal)	3.82		3.82	
23.4-4	Insert part (10 14-terminal)	6.42		6.42	
23.4-4	Insert part (10 16-terminal)	7.28		7.28	
23.4-3	Install insulator (6 insulators)	.88		.88	
23.4-5A	Machine crimp (40 parts)	1.84		1.84	
23.4-5B	Machine advance (50 parts)	1.29		1.29	
23.4-6	Inspect board	1.001		1.00	
	Total			61.02	.5
	Lot estimate	15.76 hr			

Light-Directed Component-Selection Insertion Machines

TABLE 23.4 LIGHT-DIRECTED COMPONENT-SELECTION INSERTION MACHINES

Setup .50 hr

Operation elements in estimating minutes

1. Load board and unload board 1.08

2. Reach for part(s)

No. Parts	Min	No. Parts	Min	No. Parts	Min
30	1.51	44	1.79	64	2.20
33	1.57	48	1.87	71	2.34
36	1.63	53	1.98	78	2.48
40	1.71	58	2.08	85	2.63
				Add'l	0.21

3. Install insulator on board

No. Insulators	Min	No. Insulators	Min	No. Insulators	Min
3	0.44	8	1.18	13	1.91
4	0.59	9	1.32	15	2.21
5	0.74	10	1.47	16	2.35
6	0.88	11	1.62	18	2.65
7	1.03	12	1.76	20	2.94
				Add'l	.15

4. Insert component

No.	\multicolumn{7}{c}{No. of Terminals on Components}

No.	2	3	4	6	8	14	16
6	0.73	0.99	1.25	1.77	2.29	3.85	4.37
7	0.85	1.15	1.46	2.06	2.67	4.49	5.10
8	0.97	1.32	1.67	2.36	3.05	5.13	5.83
9	1.09	1.48	1.87	2.65	3.43	5.77	6.55
10	1.22	1.65	2.08	2.95	3.82	6.42	7.28
11	1.34	1.81	2.29	3.24	4.20	7.06	8.00
12	1.46	1.98	2.50	3.54	4.58	7.70	8.74
13	1.58	2.14	2.71	3.83	4.96	8.34	9.47
15	1.82	2.47	3.12	4.42	5.72	9.62	10.92
16	1.94	2.63	3.33	4.72	6.10	10.26	11.65
18	2.19	2.97	3.75	5.31	6.87	11.55	13.11
20	2.43	3.30	4.16	5.90	7.63	12.83	14.56
22	2.67	3.63	4.58	6.49	8.39	14.11	16.02
24	2.92	3.96	5.00	7.08	9.16	15.40	17.48
27	3.28	4.45	5.62	7.96	10.30	17.32	19.66
29	3.52	4.78	6.04	8.55	11.06	18.60	21.17
32	3.89	5.27	6.66	9.43	12.21	20.53	23.30
35	4.25	5.77	7.29	10.32	13.35	22.45	25.49
39	4.74	6.53	8.12	11.50	14.88	25.02	28.40
43	5.22	7.09	8.95	12.68	16.40	27.58	31.31
47	5.71	7.75	9.78	13.85	17.93	30.15	34.22
52	6.32	8.57	10.83	15.33	19.84	33.36	37.86

	No. of Terminals on Components						
No.	2	3	4	6	8	14	16
57	6.93	9.39	11.87	16.81	21.75	36.57	41.51
63	7.65	10.38	13.11	18.57	24.03	40.42	45.87
69	8.38	11.37	14.36	20.34	26.32	44.26	50.24
76	9.23	12.53	15.82	22.41	29.00	48.75	55.34
84	10.20	13.85	17.49	24.77	32.05	53.89	61.17
Add'l part	0.12	0.16	0.21	0.30	0.38	0.64	0.73

5. Cut and crimp leads, advance machine

   A. Cut, crimp and advance

No. Parts	Min	No. Parts	Min
28	1.25	55	2.47
31	1.39	60	2.69
34	1.52	66	2.96
37	1.66	73	3.27
41	1.84	80	3.59
45	2.02	Add'l	.046
50	2.24		

   B. Advance only

No. Parts	Min	No. Parts	Min
28	0.72	50	1.29
31	0.80	55	1.42
34	0.88	60	1.55
37	0.96	66	1.71
41	1.06	73	1.89
45	1.16		

6. Inspect board   1.00

23.5 Printed Circuit Board Stuffing

DESCRIPTION

The estimating information included here is for a manual bench assembly. The data are not associated with fundamental motions, and no machines are used in the process, but a few hand tools are available when needed. A worker completes an entire printed circuit (PC) board completely; that is, there is no conveyor line. The worker starts by adding each element to the PC board one at a time. Every element is carefully checked for its identification number. After the element is in place on the board, it is soldered to the board. One or more parts may be soldered to the board at the same time.

Once all of the parts are attached to the board, the worker cleans it by brushing it with alcohol. The board is then dried with an air hose. Next the PC board is sent to the inspector where it must pass an inspection. If any errors are found, they are fixed immediately and it is reinspected. When the board has passed the inspection it is chemically sealed with hysol.

Those factors not considered are the drying time of the chemical seal, an overall final inspection, and electronic and environment testing.

Every item on the bench is easily accessible. The only times the worker needs to move from the desk is when hysoling, cleaning, or taking a PC board to the inspector.

ESTIMATING DATA DISCUSSION

In these data a setup time is not allowed.
Because of the high quality of a board, a handle is

needed for each part placed on the board. The parts (i.e., chips, resistors, electronic components, etc.) are carefully inspected for their identification number so that it matches the one given on the blueprint. The parts used in the PC board stuffing can be broken into two categories: components and chips. The data treats these as identical parts. The data separate the parts by the number of pins or wires (called electronic parts in this discussion) being soldered to the board. Separating the electronic parts in this fashion eliminates the problem of separating these by part size.

The data are given in minutes per occurrence. As the data are multiplied by the frequency of occurrence, rounding-off errors may result.

EXAMPLES

A. A PC board has five 2-pin elements, twelve 6-pin elements, a connector with 36 pins, a cleaning, and a single inspection. Determine a unit estimate. Assume only one board is to be fabricated for the lot.

Table Number	Process Description	Table Time	Adjustment Factor	Cycle Minutes	Setup Hours
23.5-1	Handle and inspect parts	.40	17	6.80	
23.5-2	Adjust part to holes				
	2 pin	.40	5	2.00	
	6 pin	.54	12	6.48	
23.5-3	Solder and flux				
	2 pin	.44	5	2.20	
	6 pin	.75	12	9.00	
23.5-5	Insert and solder connectors, 36 pin	3.94		3.94	
23.5-6	Screw on connectors	1.86		1.86	
23.5-4	Cleaning and drying	.73		.73	
23.5-7	Inspection, one board only	28.13		28.13	
	Unit estimate			61.31	

B. Twenty PC boards will be made with the following: three 2-pin parts, five 4-pin parts, two 10-pin parts, three connectors with 12, 14, and 26 pins respectively. Soldering is done after all parts are connected. A single cleaning, inspection, hysoling, and another inspection are required. Determine the unit estimate for this process. Find the lot estimate. The work will be done on a pass-along assembly line. Element 1 of Table 23.5 is not used in this case. Only the first board receives a full inspection.

Table Number	Process Description	Table Time	Adjustment Factor	Cycle Minutes	Setup Hours
23.5-2	Adjust part to holes				
	2 pin	.40	3	1.20	
	4 pin	.44	5	2.20	
	10 pin	.65	2	1.30	
23.5-3	Solder and flux, 2 (3) + 4 (5) + 10 (2) = 41 locations	4.27		4.27	
23.5-5	Insert and solder connectors				
	12 pin	2.96		2.96	
	14 pin	2.96		2.96	
	26 pin	3.58		3.58	
23.5-6	Screw on connectors	1.86	3	5.58	
23.5-4	Cleaning and drying	.73		.73	
23.5-7	Inspection. First board prorated over 20	28.3	1/20	1.42	
	Each additional board	5.40		5.40	
23.5-8	Hysoling	2.32		2.32	
	Cycle estimate			33.92	

C. A board is planned to have the following elements: nineteen 2-pin, twenty-six 3-pin, and seventeen 8-pin. A cleaning is to be done after each set of pins is attached to the board. A connector is then added (36 pins) with another cleaning. This is followed by an inspection and hysoling. Find the unit estimate. A lot of 30 boards is needed.

Table Number	Process Description	Table Time	Adjustment Factor	Cycle Minutes	Setup Hours
23.5-1	Handle and inspect 62 parts	.40	62	24.80	
23.5-2	Adjust part to holes				
	2 pin	.40	19	7.60	
	3 pin	.44	26	11.44	
	8 pin	.59	17	10.03	
23.5-3	Solder and flux				
	2 pin	.44	19	8.36	
	3 pin	.48	26	12.48	
	8 pin	.94	17	15.98	
23.5-5	Insert and solder connectors, 36 pin	3.94		3.94	
23.5-6	Screw on connectors	1.86		1.86	
23.5-4	Cleaning and drying	.73	4	2.92	
23.5-7	Inspection, 28.3/80, first part	28.30	1/80		
	Each subsequent board	5.40		5.40	
23.5-8	Hysoling	2.32		2.32	
	Total			107.13	
	Lot estimate	53.57 hr			

TABLE 23.5 PRINTED CIRCUIT BOARD STUFFING

Setup

Operation elements in estimating minutes   0

1. Handle and inspect a part   .40

2. Adjust an electronic element for a hole and insert in board

No. of Pins	Min
2	.40
4	.44
5	.49
7	.54
9	.59
11	.65
13	.71

3. Solder and flux an element to board

No. of Pins	Min
2	.44
3	.48
4	.59
5	.71
6	.78
7	.86
8	.94
9	1.04
10	1.14
12	1.26
13	1.38
14	1.52
16	1.67
Average	.104 each

4. Clean and air dry board .73

5. Insert and solder connectors

No. of Pins	Min
15	2.96
22	3.26
30	3.58
38	3.94
47	4.33
57	4.78

6. Connector to the board with screws 1.86

7. Inspect board
   First time 28.3
   Each additional time 5.4

8. Hysol the board 2.32

23.6 Wave Soldering Machines

DESCRIPTION

A wave soldering machine applies a layer of molten solder to a printed circuit board. Solidification of the metal electrically secures the inserted components into place. The machine consists of a conveyor that mechanically transports the circuit boards through the system. Machines vary, but a typical process for the board is:

1. Flux, which applies a liquid layer to remove surface oxides and to give the filler metal the fluidity to wet the joint surfaces completely;
2. Heating elements, which raise the temperature for proper solder adhesion;
3. Air knives, used to remove excess flux;
4. A pass-over the crest of the liquid solder wave at a set height and;
5. The washing cycle, which may vary from a series of washes, rinses and air knives to a simple dishwasher.

The machines are able to solder different sizes of boards at variable rates of speed. Maximum and minimum rates are determined by the design of the machine and the application limitations. One or two operators may perform the loading and unloading operations, which is important in time estimating. Machine types are variable. Smaller machines, used for smaller lot quantities, make use of conveyor racks which can be adjusted to different board sizes. Larger machines have adjustable width conveyors and are used for large lots. Sometimes they run continuously for one or two shifts.

Figure 23.6A is a sketch of a large unit that is intended for a large quantity of the same size boards. Two operators are required. Figure 23.6B is a sketch of a semiautomatic wave soldering machine intended for short runs or occasional use. It uses either one or two operators. It handles variable sized boards, which are first inserted into fixtures or racks. The racks are returned on a return carriage. A dishwasher may be used for cleaning and flux removal. Figure 23.6C is a photograph of a unit that can solder various sized boards, fine lines, large lands, and multilayer boards. A return conveyor system is also available.

ESTIMATING DATA DISCUSSION

These estimating data cover the wave soldering operation of a continuously operating automatic self-washing machine and semiautomatic machine. These data describe the limits of wave soldering machines, so time estimates should be possible for a variety of applications.

The setup elements are given in hours. Element 1 includes the time required to fill the flux tank, check the level, and ensure that the foamer is working prop-

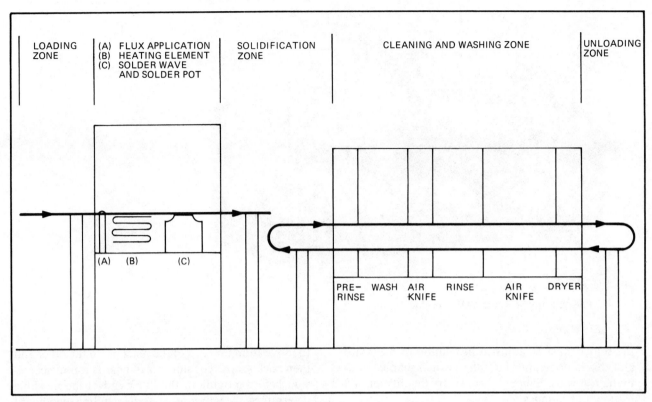

FIGURE 23.6A Automatic wave soldering machine, variable board sizes and speed rates.

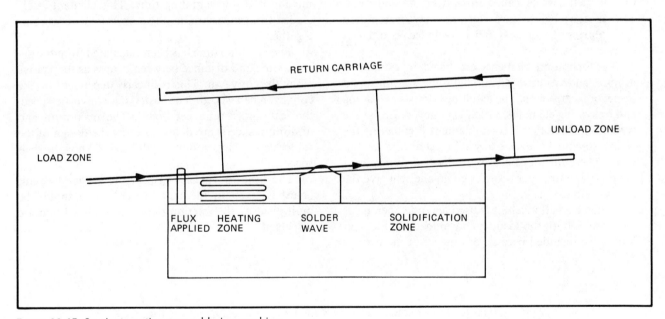

FIGURE 23.6B Semiautomatic wave soldering machine.

erly. The height must be within limits, and this is also included.

Some wave soldering machines use racks which hold the board and transport it along the conveyor. Element 2 lists the time required to adjust each rack to the correct width of the board. The time must be multiplied by the number of conveyor racks that will be used.

A setup accounts for the time to adjust the width of the conveyor for board width. This may require a few fine adjustments and these have been included in the value. To ensure that the board is being accurately soldered, it is moved over the solder wave manually and then inspected. Then it is run automatically to ensure that the belt speed is producing a good solder finish. Elements 5 and 6 give these values. The height of

Figure 23.6C Electrovert Ultrapak Wavesolder System.

the solder must be adjusted in relation to the board, and this is done after both the normal manual automatic test runs. Element 5 considers the different adjustment techniques.

Element 6 is the time required to add solder and to remove the waste, called dross, from the solder pot. This is done before each run and may be repeated again during the process and is included as often as necessary.

The operation elements are listed in estimating minutes and are arranged into handling, machine, and inspection categories. The handling consists of loading and unloading elements, which vary depending on the type of machine that is used. Element 1, gathering the boards, includes the time to walk to and from the position where the lot of boards are located. When trays are used, the time per board is given and will give the correct unit estimate.

When a small machine is used and one operator is performing both the loading and unloading, Element 5 must be included because of the extra time it takes to move from one end of the machine to the other and then back again. The number of boards run in succession before moving to the other end of the machine determines the amount of time required per board unit. Determine the correct value from the table and include it in the handling time. This element is excluded from the estimate if two operators are working together.

The machine time has been calculated from the average time and distance between boards as they move along the conveyor. This is linearly dependent on the conveyor belt speed. The faster the conveyor speed, the less solder time per board. The maximum and minimum speeds are determined by the design of the solder wave. The estimator will need to know the belt speed.

The time to visually inspect the soldered circuit board is given in Element 11. This time should be multiplied by the fraction of boards soldered that are inspected.

EXAMPLES

A. A lot of 72 circuit boards is to be soldered. A machine using conveyor racks will be used and will be run at a rate of 4 ft per min. One operator will load and unload so four conveyor racks will be used. Each board will be inspected. This will be the only lot of the day so the setup is extensive. Find unit and lot estimates.

Table Number	Process Description	Table Time	Adjustment Factor	Cycle Minutes	Setup Hours
23.6-S2	Adjust conveyor racks to correct board width, 4 boards × .01	.04			.04
23.6-S1	Fill flux tank	.12			.12
23.6-S5	Run first board manually	.06			.06
23.6-S7	Adjust height of solder wave, table leg adjust	.07			.07

Table Number	Process Description	Table Time	Adjustment Factor	Cycle Minutes	Setup Hours
23.6-S6	Run second board automatically	.03			.03
	Setup total				.32
23.6-1	Gather tray of boards	.13		.13	
23.6-2	Position board in conveyor rack	.09		.09	
23.6-3	Place rack on conveyor	.04		.04	
23.6-5	Operator moves to opposite end (4 boards run in succession)	.07		.07	
23.6-6	Remove rack from conveyor, remove board	.03		.03	
23.6-7	Place rack on return carriage	.05		.05	
23.6-9	Place board on tray	.04		.04	
23.6-10	Machine time (4 ft per min)	.49		.49	
23.6-11	Inspection, visual; each board	.06		.06	
	Cycle total			1.00	
	Lot estimate	1.52 hr			

B. Using an automatic wave soldering machine, 1100 circuit boards will be soldered. Two operators are required and this will be included in the handling time needed in the estimate. Assume this will be the first run of the day and the setup will be extensive. Conveyor speed is 9 ft per min. One out of 25 boards will be inspected. Find the unit and lot estimates.

Table Number	Process Description	Table Time	Adjustment Factor	Cycle Minutes	Setup Hours
23.6-S1	Fill flux tank	.12			.12
23.6-S4	Fill automatic washers with soap	.03			.03
23.6-S8	Add solder (four 5-pound bars)	.15			.15
	Clean dross from solder pot				
	Adjust solder temperature				
23.6-S3	Adjust conveyor to correct board width	.04			.04
23.6-S5	Run first board manually	.06			.06
23.6-S6	Run second board automatically	.03			.03
23.6-S7	Adjust height of solder wave, crank adjustment	.03			.03
	Setup total				.46
23.6-1	Gather stack of boards, tray	.13		.13	
23.6-4	Place and position boards on conveyor	.07		.07	
23.6-8	Unload board from conveyor	.02		.02	
23.6-9	Place board on tray	.04		.04	
23.6-10	Machine time (9 ft per min)	.34		.34	
23.6-11	Inspection, visual; one out of 25 boards	.06	1/25	.002	
	Cycle total			.602	
	Lot estimate	11.56 hr			

C. A wave soldering machine that uses conveyor racks has a small lot to run. There are 20 boards of one size, six of another, and then two of a small size. The conveyor speed is set at 5 ft per min, and 5 racks will be run in succession. Two operators will be used to load and unload the boards. This is not the first lot run of the day so setup will be average.

Table Number	Process Description	Table Time	Adjustment Factor	Cycle Minutes	Setup Hours
23.6-S	Adjust conveyor racks to correct board width, 5 × .01 (three times)	.01	5 × 3		.15
23.6-S	Run first board manually	.06			.06
23.6-S	Run second board automatically	.03			.03
23.6-S	Adjust height of solder (table leg adjustment)	.07			.07
23.6-S	Clean dross from solder pot	.05			.05
	Setup total				.36
23.6-1	Gather stock, one board at a time	.19		.19	
23.6-2	Position board in conveyor rack, tighten	.09		.09	

Table Number	Process Description	Table Time	Adjustment Factor	Cycle Minutes	Setup Hours
23.6-3	Plack rack in conveyor	.04		.04	
23.6-6	Remove rack from conveyor, remove board	.03		.03	
23.6-7	Place rack on return carriage	.05		.05	
23.6-9	Place board on tray	.04		.04	
23.6-10	Machine time, 5 ft per min	.45		.45	
23.6-11	Inspection, visual one out of two boards	.06	1/2	.03	
	Unit estimate			1.02	
	Lot estimate	.936 hr			

D. A lot of 750 circuit boards is to be soldered, and an automatic wave soldering machine is used. The rate of the conveyor is 7 ft per min, and one out of every ten boards is inspected during the operation. The second shift will begin with this lot so the setup will include most elements. Find the unit and lot estimates.

Table Number	Process Description	Table Time	Adjustment Factor	Cycle Minutes	Setup Hours
23.6-S	Setup total	.46			.46
23.6-1	Gather stack, tray of boards	.13		.13	
23.6-4	Place and position board on conveyor	.07		.07	
23.6-8	Unload board from conveyor	.02		.02	
23.6-9	Place board on tray	.04		.04	
23.6-10	Machine time, 7 ft per min	.39		.39	
23.6-11	Inspection, one out of ten boards	.06	1/10	.006	
	Total			.656	.46
	Lot estimate	8.66 hr			

E. One operator will run the weekly lot of circuit boards for a small company. There are 65 boards of one size and 47 of another that must be soldered. Conveyor racks are used on this type of machine and the belt speed is set at 4.5 ft per min. The operator will perform all of the loading and unloading, plus the setup. Three conveyor racks are run in succession before the operator moves to the opposite end. Find the lot estimate and pc/hr.

Table Number	Process Description	Table Time	Adjustment Factor	Cycle Minutes	Setup Hours
23.6-1	Gather stack, tray of boards	.13		.13	
23.6-2	Position board in conveyor rack	.09		.09	
23.6-3	Place rack on conveyor	.04		.04	
23.6-5	Operator moves to opposite end (3 boards run in succession)	.09		.09	
23.6-6	Remove rack from conveyor, remove board	.03		.03	
23.6-7	Place rack on return carriage	.05		.05	
23.6-9	Place board on tray	.04		.04	
23.6-10	Machine time, 4.5 ft per min	.45		.45	
23.6-11	Inspection, one out of two boards	.06	1/2	.03	
	Cycle total			.95	
	Lot estimate	2.26 hr			
	Pc/hr	63			

TABLE 23.6 WAVE SOLDERING MACHINES

Setup

S1.	Fill flux tank and check specific gravity	.12 hr
S2.	Adjust conveyor racks to correct width, per rack	.01
	Adjust conveyor to correct board width	.04

| S3. | Fill automatic washers with soap | .03 |

| S4. | Run first board manually | .06 |
| | Run second board automatically | .03 |

S5. Adjust height of solder wave:
    crank adjustment   .03
    table leg adjustment   .07

S6. Add solder, clean dross, and adjust temperature   .15

Operation elements in estimating minutes

1. Gather stack, move to machine:
       one board at a time   .19
       tray of boards   .13

2. Position board in conveyor rack, tighten   .09

3. Place rack on conveyor   .04

4. Place and position board on conveyor   .07

5. Operator moves to opposite end of conveyor and then back again, for one operator loading and unloading the boards

Boards Run in Succession	1	2	3	4	5
Time per Board, Min	.14	.11	.09	.07	.06

6. Remove rack from conveyor, remove board   .03

7. Place rack on return carriage   .05

8. Unload board from conveyor   .02

9. Place board on tray   .04

10. Wave solder

Ft. per Min	1	2	3	4	5	6	7	8	9	10
Solder Time per Board, Min	.64	.58	.53	.49	.45	.42	.39	.36	.34	.32

11. Inspect (visual), time per inspection   .06

23.7 Vertical Solder Coating Machines

DESCRIPTION

The vertical solder coating machine applies a solder coating to selected areas on a printed circuit board. Solder dip and hot air leveling is claimed to give better solderability and shelf life to simple print and etch boards. Machines of this class are of modular design and are available in different combinations. One style of machine will have four carousels for different steps. The control features minimizes the role of the operator. Everything but the load and unload of the panel are predetermined by the type of material being processed as the specific times and temperatures are entered in an instrument panel on the front of the machine. Figure 23.7 has automatic adjustments for panel thickness from 0+ to 0.187 in. The solder dwell adjustment ranges from 0+ to 99 sec.

ESTIMATING DATA DISCUSSION

Setup includes the time for the machine to reach operating temperature, to check the fluid levels, and to perform various minor maintenance operations.

The following operation elements make up the cycle for a four-carousel machine. The estimating data can be applied to other machines.

Element 1 is the operator load. It consists of the operator taking the board from the solution tank in front of the machine and placing it in the clamp. A waiting period follows that includes time for the cycle in process to finish and the carousel to rotate 90 degrees and lower. When the carousel comes to rest at the bottom, Element 3, the flux station, begins. The carousel raises and places the printed circuit board in the preheat stage. The board is then placed in the solder bath. The air knives are activated at the end of this period as the carousel raises. The raising time is set in accordance with the air knives. Leveling, the next element, is important to the entire operation. The printed circuit board in the clamp rotates to a position slightly greater than horizontal. Small fans cool the solder preventing unevenness. The rest of the cycle is devoted to cooling. When the operator unloads the finished

FIGURE 23.7 Vertical solder coating machine.

board, it is hot enough so that warping may occur easily. The final element includes the unload, inspection, which is visual, and placing the circuit board in the cooling water.

EXAMPLES

A. A printed circuit board is to be solder coated. Find the unit estimate for a board of thickness .032 in. with total plating area of 186.30 sq. in. and a solder area of 31.09 sq. in.

Table Number	Process Description	Table Time	Adjustment Factor	Cycle Minutes	Setup Hours
23.7-S	Setup	2.00			2.00
23.7-1	Load board	.13		.13	
23.7-2	Machine insert	.82		.82	
23.7-3	Flux	.19		.19	
23.7-4	Preheat	.25		.25	
23.7-5	Solder	.15		.15	
23.7-6	Level	.25		.25	
23.7-7	Dry	.15		.15	
23.7-8	Wait	.41		.41	
23.7-9	Unload and inspect	.16		.16	
	Total			2.51	2.00

B. A printed circuit board is to be soldered using a vertical soldering technology. Find the unit estimate and the hours per 100 units for a board thickness of .016 in.

Table Number	Process Description	Table Time	Adjustment Factor	Cycle Minutes	Setup Hours
23.7-S	Setup	2.00			2.00
23.7-1	Load board	.13		.13	
23.7-2	Machine insert	.82		.82	
23.7-3	Flux	.19		.19	
23.7-4	Preheat	.25		.25	
23.7-5	Solder	.15		.15	
23.7-6	Level	.25		.25	
23.7-7	Dry	.15		.15	
23.7-8	Wait	.41		.41	
23.7-9	Unload and inspect	.16		.16	
	Total			2.51	2.00
	Hr/100 units	4.18			

C. Vertical soldering technology is planned for a PCB. With a board thickness of .062 in., find the unit estimate, setup time, hr/100 units, pc/hr, and lot time. The order quantity is 271 units.

Table Number	Process Description	Table Time	Adjustment Factor	Cycle Minutes	Setup Hours
23.7-S	Setup	2.00			2.00
23.7-1	Load board	.13		.13	
23.7-2	Machine time	.82		.82	
23.7-3	Flux	.21		.21	
23.7-4	Preheat	.27		.27	
23.7-5	Solder	.18		.18	
23.7-6	Level	.27		.27	
23.7-7	Dry	.20		.20	
23.7-8	Wait	.50		.50	
23.7-9	Unload and inspect	.21		.21	
	Total			2.79	2.00
	Unit estimate	2.79 min			
	Setup	2.00 hr			
	Hr/100 units	4.655			
	Pc/hr	21			
	Lot hr	14.62			

TABLE 23.7 VERTICAL SOLDER COATING MACHINES

Setup   **2.00 hr**

Operation elements in estimating minutes

1. Load part   .13

2. Machine insert   .82

3. Flux

Thk	.047	.065	.085	.106	.130
Min	.19	.21	.23	.25	.28

4. Preheat

Thk	.047	.075	.106	.140
Min	.25	.27	.30	.33

5. Solder

Thk	Min	Thk	Min
.047	.15	.082	.22
.054	.16	.092	.24
.063	.18	.104	.27
.072	.20	.118	.29
		.132	.32

6. Level

Thk	.047	.065	.086	.108	.132
Min	.25	.27	.30	.33	.37

7. Dry

Thk	Min	Thk	Min
.047	.15	.084	.25
.053	.16	.093	.27
.060	.18	.104	.29
.067	.20	.116	.32
.075	.22	.129	.35

8. Wait

Thk	Min	Thk	Min
.047	.41	.094	.54
.061	.45	.113	.60
.076	.50	.133	.66

9. Unload and Inspect

Thk	Min	Thk	Min
.047	.16	.089	.29
.052	.17	.099	.31
.058	.19	.109	.34
.065	.21	.121	.37
.072	.23	.134	.41
.084	.25		

23.8 Wire Harness Bench Assembly

DESCRIPTION

Assembly of a harness is performed at the bench having a board with an attached harness layout. Nails or pins are located along the track and benches. These pins confine the loose wires within the layout. Associated with the laying of the wires is lacing or strapping, which requires a roll of lacing cord or plastic straps. These cords or straps are secured around the bundle with a special strapping tool. Other common elements are the cutting of the wire to an exact length per the layout, which strips insulation from the end of the wire to a predetermined length, and crimping lugs to the ends of the wires. Each of the elements requires a tool.

Wires that interconnect various electrical and electronic components are often routed together. This group of wires, tied into a neat bundle, is known as a wire harness or cable. A wire harness may be prefabricated on a jig or bench and, sometimes, in a chassis. A wire harness is similar to multiple-wire cable. However, a wire harness is not usually a linear cable, but provides for the branching of individual wires throughout its length as required by the circuits associated with the electrical circuitry. In the production of the harness, a wire harness ends at a terminal board or at a cable connector.

Harnesses vary in length a few inches to many feet. The intent of these data is to present data that represents the size that will be assembled on a bench-style station. Automatic assembly methods are not considered. Figure 23.8 is a sample of a harness. Other terms refer to this operation as "cabling."

ESTIMATING DATA DISCUSSION

Spot ties are individual ties produced with linear lacing cord or with nylon lacing tape. Various knots such as the square knot secure the tie. Cable lacing is the term applied to the production of many ties with a single length of lacing cord or tape. In the lacing of the cable, care is exercised to have each tie as snug and self-locking; otherwise, the lacing is inefficient. Nylon cable ties are small belts or straps used to bind a wire harness individually. The spacing between adjacent ties of a wire harness is ½ in. or more, but the maximum spacing is determined by the diameter of the harness. A harness should be neat and orderly with all wires running parallel to each other. Wires should not weave in and out as illustrated by the dark wire in Fig. 23.8. All spot tying and lacing should also be uniform.

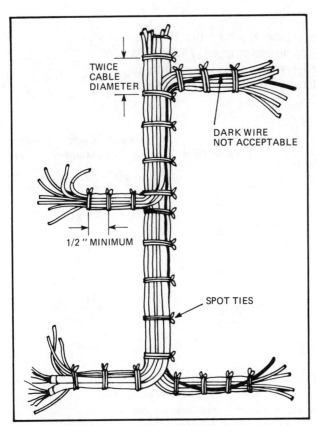

FIGURE 23.8 The spacing between spot ties should be at least ½ in., and the ties should be snug and self-locking.

A harness that leans against a sharp corner or abrasive surface should be protected against abrasion. A plastic sleeving is placed around the harness and held in place with spot ties.

A great variety of cable connectors are encountered in electronics manufacture. Their elaboration is found in other locations of this *AM Cost Estimator*.

Setup is provided for loading bundles of wires into a wire rack and also deals with the usual chores of starting a new job.

Cycle data assume a harness board with nails along the pattern to guide the assembler. Each end of the pattern has a spring holding device to hold the wire. Identification of harness branches are provided on the layout. A written instruction is available to aid where each wire is laid.

The operation elements are provided as a consolidated grouping or as single elements. The consolidated elements are a group of several individual elements, which are also provided. The consolidated elements do not require grouping and selection. On the other hand, an estimator may separately identify those elements and build up an estimate. Both opportunities are available. Our discussion first describes the individual elements, as these are the items that are grouped for a consolidated element.

Element 1 includes several items. Get a wire includes obtaining a wire from a predetermined location in a wire rack and a move of the wire to the harness board. It assumes that locations are memorized. Start a wire begins with wire at the harness board and moves an end of the wire to a starting point, place in holding device, and regrasp wire preparatory to laying wire. The holding device is pressure type such as a spring. Lay a wire on a harness board begins with grasp of a loose end of wire, and lay wire on the harness board between pins along a directed route to a termination point of the wire. End the lay in wire (time also includes start of a wire) involves the wire at the end of the trunk or branch and place end of wire into a holding device and hand releases wire. The consolidated element involves a get, start, lay, and end wire.

In Element 2, the operator assembles a pre-wired connector to board and end-run wires. It also includes obtaining a pre-wired connector from a bench and placing it on a harness board between posts. Get back wire and lay on board between pins along described path to the termination point on the bench is also included. It concludes with the placement of the wire end into the holding device.

Element 3, lacing a harness, is described by the following work. It involves manually securing the harness with cord prior to removing the harness from the board. The steps are start and tie. Starting and ending ties are combined since one always involves the other. The value also includes time to varnish two ties and changing the position of the board for ease of lacing. A lock stitch is an element between the ties that secures the body of the harness. Each stitch is spaced three inches. Congestion of these elements is possible due to branches and pins on the board. A dress and remove element separates and straightens the wires of the harness for a neat-appearing harness. A time is provided to remove the completed harness from the harness board and place the harness aside.

Strapping of a harness or branch is identified as Element 4. It may be more economical to strap instead of lace, and sometimes strapping is used in addition to lacing. Strapping involves placing a strap around the harness every three inches. The time includes getting and placing strap and tightening. A tool is used. Congestion may occur because of branches in the harness or interfering pins on the board. The action of dressing and removing is used to separate and straighten end compressed wires for a neater finished appearance.

Element 5 consists of consolidated work and single elements. This element is collectively referred to cut, strip, and lug. Usually some, if not all, of the wire end need to be cut and stripped to a prescribed length. Sometimes stripping is eliminated which prevents fraying of wire ends. Lugging, when required, is usually over a stripped end, although not always. This action is determined by the type of lug. Cut wire includes getting the cutter and placing aside. Each additional cut time value is without the get and aside cutter. The cut dimension is determined by a mark on the harness board for each wire. Similar instructions are noted for stripping. Lugging the first end involves a get and aside of a lugging tool. Additional lugged ends are without this tool time. Labeling effort includes placing and securing a marker on the harness. Time is indicated for removing the harness from the board. Observe that time is associated with the number of branches. The time allows for care to ensure that lugs and connections are disengaged.

Miscellaneous elements are identified in Element 6. The end of a shielded wire is taped, which includes obtaining a roll of tape, start taping and wrapping to a specific location, cutting tape, replacing wire harness into position, and dropping of tape. A length may be taped, and the length of run is indicated. It includes the usual get and aside of the tape. A tube, which is pre-split along the length, is forced over the harness and includes the tieing. A tube may be guided over several wires by sliding the tube to a location. The time is determined by the length of sleeving.

EXAMPLES

A. A 24-in. long harness having five branches is to be estimated for a lot of 10 units. Branch 1, 2, 3, 4 and 5 have 12, 6, 4, 10, and 4 ends which are cut and stripped. The harness has 18 wires. The main trunk is laced and branches 2, 3, and 4 are spot tied. The main trunk is 24 in., and branches 2, 3, and 5 are 6, 18, and 8 in. in length. Find the hr/100 units and lot estimate.

Table Number	Process Description	Table Time	Adjustment Factor	Cycle Minutes	Setup Hours
23.8-S	Setup	.05			.05
23.8-1	Lay 18 24-in.	4.00		4.00	
23.8-3	Lace 24 in. of main trunk	1.84		1.84	
23.8-4	Spot tie, 13 places, branches 2, 3, and 5	.36	13	4.68	
23.8-5A	Cut 36 ends; .78 + 11 × .03	1.11		1.11	
23.8-5A	Strip 36 ends of wires; 1.03 + 11 × .04	1.47		1.47	
23.8-5B	Remove harness from board, use six branches	.70		.70	—
	Total			13.80	.05
	Hr/100 units	23.00			
	Lot estimate	2.35			

B. A harness is to be produced on a board. Features of the harness are as follows: There are 26 wires, one main trunk, four branches, and one branch less than 3 in. There is one connector with 10 wires having two mounting holes. Pins and nails are positioned on the board according to drawing specifications. The wires are cut to indicated drawing length, and the ends are stripped ¼ in. The main trunk and branch 2 is laced per specification.

Branch 4 is spot tied. Branches 1, 3, and 5 are strapped tied per specification. One end has a crimped lug. Wires are numbered for identification. A lot of 27 harnesses is to be estimated. Find the unit estimate, hr/100 units, and lot time.

Table Number	Process Description	Table Time	Adjustment Factor	Cycle Minutes	Setup Hours
23.8-S	Setup	.05			.05
23.8-2	Load pre-wired connector, lay in 10 wires, and distance is 30 in.	1.50		1.50	
23.8-1	Lay in 16 wires with average length of 24 in., 4.05 + .23	4.28		4.28	
23.8-3	Lace main branch, 30 in.	2.05		2.05	
23.8-3	Lace branch 2, 18 in.	1.63		1.63	
23.8-4	Spot-tie branch 4	.36		.36	
23.8-4	Strap tie branches 1, 3, 5				
	Branch 1 is 12 in.	.91		.91	
	Branch 3 is 8 in.	.91		.91	
	Branch 5 is 10 in.	.91		.91	
23.8-4	Cut 42 wires to length, .78 + 17 × .03	1.29		1.29	
23.8-4	Strip 42 wires, ¼ in., 1.03 + 17 × .04	1.71		1.71	
23.8-4	Lug one wire	.24		.24	
23.8-5B	Remove harness, 7 branches	.80		.80	
23.8-5B	Label cable with marker	.24		.24	
	Total			16.91	.05
	Hr/100 units	28.184			
	Lot time	7.660 hr			

C. A harness is to be bench assembled. The harness has 48 wires, and 24 are preassembled to a connector. The main branch is 60-in. long. Branches 2, 3, 4, 6, 7, and 8 are 12 in., 12 in., 10 in., 8 in., 15 in., and 10 in. in length. All ends are cut and stripped. Six ends are lugged. The main branch is laced every three inches and the branch is strapped. Branches 2, 3, 4, 6, and 7 are spot tied. The harness is taped at a bend near branch 3 and is 3-in. long. Find the unit and shop estimate. (The unit and the cycle estimate are similar terms.)

Table Number	Process Description	Table Time	Adjustment Factor	Cycle Minutes	Setup Hours
23.8-S	Setup	.05			.05
23.8-2	Assemble pre-wire connector to board, 24 wires with average length of 60 in., 3.98 + 4 × .18	4.70		4.70	
23.8-1	Lay wires, 10 are 60 in.,	2.40		2.40	
	10 are 36 in.,	2.10		2.10	
	4 are 24 in.	.80		.80	
23.8-3	Lace 60 in.	3.10		3.10	
23.8-4	Spot tie 24 locations	.36	24	8.64	
23.8-4	Strap 3 places for 6 in., use minimum	.91		.91	
23.8-6A	Tape 3 in. long	1.26		1.26	
23.8-5A	Cut 72 ends, .78 + 47 × .03	1.41		1.41	
23.8-5A	Strip 72 ends 1.03 + 47 × .04	1.88		1.88	
23.7-5A	Lug 6 ends, .48 + .06	.54		.54	
23.8-5B	Remove harness from board	.90		.90	
	Total			28.64	.05
	Pc/hr	2.1			
	Hr/100 units	47.734			

TABLE 23.8 WIRE HARNESS BENCH ASSEMBLY

Setup .05 hr

Operation elements in estimating minutes

1. Get wire, start, lay, and end wire

No. of Wires	Distance					
	12	24	36	48	60	72
1	.19	.20	.21	.22	.24	.26
2	.38	.40	.42	.44	.48	.52
3	.57	.60	.63	.66	.72	.78
4	.76	.80	.84	.88	.96	1.04
5	.95	1.00	1.05	1.10	1.20	1.30
6	1.14	1.20	1.26	1.32	1.44	1.56
7	1.33	1.40	1.47	1.54	1.68	1.82
8	1.52	1.60	1.68	1.76	1.92	2.08
9	1.71	1.80	1.89	1.98	2.16	2.34
10	1.90	2.00	2.10	2.20	2.40	2.60
12	2.28	2.40	2.52	2.64	2.88	3.12
15	2.85	3.00	3.15	3.30	3.60	3.90
20	3.80	4.00	4.20	4.40	4.80	5.20
25	4.75	5.00	5.25	5.50	6.00	6.50
28	5.32	5.60	5.88	6.16	6.72	7.28
30	5.70	6.00	6.30	6.60	7.20	7.80
40	7.50	8.00	8.40	8.80	9.60	10.40
50	9.50	10.00	10.50	11.00	12.00	13.00
75	14.25	15.00	15.75	16.50	18.00	19.50
100	19.00	20.00	21.00	22.00	24.00	26.00
Add'l Wire	.22	.23	.24	.25	.27	.29

Get wire from holder, per wire .04
Start wire and end run, per wire .09
Run wire in cable, per wire .08

2. Assemble pre-wired connector to board, lay and end wire

No. of Wires	Distance					
	12	24	36	48	60	72
1	.46	.48	.50	.53	.56	.59
2	.54	.58	.62	.68	.74	.80
3	.62	.68	.74	.83	.92	1.01
4	.70	.78	.86	.98	1.10	1.22
5	.78	.88	.98	1.13	1.28	1.43
6	.86	.98	1.10	1.28	1.46	1.64
7	.94	1.08	1.22	1.43	1.64	1.85
8	1.02	1.18	1.34	1.58	1.82	2.06
9	1.10	1.28	1.46	1.73	2.00	2.27
10	1.18	1.38	1.58	1.88	2.18	2.48
20	1.98	2.38	2.78	3.38	3.98	4.58
25	2.38	2.88	3.38	4.13	4.88	5.63
30	2.78	3.38	3.98	4.88	5.78	6.68
40	3.58	4.38	5.18	6.38	7.58	8.78
50	4.38	5.38	6.38	7.88	9.38	10.88
75	6.38	7.88	9.38	11.63	13.88	16.13

No. of Wires	Distance					
	12	24	36	48	60	72
100	8.38	10.38	12.38	15.38	18.38	21.38
Add'l Wire	.08	.10	.12	.15	.18	.21

Assemble pre-wired connector to cable board	.38
Lay wire, each	.08
End wire, each	.04
Mount wired connector to holding fixture	1.29
Mount wiring plug to plate with mounted plug	1.01

3. Lacing of harness trunk or branch

Distance	12	18	24	30	36	42	48
Min	1.42	1.63	1.84	2.05	2.26	2.47	2.68

Distance	54	60	66	72	Add'l ft
Min	2.89	3.10	3.31	3.52	.42

Start and end tie	1.09
Lock stitch every 3 in.	.09
Correct for congestion of cable every 12 in.	.04
Dress harness, every 12 in.	.02

4. Strapping of harness trunk or branch

Distance	12	18	24	30	36	42
Strapping Min	.85	1.19	1.70	2.89	4.08	4.42
Congestion Min	.04	.06	.08	.10	.12	.12
Dressing Min	.02	.03	.04	.05	.06	.07
Consolidated	.91	1.28	1.82	3.04	4.26	4.61

Distance	48	54	60	66	72	Add'l Ft
Strapping Min	4.76	5.10	5.44	5.78	6.12	.68
Congestion Min	.16	.18	.20	.22	.24	.04
Dressing Min	.08	.09	.10	.11	.12	.02
Consolidated	5.00	5.37	5.74	6.11	6.48	.74

Strapping a bundle with plastic tie wrap every 3 in.	.17
Correct for congestion of interference with posts or branch per ft	.04
Dress trunk or branch every 12 in.	.02
Spot tie for branch from trunk, per tie	.36
Spot tie for branch from trunk with varnish	.42

5.A. Cut, strip, and lug wires

	Wires								
	1	2	3	4	5	10	15	25	Add'l
Cut	.06	.09	.12	.15	.18	.33	.48	.78	.03
Strip	.07	.11	.15	.19	.23	.43	.63	1.03	.04
Lug	.24	.30	.36	.42	.48	.78	1.08	1.68	.06

Cut wire,	.06 for first,	.03/add'l	
Strip wire,	.07 for first,	.04/add'l	
Lug wire,	.24 for first,	.06/add'l	
Label harness with marker			.24

B. Remove Harness from Board

	6	7	8	9	10	Add'l
Min	.70	.80	.90	1.00	1.10	.10

6. Miscellaneous

A. Tape Ends of Shielded Wire

	1	2
Min	.40	.80

B. Tape Start and Finish Length

	2	3	4	5	6	7	8	Add'l
Min	.42	1.26	1.68	2.10	2.52	2.94	3.36	.42

Roll tape around harness, per in. .22

C. Taping Around Branch, Each

	1	2	3	4	5	Add'l
Min	.27	.54	.81	1.08	1.35	.27

D. Split Tubing, Lay Over Harness, Tie Tubing

	1	2	3	4	5	Add'l
Min	.74	1.48	2.22	2.96	3.70	.74

E. Sleeve Over Several Wires

	1	2	3	4	5	Add'l
Min	.39	.78	1.17	1.56	1.95	.39

F. Route wire through aparture
   Easy (no force required) up to 5 in. of wire   .03
   Difficult (force required) up to 5 in. of wire   .06
   Each additional 5 in. of wire   .01

23.9 Flat Cable Connector Assembly

DESCRIPTION

Growing industrial acceptance of flat ribbon cable has led to increased production. The demand has resulted from advantages of faster insulation stripping and mass termination on conductors offered by the connectors. In utilizing these machines, the production rate improves over hand labor requirements. Connectors are assembled to flat ribbon cable leading to reduced costs and improved quality.

Early development of connector/flat ribbon cable was initiated about 1960. This concept was based on the "U" shape contact principle, which strips insulation from round conductors during mating. Later, the U-contact principle was applied to develop a connector/flat ribbon cable system with characteristics demanded by the electronic industry. The design solved a number of problems including elimination of wiring errors and simplification of harness assembly. The U-contacts are formed with parallel legs, which displace insulation as the conductor is forced into the slot. Deflection of these legs during termination results in a permanent, gas-tight grip on the cable conductor.

The mass termination machines are simple and accurate. We consider three types of equipment: manual press, pneumatic press, and semi-automatic machine. The assembly procedures utilizing the first two types of equipment are as follows.

The assembler begins by removing a connector cover from the transfer adhesive strip. The cover snaps down against the adjacent cover, breaking it from the liner. The cover is pulled laterally away from the strip, leaving the adhesive with a smooth clean edge. The cable is placed on the cover with the ribbed cable contour matching. The cable and cover subassembly are next placed in a steel locator plate, and the mating connector body is positioned over them. The press is operated to force U-contacts over the flat cable conductors ensuring a reliable electrical connection. This is the standard assembly procedure for all mass termination connectors. Only the locator plate need be changed to accommodate differences in connector format.

Following is the assembly procedure utilizing the semi-automatic machine for manufacturing mass termination flat cables. The feed followers are removed, the channel widths are adjusted to connector size, the connector covers are placed into the channel, and pressure sensitive tape is removed. The width of feed follower is adjusted by sliding a spring loaded adjustable rail to appropriate connector width. The above steps should be repeated for connector. Cable is fed through the magazine until it hits the stop. The foot pedal is depressed.

EXAMPLES

A. A cable assembly consists of 6 in. of cable and two connectors. Find the unit estimate for the manual press operation to terminate the cable with two connectors for both ends. Also find the estimated time for a lot of 25 cables.

Table Number	Process Description	Table Time	Adjustment Factor	Cycle Minutes	Setup Hours
28.9-S	Setup	.02			.02
23.9-1	Remove top covers	.05	2	.10	
23.9-1	Place cover on cables	.07	2	.14	
23.9-1	Place cover-cables in locator plate	.03	2	.06	
23.9-1	Remove connectors	.01	2	.02	
23.9-1	Place connectors in locator plates	.06	2	.12	
23.9-1	Press down	.04	2	.08	
23.9-1	Remove the assembly	.04		.04	
	Total			.56	.02
	Lot estimate	.25 hr			

B. Repeat Example A utilizing a pneumatic press.

Table Number	Process Description	Table Time	Adjustment Factor	Cycle Minutes	Setup Hours
23.9-S	Setup	.02			.02
23.9-2	Remove top covers	.05	2	.10	
23.9-2	Place covers on cables	.07	2	.14	
23.9-2	Place cover cables in locator plate	.03	2	.06	
23.9-2	Remove connectors	.01	2	.02	
23.9-2	Place connectors in locator plates	.06	2	.12	
23.9-2	Press down	.02	2	.04	
23.9-2	Remove the assemblies	.04		.04	
	Total			.52	.02
	Lot estimate	.25 hr			

C. Repeat Example A utilizing a semi-automatic machine.

Table Number	Process Description	Table Time	Adjustment Factor	Cycle Minutes	Setup Hours
23.9-S	Setup	.03			.03
23.9-3	Feed cable	.01	2	.02	
23.9-3	Terminate cable	.01	2	.02	
23.9-3	Remove cable	.01	2	.02	
	Total			.06	.03
	Lot estimate	.06			

TABLE 23.9 FLAT CABLE CONNECTOR ASSEMBLY

Setup
Manual or pneumatic press	.01 hr
Semiautomatic	.03 hr

Operation elements in estimating minutes

1. Manual press
    - Remove top cover .05
    - Place cover on cable .07
    - Place cover cable assembly in locator plate .03
    - Remove connector .01
    - Place connector in locator plate .06
    - Press down .04
    - Remove the assembly .02

2. Pneumatic press
    - Remove top cover .05
    - Place cover on cable .07
    - Place cover cable assembly in locator plate .03
    - Remove connector .01
    - Place connector in locator plate .06
    - Press down .02
    - Remove the assembly .02

3. Semiautomatic
    - Feed cable .01
    - Terminate cable .01
    - Remove cable .01

23.10 D-Subminiature Connector Assembly

DESCRIPTION

D-subminiature connectors are designed to accommodate cable-to-panel and cable-to-cable applications. These are pin and socket devices which employ contacts encased in a molded dielectric insert surrounded by D-shaped shells for polarization. These connectors are designed for applications where space and weight are prime considerations. These high-contact density connectors are especially suited for connecting a computer to a terminal or a printer. Other applications include circuit-termination, computers, and computer peripheral equipment, data terminal, telecommunications systems, medical equipment, military systems, instrumentation, and cable assemblies.

Four sizes of D-subminiature connectors are considered here. These are 9, 15, 25, and 37 contacts. The most popular of these connectors are the 25 and 37 positions. The 25-position is called RS232, and the 37-position is called RS449. These connectors are used as communication input/output connecting devices. Four types of D-subminiature connectors are discussed: insulation displacing (discrete wires), solder, solder with linear, and crimp-contact type.

Only operation time for 25 and 37 positions are provided. For operation data, mass termination and the manufacturing procedure on mass termination, see the flat cable section, Table 23.9.

ESTIMATING DATA DISCUSSION

After mass termination, the operation of terminating D-subminiature connectors is a cost-effective method. Using a semi-automatic machine, two wires are simultaneously inserted in connectors' contacts. The wire is pierced and cut.

Solder types of connectors are terminated by hand stripping the wires, pre-soldering connector contacts, and soldering the wires to one side of connector and then to the other side. A vise is used to hold the connector.

A fixture is used to terminate a solder with linear connector. All the wires are selected, inserted, and soldered to one side of the connector. Then the other side is terminated.

The method of crimp contacts is described as follows. First the contacts are crimped to the wires. Then the contacted wires are inserted into the connector. The advantage of this type of connector is significant when few wires are terminated.

EXAMPLES

A. Estimate the time for RS232 cable assembly connecting a computer terminal to a printer. All the 25 contacts of D-subminiature connector are to be wired. A lot quantity of 50 cables will be manufactured. Use the method of insulation displacing connector.

Table Number	Process Description	Table Time	Adjustment Factor	Cycle Minutes	Setup Hours
23.10-S	Setup cables and load connectors	.03			.03
23.10-1	Place connector in the nest	.03		.03	
23.10-2	Spread and fan wires	.11		.11	
23.10-3B	Connectorizing 25 pins	.68		.68	
23.10-4	Remove connector	.03		.03	
	Total			.85	.03
	Lot estimate	.73 hr			

B. Find the unit time, hr/100 units, and lot estimates using solder connector for the problem described above. Assume a quantity of 500 cable assemblies.

Table Number	Process Description	Table Time	Adjustment Factor	Cycle Minutes	Setup Hours
23.10-S	Setup	.05			.05
23.10-5	Strip wires, 25 wires	.03	25	.75	
23.10-6	Place the connector in the vise	.42		.42	
23.10-7	Get solder iron, fill solder cup, 25 contacts	.06	25	1.50	
23.10-7	Solder wires to the connector, 25 wires	.07	25	1.75	
23.10-8	Turn the connector in the vise	.42		.42	
23.10-9	Remove cable and connector	.09		.09	
	Total			5.28	.05
	Hr/100 units	8.800			
	Lot estimate	44.05 hr			

C. Find unit time using solder linear connector for the problem described above. The production quantity is 2000 cable assemblies.

Table Number	Process Description	Table Time	Adjustment Factor	Cycle Minutes	Setup Hours
23.10-S	Setup	.03			.03
23.10-10	Place wires	.14		.14	
23.10-11	Insert wires into fixtures	.01		.01	
23.10-11	Strip and cut	.03		.03	
23.10-12	Handle solder iron and solder pins	.87		.87	
23.10-13	Turn connector and release cable	.02		.02	
	Total			1.15	.03
	Hr/100 units	1.197			
	Lot estimate	23.97 hr			

TABLE 23.10   D-SUBMINIATURE CONNECTOR ASSEMBLY

Setup
Insulation displacing	.03 hr
Solder	.05 hr
Solder linear	.03 hr

Operation elements in estimating minutes

1. Place connector in the nest .03

2. Spread and fan wires .11

3.A. Select and insert wires

Pair	Min	Pair	Min
1st	.080	8th	.046
2nd	.074	9th	.041
3rd	.071	10th	.038
4th	.068	11th	.034
5th	.062	12th	.034
6th	.057	13th	.031
7th	.052	Total	.688

B. Selecting paired wires out of 13th pair cable to connectorize 25 pin connector

Wires in Pair	Min	Wires in Pair	Min
1st	.08	1st–8th	.51
1st + 2nd	.15	1st–9th	.55
1st + 2nd + 3rd	.22	1st–10th	.58
1st–4th	.29	1st–11th	.62
1st–5th	.35	1st–12th	.65
1st–6th	.41	1st–13th	.68
1st–7th	.46		

C. Selecting paired wires out of 19th pair cable to connectorize 37 pin connector

Wires in Pair	Min	Wires in Pair	Min
1st	.083	11th	.049
2nd	.080	12th	.046
3rd	.077	13th	.043
4th	.074	14th	.041
5th	.071	15th	.039
6th	.068	16th	.038
7th	.062	17th	.034
8th	.060	18th	.034
9th	.057	19th	.031
10th	.052		

D. Selecting paired wires out of 19th pair cable to connectorize 37 pin connector

Wires in Pair	Min	Wires in Pair	Min
1st	.08	1st–11th	.72
1st + 2nd	.16	1st–12th	.77
1st + 2nd + 3rd	.24	1st–13th	.81
1st–4th	.31	1st–14th	.85
1st–5th	.38	1st–15th	.89
1st–6th	.45	1st–16th	.93
1st–7th	.51	1st–17th	.96
1st–8th	.57	1st–18th	1.00
1st–9th	.63	1st–19th	1.03
1st–10th	.68		

4. Remove connector .03

5. Strip wires, solder method .03 ea

6. Place the connector in the vise .42

7. Get solder iron .30
   Fill solder cup with solder .06 ea contact
   Solder wires to the connector .07 ea wire

8.	Turn the connector in the vise	.42
9.	Remove the cable and connector	.09
10.	Place wires, solder linear method	.14
11.	Insert wire into fixture, solder linear method	.01
	Strip and cut, solder linear method	.03
12.	Handle solder iron and solder, solder linear method:	
	25 pins	.87
	37 pins	1.26
13.	Turn connectors, solder linear method	.02
14.	Release	.08

23.11 Single-Wire Termination Machines

DESCRIPTION

Wire terminal applicator machines apply strip terminals used in the manufacture of electrical connections on cables, jumpers, and related assemblies. Most machines will accommodate a range of wire gages and specify a necessary crimp height for each size. The crimp height refers to the thickness of the terminal at the crimp location after crimping (see Figure 23.11A).

The applicator machines power a vertical action ram which drives the crimping die. The die performs three specific functions:

1. It crimps the terminal at two locations, folding the terminal ears around the wire insulation and crimping the terminal barrel around the bare wire;
2. It cuts the terminal from its carrier strip and;
3. It advances the carrier strip, placing the next terminal in the crimping area.

The terminal carrier strip is fed to the left side of the crimping die from a reel mounted on top of the machine. Termination is accomplished by inserting a stripped wire end into a terminal on the crimping anvil of the die and actuating the ram with a foot switch. Operation is simple and specially trained operators are not required.

ESTIMATING DATA DISCUSSION

Basic setup elements for single wire termination include locating wires near the machine, cleaning the feed track to remove the metal cut from between the terminals, and testing the machine's operations. The operation test involves terminating a few wires, inspecting the crimp, and measuring the crimp height with a micrometer. Two setup elements, changing the terminal reel and adjusting the die, are not necessarily performed each time with the basic elements. For example, the machine is often equipped with sufficient

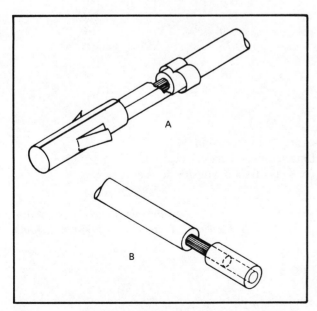

FIGURE 23.11 (A) Crimp height; (B) pre-cut insulation.

correct terminals and does not require adjustment, in which case one should not include these elements in the setup time.

There are six general operation elements. Certain elements are always included in the cycle time while others are only included in special instances.

Element 1 involves securing a handful of wire from the workbench and turning to face the machine. This element is constant.

Element 2 involves turning the wires end for end to organize common ends. This element should only be included when the two ends of the wire are stripped different lengths or a terminal is already attached to one end.

When the wires are supplied with pre-cut insulation (see Figure 23.11B), so that the insulation piece may be removed by hand, Element 3 should be included.

Element 4 involves placing the stripped wire end into the crimping die, actuating the crimping mechanism with a foot switch, and removing the terminated wire. This element is constant.

Element 5 includes brief inspection of the handful of termination wires and placing them on the table. This inspection is a quick check to make sure that all the terminals are securely fastened to the wires. Bad wires are set aside.

Element 6 involves sorting the finished wires into piles of 100 and taping them into bundles to be sent for final inspection.

Note that the handling times for Elements 2, 5, and 6 are a function of wire length. This is a result of increased handling difficulty for the longer wires due to the tangling and increased bulk.

It is recommended that the following three rules be used to obtain values from the estimating table.

1. If the value of the time driver (in our case wire length) is less than the minimum table value then use the smallest value given.
2. If the time driver value falls between values in the table use the largest value.
3. If the time driver value is greater than the maximum table value use the maximum value.

EXAMPLES

A. A lot of 6-in., 18-gage wires are to be terminated on one end. The wires are supplied with the insulation on each end pre-cut to different lengths. The machine is already equipped with a reel of the correct terminals and the die does not require adjustment. Lot size is 5000 wires. Estimate the unit and lot times.

Table Number	Process Description	Table Time	Adjustment Factor	Cycle Minutes	Setup Hours
23.11-S1	Locate wires near machine	.04			.04
23.11-S3	Clean feed track	.01			.01
23.11-S5	Test operation	.06			.06
	Setup subtotal				.11
23.11-1	Grab wires	.003		.003	
23.11-2	Arrange wires	.006		.006	
23.11-3	Remove insulation	.010		.010	
23.11-4	Terminate	.028		.028	
23.11-5	Inspect and place aside	.005		.005	
23.11-6	Count and bundle	.009		.009	
	Total			.061	.11
	Lot estimate	5.19 hr			

B. A lot of 16-gage wires are to be terminated. The wires are 17-in. long, and both ends have the insulation pre-cut to the same length. The machine requires a reel change and adjustments along with the other setup elements. The lot consists of 8000 wires. Find the unit and lot estimates.

Table Number	Process Description	Table Time	Adjustment Factor	Cycle Minutes	Setup Hours
23.11-S1	Locate wires near machine	.04			.04
23.11-S2	Change terminal reel	.05			.05
23.11-S3	Clean feed track	.01			.01
23.11-S4	Adjust die	.02			.02
23.11-S5	Test operation	.06			.06

Table Number	Process Description	Table Time	Adjustment Factor	Cycle Minutes	Setup Hours
	Setup subtotal				.18
23.11-1	Grab wires	.003		.003	
23.11-3	Remove insulation	.010		.010	
23.11-4	Terminate	.028		.028	
23.11-5	Inspect and place wires aside	.011		.011	
23.11-6	Count and bundle	.016		.016	
	Total			.068	.18
	Lot estimate	9.25 hr			

C. A lot of 20-gage wires are to be terminated. The wires are 11-in. long with one end already terminated. All setup elements are required. The lot is 3000 wires. Estimate the unit and lot times.

Table Number	Process Description	Table Time	Adjustment Factor	Cycle Minutes	Setup Hours
23.1-S	Setup (1–5)	.18			.18
23.1-1	Grab wires	.003		.003	
23.11-2	Arrange wires	.009		.009	
23.11-3	Remove insulation	.010		.010	
23.11-4	Terminate	.028		.028	
23.11-5	Inspect and place aside	.008		.008	
23.11-6	Count and bundle	.012		.012	
	Total			.070	.18
	Lot estimate	3.68 hr			

TABLE 23.11 SINGLE-WIRE TERMINATION MACHINES

Setup   **Hr**
- S1. Locate wires near machine — .04
- S2. Change terminal reel — .05
- S3. Clean feed track — .01
- S4. Adjust die — .02
- S5. Test operation — .06

Operation elements in estimating minutes

1. Grab handful of wires — .003

2. Arrange wires in hand

In.	6	7	9	11	14	17
Min	.006	.007	.008	.009	.011	.013

3. Remove pre-cut insulation — .010

4. Terminate — .028

5. Inspect and place aside

In.	6	7	9	11	14	17
Min	.005	.006	.007	.008	.010	.011

6. Count and tape in bundles of 100

In.	6	7	9	11	14	17
Min	.009	.009	.011	.012	.014	.016

23.12 Component Lead Forming Machines

DESCRIPTION

The component lead former is a machine which can cut and bend leads to a specified geometry. These machines allow interchangeable dies. Each die set creates a different geometry for the component. A die set is composed of two elements. The first element is attached to the body or case of the machine and the second element is attached to a press. The press can be propelled pneumatically. Each component is placed manually between the set of dies (either individually or in small groups) and formed between the die set where it is sheared and bent to its desired geometry.

ESTIMATING DATA DISCUSSION

The setup time provides for a die change and obtaining materials. The die change is the time required to remove the current die set and replace it with the desired set. Obtain materials includes the time required to obtain the components from inventory or storage.

Element 1 provides total handling on a bulk basis for a specific lot. Handling and preparation is manual. Note that the initial condition of the components (either in bulk or in ordered strips) is insensitive to handling and preparation time. These lot minutes are divided by the number in the lot to find the cycle time. The number of components in the lot is entered as the adjustment factor entry.

Element 2 encompasses the machine forming time. This includes the manual placement of the part between the dies, triggering of the press, forming the components, and obtaining the next component. Entry to Element 2 is lot quantity. The adjustment factor reduces the time to a cycle value. In a way, the data for this machine reflects operator knowledge.

EXAMPLES

A. A lot of 100 resistors need to be formed to a hook shape. Find the estimates.

Table Number	Process Description	Table Time	Adjustment Factor	Cycle Minutes	Setup Hours
23.12-S	Setup	.15			.15
23.12-1	Handle and preparation	1.36	1/100	.014	
23.12-2	Machine time	5.10	1/100	.051	
	Total			.065	.15
	Lot estimate	.158 hr			

B. A lot of 70 resistors and 41 diodes are to have their leads formed. The parts are provided in separate bulk quantities but are to be produced jointly. Two tool changes are necessary. Find the unit, setup, lot and pc/hr estimates. Total quantity assumed to 111 units.

Table Number	Process Description	Table Time	Adjustment Factor	Cycle Minutes	Setup Hours
23.13-S	Setup	.15	2		.30
23.13-1	Handle resistors	1.36	1/70	.019	
23.13-2	Form resistor lead	3.49	1/70	.050	
23.13-1	Handle diodes	1.36	1/41	.033	
23.13-2	Form diode lead	2.16	1/41	.053	
	Total			.155	.30
	Hr/100	.258			
	Lot estimate	.59 hr			
	Pc/hr	387			

Note: There would be no difference if the parts had been estimated separately.

TABLE 23.12 COMPONENT LEAD FORMING MACHINES

Setup
    Die change and supplies   .15 hr

Operation elements in estimating minutes

1. Bulk handling, preparation of quantity   1.36
   (Divide by lot quantity)
   Handle, one at a time   .03

2. Form lead

No. of Components	Lot Min	No. of Components	Lot Min	No. of Components	Lot Min
30	1.47	48	2.38	77	3.84
33	1.62	53	2.62	84	4.22
36	1.78	58	2.89	93	4.64
40	1.97	64	3.17	102	5.10
44	2.16	70	3.49	Add'l	.051

23.13 Printed Circuit Board Drilling Machines

DESCRIPTION

The printed circuit board (PCB) drill machine is a paper-tape programmable drilling machine used for production drilling of one or several PCBs simultaneously. The drill needs the appropriate data for each board. Hole locations, according to a specific size, are stored in a pattern which is efficient as well as acceptable for the drill press. Once the data have been loaded into the machine, the operator changes the appropriate carbide drills and oversees the process.

Figure 23.13 is a five-spindle drilling machine. Not all jobs will require the use of all five spindles. The operator selects which of the spindles will be active.

FIGURE 23.13 Printed circuit board drilling machine (Excellon)

Once the process has started only those spindles selected for the operation are lowered to drill the board. Depth and clearance of the drill can be adjusted. Larger bits require more clearance and the spindle upper limit must be adjusted to allow for this clearance, which increases the process time. Feed and drill speed are chosen. These variables have the largest effect on the drilling time. The larger carbide bits require a slower feed as well as drill speed. The combination of these two variables leads to longer drill time. Adjustment of the feed and drill speed is handled by software. Depth controls are adjusted by the operator to accommodate for one or several boards being simultaneously drilled.

Drills are only loaded once or twice during the run. The operator starts the drilling cycle and the machine keeps track of which drill is required. Once loaded, the drilling process begins. The operator is able to prepare for the next job. This reduces most of the job time to drilling. Drilling time is usually on the order of a few minutes for each drill size. The operator can perform non-drill time tasks during machining, which may include changing worn out, non-active drills.

The machine tool is controlled by a microprocessor to position the table according to the data for a particular circuit board. Each hole is programmed with an x and y coordinate. The table moves to this position and signals the processor that the table is ready for that hole to be drilled. The processor loads the machine tool control with the next position, and when the spindle control signals that the hole has been drilled, the table moves to the new drilling location and the process continues until all the holes for that particular bit size have been drilled. At this point, the drill is unloaded and the machine awaits instructions to proceed with the next hole. Time to drill is determined more by the number of holes than by their depth.

Tool changes are required every 3000 holes or so. After 3000 hits the machine parks the table and signals the operator, via the terminal screen, that a tool change is required. Also, should something go wrong during the drill cycle, such as an undefined table position due to an error in reading the data, the operator is signaled and appropriate actions are taken.

ESTIMATING DATA DISCUSSION

Setup requires inspection of the machine and loading the machine operating system. Cycle elements include loading the machine with the circuit boards, drilling of the holes, and the unloading. Loading the machine consists of selecting the particular spindles for drilling as well as positioning and securing the boards. Positioning and securing the boards consist of pounding nails into guides on the bench and then taping the boards in place. The time taken to load the boards from the job rack depends on the number of boards loaded.

Drilling the holes consists of tool selection and drilling. The time to drill the holes depends on the number of holes and on the diameter of the drill. Due to the variety of sizes, table choices are limited within a range. To use the tables determine the range of the bit diameter and the number of holes. Use the higher value if the number of holes falls between two choices. If the number of total hits exceeds 3000 hits, a tool change time is necessary.

Unloading the boards consists of removing the tape which secures the boards and placing them on a table so the boards can be separated.

EXAMPLES

A. A stack of one board is to be drilled with 49 holes of .040-in. OD, 912 holes of 0.057-in. OD, and 6 .072-in. holes. Find the cycle and setup estimates.

Table Number	Process Description	Table Time	Adjustment Factor	Cycle Minutes	Setup Hours
23.13-S	Setup	.5			.5
23.13-1	Handle board	1.60		1.60	
23.13-2B	Drill 40 .040-in. holes	.8		.80	
23.13-2C	Drill 912 .057-in. holes	5.3		5.3	
23.13-2D	Drill 6 .072-in. holes	.40		.40	
	Total			8.10	.5

B. Two sets of boards, two deep, are loaded on the table. Each board will require 2617 holes of .028 in., 97 holes of .040 in., and 73 holes of .0635 in. Note that drilling by two spindles of the same size is simultaneous. Also a stack has two boards. Find the unit and setup estimates.

Table Number	Process Description	Table Time	Adjustment Factor	Cycle Minutes	Setup Hours
23.13-S	Setup	.50			.50
23.13-1	Load and unload two board stack	1.70	½	.85	
23.13-2A	Drill 2617 holes, .028-in. OD	15.1	¼	3.78	
23.13-2B	Drill 97 holes, .040-in. OD	1.1	¼	.28	
23.13-2D	Drill 73 holes, .0635-in. OD	.9	¼	.23	
	Total			5.14	.50
	Floor-to-floor time for four boards	20.5 min			

C. Three sets of boards, three deep, are to be drilled with the following quantities and diameters: 108 0.033-in. holes, 2545 0.035-in. holes, 897 0.040-in. holes, 145 0.0465-in. holes, 41 0.055-in. holes, and 160 0.129-in. holes. Find the estimates. The lot has 848 boards.

Table Number	Process Description	Table Time	Adjustment Factor	Cycle Minutes	Setup Hours
23.13-S	Setup for three-spindle drilling	.5			.5
23.13-1	Load three-board stack	1.9	⅓	.63	
23.13-2A	Drill 108 holes 0.033-in. diameter	1.0	⅑	.11	
23.13-2B	Drill 2545 holes 0.035-in. diameter	12.4	⅑	1.38	
23.13-2B	Drill 897 holes 0.040-in. diameter	5.3	⅑	.59	
23.13-2C	Drill 145 holes 0.0465-in. diameter	1.3	⅑	.14	
23.13-2C	Drill 41 holes 0.055-in. diameter	0.8	⅑	.09	
23.13-2E	Drill 160 holes 0.129-in. diameter	1.8	⅑	.20	
	Total			3.14	.5
	Floor-to-floor	28.3 min			
	hr/100			5.241	
	Lot estimate			44.94 hr	

TABLE 23.13 PRINTED CIRCUIT BOARD DRILLING MACHINES

Setup in estimating hours 0.5 hr

Operation elements in estimating minutes

1. Load and unload

Stack of Boards	Min
1	1.60
2	1.70
3	1.90
4	2.00

2. Drill

   A. Drill diameter 0–0.033 in.

No. of Holes	Min	No. of Holes	Min	No. of Holes	Min
4	.5	276	1.8	1232	6.5
20	.6	333	2.1	1433	7.5
39	.7	398	2.4	1663	8.6
60	.8	474	2.8	1929	9.9
85	.9	560	3.2	2233	11.4
113	1.0	660	3.7	2584	13.1
146	1.2	775	4.3	2987	15.1
183	1.4	906	4.9	3451	17.4
226	1.6	1058	5.6		

   B. Drill diameter 0.034–0.040 in.

No. of Holes	Min	No. of Holes	Min	No. of Holes	Min
15	.6	371	2.3	1625	8.1
36	.7	446	2.6	1887	9.4
61	.8	532	3.0	2189	10.9
89	1.0	631	3.5	2537	12.4
121	1.1	744	4.0	2936	14.3
158	1.3	875	4.6	3396	16.4
201	1.5	1025	5.3	3924	18.9
250	1.7	1198	6.1		
306	2.0	1396	7.1		

C. Drill diameter 0.041–0.060 in.

No. of Holes	Min	No. of Holes	Min	No. of Holes	Min
15	0.6	314	2.0	1227	6.1
37	0.7	380	2.3	1430	7.1
62	0.8	457	2.6	1664	8.1
91	1.0	545	3.0	1933	9.4
124	1.1	646	3.5	2243	10.8
162	1.3	762	4.0	2599	12.4
206	1.5	896	4.6	3008	14.3
256	1.7	1050	5.3		

D. Drill diameter 0.061–0.10 in.

No. of Holes	Min	No. of Holes	Min
2	.3	112	1.2
10	.4	137	1.3
20	.5	165	1.6
42	.6	197	1.8
56	.7		
72	.9		
91	1.0		

E. Drill diameter 0.11–0.252 in.

No. of Holes	Min	No. of Holes	Min	No. of Holes	Min
7	.4	135	1.6	430	4.3
15	.5	161	1.8	501	4.9
34	.6	192	2.1	582	5.6
45	.8	227	2.4	676	6.5
58	.9	268	2.8	784	7.5
74	1.0	314	3.2	908	8.6
91	1.2	368	3.7	1051	9.9
111	1.4				

3. Change drill per size   3.00
   (Prorate over no. of boards)

23.14 Shearing, Punching, and Forming Machines

DESCRIPTION

These data describe shearing, punching, and forming machines used with electronic products that require these processes. These data would customarily be used with electronic parts in electronic fabrication operations. Similar equipment would be used for these operations provided the proper dies are available. The machine used to drive the die is an upright punch press, and a sketch is given by Figure 23.14.

The power punch press is a simple machine. Through a ram, it transfers power to a die which in turn forms the part. The machine is used widely in industry for operations that require high concentrated force on one surface of an object.

With the exception of manual energy, these data can be applied for a variety of power sources (electrical, mechanical, hydraulic, or pneumatic) and applied

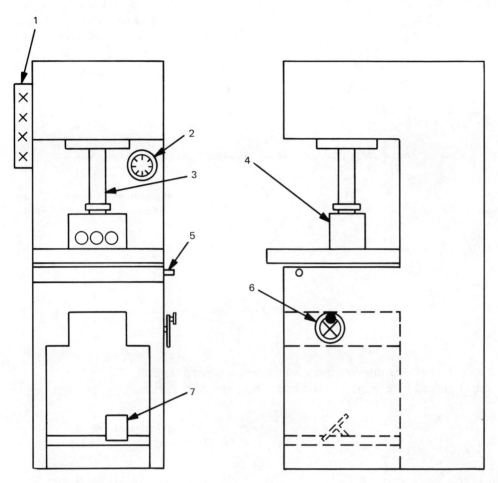

1 — Start/Stop Buttons
2 — Pressure Gage (0–30,000 P.S.I.)
3 — Ram
4 — Die
5 — Outlet for Air Blower Nozzle
6 — Crank for Raising and Lowering Die Surface
7 — Adjustable Foot Pedal

FIGURE 23.14 Sketch of an upright punch press.

to estimate cycle and setup times. Application of these data is limited to machines that use the following systems of power transfer: crank, cam, eccentric, or screw. These estimates were compiled for a single-die process and would not be applicable for multiple die systems.

The punch press is activated by either palm buttons or a foot pedal. It is possible to shear, punch, form, coin, or crimp by adjusting the force with the appropriate die.

ESTIMATING DATA DISCUSSION

Element 1 loads the part from the bins or bin to the die. Positioning of the component in the die is covered in Element 2. Once the operator has the part in the die the second element begins. With a component that requires a minor assembly such as staking or crimping, the given estimate is for insertion and activating the press with either palm buttons or a foot pedal. Stroke time is for one hit.

The third element begins as soon as the punch has returned to its retracted position. The operator takes the part from the die to the "finished" bin completing the entire process. In a few cases the component will have a stray ground wire. This must be cleared before the stroke.

EXAMPLES

A. It is necessary to crimp a header onto a Schotky DO-4 diode. The part weight is .16 oz. Find the estimates.

Shearing, Punching, and Forming Machines

Table Number	Process Description	Table Time	Adjustment Factor	Cycle Minutes	Setup Hours
23.14-S	Setup	1.5			1.5
23.14-1	Pick up two parts	.033		.033	
23.14-2	Crimp	.010		.010	
23.14-3	Lay aside	.016		.016	
	Total			.059	1.5

B. Form a lead on to a 8.5 oz. electronic chip. There are 8010 assemblies necessary. Find the estimates.

Table Number	Process Description	Table Time	Adjustment Factor	Cycle Minutes	Setup Hours
23.14-S	Setup	1.5			1.5
23.14-1	Two parts to die	.033		.033	
23.14-2	Form	.070		.070	
23.14-3	Lay aside	.016		.016	
	Total			.119	1.5
	Hr/100	.198			
	Pc/hr	504			
	Lot estimate	17.39 hr			

C. A header is formed on a 3.6-oz. electronic assembly, which behaves as an integral single part. The part has a stray ground wire. Find the estimates. There are 1936 units in the lot.

Table Number	Process Description	Table Time	Adjustment Factor	Cycle Minutes	Setup Hours
23.14-S	Setup	1.5			1.5
23.14-1	Pick up part, move to die	.015		.015	
23.14-2	Form (two strokes)	.031	2	.062	
23.14-2	Separate wire	.038		.038	
23.14-3	Lay aside	.016		.016	
	Total			.131	1.5
	Hr/100	.218			
	Lot estimate	5.73 hr			

TABLE 23.14 SHEARING, PUNCHING, AND FORMING MACHINES

Setup   1.5 hr

Operation elements in estimating minutes

1. Pick up part and move to die

    One part

Wt (Oz.)	Min
.2	.023
2.5	.018
4.4	.015
6.3	.012
7.5	.009
8.8	.007

    Two parts and move to die   .033

2. Shear, punch, form, crimp (one stroke)

One part	Wt (Oz.)	Min
	.2	.010
	.6	.013
	1.1	.015
	1.8	.018
	2.6	.022
	3.5	.026
	4.5	.031
	6.0	.038
	7.4	.045

Two-part minor assembly	.070
Separate wire or wires	.038

3. Remove part and lay aside

Small assembly	.016
Single part	.006

23.15 Resistance Micro-Spot Welding

DESCRIPTION

Resistance micro-spot welding (RMSW) welds two pieces of metal together by passing a current through the metal. Pressure is necessary at a high amperage. The metal is raised to a temperature sufficient to lightly melt the metal. Two electrodes press the metal together, thus completing the weld. The two electrodes are then removed and the metal returns to room temperature.

There are two types of RMSW considered here. One is a foot-pedal operated press in which an upper electrode is lowered by a foot pedal and brought into contact with the metal. The lower electrode is positioned below the metal. The other setup is a hand-held weld clamp in which the electrodes are mounted to the tips of the C-clamp. Two electrodes bear on the metal by squeezing the clamp. In both setups, the current is actuated by the pressure between the electrodes reaching a pre-set level. The purpose of RMSW is not only in ensuring a reliable spot-weld joint. It is also used in place of other operations such as soldering when the design requires the absence of foreign material.

ESTIMATING DATA DISCUSSION

The setup required for RMSW consists of setting pressure to activate the current, the power level to pass through the metal, and obtain parts. For the weld press, it is necessary to set the gap for the electrodes.

Element 1 consists of loading the parts to be welded into the welding fixture or onto the welding machine. Usually there are several parts loaded before each pair of parts is welded. The independent variable is the number of parts loaded.

Element 2 consists of the actual welding process. The independent variable is the number of welds.

Element 3 is the removal of the part after welding. The time required to perform this element is generally a constant.

The process of data adjustment for different RMSW operations is simply done by counting the number of parts loaded and the number of welds required. There may be several RMSW operations performed before the product is completed and setup time is required for each operation.

EXAMPLES

A. To assemble the mandrel, one places two end pieces or caps to each end of a cylindrical sheet of metal. Two identical operations are performed, consisting of placing the tube and one endpiece into a fixture, and then welding using 11 welds around the tube. The weld press is used.

Table Number	Process Description	Table Time	Adjustment Factor	Cycle Minutes	Setup Hours
23.15-S1	Setup	.15			.15
23.15-1	Load parts	.26		.26	
23.15-2	Spot weld 11 times	.28		.28	2
23.15-3	Remove piece	.08		.08	
	Repeat 1–3	.62		.62	
	Total			1.24	.15
	Shop estimate	48 pc/hr			

B. Two small interpieces are welded to two pins. The interpieces are placed into a custom made fixture which holds the header. The fixture is then placed between the electrodes of the weld press. This requires two welds and three parts to be loaded. Find the unit and shop estimate.

Table Number	Process Description	Table Time	Adjustment Factor	Cycle Minutes	Setup Hours
23.15-S1	Setup	.15			.15
23.15-1	Load parts	.33		.33	
23.15-2	Spot weld two times	.08		.08	
23.15-3	Remove piece	.08		.08	
	Total			.52	.15
	Shop estimate	115 pc/hr			

TABLE 23.15 RESISTANCE MICRO-SPOT WELDING

Setup

    S1. Machine press   .15 hr

    S2. C-clamp style   .05 hr

Operation elements in estimating minutes

1. Load parts

No. of Parts	Machine Press	C-clamp
2	.24	.14
3	.33	.16
4	.42	.18
5	.52	.23
6	.61	.27
8	.80	.32
10	.99	.39

2. Spot weld

No. of Welds	Machine Press	C-clamp
2	.07	.19
3	.09	.22
4	.11	.25
5	.14	.31
6	.16	.37
7	.18	.46
9	.22	.55
11	.26	.55
14	.32	.67

No. of Welds	Machine Press	C-clamp
17	.38	.82
21	.40	1.00
26	.57	1.23

3. Unload
       Machine press   .07
       C-clamp part   .19

23.16 Turret Welding Machines

DESCRIPTION

The turret welder is a semiautomatic process that welds two components between metal electrodes. Normally the welded surfaces of the electronic components are steel, small sized, and related to electronic fabricated products; however, this is not a dominant requirement. The major operational feature is the six stations, known as turrets, on a rotating turn table. Note Figure 23.16. Each station has two electrodes for welding two parts. The electrodes are air and water cooled.

ESTIMATING DATA DISCUSSION

Four elements are performed simultaneously. The limiting or longest time element is usually loading. Turntable rotation is limited by how fast the loader can seat two pieces into the base electrodes. As the operator is loading two parts the machine is performing three other functions. Sensor switches check seating of the parts into the base electrodes. Header electrodes drop onto the parts and apply pressure while welding the two pieces together. Finally, the ejectors pluck the parts out of the base electrode and eject them down a ramp to a collector tray. The sensor, welding, and ejector stations are located at 60-degree intervals above the back half of the turntable. The turntable rotates at 60-degree intervals stopping long enough for the operator to load two parts. The operator must stop the machine occasionally to pick up dropped pieces and change collector trays.

Turret welder setups for different parts remain essentially constant. The setup process consists of checking the quality of the weld.

The operational elements are loading, sensor stroke, pressure and weld stroke, ejector stroke, and turntable rotation time. Loading is the only stage where the operator handles the part. The operator must get the pieces from a holding tray and load them into the base electrode, usually two pieces at a time. The sensor stroke is contained on the sensor station. Two spring-loaded sensors drop down onto the lower

FIGURE 23.16 Turret welding machine.

electrode checking for two pieces. If there are no pieces or only one piece, the welding process is automatically interrupted. Element 3, pressure and welding, is the critical part of the welding process. The upper set of electrodes drop down onto the lower electrodes. As the upper electrodes are seated, they apply pressure and current on the pieces. Element 4, unloading, is the final element. A set of ejectors clamp onto the two pieces in the lower electrode and ejects them down a ramp into a holding tray.

The turret welding operation is governed by the size of the pieces to be welded, $L + W$ is the seating space of the parts and the largest size governs the selection.

Irregular shaped objects can be dimensioned quickly for the $L + W$ squat size. A round washer, for example, will have $L + W = 2$ dia. A cable, wire, or lead will be the length + diameter of the wire. A resister will be the length of the body + the diameter of the resistor. This simple scheme for using the overall dimensions encourages speed in estimating and consistency. The larger parts dominates for two parts. The rules for using tables apply to these data. If the $L + W$ factor is less than any table value, the minimum value is adopted. If the $L + W$ falls between two values, the policy is adopt the higher value. If the $L + W$ exceeds the upper value, the additional time factor is employed. The cycle time is controlled by the longest station time.

EXAMPLES

A. Estimate the time to produce one tray of 130 DO-5 series glass diodes with a dimensional standard of .940-in. header and mechanical .687-in. hex ½-in. long.

Table Number	Process Description	Table Time	Adjustment Factor	Cycle Minutes	Setup Hours
23.16-S	Setup	.95			.95
23.16-1	Load, $L + W = 1.19$	.082		.082	
23.16-2	Sensor stroke	.087			
23.16-3	Weld, $L + W = 1.19$	.044			
23.16-4	Unload, $L + W = 1.19$	.06		—	—
	Total			.082	.95
	pc/hr	731			

B. Estimate the time to produce 1600 DO-5 series diodes with a dimensional criteria of .940-in. header × 1.5 in. and a .687-in. hex × .48-in. steel header and copper base. Inspection is required.

Table Number	Process Description	Table Time	Adjustment Factor	Cycle Minutes	Setup Hours
23.16-S	Setup	1.5			1.5
23.16-1	Load, $L + W = 2.44$	.097		.097	
23.16-2	Sensor stroke	.0087		.0087	
23.16-3	Weld, $L + W = 2.44$	.042		.042	
23.16-4	Unload, $L + W = 2.44$	.112		.112	—
	Total longest turret			.112	1.5
	hr/100 units	.187			
	Lot estimate	4.47 hr			
	pc/hr	535			

TABLE 23.16 TURRET WELDING MACHINES

Setup

Mechanical turret and tooling	.95 hr
Weld check and inspection	.55 hr

Operation elements in estimating minutes

1. Load part into electrodes on turret

$L + W$	Min	$L + W$	Min
1.15	.080	1.84	.089
1.25	.082	1.90	.090
1.63	.087	Add'l 1	.013
1.70	.088		

2. Sensor stroke .0087

3. Weld

L + W	Min
1.15	.045
1.25	.044
1.63	.043
1.90	.042

4. Unload welded part

L + W	Min	L + W	Min
1.25	.06	1.75	.033
1.63	.024	1.84	.043
1.65	.025	1.90	.050
1.70	.029	2.0	.064
		Add'l 1	.108

23.17 Wafer Dicing Machines

DESCRIPTION

The wafer dicing process involves the division of a silicon wafer into dice or chips. Typically the die dimensions are predetermined and inherent to the design of the wafer. Once cut, breaking, unmounting the dice, and sorting are considered in other operations. That data are provided in Table 21.1.

Two different types of machines are considered: sawing and laser scribing. The discussion and data are identified for those two types of equipment. The sawing operations apply to a range of die sizes and saw speeds. Non-sawing operations apply to a variety of mounting techniques and time of sawed-wafer cleaning. Data included for sawing operations are generally applicable to semi- and fully-automatic dicing saws. Semi-automatic saws include features which align streets (see part Figure 23.17) by remote on an x-y horizontal axis, have a television display of street magnification, and digital speed and index settings. Fully-automatic saws consider in addition to those features suggested for the semi-automatic saw, a function to remotely align and rotate the streets to angles between 0 and 90 degrees. Preprogrammable options provide for speed, index, and depth settings. Essentially the data provided are reasonable for machines that (1) self advance (step); (2) may be aligned by remote and television display; (3) that must be rotated either manually or remotely after cutting all streets parallel to cut streets perpendicular; and (4) are programmable. This does not include observing and aligning the streets through a microscope and manually operated machines commonly used for lab work.

The laser scriber also cuts silicon wafers into individual elements. The machine requires one operator to load the wafers and manually "scribe" them by manipulating controls while viewing a particular wafer through a magnifier. The laser scriber is approximately the size of an average business desk and seats an operator at the center. Refer to section 4.3 for another discussion on laser marking machines.

For a particular job x and y coordinate controls specify the dimensions of the particular wafer chips so that channels or streets may be cut through a chip at various separations and angles. This mode of operation is prespecified for a particular run. Once a mode is selected the operator positions a wafer on the scriber surface where it is held by a vacuum. Now in the proper position the controller aligns the channels with a guide which is located on the viewer through the use

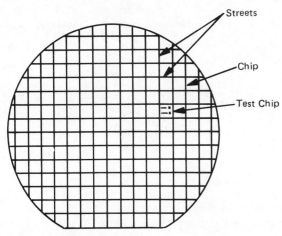

FIGURE 23.17 Typical wafer.

of the table control. It is necessary to focus the viewer at times due to differing channel thickness. After a wafer has been aligned the doors over the scriber surface are closed to ensure operator safety. At this point the run button is actuated to begin the laser scribing procedure. When the end light indicated the operator opens the doors and removes the wafer. The element is sometimes inspected and has its edges removed at this time. The finished product is placed in the finished box and the trimmed corners in the scrap box.

ESTIMATING DATA DISCUSSION

Machine sawing setup typically includes warm up, blade dressing, and program initiation. Fifteen minutes is allowed for warm up. In particular, setup turns on the air, water, and power supplies, television and the machine fan. Selecting program choices may offer six to thirteen buttons to choose, depress, and visually verify. Setup will include turning on the wafer cleaning and mounting apparatus. As part of setup time provided, cleanup reverse the preparation process. Included in clean up, however, the chuck and surroundings are expected to be blown dry and there is no warm down. Blade dressing is an optional choice but is necessary whenever large lots or special conditions are encountered.

The setup time required for the scriber involves only reprogramming the processor for differing chip dimensions. The cycle elements are identified for either sawing or laser. Most of the work is for the wafer, and the time must be dejointed using the adjustment factor. The wafer is considered a "multiple," i.e., has many finished chips. The product of the adjustment factor and the table time is a unit elemental time. The following is an example of the output for either saw or laser scribing.

Wafer Dia	Chip Size	No. Channels	No. Chips
4.5 in.	.238 in.	38	210
4.5 in.	.150 in.	47	340

A firm may have other wafer dimensions and the number of chips will vary. Usually the number of chips chosen does not account for yield losses. The effects of yield are adjusted elsewhere in an estimate.

After preparation sawing operations may involve several variations but are linked by mounting, cutting, rotating, and cleaning. Variations are likely in mounting techniques given different apparatus. Odd elements for dicing will include changing the blade and detailed microscope inspections.

Different time elements are necessary for laser scribing. Pick up and insert reflects the operator's positioning of a silicon wafer on the scribing surface where the wafer is held by vacuum. Centering is achieved by jogging of the table control while aligning the guide to the channel in the viewer. It is essential that proper alignment be obtained before initiation of the scribing process. Some runs may require refocussing for each element while others require refocussing for only the first element of a particular job where subsequent elements are identical. It is left to the estimator to decide to include the focussing step depending on the process being performed. The run operation is self explanatory. The end light tells the operator when the cycle is complete. The removal of the chip is covered by the side element. If corners are to be removed from the chip, the estimator should increase the side value by a factor of 4.

EXAMPLES

A. A batch of 830 chips is desired. Chip size is .230 in. and the number of streets is 38 to be cut on a 4.5-in. OD wafer. Part feed is assumed to be .512 in./second for a sawing operation. Estimate the lot time. There are 210 chips on a 4.5-in. OD wafer.

Table Number	Process Description	Table Time	Adjustment Factor	Cycle Minutes	Setup Hours
23.17-S	Setup	.58			.58
23.17-1	Precut tape and mount	.44	1/210	.002	
23.17-2	Mounting	.11	1/210	.001	
23.17-3	Align streets	.58	1/210	.003	
23.17-4	Saw wafer	12.3	1/210	.059	
23.17-3	Rotate 90 degrees	.34	1/210	.002	
23.17-5	Unmount	.31	1/210	.001	
	Total			.067	.58
	Lot estimate	1.51 hr			

B. A batch of DO5 wafers is to be sawed and cleaned. Five thousand chips are necessary and a speed of .35 in./second is appropriate. Assume a new blade is necessary which will last 20,000 cuts. Inspection is appropriate. Each wafer has 210 chips. Find the estimates.

Table Number	Process Description	Table Time	Adjustment Factor	Cycle Minutes	Setup Hours
23.17-S1	Setup	.58			.58
23.17-S1	Blade dress	2.5			2.5
23.17-1	Mount wafer	.56	1/210	.003	
23.17-2	Mount ring	.11	1/210	.001	
23.17-3	Align streets	.58	1/210	.003	
23.17-4	Saw at .35 in./second	15.8	1/210	.075	
23.17-3	Rotate 90 degrees	.34	1/210	.002	
23.17-5	Unmount	.31	1/210	.001	
23.17-6	Ultrasonic clean	6.3	1/210	.03	
23.17-7	Blade change	.0006		.006	
23.17-8	Mike inspection	4.00	1/1000	.004	
	Total			.119	3.08
	Hr/100	.198			
	Pc/hr	504			
	Lot estimate	12.99 hr			

C. A .304 × .304-in. wafer is to be laser scribed. Corners will broken and refocussing is thought necessary. Find the unit and lot estimates for a quantity of 50. Excess chips from a wafer size of 4.5 in. will be scrapped.

Table Number	Process Description	Table Time	Adjustment Factor	Cycle Minutes	Setup Hours
23.17-S2	Setup	.1			.1
23.17-1	Load for laser	.10	1/50	.002	
23.17-2	Laser center	.25	1/50	.005	
23.17-3	Focus	.10	2/50	.004	
23.17-4	Laser cut	.49	1/50	.01	
23.17-5	Unload	.05	4/50	.004	
	Total			.018	.1
	hr/100	.030			
	Lot hr	.12			

D. Re-estimate Example B using a laser scriber.

Table Number	Process Description	Table Time	Adjustment Factor	Cycle Minutes	Setup Hours
23.17-S2	Setup	.1			.1
23.17-1	Load wafer	.10	1/210	.0005	
23.17-3	Laser center	.25	1/210	.0012	
23.17-3	Adjust focus	.10	1/210	.0005	
23.17-4	Laser cut	.49	1/210	.0023	
23.17-5	Unload laser	.05	1/210	.0002	
	Total			.0047	.1
	Hr/1000 units	.0786			
	Pc/hr	12,727			
	Lot estimate	.49 hr			

TABLE 23.17   WAFER DICING MACHINES

Setup
S1. Setup for sawing .58 hr
    Blade dress for sawing 2.5
S2. Setup for laser scriber .1 hr

Operation elements in estimating minutes

1. Load wafer, per wafer
   - Mount for sawing .56
   - Precut tapes and mount for sawing .44
   - Load for laser scriber .10

2. Mounting and blowoff for sawing, per wafer .11

3. Machine control, per wafer
   - Align streets for sawing .58
   - Rotate chuck for sawing, 90 degrees .34
   - Laser center and close door .25
   - Adjust focus for laser to secure alignment .10

4. Dice wafer, per wafer
   Saw and cut

Feed	.2	.29	.33	.40	.46	.53	.61	.70	.81
Min	16.6	15.8	14.9	13.6	12.3	10.9	9.3	7.4	5.2

   Laser cut .49

5. Unload per wafer
   - Unmount and blowoff for saw .31
   - Laser .05

6. Ultrasonic clean for sawing, per wafer
   - 3-min clean 5.3
   - 4-min clean 6.3
   - 5-min clean 7.3

7. Saw blade change, per die prorated .0006

8. Inspection by micrometer
   - Per wafer 4.00

23.18 Coil Winding Machines

DESCRIPTION

A coil winder provides a means to wind different shapes and types of coil quickly. Used in production of electromagnetic devices such as deflection coils, focus coils, inductors, and toroids, etc., the machines are capable of winding coils with segment windings of up to 180 degrees or 360 degrees continuous windings or bobbin windings. Wire used on the machine as reported here is between 16- and 45-gage, and cores up to 1-in. diameter can be accommodated. The winder contains a pre-set electronic counter assembly which controls the number of turns wound on the core. The motor provides a 0–400 rpm output to the winding head assembly. The core holding fixture rotates and the interaction between it and the rotating winding head yields correct winding of the coil. The winding process is semi-automatic, requiring the operator to thread the winding head and the core fixture. Constant supervision of the machine is needed to ensure that there is no disalignment of the shuttle and core. The machine is best suited for small quantity jobs. Now consider the general classification of coil winding machines.

ESTIMATING DATA DISCUSSION

Since the coil winder can produce coils of many different sizes and dimensions, a distinction is made concerning coil type. Distinctions are made in some of the cycle elements between bobbins or toroids. For the toroidal pieces, the setup time is .10 hr. This allows time to mount the winding head, install the shuttle, attach the wire spool and core holding fixture, and the core itself. Cleanup and other chores are included. This basic setup also assumes that the choice of winding head, shuttle size, and core size are all known. Experimentation in fitting shuttle size to core proportions can result in a much longer setup time and is not accounted for here. For the bobbins setup time include time for mounting the chuck, inserting the bobbin, and mounting the thread spool, and cleaning up.

The cycle time for toroids is arranged into six main elements. Element 1 consists of picking up the spool thread, leading it over the wire tension guide, threading it through the shuttle, and securing. The wire is wrapped around once and the free end clipped short.

In Element 2 the process is automatic. The operator initially sets the load/wind switch, the digital footage counter, and the speed control. The speed control regulates the feed rate of the shuttle and is set at approximately 160 rpm. The operator then flips the power switch, and the machine loads the shuttle. This element has both a constant and variable portion to it.

Element 3 consists of mainly wrapping the thread through the core and securing it before the winding process. Occasionally adjustment of the core's position is needed.

Element 4 deals with the actual winding of the toroids by the machine. The number of turns desired and the wire size used are the variable factors in determining the element time. In Element 5 the core and shuttle are removed from the machine and the extra wire is cut from the shuttle. The toroid's wire ends are wrapped around by hand and trimmed. The coil is then inspected for even turns of wire and to ensure that none of the wire insulation has been scraped off the winding process.

The cycle elements for the production of bobbins are simpler than for a toroid. Element 1 consists of leading the wire over the tension guides and attaching it to the bobbin. The winding time, Element 4, varies with the number of turns required for the bobbin. It includes operator start and stop time. Elements 5 and 6 give the estimating times for removing the bobbin from the holder, clipping the wire, and inspecting for even turns. In Element 4, the correct number of turns is found on the table and the wire gage bracket is located at the table's head. The intersection of that column and row is the elemental time. If any variable is less than the minimum shown in the table, the minimum value listed is chosen. If any variables fall between the listed table values, the next higher value is used. If any exceed the tabular listed value, the maximum of the values is used. These rules aid consistency in estimates.

EXAMPLES

A. A toroid is needed for a high voltage power supply. It requires 1500 turns of 36-gage wire. The core diameter is small, requiring precise alignment. Find the unit estimate and hr/100 units for the job.

Table Number	Process Description	Table Time	Adjustment Factor	Cycle Minutes	Setup Hours
23.18-S	Setup	.10			.10
23.18-1	Thread shuttle	.50		.50	
23.18-2	Operator start and stop	.25		.25	
23.18-2	Machine load 40 ft	.32		.32	
23.18-3	Thread wire through core	.57		.57	
23.18-4A	Machine wind wire on core	21.96		21.96	
23.18-5	Stop, removal of core and shuttle	.78		.78	
23.18-6	Inspect insulation and windings	.60		.60	
	Total			28.60	.10
	Hr/100 units	47.667			

B. An inductor for a low voltage power supply has been ordered. It requires 1000 turns of 41-gage wire. Find the unit and lot estimate if 50 are ordered.

Table Number	Process Description	Table Time	Adjustment Factor	Cycle Minutes	Setup Hours
23.18-S	Setup	.10			.10
23.18-1	Thread shuttle	.50		.50	
23.18-2	Operator start and stop	.25		.25	
23.18-2	Machine load 30 ft	.24		.24	
23.18-3	Thread wire through core	.57		.57	
23.18-4A	Machine wind wire on core	14.88		14.88	
23.18-5	Stop, removal of core, shuttle	.78		.78	

Coil Winding Machines

Table Number	Process Description	Table Time	Adjustment Factor	Cycle Minutes	Setup Hours
23.18-6	Inspect insulation and windings	.60		.60	
	Total			21.42	.10
	Hr/100 units	35.70			
	Lot estimate	17.95 hr			

C. A 7/32-in. diameter bobbin coil is assembled. It has 400 turns of 38-gage wire. Estimate the unit time and pc/hr.

Table Number	Process Description	Table Time	Adjustment Factor	Cycle Minutes	Setup Hours
23.18-S	Setup	.05			.05
23.18-3	Lead wire over guides, thread bobbin	.25		.25	
23.18-4B	Machine wind thread onto bobbin	.83		.83	
23.18-5	Stop, remove bobbin from chuck	.17		.17	
23.18-6	Inspection of windings	.58		.58	
	Total			3.03	.05
	pc/hr	19.8			

D. A special toroid requires 478 turns of 38-gage wire. A lot of 400 has been ordered. Find the unit and lot estimate for the job.

Table Number	Process Description	Table Time	Adjustment Factor	Cycle Minutes	Setup Hours
23.18-S	Setup	.10			.10
23.18-1	Thread shuttle	.50		.50	
23.18-2	Operator start and stop	.25		.25	
23.18-2	Machine load, 20 ft	.16		.16	
23.18-3	Thread wire through core	.57		.57	
23.18-3	Adjust core position with shuttle	.67		.67	
23.18-4A	Machine wind the core	5.98		5.98	
23.18-5	Stop, removal of core, shuttle	.78		.78	
23.18-6	Inspect windings, insulation	.60		.60	
	Total			13.11	.10
	Hr/100 units	21.850			
	Lot estimate	87.500 hr			

TABLE 23.18 COIL WINDING MACHINES

Setup

Toroids	.10 hr
Bobbins	.05 hr

Operation elements in estimating minutes

1. Thread wire over guides and through the shuttle .50

2. Load wire onto shuttle, operator start and stop .25
   Machine loading

Feet	10	15	20	25	30	35	40	45
Min	.08	.12	.16	.20	.24	.28	.32	.36

3. Thread wire through core and secure .57
   Lead wire over guides and through bobbin .25
   Adjust core position with shuttle .67

4. Machine wind
   A. Torrids:

No. of Turns	Wire Size (Gage)		
	24–32	34–36	38–42
300	2.17	2.25	2.42
400	3.78	3.90	4.20
500	5.38	5.56	5.98
600	6.98	7.21	7.76
700	8.59	8.87	9.54
800	10.19	10.52	11.32
900	11.79	12.18	13.10
1000	13.39	13.83	14.88

No. of Turns	Wire Size (Gage)		
	24–32	34–36	38–42
1000	13.39	13.83	14.88
1100	15.00	15.49	16.66
1200	16.60	17.15	18.44
1300	18.18	18.78	20.22
1400	19.80	20.46	22.00
1500	21.26	21.96	23.78
1600	23.01	23.77	25.56
1700	24.61	25.42	27.32
1800	26.21	27.08	29.12

   B. Bobbins:

No. of Turns	200	300	400	500	600	700	800	900	1000
Min	.42	.63	.83	1.04	1.25	1.46	1.75	1.97	2.19

5. Removal of coil from winder
   Removal of core and shuttle, wire is stripped from shuttle, core wire is wrapped and trimmed .... .78
   Removal of bobbin from chuck, trim wire ends .... .17

6. Inspection
   Toroids, check windings and insulation .... .60
   Bobbins, check windings .... .58

PACKAGING

24.1 Packaging Bench and Machines

DESCRIPTION

These estimating data are intended for temporary or final packaging. The purpose of temporary packaging may be for interim inventory and for protection against corrosion, moisture, abuse, or loss. The intention of final packaging may be for consumer or industrial pack, box, envelope, and/or bag. The materials for packaging are wood, paper, plastic, and metal. Cushioning for transport or stability is also considered. The items to be packaged are industrial and are not normally foodstuffs or pharmaceuticals.

The information is for low- to moderate-lot quantities using bench work. Semi- or fully automatic machines are necessary for moderate-to-large quantities. Individual, multiple, and bulk quantities are factors of choice in using the tables.

Figure 24.1 is a universal wrapping machine for short or long runs. It employs shrink or non-shrink wrapping materials. The product infeed and discharge are on the same end of the machine which allows one operator to feed the product and pack off where higher speeds are not required, which is about 40 units/min.

ESTIMATING DATA DISCUSSION

The data are organized differently than other tables. In addition to the setup, operational elements are associated to (1) handling of one, several, or bulk, (2) order time per occurrence, and (3) a unit time for pack. Thus, the data are pro-rated against the lot, order, or quantity restriction of the container, as well as the unit time for the pack. Single or multiple packaging must be evaluated. In addition to a unit or lot estimate, we may want to know an order estimate, as the order quantity may be less than the lot quantity.

The description of packaging direct labor can be diverse; it can be bench work only, bench and machine alternately or in some combined way, or machine tending exclusively. Element choice is provided for this job diversity. The final package can consist of a variety of intermediate packages, each with different quantities of the same or different parts.

Order paperwork is optional, but it relates to invoice or billing paper, inventory slips, routings, guarantees, or computer cards added to the order, or marked in accordance with shipping instructions either to an interim internal or external inventory or directly to the customer. Shipping labels and general package instructions are included in this element. If there are several units for each order, the value is reduced by the adjustment factor.

Element 1 is usually independent of the type of pack and the quantity packaged.

Element 2 is manual get and position part for packaging. Conceivably, the part(s) get and position could be from a cart, skid, tote box, hanger, conveyor line, rack, moveable cage, pan, rod, jar, carton, etc. The three entry variables are maximum dimension, weight, and the nature of the arrival of the units. While there may be a contradiction between the maximum dimension and weight range, these suggestions

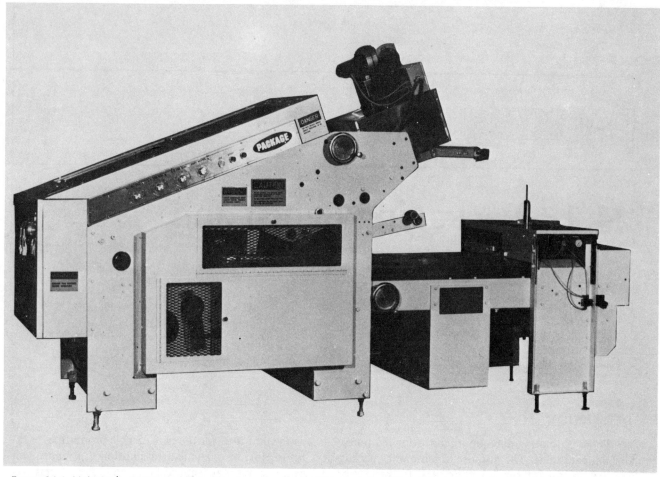

FIGURE 24.1 Universal wrapping machine for size range up to 16-in. length, 9-in. width and 5-in. height *(Package Machinery Company)*

allow for a consistent selection. Simple implies easy reaches, object handlings, and moves underload. Stack implies interference, while several suggest that several can be handled simultaneously. Bulk remove is a containerized transfer and dumping is possible. All manual packaging and many machine packaging operations will include Element 2.

Elements 3 through 16 deal with bench packaging. The estimator should carefully note what the element is "per." Unit differs from units; and bag, package, carton, or box implies one or more than one unit inasmuch as one or more may be jointly packaged in these containers. Some elements give a constant time per the package type plus a time for the part being loaded. Thus the order or lot (and they may differ) quantity is important. Some work would be done for the bag, package, etc., as a finish wrap. The procedure is to find the unit estimate, and multiply by the lot or order (or both) quantity to establish the estimate.

Element 3 constant time is per straw, and the number of parts used with each straw must be determined from the dimensions. The .02 min refers to part loading.

Two types of loading are considered in Element 6. In the nonbulk pack, the entry variable is the smallest x-y-z coordinate dimension. With 200+ bulk pack, the time is per the number loaded into the bag. No distinction is made to the bag seal, i.e., fold and staple, heat seal, or pressure close. In Element 7, the bag units are the divisor.

For Element 9, the .83 min covers preparation of the plastic or glass jar, and it is prorated against the number of units placed with the jar. Optional cushion or inner layer time is provided. But each part, singly or bulk, is placed within the jar and is added to the time for this pack. A lid means a turned and pressure cap.

The foam is preslitted in Element 11, and the slit length is the entry variable.

Element 12, or inner wrap, is wrapping material laced between two or more units. Distinctions are made with respect to materials used. Thus, if several units are packaged, the time includes both inner wrap package and minutes per unit (min/unit). Overwrap, Element 13, has the wrapping material around the outside of one or more units. It includes a constant time for material related to the package units and a variable wrapping time related to girth $L + W + H$. A third part of the overwrap is the number of units, and is expressed as a unit time.

Elements 14 and 15 are organized similarly. A container constant plus a unit or bulk load is stated. A final total time is divided by the number of units in the container.

Element 17 is for automatic packaging. Entry variable is the package per minute and units per package. The time is per unit of final product and is labor time for machine watching. If the operator watches more than one machine, the minutes per part is divided by the number of machines.

EXAMPLES

A. Parts that are easily removed from a hook-plating rack are placed into a tote box. Find the unit estimate. No order paperwork is required. Dimension is 4 in.

Table Number	Process Description	Table Time	Adjustment Factor	Cycle Minutes	Setup Hours
24.1-2	Unload from rack	.03		.03	
24.1-5	Place in tote box by stacking	.01		.01	
	Total			.04	

B. Painted parts that have a girth of $L = 4$, $W = 2$, and $H = 0.6$ in. are innerwrapped. Ten parts are placed in a corrugated paper carton for shipment. The parts are removed from a belt conveyor and the order size is moderate. Find the carton time and the unit estimate. Order paperwork is required for each carton.

Table Number	Process Description	Table Time	Adjustment Factor	Cycle Minutes	Setup Hours
24.1-1	Order paperwork	2.00	1/10	.20	
24.1-2	Remove from conveyor, each	.09		.09	
24.1-12A	Brown paper innerwrap, and 10 units will have an approximate box dimension of 5 high and 2 across for a 10 × 8 × 10-in. dimension	.37	1/10	.037	
24.1-12A	The sum of the two smallest major dimensions is 6	.33		.33	
24.1-10	Corrugated carton	1.51	1/10	.15	
24.1-10	Nonbulk pack	.06		.06	
	Total			.87	
	Order estimate	8.67 min			

C. The part in B is from a lot of 600, 200 of which are packaged as in B above, and the remainder will be Kimpak overwrapped for cartons containing 6 units. Each carton has order-processing. Find the lot estimates for B and C.

Table Number	Process Description	Table Time	Adjustment Factor	Cycle Minutes	Setup Hours
24.1-S	Setup	.15			.15
24.1-1	Order processing, 2.00/6	.33		.33	
24.1-2	Remove from conveyor, each	.09		.09	
24.1-13	Kimpack material	1.06	1/6	.18	
24.1-13	Package of six has $L + W + H = 26$,	.01	26	.26	
24.1-13	Unit time	.04		.04	
24.1-10	Corrugated carton	1.51	1/6	.25	
24.1-10	Nonbulk pack	.06		.06	
	Total			1.18	.15
	Order estimate	7.08 min			
	Lot estimate, .15 + 200 × .87/60 + 400 × 1.18/60	10.92 hr			

D. Two little parts are inserted into a coin envelope. Ten coin envelopes are placed in a metal-edge paper box. Find the unit estimate.

Table Number	Process Description	Table Time	Adjustment Factor	Cycle Minutes	Setup Hours
24.1-S	Setup	.15			.15
24.1-2	Get and position unit	.03		.03	
24.1-4	Load two units in coin envelopes, .14 + .03 = 17	.17	½	.09	
24.1-15A	Metal-edge paper box	.75	³⁄₁₀	.04	
24.1-15A	Insert envelope and unit is ½ × 3 × 5-in. size	.01	½	.01	
	Total			.17	.15

E. A part is to be packaged four to a corrugated box. The lot order is 7075 units.

Table Number	Process Description	Table Time	Adjustment Factor	Cycle Minutes	Setup Hours
	Package four in box				
24.1-S	Setup	0.15			0.15
24.1-1	Order paperwork; each box gets paper work, divide by 4	2.00	¼	0.50	
24.1-2	Get and position	0.08		0.08	
24.1-10	Paper carton corrugated carton where four in box	1.51	¼	0.38	
24.1-10	Paper separates four parts	0.22	½	0.11	
24.1-16	Miscellaneous adhesive label on box	0.26	¼	0.07	
	Total			1.14	0.15
	Lot estimate	134.58 hr			

TABLE 24.1 PACKAGING BENCH AND MACHINES

Setup

    Bench packaging   .15 hr
    Machine packaging   2.0 hr

Operation elements in estimating minutes

1. Order paperwork. Prorate over quantity per order paperwork   2.00

2. Get and position unit, manual

Maximum Dimension	Weight Range	Simple	Stack	Several	Bulk Remove
Difficult grasp less than 3 in.	Under .2 lb	.03 .02	.04 .04	.005 .01	.001–.0005 .001–.0005
3 to 9 in. 9 to 20 in.	Under 3 lb Under 8 lb	.03 .08	.05 .09	.01 .02	.02 .04
20 to 36 in. 36+ in.	Under 12 lb Under hoist load		.14 .25		

3. Soda straw or pipe cleaner   .28/straw + .02/unit

4. Coin envelope   .14/first unit + .03/add'l unit

5. Metal, wood, plastic tote box
    Dump   .005–.0001
    Stack, place   .01

6. Plastic bag

   A. Non bulk pack

Smallest x, y, or z Dimension	.1	.5	1	1.5	2	Add'l
Min/Unit	.08	.11	.17	.22	.29	.13

B. Bulk pack

1–200 Units	.01/Unit
200 or More Units	.58/bag units

7. Aluminum foil bags

10 or Less Bags	.97/Bag Units
11 or More Bags	.85/bag units

8. Paper sack

Units in Sack	1	3	6	9	Add'l	Bulk
Min/Sack	.78	.85	.98	1.05	.03	1.05/sack

9.A. Jar with lid
    Optional tissue cushion   .83/jar units
    Optional tissue innerlayer   .21/jar units

No. of Layers/Jar	2	3	4	5	Add'l
Min/Jar Units	.18	.36	.53	.71	.18

B. Nonbulk pack

$L + W + H$	1	2	3	4	Add'l
Min/Unit	.01	.02	.03	.04	.01

C. Bulk pack

1–200 Units	.01/Unit
200 or More Units	.58/jar units

10. Corrugated paper carton (ctn)   1.51/ctn units
    Optional paper cushion or layer   .22/layer units
    Coin envelope, or bag, or sack placed within carton   .01/bag
    Nonbulk pack   .06/unit
    Bulk pack

1–200 Units	16/Ctn + .01/Unit
200 or More Units	.58/units

11. Slitted foam   .21/foam units

Slit Length, In.	1	2	3	4	Add'l
Min/Unit	.13	.16	.20	.23	.03

12A. Innerwrap
     Brown paper

Largest Major Dimension	2	7	12	19
Min/Innerwrap Package (pkg)	.30	.33	.37	.41

B. Kimpack   .50/innerwrap pkg
   Tissue   .26/innerwrap pkg

Sum of 2 Smallest Major Unit Dimensions	1	2	3	4	Add'l
Min/Unit	.05	.07	.08	.10	.02

13. Overwrap
     Brown paper   .87/overwrap pkg + .02/(pkg $L + W + H$)

Kimpack   1.06/overwrap pkg + .01/(pkg $L + W + H$)
   Tissue   .45/overwrap pkg + .01/(pkg $L + W + H$)
   No. of units/overwrap pkg   .04/unit

14. Plastic tube: .63/tube units + following value

No. of Units Placed in Plastic Tube	1	20	80	140
Min/Units	.08	1.26	1.33	1.38

   Optional cushion   .16/tube units

15A. Metal-edge paper box: .75/box units + following value
    Nonbulk pack

Smallest Major Dimension of Unit Placed in Box	.1	.3	.8	1.3	Add'l
Min/Unit	.01	.02	.03	.04	.02

   B. Bulk pack

1–200 Units	.01/Unit
200 or More Units	.58/box units

   Optional tissue cushion   .36/box units

16. Miscellaneous bench packaging elements
     Stamp part with rubber stamp   .11
     Apply adhesive label   .26
     Add desiccant   .14

17. Automatic machine packaging, min/part

Units/Pkg	Pkg/Min												
	10	20	30	40	50	60	70	80	90	100	150	200	300
1	.10	.05	.03	.03	.02	.02	.01	.01	.01	.01	.007	.005	.003
2	.05	.03	.02	.01	.01	.01	.007	.006	.006	.005	.003	.003	.002
3	.03	.02	.01	.01	.007	.006	.005	.004	.004	.003	.002	.002	.001
4	.03	.01	.01	.006	.005	.004	.004	.003	.003	.003	.003	.001	.001
5	.02	.01	.007	.005	.004	.003	.003	.003	.002	.002	.001	.001	.001
6	.02	.008	.006	.004	.003	.003	.002	.002	.002	.002	.001	.001	.001
8	.01	.007	.004	.003	.003	.002	.002	.002	.001	.001	.001	.001	.0004
10	.01	.006	.003	.003	.002	.002	.002	.001	.001	.001	.001	.001	.0003

24.2 Corrugated-Cardboard Packaging Conveyor-Assembly

DESCRIPTION

These data are for the final packaging of industrial products in corrugated cardboard cartons. The items have a girth which ranges from 75 to 300 and are to be packaged one item per carton. Materials used in the packaging are styrofoam, foam rubber, and polyurethane strapping in addition to the corrugated paper carton elements. Cushioning for the stability and protection of delicate products during transport is also considered.

The packaging assembly uses a mechanical roller

conveyor, which is belt driven to transport the unit from final assembly and inspection. There is no power drive through the packaging assembly area, however, allowing the worker to move the product through the work stations at his or her speed. The information provided is for low to moderate lot quantities using bench work for the majority of the elements with machine assistance for some elements.

ESTIMATING DATA DISCUSSION

Setup includes time for punch in and out and to queue the product to be processed at the beginning of the roller conveyor. The setup involves stocking of packaging materials in bins or on shelves, where they are accessible to the worker. This can be done manually where the material is removed from its shipping carton and placed on the shelves, or alternatively the carton may be cut and placed directly on the shelving with a forklift.

Element 1 is the paperwork that may be necessary, including inventory slips, routing, shipping labels, etc. If it is used, the element time is divided by the quantity in the lot.

Element 2 is the manual get, fold, and position of the previously cut corrugated cardboard parts. The parts are located on the shelving and the transport of the work down the conveyor to the next part storage is included. Tabulated data include constant time related to the size of the material and a variable folding time related to the number of folds in each part. Element 2 is included in all manual packaging operations.

Element 3 is the manual get, position, and secure of any styrofoam or foam rubber cushioning materials. As in Element 2, the materials are presumed to be prefabricated to the proper dimensions and transport time is included. The padding may be molded to the product and thus requires fitting or it may be placed loosely on or around the product.

Element 4 is the manual placing of the carton over the completed product or work station. It is assumed that the large dimensions of the product make it easier to place the carton over the product and secure it rather than picking up the product and placing it into a carton with a lid. The relative clearance between the product and the inner packaging and the outer carton is the major time driver for Element 4.

Element 5 is the securing of the carton with polyurethane straps. Strapping can be done either manually or with an automatic strapping machine. If the machine is used, it is assumed that it is integrated directly into the conveyor system, and therefore no special handling is required. Element 5 includes a variable strapping time related to the perimeter of the box encircled by the strap (strap length). Provisions for any special positioning of the carton are also considered. Taping of the strap ends may be required in the case of automatic strapping to prevent the loose ends from catching on other cartons.

Element 6 considers plastic and paper sacks that may be a part of the inner packaging.

EXAMPLES

A. A unit is to be packaged that uses three separate cardboard parts requiring 4, 8, and 16 folds respectively. The unit is to be padded with two molded styrofoam inserts. The outside box fits tightly over the unit and is to be strapped four times with 180-in. straps. When strapping, it is necessary to rotate the unit 90 degrees for each strap. Find the unit estimate and lot estimate for 55 units.

Table Number	Process Descriptio	Table Time	Adjustment Factor	Cycle Minutes	Setup Hours
24.2-S	Setup, manual	.80			.80
24.2-2	Fold 1st box	.08		.08	
24.2-2	Fold 2nd box	.65		.65	
24.2-2	Fold 3rd box	1.73		1.73	
24.2-3	Cushioning (2 inserts)	.04	2	.08	
24.2-3	Secure inserts	.12		.12	
24.2-4	Fit outside carton	2.40		2.40	
24.2-5A	Position carton on strapping machine	.03		.03	
24.2-5A	Strap with machine (4 straps)	.50	4	2.00	
24.2-5B	Rotate four times	.04	4	.24	
	Total			7.33	.80
	Lot estimate	7.51 hr			

B. A large standing unit is to be rolled into a box requiring eight folds. Loose padding is placed inside the box. Because of the bulk (250-in. perimeter) and weight, hand strapping is required. Three straps are needed. Find the unit estimate and lot estimate for 15 units.

Table Number	Process Description	Table Time	Adjustment Factor	Cycle Minutes	Setup Hours
24.2-S	Setup, manual	.80			.80
24.2-4	Fit outside, moderate clearance	1.05		1.05	
24.2-2	Fold carton	.65		.65	
24.2-3	Padding	.06		.06	
24.2-5C	Three straps, 250 in.	4.76	3	14.28	—
	Total			16.04	.80
	Lot estimate	4.81 hr			

C. Four small boxes (8 folds) are placed in a large carton (8 folds) and are secured with four pieces of molded cushioning. Two 120-in. straps are manually secured around the carton. A forklift is used for setup.

Table Number	Process Description	Table Time	Adjustment Factor	Cycle Minutes	Setup Hours
24.2-S	Setup	.47			.47
24.2-2	Fold four inner boxes	.65	4	2.60	
24.2-3	Tape four boxes	.06	4	.24	
24.2-3	Position padding	.16		.16	
24.2-2	Fold outer carton	.65		.65	
24.2-4	Place outside carton over inner pack	.25		.25	
24.2-5C	Strap manually	2.56	2	5.12	—
	Total			9.02	.47

D. Repeat Example A, using automated setup and consider order paperwork.

Table Number	Process Description	Table Time	Adjustment Factor	Cycle Minutes	Setup Hours
24.2-S	Setup, automated	.47			.47
24.2-1	Order paperwork	2.00	1/55	.04	
	Unit estimate from Example A	7.33		7.33	—
	Total			7.37	.47
	Lot estimate	7.23 hr			

E. Fold outer carton (6 folds) and place product inside with minimum clearance. Both ends of the carton are taped shut. Find unit estimate.

Table Number	Process Description	Table Time	Adjustment Factor	Cycle Minutes	Setup Hours
24.2-2	Fold outer carton	.37		.37	
24.2-4	Place product in carton	2.40		2.40	
24.2-3	Secure ends with tape	.12		.12	
	Total			2.89	

TABLE 24.2 CORRUGATED-CARDBOARD PACKAGING CONVEYOR-ASSEMBLY

Setup

Manual stocking	.80 hr
Stocking with forklift	.47 hr

Operation elements in estimating minutes

1. Order paperwork. Prorate over quantity per order paperwork   2.00

2. Get, fold and position cardboard piece, manual

No. of Folds	Min	No. of Folds	Min
4	.08	10	.92
5	.24	12	1.20
6	.37	14	1.46
7	.49	16	1.73
8	.65	18	2.01
9	.78	20	2.28

3. Get and secure cushioning materials, manual
   - Molded fit — .04
   - Loosely placed — .015
   - Secure with tape — .06

4. Place carton over unit with inner packaging, manual
   - Minimum clearance — 2.40
   - Moderate clearance — 1.05
   - Clearance not a factor — .25

5. Secure carton with straps
   A. Automatic strapping machine

Perimeter (In.)	Min.
80	.25
100	.30
115	.34
130	.37
145	.41
160	.45
180	.50
200	.56

   Position carton on machine — .03

   B. Rotate carton 90–180 degrees

Large	.06
Small	.08

   C. Manual strapping, thread strap and crimp metal fastener

Perimeter (In.)	Min
130	2.56
150	2.92
170	3.29
190	3.66
210	4.03
230	4.40
250	4.76

6. Inner package
   - Plastic bag — .11
   - Paper sack — .09

24.3 Case Packing Conveyor

DESCRIPTION

The side loading case packer often constitutes the final stage of packaging before shipment. The type of work is continuous, fully automatic packing of boxes in a case. The machine is directly related to the previous operation within a conveyor system. Due to the varied output of the boxing operations, delays in the casing operation take place. The usual operation on this machine is 14 cases per minute, maximum output, and 12 boxes per case. The data, however, allow for other conditions, should they vary from these experiences. Operations peculiar to this machine are the reloading of empty cases, the monitoring of both the sealer glue, and the ink roller that labels each case. This machine may not require an operator. Now consider the general industrial operation of loading boxes into a case on a conveyor.

ESTIMATING DATA DISCUSSION

The setup time for this machine is considered an overhead cost.

Element 1 of the packing process is the queuing of boxes on the conveyor, or just enough to set off the pressure needle that pushes a load of boxes to Element 2.

Element 2 consists of stacking rows of boxes vertically onto a horizontal plate and inserting them into an opened case. For example, there may be 3 rows of 4 boxes loading into a case.

Element 3 is merely the closing and sealing of the cases. Elements 1 and 2 are internal to the workings of the machine so their times will be 0. Element 3 alone determines the rate at which the machine packs cases. The time provided for Element 3 is the unit time for one box which depends upon the number loaded and the conveyor rate of cases per min.

EXAMPLES

A. A lot of 6000 boxes is to be prepared for shipment. Determine the unit and lot estimates for 12 boxes per case and a rate of 5.7 cases per minute.

Table Number	Process Description	Table Time	Adjustment Factor	Cycle Minutes	Setup Hours
24.3-1	Line up boxes on conveyor	0.00		0	
24.3-2	Stack boxes and load into case	0.00			
24.3-3	Close and seal case	0.015		.015	
	Total			.015	
	Lot estimate		1.5		

B. Using a rate of 8 cases per minute and 10 boxes per case, determine the unit estimate, lot estimate, and hr/100 for the casing of 4000 boxes.

Table Number	Process Description	Table Time	Adjustment Factor	Cycle Minutes	Setup Hours
24.3-1	Line up boxes on conveyor	0.00		0	
24.3-2	Stack boxes and load into case	0.00		0	
24.3-3	Close and seal case	0.013		.013	
	Total			.013	
	Lot estimate	0.87 hr			
	Hr/100	.022			

TABLE 24.3 CASE PACKING CONVEYOR

Setup   0.00

Operation elements in estimating minutes

1. Line up boxes on conveyor   0

2. Stack boxes and load into case   0

3. Close and seal cases

No. of Boxes	Cases/Min						
	2	4	5	5.7	6	8	10
9	.058	.029	.023	.020	.019	.015	.012
10	.053	.026	.021	.018	.018	.013	.011
12	.044	.022	.018	.015	.015	.011	.009
16	.033	.016	.013	.012	.011	.008	.007
20	.026	.013	.011	.009	.009	.007	.005

24.4 Foam-Injected Box Packaging Bench

DESCRIPTION

This packaging bench data involve the injection of a chemically controlled mixed liquid that expands to become a soft foam when exposed to air. This process provides an ideal medium for packing and shipping fragile materials. Drums containing the two chemicals have pumps and hoses which combine and mix the chemicals at a trigger gun. A thermal heater is used to heat the chemicals in the hoses while they are pumped from the drums to the gun. The sequence in using the foam is as follows: inject a layer on the bottom of the box; cover the rising foam with a thick polyethelene wrap; place the object to be shipped on top of the poly; cover the object with another sheet of poly; and inject more foam around and on top of the object. Because the foam expands, a general rule of thumb is to inject two inches in the bottom of the box. This will increase in volume between three and five times creating the mechanically insulated effect desired to protect the object being shipped.

ESTIMATING DATA DISCUSSION

The thermal heater is the first element in setup because it takes the longest time (approximately seven minutes) to preheat. Setup time is constant because parts do not affect the setup.

Although the packaging process is relatively the same procedure each time, some of the cycle elements are variable because of the different sized boxes used. Box and part handling involve obtaining the correct size box and folding to form. Preparing the box includes taping the bottom. The operator must select the appropriate length of tape from the tape machine.

Foam injection begins with the element of actual injection. Putting the object in the box on top of the poly is a step followed by a second sheet of poly to cover the object. The object is foamed with a second layer. The box is closed to give the rising foam a top form. A shipping order may be placed inside the box, if necessary. By entering the first three numbers of the zip code a scale electronically determines the amount of postage required.

EXAMPLES

A. Estimate the setup hours and cycle minutes to foam package on electronic meter of box dimensions of 12 × 6 × 4 in. An approximate corrugated carton of size 14 × 8 × 8 in. is used. Lot quantity is 1.

Table Number	Process Description	Table Time	Adjustment Factor	Cycle Minutes	Setup Hours
24.4-S	Basic setup	.30			.30
24.4-S	Postage requirement	.20			.20
24.4-1	Get box and fold	.13		.13	
24.4-1	Tape bottom of box	.083		.083	
24.4-1	Polyethelene in box bottom, volume of box = 896 in.	.20		.20	
24.4-2	Fill foam on box bottom	.11		.11	
24.4-2	Place polyethelene on top	.23		.23	
24.4-2	Load part	.25		.24	

Table Number	Process Description	Table Time	Adjustment Factor	Cycle Minutes	Setup Hours
24.4-2	Place polyethelene on top	.14		.14	
24.4-2	Inject foam	.33		.33	
24.4-2	Close box	.28		.28	
24.4-3	Weigh box and stamp	.29		.29	
	Total			2.04	.40
	Lot estimate	.53 hr			

B. An electronic assembly of box size 3 × 5 × 8 in. is to be packaged one per box in a carton of 9½ × 8½ × 7 in. There are 10 units. The box is to be foam filled. Find the lot hours.

Table Number	Process Description	Table Time	Adjustment Factor	Cycle Minutes	Setup Hours
24.4-S	Basic setup	.30			.30
24.4-1	Get box and unfold	.13		.13	
24.4-1	Tape bottom of box	.083		.083	
24.4-1	Poly on box bottom, volume = 565 in.^{3}	.20		.20	
24.4-2	Inject foam on bottom	.11		.11	
24.4-2	Place poly on top of foam	.23		.23	
24.4-2	Load part	.25		.25	
24.4-2	Place poly on part	.14		.14	
24.4-2	More foam	.33		.33	
24.4-2	Close box	.28		.28	
	Total			1.753	.30
	Hr/100 units	2.922			
	Lot estimate	.59 hr			

C. Estimate the time to package 300 electronic meters. The size of the meter is 12 × 8 × 6 in. and the box is 20 × 14 × 14 in. Two meters are placed in each box. Individual box weighing is necessary.

Table Number	Process Description	Table Time	Adjustment Factor	Cycle Minutes	Setup Hours
24.4-S	Basic setup	.3			.3
24.4-S	Postage equipment	.2			.2
24.4-1	Get box and unfold	.13	½	.065	
24.4-1	Tape bottom	.083	½	.04	
24.4-1	Box bottom poly	.34	½	.17	
24.4-2	Inject foam on floor 3920 in.^{3}	.17	½	.09	
24.4-2	Place poly on top for one unit	.23		.23	
24.4-2	Load part	.25	2	.50	
24.4-2	Place poly on part	.23	2	.46	
24.4-2	Inject foam on part	.17		.17	
24.4-2	Inject foam in top area	.47	½	.24	
24.4-2	Insert ship order	.17	½	.09	
24.4-2	Close box	.28	½	.14	
24.4-3	Weigh	.12	½	.06	
	Total			2.26	.5
	Hr/100 units	3.758			
	Lot estimate	11.78 hr			

TABLE 24.4 FOAM-INJECTED BOX PACKAGING BENCH

Setup
   Basic   .3 hr
   Foam supply, large usage   1.5 hr

Stocking of boxes .6 hr
Postage equipment .2 hr
Heater wait .15 hr

Operation elements in estimating minutes

1. Box handling
   Get from bin and unfold .13
   Tape bottom of box .083
   Polyethelene in box bottom

Vol	560	1400	3000	7330
Min	.18	.20	.24	.34

2. A. Filling and part handling in box
      Filler gun and inject foam on floor of box

Vol	560	1400	3000	7330
Min	.07	.11	.13	.17

   B. Place poly on top of foam .23
      Load part on soft foam and poly and wait for foam to rise .25
      Place poly on top of part and foam

Vol	1400	3000	7330
Min	.14	.19	.23

   Filler gun and inject top area foam on poly-protected part

Vol	560	1400	3000	7330
Min	.25	.33	.43	.47

   Insert shipping order .17
   Close box, hand distribute foam .28

3. Weighing and postage and stamping
   Weigh box .12
   Invoice control .15
   Stamp box .17

TOOLING

25.1 Dies, Jigs and Fixtures

DESCRIPTION

The following estimating data[1] are for the construction of dies, jigs, and fixtures. These tool building data are for tool makers, model builders, or journeyman tool craftsmen. The work is prototype, or single quantity, and the estimating is done prior to a tool design or tool construction. Estimating may be done with either a part print or verbal instructions. A tool design is not mandatory for these data. But in general the estimating of dies, jigs, and fixtures requires a part print and the elemental setup tables given in the *AM Cost Estimator*. The data can be used to determine the time for making tooling using several methods. Essentially this method estimates tool cost in terms of hours, but once the shop rate multiplies the time, a tool cost for the job results. Using other equipment and data throughout the *AM Cost Estimator*, it is possible to find piece part costs. There are no cycle minutes given in Table 25.1 as all the time values are given in hours for the construction of one final jig, die, or fixture. This method is a shorthand guide for people who may be required to give on-the-spot estimates to part designers.

When first estimating, the tool building times may be on the high or low side when compared to the estimates, but the deviations will be consistent. All that's necessary is to determine a suitable productivity factor based on your conditions. Note the Introduction for discussion about the productivity factor. The Introduction will state the practices for adjusting your estimate to actual time and cost.

ESTIMATING DATA DISCUSSION

The time data are in units of hour. No units of minute are given. Time values given in the tables are based on the construction of a die to these specifications.

1. The punch and die will be made of oil-hardening tool steel.
2. Punch and die blocks will be machined squared, then ground before and after heat treatment.
3. These punch and die blocks will be mounted with screws and positioned with dowels.
4. The punch and die blocks will be assembled on the surface of the die shoe and punch holder of the die set.
5. Time values are included in the estimate for die accessories such as gages, nests, strippers, punch pads, locating pins, spring pins, stops, knockouts, and pressure pads.
6. The minimum tolerance allowed for all dimensions will be ±0.002 in.

[1]*American Machinist*, September 4, 1961, "How to Estimate Dies, Jigs, and Fixtures from a Part Print," L. Nelson.

7. Material thickness will range from 0.010 to 0.250 in.
8. Physical limits of the die set will not exceed 1½ in. in thickness for die shoe and punch holder, and the assembling space will not exceed 6 in. in width by 8 in. in length.

Additional time is necessary under these circumstances:

1. Add 20 percent when high-carbon, high-chrome steel is used instead of oil-hardening tool steel. The extra time is required because of the difficulty encountered in the drilling, machining, and grinding.
2. Add 25 percent when the blank thickness is less than 0.010 in. More time is needed because the clearance between punch and die is extremely close, and extra care is required in the fitting, filing, and stoning of the punch and die.
3. Add 10 percent when the die set is milled to fit the die blocks. Usually, milling is employed when material thickness is greater than ¼ in., to assure stability of the die blocks.
4. Add 10 percent when the physical limits of the die set, as stated above, are exceeded. More time must be allowed to compensate for the extra handling and drilling of the die set.

Hole-piercing dies can be classified into three types, according to hole diameter.

Type	Size Range
1. Small	0.0156–0.1245 in.
2. Medium	⅛–½ in.
3. Bored	Over ½ in.

The time values in Element S1 are based upon the diameter of the hole. Holes from a ¹⁄₆₄ to ⅛ in. are designated as "small pierce holes." Time values are greater than for medium pierce holes because of the extra care in drilling the die block, to avoid drill breakage, and the extra time for grinding and making the pierce punch. Holes with diameters of ¹⁄₆₄–³⁄₆₄ in. require a special quill-type punch and holder.

Holes from ⅛–¼ in. diameter will be designated as "bored holes." Time values are based on the need to bore the hole in the die block, coupled with the extra time for grinding and machining the larger pierce punch.

Basic blanking dies are estimated according to the periphery of the blank. The contour of the blank will be made up of one or a combination of the four different types of contours:

1. Straight lines. A rectangle would have four straight lines.

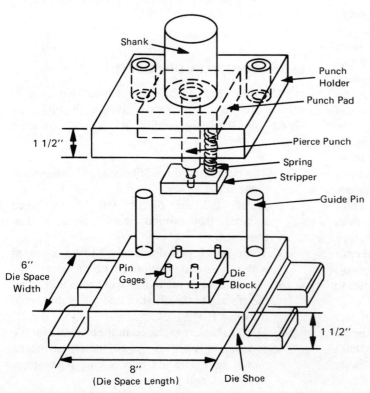

FIGURE 25.1-1 Time data for a basic pierce hole die are given for a tool that does not exceed these physical limits. If the required die is larger, add 10 percent to compensate for extra handling.

2. Angular lines. A triangle is an example.
3. Curved lines that cannot be bored. An ellipse illustrates.
4. Irregular lines. Compound curves are typical. Time values for basic blanking dies are given in Element S2. Note that times are based on use of oil-hardening tool steel. Add 20 percent if high-carbon, high chromium steel is selected.

Basic extruding dies can be estimated according to the hole diameter of the extrusion and the type of the extrusion. Consider three types of extrusions given by Figure 25.1-2.

1. An extrusion without a pierced hole.
2. An extrusion with a pierced hole requiring two stages for completion. First-stage piercing is followed by an extruding stage.
3. An extrusion with a pierced hole done in one stage.

Time values for basic extrusion dies are given in Element S3.

A basic bending or forming die can be estimated according to the width, length, and type of bend. Types of bends are classified in Figure 25.1-3. Most forms are 90-degree bends. The L, U, and V forms are

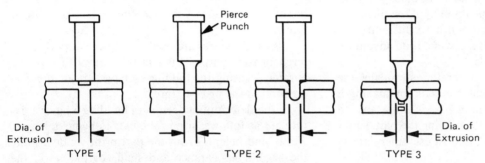

FIGURE 25.1-2 Basic extrusion dies may push the metal part way out of the sheet, Type 1, or pierce a hole and extrude a collar or eyelet, Types 2 and 3.

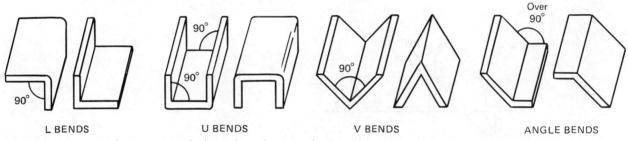

FIGURE 25.1-3 Time values are given for basic form dies to make these four types of bends.

variations of the 90-degree bends. An angle bend is a bend other than 90 degrees. Time values are based on a standard form die for a part that contains a bend 1-in. wide by 1-in. long. Notice Figure 25.1-4. All 90-degree bends that are less than 1-in. wide and 1-in. long will use the same value as the standard form die. For bends wider than 1 in. or longer than 1 in., an additional 2.5 hr per in. of width or length is added for additional machining grinding, drilling, and assembly of larger form blocks. For angle bends, an additional 3.5 hr per in. of width or length will be added. Element 25.1-S4 applies to one bend only of each type. If there are more than one, use Element 25.1-S8.

Combination dies perform more than one function. For dies that perform more than one function, it is necessary to estimate the time for each function and

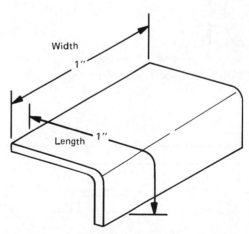

FIGURE 25.1-4 Tabulated data are given in terms of variation from this standard size of 90 degrees or "el" bend.

Dies, Jigs and Fixtures 533

then add the figures obtained for the various functions. See the Index of Elements.

Using this approach it is possible to compare different tool costs. Notice Figure 25.1-5 and Examples E and F. Example E employs a progressive die. Tool construction time is 132.75 hr from Example E.

Example F involves use of three dies: pierce and blank, form two ears and the end, and final form. Here the tool-building time is 48.5 + 45 + 30 = 123.5 hr.

Thus, the progressive die requires only 9.5 hr additional to build and may well provide the lowest over-all piece-part cost. Piece part estimating would use the data from other parts of the *AM Cost Estimator,* such as Presswork, Chapter 3.

Example G discusses a problem of piecing, blanking, and forming that can be done by three methods. Figure 25.1-6 is the part that will have different tooling. The second method has a distinct time advantage over the other two proposals.

Availability of time data therefore stimulates the estimator to consider several methods for making a part. And within a few minutes he is often able to show the cheapest method for making the part, so far as tooling cost is concerned, and to use other data to find piece-part cost.

Rules for estimating jigs and fixtures follow. This section is a short hand guide for determining the number of hours necessary for a toolmaker to build a jig or fixture.

Jigs and fixtures are used to produce duplicate, interchangeable parts. This aim is accomplished by accurately locating and holding each piece in an identical position in the jig or fixture.

Time-determining factors in building a jig or a fixture are as follows: size of the part to be held, length, width, and height. From the part print we determine the general appearance and shape of the work, the number of operations to be done, and the approximate dimensions of the jig or fixture.

Index of Elements

Type of Die	Use Elements			
	S5	S6	S7	S8
Pierce and blank	x	x		
Pierce, extrude and blank	x	x	x	
Pierce, extrude, blank and form	x	x	x	x
Pierce, cutoff and form	x	x		x
Pierce and cutoff	x	x		
Form (more than one bend)				x
Pierce, extrude & cutoff	x	x	x	

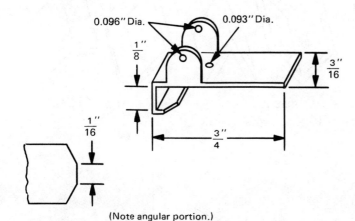

FIGURE 25.1-5 Small sheet metal part which is estimated several ways.

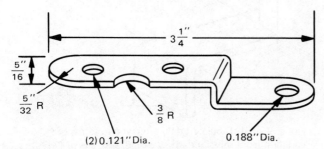

FIGURE 25.1-6 Part estimated by several dies; material is .031-in. brass.

Five basic components of a jig or a fixture are as follows: (See Figure 25.1-7.)

1. A baseplate or mounting plate. The baseplate is comparable to the die set in the die making. All fixture blocks are assembled on the baseplate by means of screws and dowels.
2. A locating means. The part is located accurately by pins, edge-locating blocks or nests, or V-blocks.
3. A clamping means. Several kinds of clamps are used: thumbscrews, knobs, quick-acting hand-operated clamps.
4. A positioning means. Desired relationship between the cutting tool and a jig or fixture is established by a drill plate; in a fixture a set block would be the positioning means.
5. Tool guides or bushings. These are used in jigs for drilling and reaming holes accurately.

Information for classification of jigs and fixtures is as follows:

As many as 40 or more types of fixtures will be used in a large plant. They are named according to the operation performed, for instance:

Assembly, staking, testing, locating, holding, positioning, alignment, drill, lathe, milling, reaming, tapping, grinding, turning, etc. For our purposes, let's put fixtures into four groups:

1. Drilling fixtures   Drill press
2. Milling fixtures   Milling machines
3. Bench fixtures   Bending and burring
4. Turning fixtures   Lathes, turret lathes

Time allowance to machine a purchased faceplate, to make a turning fixture, is 6 hr.

Physical limits of jigs and fixtures are as follows:

All tabulated time values are based upon construction of a jig or a fixture to the following specifications:

1. The steel employed for the baseplate is either hot or cold-rolled steel (HRS or CRS).
2. Locators, clamps, positioner, and tool guides (with the exception of the drill plate and supporting blocks) will be made of oil-hardening tool steel (OHTS).
3. All blocks will be machined squared and ground.
4. All blocks will be positioned by dowels and fastened with screws.
5. Part-size limits will be 6-in. wide by 6-in. long, or 6-in. diameter by 3-in. high, and these limits apply to parts that are flat, round or a combination of both and tubing.
6. Physical limits of the mounting plate will not exceed 1-in. thick by 12-in. wide by 12-in. long
7. The minimum tolerance for jig boring will be $\pm$ 0.0002 in.

Additional time allowances are provided when the jig or fixture exceeds the above physical limits. For instance:

1. An allowance of 10 percent is made when "standard" size limits of the baseplate are exceeded. The additional time compensates for extra handling and drilling of the plate.
2. An allowance of 25 percent is added when the standard physical limits of the part are exceeded.
3. An allowance of 10 percent is necessary when a trunnion jig or fixture is needed. More time must be allotted for extra handling and assembly of the intricate fixture.

Baseplates or mounting plates fall in three categories. Notice Figure 25.1-8.

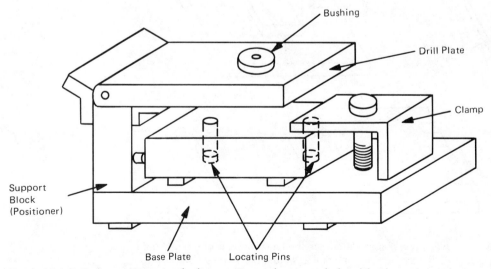

FIGURE 25.1-7 Basic components of a fixture. Time values are tabulated for the various parts.

Dies, Jigs and Fixtures   535

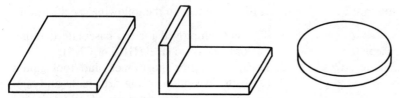

FIGURE 25.1-8 Baseplates for drilling and milling fixtures are of various shapes.

The selection of the proper mounting plate depends upon the "base" dimensions of the part, the general shape (rectangular, round, a combination of both, or tubing) and the type of machine to be used. The base dimensions of a part are the two linear bounds of the base of the part which locates itself on the mounting plate. Baseplates are customarily specified to be made of hot-rolled or cold-rolled steel plates of various shapes to suit the job.

Each baseplate of the four general types is different in construction. The size of the baseplates will be classified into small, medium and large.

A small baseplate will house a part with dimensions up to and including 1×1 in.

A medium baseplate will hold a part with dimensions greater than 1×1 in. and up to and including 3×3 in.

A large baseplate will encompass a part with dimensions greater than 3×3 in. and up to and including 6×6 in.

The baseplate will be larger than the part in base dimension; usually from 1 to 3 in. is allowed on each side. This space is needed to assemble the next or the fixture blocks.

Time values included in Element S9:

1. For drilling, milling, and bench fixtures. Cutoff, squaring, grinding, layout, plus drilling, reaming and tapping for locators, clamps form blocks or vise grip as required, and assembly.
2. For drilling-fixture baseplates. Making four feet for the baseplate and four rest buttons for the work. Notice Figure 25.1-9.
3. For milling fixture baseplates. Milling four bolt slots and two keyways, making and fitting two keys, and making four rest buttons.
4. For bench-fixture baseplates. Making the vise grip.

Turning fixture-time values include (a) drilling, reaming and tapping the purchased faceplate for installation of locators, clamps, set blocks, (b) providing bolt holes to fasten faceplate to chuck, (c) milling slots for guiding clamps, and (d) assembly of faceplate to lathe chuck.

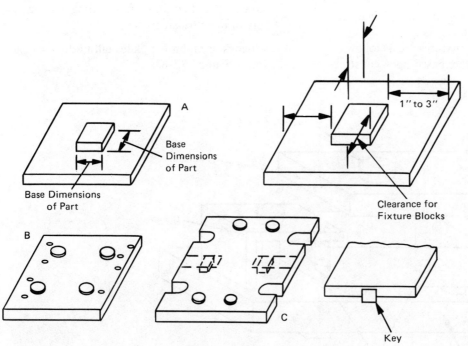

FIGURE 25.1-9 (A): In selecting a baseplate, two considerations are the linear bounds of the part (left) and clearance for fixture blocks (right). (B): A baseplate for a drilling fixture has four feet and rest buttons. (C): Milling-fixture baseplates require four rest buttons, two keyways, and two keys.

Baseplates for bench fixture (tube bending) are shown by Figure 25.1-10. Many bench fixtures are used for bending tubing. In order to determine the baseplate size of a bending fixture, the outline of the completed part is the key. The width and length of the bend are comparable to the base dimensions, as shown in Figure 25.1-10.

A small baseplate will house a tubing part up to and including 1-in. wide by 4-in. long. A medium baseplate will house a tubing up to and including 3-in. wide by 6-in. long. A large baseplate will house a tubing part up to and including 6-in. wide by 10-in. long.

Angular baseplates for circumferential operations are used in conjunction with a regular baseplate when the part is round and requires work done on its circumference. The shape of the part determines use of an angular baseplate in drilling, milling and turning fixtures. Notice Figure 25.1-11.

Time values are based on the use of the angle plate as a base or in combination with another baseplate. The C-shaped angle plate is also used in the making of a trunnion fixture. Time values include the complete machining of the angle plate along with the drilling, reaming and tapping for locators, clamps and set blocks. "In combination" refers to use of an angle plate with a flat baseplate.

Special baseplates for two or more operations are discussed for Element S11. Notice Figure 25.1-12. Baseplates discussed so far are used for one operation—in one plane. When two or more operations are required on a part, they are ordinarily done by two separate fixtures. With a special fixture, however, the two operations can be performed in the same fixture with minimum of movement on the part of the operator. Examples of such fixtures are as follows:

1. A box or tumble jig is used for drilling opposed holes. The support legs are of such length that all holes of the same diameter can be drilled with the same spindle stop setting. Both holes can be drilled with only one positioning for depth, which is made during setup. In a box jig, two baseplates are necessary. See Table 11 for time values.

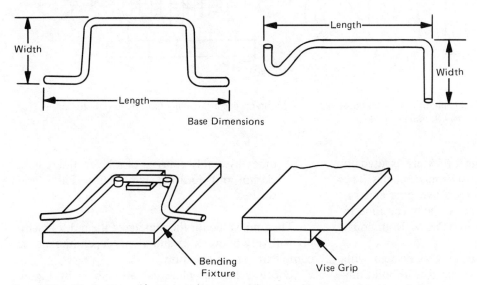

FIGURE 25.1-10 Outline of bent part determines the size of baseplate for a bench figure.

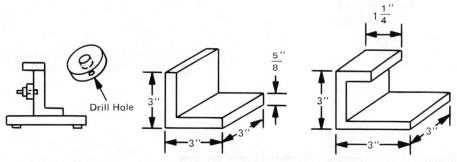

FIGURE 25.1-11 (A): Angular baseplates are mounted on flat baseplates to hold round parts requiring work on their circumference. (B): Plain and C-shaped types are given time values for three sizes, Element S 10.

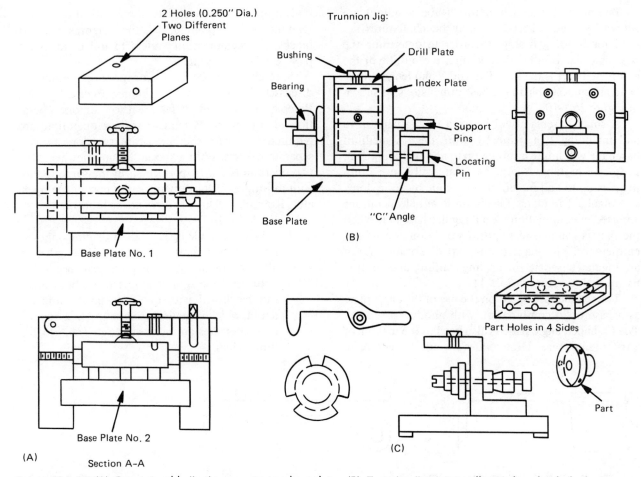

FIGURE 25.1-12 (A): Box or tumble jigs incorporate two baseplates. (B): Trunnion jigs are usually employed only for large parts. (C): Index drill jigs require 25 to 40 hours to make.

2. A trunnion-mounted jig or fixture is used to drill opposed holes in two or more planes. The device is mounted between two pins and rotated by means of an index plate. The time to make the baseplate (12-in. wide × 18-in. long) is approximately 14 hr.

   Trunnion jigs are usually associated with large parts. It is not cost effective to build this type of jig for small parts.

3. An index drill jig. When the part to be drilled is round and more than one hole is to be drilled on the circumference, an index drill jig is used. The device is mounted on an angle plate and uses an index assembly to rotate the part. The index assembly consists of a lever, index hub, clamp and locating bar, clamp knob and washer.

   The time value for the making of the index assembly will vary between 25 and 40 hr, depending on the complexity of the assembly.

   Locators for baseplate use are shown by Figure 25.1-13. Every part has to be located on the baseplate to assure identical parts. A minimum of two-point location is necessary. There are many types of locators:

1. Pin
2. Block

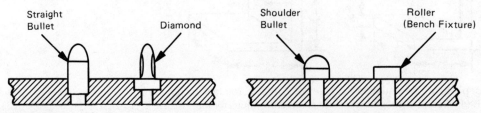

FIGURE 25.1-13 Pin locators are used to position parts having machined holes.

3. Edge block
4. Nest
5. V-block

Parts with machined holes can use pin locators effectively—either straight pin or diamond pin. Small pin locators are up to ½-in. diameter by 2-in. long; large pin locators are 1½-in. diameter up to 2-in. long. Time values are given in Element S12.

Block locators for tubing are used in bench fixtures, especially for bending. Time values include the cutoff, squaring, grinding, drilling and assembly of these blocks on the fixture.

Size	Est. Hr
Small	2.5
Medium	3.5
Large	4.5

Edge blocks and nests are an effective means for locating a flat sheet metal part. Time values include the cutoff, squaring, grinding, and assembling the blocks. The estimated time for one nest or edge is 2.5 hr. Notice Figure 25.1-14.

V-block locators are an effective means of locating round parts. Time values include the cut-off, squaring, grinding and assembling the blocks.

Size	Est. Hr
Small	3.5
Medium	4.5
Large	5.5

Drill plates are evaluated by Element S13.

The size of the drill plate is roughly the size of the baseplate. The support block will be the width of the baseplate and the height will be approximately ½ in. more than the height of the part. The ½-in. space is necessary for chip clearance and room for the work to be supported on rest buttons. Therefore, the support-block dimension will depend upon the height of the part. Notice Figure 25.1-15.

Index plates are similar to drill plates. They are used on trunnion jigs for large parts only.

In the making of a milling or turning fixture, a set block is employed for positioning the milling cutter or the toolbit. The time value for a set block is 2.5 hr. Notice Figure 25.1-16.

Bending levers are shown by Figure 25.1-17. In the making of a bending fixture, the most common type of bending lever requires 3 hr for cutoff, squaring, grinding, drilling, locating, bending, and assembly. This device is shown as being 7-in. long.

Figure 25.1-18 shows knobs and thumbscrews.

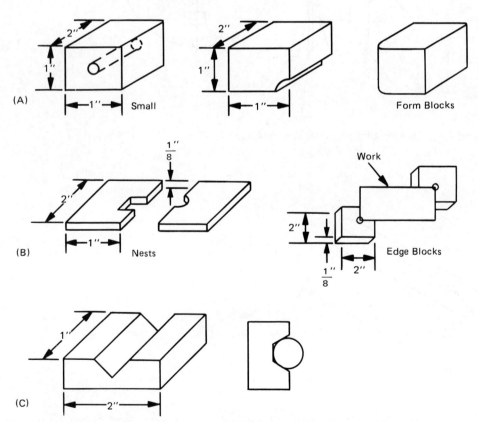

FIGURE 25.1-14 (A): Edge blocks and form blocks for bench figures require from 2.5 to 4.5 hours to make depending on size. (B): Nests and edge block locate flat parts. (C): Round parts are customarily located in V blocks.

Dies, Jigs and Fixtures   539

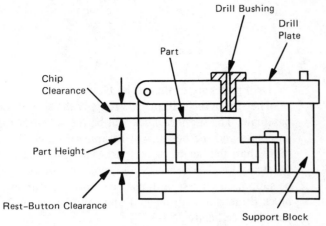

FIGURE 25.1-15 Support blocks position workpieces in fixtures, and drill plates position and guide the tools.

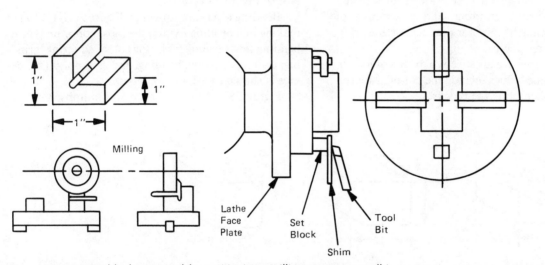

FIGURE 25.1-16 Set blocks are used for positioning a milling cutter or a toolbit.

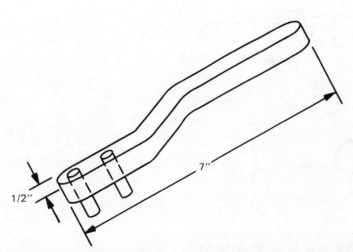

FIGURE 25.1-17 Bending levers require 3 hours to make and assemble to the fixture.

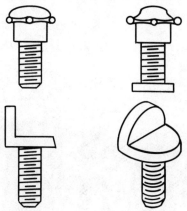

FIGURE 25.1-18 Time for drilling and assembly of fixture parts must be considered when knobs and thumbscrews are used.

There are many types of clamps used in the making of jigs and fixtures. Time data are given for the four most common clamps.

1. Hand knob clamps
2. Strap clamps with heel
3. Strap clamps with special heel and pin for guiding on turning fixtures
4. Quick-acting milling clamps. Refer to Figure 25.1-19

Knobs and thumbscrews are stock items. See Element S14 for time for drilling and assembly of fixture part.

For strap clamps with heel, time values include cutoff, squaring, slotting, grinding and assembly of the strap clamp and heel block:

Size	Est. Hr
Small	3 hr
Medium	7 hr
Large	5 hr

Guided strap clamps have a heel and pin for easy movement in the milled slots of milling or turning fixtures. Time values are given as:

Size	Est. Hr
Small	6 hr
Medium	7 hr
Large	8 hr

Quick-acting milling clamps are purchased. Allow 1.0 hr for assembly.

Guide bushings are estimated by Element S15.

Time values are assigned for jig boring the hole for each drill bushing. Separate time values are necessary for setup and layout on the jig boring machine. This value is influenced by the size of the part and the type of the jig.

A drill plate is jig bored for press-fit shoulder bushings; an index plate is jig bored for headless bushings.

Jig boring-setup and inspection are estimated by Element S16. A typical jig machine tool is shown by Figure 25.1-20.

Additional time must be allowed for the setting up

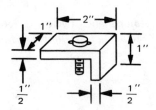

Strap Clamp with Heel
for Drilling and Milling Fixtures

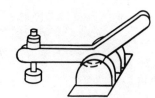

Quick-Acting Clamp

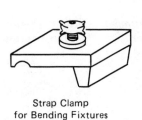

Strap Clamp
for Bending Fixtures

Guided Strap Clamp

FIGURE 25.1-19 These four types of clamps are commonly used in making jigs and fixtures.

Dies, Jigs and Fixtures

the jig or fixture on the jig-boring machine for boring the drill plate, index plate and the baseplate (for the accurate position of the locators). At the same time the jig borer inspects the relative parts of the fixture. The time allowance is greater for special jigs, box and trunnion, because of the complexity.

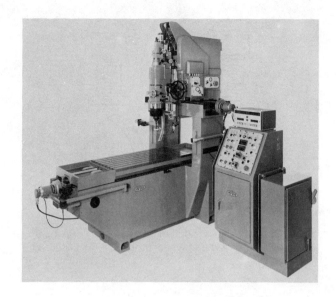

FIGURE 25.1-20 A jig grinder. *(Moore)*

EXAMPLES

A. Estimate the hours to pierce five 0.031-in. diameter holes in .009-in. stock. Oil hardening tool steel is used. A basic piercing die is planned.

Table Number	Process Description	Table Time	Adjustment Factor	Cycle Minutes	Setup Hours
25.1-S1A	Pierce hole die,	21			21
25.1-S1A	Add'l holes	3.5	4		14
25.1-S1D	Material thickness, 21 + 14 = 35 hr	35	.25		9
	Total				44

B. Estimate the hours to pierce five .250-in. holes in .031-in. thickness stock. Die material is high carbon high chrome steel. A basic piercing die is planned.

Table Number	Process Description	Table Time	Adjustment Factor	Cycle Minutes	Setup Hours
25.1-S1B	Pierce hole die	16			16
25.1-S1B	Add'l holes	2.5	4		10
25.1-S1D	Add'l for chrome steel, 16 + 10 = 26 hr	26	.20		5
	Total				31

C. Estimate the hours to construct a basic piercing die for a part having five .745-in. holes in .188-in. stock. Material is high-carbon high-chrome steel. The die set is extra large as each side is 10 in.

Table Number	Process Description	Table Time	Adjustment Factor	Cycle Minutes	Setup Hours
25.1-S1A	Bored hole	21			21
25.1-S1A	Add'l holes	9	4		36
	Subtotal				57
25.1-S1D	Add'l for HCHC	57	.20		11.4
25.1-S1D	Add'l for die set	57	.10		5.7
25.1-S1D	Add'l for extra handling	57	.10		5.7
	Total				79.8

D. Find the hours to construct a basic piercing die which will pierce one .031-in. hole, one .250-in. hole, and one 1.250-in. hole in .031-in. material. The die is to be constructed from oil hardening tool steel.

Table Number	Process Description	Table Time	Adjustment Factor	Cycle Minutes	Setup Hours
25.1-S1A	Small pierce hole	21			21
25.1-S1A	One add'l pierce hole	2.5			2.5
25.1-S1A	One add'l bored hole	9			9
	Total				32.5

E. Estimate the hours for a progressive die to make the part given by Figure 25.1-5. For dies that perform more than one function, it is necessary to find the time for each function and then add the figures obtained.

Table Number	Process Description	Table Time	Adjustment Factor	Cycle Minutes	Setup Hours
25.1-S5	Pierce hole, 3 small	7	3		21
25.1-S5	8 small pilot holes	3.5	8		28
25.1-S6	Straight outline 1½ in.	5	2		10
25.1-S5	Pierce 2 tab holes	5	2		10
25.1-S6	Angle, ½ in.	7.5	.5		3.75
25.1-S8	90-degree bends, 4 times	15	4		60
	Total				132.75

F. Estimate Figure 25.1-5 using three dies, a pierce and blank die, form two ears and end die, and final form die. This can be compared to Example E.

Table Number	Process Description	Table Time	Adjustment Factor	Cycle Minutes	Setup Hours
	Pierce and blank die,				
25.1-S5	3 small pierce holes	7	3		21
25.1-S5	1 small pilot	3.5	1		3.5
25.1-S6	Straight, 1½ in.	5	1.5		7.5
25.1-S5	Pierce tab holes, 2	5	2		10
25.1-S6	Angle, ½	7.5	.5		3.75
	Subtotal				45.75
	Form 2 ears and end die				
25.1-S8	Form 90-degree bends, three times	15	3		45
	Subtotal				45
	Final form die				
25.1-S4	Basic form die	30	1		30
	Subtotal				30
	Total for 3 dies				120.75

G. Estimate the construction for Figure 25.1-6. Part material is 0.031-in. brass. This part will be analyzed for three different tooling systems. In Example G we estimate construction setup hours for a blank, pierce, and form die.

Table Number	Process Description	Table Time	Adjustment Factor	Cycle Minutes	Setup Hours
25.1-S5	2 small holes	7	2		14
25.1-S5	4 small pilots	3.5	4		14
25.1-S5	3 pierce holes	5	3		15
25.1-S6	½ in. of curve	10	.5		5
25.1-S6	6 in. of straight	5	6		30

Dies, Jigs and Fixtures 543

Table Number	Process Description	Table Time	Adjustment Factor	Cycle Minutes	Setup Hours
25.1-S8	2 90-degree bends	15	2		30
25.1-S8	2 in. of length	2.5	2		5
	Total				113

H. Re-estimate Figure 25.1-6 using a pierce, cutoff, and form die.

Table Number	Process Description	Table Time	Adjustment Factor	Cycle Minutes	Setup Hours
25.1-S5	2 small holes	7	2		15
25.1-S5	2 small pilots	3.5	2		7
25.1-S5	3 pierce holes	5	3		15
25.1-S6	½ in. of curve	5			5
25.1-S8	2 90-degree bends	15	2		30
25.1-S8	2 in. length	2.5	2		5
	Total				76

I. Re-estimate Figure 25.1-6 using two dies, or a pierce and blank die, and form die.

Table Number	Process Description	Table Time	Adjustment Factor	Cycle Minutes	Setup Hours
	Pierce and blank die				
25.1-S5	2 small holes	7	2		15
25.1-S5	2 small pilots	3.5	2		7
25.1-S5	3 pierce holes	5	3		15
25.1-S6	½ in. of curve	5			5
25.1-S6	6 in of straight	5	6		30
	Subtotal pierce and blank dies				71
	Form die				
	Basic die				
25.1-S8	2 90-degree bends	15	2		30
25.1-S8	2 in. length	2.5	2		5
	Subtotal form die				35
	Total for two die method				106

J. Estimate a milling fixture for a part with base dimensions of 6 × 6 in. The part has overall dimension 6 × 6 × 3. For the purpose of these estimating data, the fixture is considered large.

Table Number	Process Description	Table Time	Adjustment Factor	Cycle Minutes	Setup Hours
25.1-S9	Baseplate	14			14
25.1-S12	One shoulder locator	12			2
25.1-S12	One diamond locator	12			2
	Estimate clamps, 2 guided	16			16
	Estimate positioner set blocks	2.5			2.5
25.1-S16	Jig bore and inspection	5			5
	Total				41.5

TABLE 25.1 DIES, JIGS, AND FIXTURES

Setup elements in estimating hours

ALL SET-UP TIME
NO CYCLE TIME

S1. Pierce Hole Dies

A. Small Holes

Hole Dia In.	Hr for Basic Die (One Hole Only)
0.0156–0.1245	21
Add'l hole	3.5

B. Pierce Holes

Hole Dia In.	Hr for Basic Die (One Hole Only)
0.125–0.55	16
Add'l hole	2.5

C. Bored Holes

Hole Dia In.	Hr for Basic Die (One Hole Only)
0.501–1.000	21
Add'l hole	9
1.001–1.500	31
Add'l hole	10
1.501–2.000	33
Add'l hole	12
2.001–2.500	36
Add'l hole	15
2.501–3.000	39
Add'l hole	18
3.001–3.500	42
Add'l hole	21
3.501–4.000	45
Add'l hole	24
4.001–4.500	48
Add'l hole	27
4.501–5.000	51
Add'l hole	30
5.001–5.500	54
Add'l hole	33
5.501–6.000	57
Add'l hole	36

D. Additional

Die of RCHC steel	add 20 percent
Material thickness between 0.001–0.009 in.	add 25 percent
Die set milled to fit die blocks	add 10 percent
Die shoe and punch holder are over 1½-in. thick and die set greater than 6 × 8 in., for extra handling	add 10 percent

S2. Basic Blank Die

Periphery In.	Contour of Blank			
	Straight	Angular	Curved	Irregular
Up to 3	24	27	30	37.5
1 to 4	27	32	40	50.0
4 to 5	29	38	50	62.5
5 to 6	32	45	60	75.0

If periphery is greater than 6 in. use these values for every inch:

Straight line	5.0
Angular line	7.5
Curved line	10.0
Irregular line	12.5

S3. Basic Extruding Die

Diameter of Extrusion	Type 1		Type 2		Type 3	
	Basic Die	Each Add'l	Basic Die	Each Add'l	Basic Die	Each Add'l
0.0156 to 0.1245	20	7.5	28	15.0	24	10.0
0.125 to 0.500	16	5.0	24	10.0	20	7.5
0.501 to 1.000	20	10.0	28	20.0	24	15.0

S4. Form Die, Single Bends Only

Length of Bend	Width of Bend	90° Bends		Angle	
		L	U	V	Bend
Up to 1	Up to 1	30	30	30	40
Over 1 to 2	Up to 1	32.5	32.5	32.5	43.5
Up to 1	Over 1 to 2	32.5	32.5	32.5	43.5
Over 1 to 2	Over 1 to 2	35	35	35	47
Over 2 to 3	Over 2 to 3	40	40	40	54

S5. Pierce Holes & Pilots in Combination Die

Type of Hole	Hole Dia	Pierce Holes	Pilots
Small	0.0156–0.1245	7 hr/hole for the first 5 holes. For each add'l hole thereafter add 3.5 hr/hole	3.5
Pierce	0.125–0.500	5 hr/hole for the first 5 holes. For each add'l hole thereafter add 2.5 hr/hole	2.5
Bored	0.500–1.000	10 hr/hole for the first 5 holes. For each add'l hole thereafter add 5.0 hr/hole	5.0

S6. Blanking Stations in Combination Die

Periphery	Contour of Blank			
	Straight	Angular	Curved	Irregular
Up to 3-In.	For odd-shaped blanks regardless of contour within blank, use a minimum of 15 hr/hole; if there are more than three similar holes, use 10 hr/hole thereafter.			

Periphery	Contour of Blank			
	Straight	Angular	Curved	Irregular
Over 3 In., Add'l per In. of Periphery	5.0	7.5	10.0	12.5

S7. Extrusions in Combination Die

Extrusion Die	Type 1	Type 2	Type 3
0.0156–0.1245	7.5	15.0	10.0
0.125–0.500	5.0	10.0	7.5
0.501–1.000	10.0	10.0	15.0

S8. Bends or Forms in Combination Die

Length of Bend	Width	90° Bends	Angle Bends
Up to 1 in.	Up to 1 in.	15 hr/bend	20 hr/bend

Over 1 in. of length or width—Add for every 1 in., either length or width—2.5 hr for 90-degree bends and 3.5 hr for angle bends.

Add'l	
HCHC	Add 20 percent
Thin Material	Add 25 percent
Mill Die Set	Add 10 percent
Extra Large Die Set	Add 10 percent

S9. Flat Baseplates

Size	Thickness	Width	Length	Hr
A. Drilling Fixtures				
Small	¾–1	3	3	6
Medium	¾–1	7	7	8
Large	¾–1	12	12	10
B. Milling Fixtures				
Small	⅝–1½	3	6	10
Medium	⅝–1⅛	6	9	12
Large	⅝–1⅛	10	14	14
C. Bench Fixtures				
Small	⅜–⅝	4	6	5
Medium		6	8	7
Large		9	12	9
D. Faceplates				
Large	Machining purchased plate			6

S10. Angular Baseplates

Size		Base	In. Combination	C Angle
Small	3 × 3 × 3	10	5	7
Medium	6 × 6 × 6	15	9	11
Large	6 × 9 × 9	20	13	15

S11. Two Baseplates for Box Jig

Size of Part	Base Thickness	Base Dimensions Width Length	Hr.
Small	¾–1	3 × 3	11
Medium	¾–1	7 × 7	14
Large	¾–1	12 × 12	17

S12. Pin Locators

Type	Machining	Jig Boring	Assembly	Total
Small—to ½ × 2 in.				
Straight	0.25	1.00	0.50	1.75
Shoulder or roller	0.50	1.00	0.50	2.00
Diamond	0.50	1.00	0.50	2.00
Large—½ to 1½ dia. to 2 in. long				
Straight	0.50	1.00	0.50	2.00
Shoulder or roller	0.75	1.00	0.50	2.25
Diamond	1.00	1.00	0.50	2.50

S13. Drill Plates and Support Blocks

A. Drill Plates

Part	Drill Plates	Two Drill Plates Hr	for Box Jig	Trimming Jig
Small	⅝ × 3 × 3	3	5	
Medium	⅝ × 3 × 3	5	9	
Large	⅝ × 3 × 3	7	13	8

B. Support Blocks

Height of Part, In.	Size		Hr
1	Small	⅝ × 1½ × 3	2
3	Medium	⅝ × 3 × 7	3
5	Large	⅝ × 5½ × 12	4

S14. Hand Knobs

Size of Part	Plain	With Block	Thumb Screw
Small	1.5	2.5	1.0
Medium	1.5	2.5	1.0
Large	1.5	2.5	1.0

S15. Guide bushings

Bushing Dia.	No. of Holes	Shoulder	Headless	Jig Boring and Assembly
Up to ⅝ in.	1 per hole	2	1.5	.5
	2 to 4 per bushing	1.5	1.0	
	5 or more per bushing	0.75	.5	

S16. Jig Borer Setup and Inspection

Part Size	Hr
Small	3
Medium	4
Large	5
Special fixtures	
Box or tumble	6
Trunnion	8

APPENDIX A

Conversion Table

PRECISION POINT (handwritten annotation pointing to table)

Estimated Min/Unit	Hr/1	Hr/10	Hr/100	Hr/1000	Hr/10,000	Pc/hr
.0001				.0017	.01667	600,000
.0002				.0033	.03333	300,000
.0003			.001	.0050	.05000	200,000
.0004			.001	.0067	.06667	150,000
.0005			.001	.0083	.08333	120,000
.0006			.001	.0100	.10000	100,000
.0007			.001	.0117	.11667	85,714
.0008			.001	.0133	.13333	75,000
.0009			.002	.0150	.15000	66,667
.001			.002	.0167	.16667	60,000
.002			.003	.0333	.33333	30,000
.003			.005	.0500	.50000	20,000
.004			.007	.0667	.66667	15,000
.005			.008	.0833	.83333	12,000
.006			.010	.1000	1.00000	10,000
.007			.012	.1167	1.16667	8571
.008			.013	.1333	1.33333	7500
.009			.015	.1500	1.50000	6667
.01			.017	.1667	1.66667	6000
.02			.033	.3333	3.33333	3000
.03		.01	.050	.5000	5.00000	2000
.04		.01	.067	.6667	6.66667	1500
.05		.01	.083	.8333	8.33333	1200
.06		.01	.100	1.0000	10.00000	1000

Appendix 549

Estimated Min/Unit	Hr/1	Hr/10	Hr/100	Hr/1000	Hr/10,000	Pc/hr
.07		.01	.117	1.1667	11.66667	857
.08		.01	.133	1.3333	13.33333	750
.09		.02	.150	1.5000	15.00000	667
.10		.02	.167	1.6667	16.66667	600
.11		.02	.183	1.8333	18.33333	545
.12		.02	.200	2.0000	20.00000	500
.13		.02	.217	2.1667	21.66667	462
.14		.02	.233	2.3333	23.33333	429
.15		.03	.250	2.5000	25.00000	400
.16		.03	.267	2.6667	26.66667	375
.17		.03	.283	2.8333	28.33333	353
.18		.03	.300	3.0000	30.00000	333
.19		.03	.317	3.1667	31.66667	316
.20		.03	.333	3.3333	33.33333	300
.21		.04	.350	3.5000	35.00000	286
.22		.04	.367	3.6667	36.66667	273
.23		.04	.383	3.8333	38.33333	261
.24		.04	.400	4.0000	40.00000	250
.25		.04	.417	4.1667	41.66667	240
.26		.04	.433	4.3333	43.33333	231
.27		.05	.450	4.5000	45.00000	222
.28		.05	.467	4.6667	46.66667	214
.29		.05	.483	4.8333	48.33333	207
.30		.05	.500	5.0000	50.00000	200
.31		.05	.517	5.1667	51.66667	194
.32		.05	.533	5.3333	53.33333	188
.33		.06	.550	5.5000	55.00000	182
.34		.06	.567	5.6667	56.66667	176
.35		.06	.583	5.8333	58.33333	171
.36		.06	.600	6.0000	60.00000	167
.37		.06	.617	6.1667	61.66667	162
.38		.06	.633	6.3333	63.33333	158
.39		.07	.650	6.5000	65.00000	154
.40		.07	.667	6.6667	66.66667	150
.41		.07	.683	6.8333	68.33333	146
.42		.07	.700	7.0000	70.00000	143
.43		.07	.717	7.1667	71.66667	140
.44		.07	.733	7.3333	73.33333	136
.45		.08	.750	7.5000	75.00000	133
.46		.08	.767	7.6667	76.66667	130
.47		.08	.783	7.8333	78.33333	128
.48		.08	.800	8.0000	80.00000	125
.49		.08	.817	8.1667	81.66667	122
.50		.08	.833	8.3333	83.33333	120
.51		.09	.850	8.5000	85.00000	118
.52		.09	.867	8.6667	86.66667	115
.53		.09	.883	8.8333	88.33333	113
.54		.09	.900	9.0000	90.00000	111

Estimated Min/Unit	Hr/1	Hr/10	Hr/100	Hr/1000	Hr/10,000	Pc/hr
.55		.09	.917	9.1667	91.66667	109
.56		.09	.933	9.3333	93.33333	107
.57		.10	.950	9.5000	95.00000	105
.58		.10	.967	9.6667	96.66667	103
.59		.10	.983	9.8333	98.33333	102
.60		.10	1.000	10.0000	100.00000	100
.61		.10	1.017	10.1667		98
.62		.10	1.033	10.3333		97
.63		.11	1.050	10.5000		95
.64		.11	1.067	10.6667		94
.65		.11	1.083	10.8333		92
.66		.11	1.100	11.0000		91
.67		.11	1.117	11.1667		90
.68		.11	1.133	11.3333		88
.69		.12	1.150	11.5000		87
.70		.12	1.167	11.6667		86
.71		.12	1.183	11.8333		85
.72		.12	1.200	12.0000		83
.73		.12	1.217	12.1667		82
.74		.12	1.233	12.3333		81
.75		.13	1.250	12.5000		80
.76		.13	1.267	12.6667		79
.77		.13	1.283	12.8333		78
.78		.13	1.300	13.0000		77
.79		.13	1.317	13.1667		76
.80		.13	1.333	13.3333		75
.81		.14	1.350	13.5000		74
.82		.14	1.367	13.6667		73
.83		.14	1.383	13.8333		72
.84		.14	1.400	14.0000		71
.85		.14	1.417	14.1667		71
.86		.14	1.433	14.3333		70
.87		.15	1.450	14.5000		69
.88		.15	1.467	14.6667		68
.89		.15	1.483	14.8333		67
.90		.15	1.500	15.0000		67
.91		.15	1.517	15.1667		66
.92		.15	1.533	15.3333		65
.93		.16	1.550	15.5000		65
.94		.16	1.567	15.6667		64
.95		.16	1.583	15.8333		63
.96		.16	1.600	16.0000		63
.97		.16	1.617	16.1667		62
.98		.16	1.633	16.3333		61
.99		.17	1.650	16.5000		61
1.00		.17	1.667	16.6667		60
1.01		.17	1.683	16.8333		59
1.02		.17	1.700	17.0000		59

Estimated Min/Unit	Hr/1	Hr/10	Hr/100	Hr/1000	Hr/10,000	Pc/hr
1.03		.17	1.717	17.1667		58
1.04		.17	1.733	17.3333		58
1.05		.18	1.750	17.5000		57
1.06		.18	1.767	17.6667		57
1.07		.18	1.783	17.8333		56
1.08		.18	1.800	18.0000		56
1.09		.18	1.817	18.1667		55
1.10		.18	1.833	18.3333		55
1.11		.19	1.850	18.5000		54
1.12		.19	1.867	18.6667		54
1.13		.19	1.883	18.8333		53
1.14		.19	1.900	19.0000		53
1.15		.19	1.917	19.1667		52
1.16		.19	1.933	19.3333		52
1.17		.20	1.950	19.5000		51
1.18		.20	1.967	19.6667		51
1.19		.20	1.983	19.8333		50
1.20		.20	2.000	20.0000		50
1.21		.20	2.017	20.1667		50
1.22		.20	2.033	20.3333		49
1.23		.21	2.050	20.5000		49
1.24		.21	2.067	20.6667		48
1.25		.21	2.083	20.8333		48
1.26		.21	2.100	21.0000		48
1.27		.21	2.117	21.1667		47
1.28		.21	2.133	21.3333		47
1.29		.22	2.150	21.5000		47
1.30		.22	2.167	21.6667		46
1.31		.22	2.183	21.8333		46
1.32		.22	2.200	22.0000		45
1.33		.22	2.217	22.1667		45
1.34		.22	2.233	22.3333		45
1.35		.23	2.250	22.5000		44
1.36		.23	2.267	22.6667		44
1.37		.23	2.283	22.8333		44
1.38		.23	2.300	23.0000		43
1.39		.23	2.317	23.1667		43
1.40		.23	2.333	23.3333		43
1.41		.24	2.350	23.5000		43
1.42		.24	2.367	23.6667		42
1.43		.24	2.383	23.8333		42
1.44		.24	2.400	24.0000		42
1.45		.24	2.417	24.1667		41
1.46		.24	2.433	24.3333		41
1.47		.25	2.450	24.5000		41
1.48		.25	2.467	24.6667		41
1.49		.25	2.483	24.8333		40
1.50		.25	2.500	25.0000		40

Estimated Min/Unit	Hr/1	Hr/10	Hr/100	Hr/1000	Hr/10,000	Pc/hr
1.51		.25	2.517	25.1667		40
1.52		.25	2.533	25.3333		39
1.53		.26	2.550	25.5000		39
1.54		.26	2.567	25.6667		39
1.55		.26	2.583	25.8333		39
1.56		.26	2.600	26.0000		38
1.57		.26	2.617	26.1667		38
1.58		.26	2.633	26.3333		38
1.59		.27	2.650	26.5000		38
1.60		.27	2.667	26.6667		38
1.61		.27	2.683	26.8333		37
1.62		.27	2.700	27.0000		37
1.63		.27	2.717	27.1667		37
1.64		.27	2.717	27.3333		37
1.65		.28	2.750	27.5000		36
1.66		.28	2.767	27.6667		36
1.67		.28	2.783	27.8333		36
1.68		.28	2.800	28.0000		36
1.69		.28	2.817	28.1667		36
1.70		.28	2.833	28.3333		35
1.71		.29	2.850	28.5000		35
1.72		.29	2.867	28.6667		35
1.73		.29	2.883	28.8333		35
1.74		.29	2.900	29.0000		34
1.75		.29	2.917	29.1667		34
1.76		.29	2.933	29.3333		34
1.77		.30	2.950	29.5000		34
1.78		.30	2.967	29.6667		34
1.79		.30	2.983	29.8333		34
1.80		.30	3.000	30.0000		33
1.81		.30	3.017	30.1667		33
1.82		.30	3.033	30.3333		33
1.83		.31	3.050	30.5000		33
1.84		.31	3.067	30.6667		33
1.85		.31	3.083	30.8333		32
1.86		.31	3.100	31.0000		32
1.87		.31	3.117	31.1667		32
1.88		.31	3.133	31.3333		32
1.89		.32	3.150	31.5000		32
1.90		.32	3.167	31.6667		32
1.91		.32	3.183	31.8333		31
1.92		.32	3.200	32.0000		31
1.93		.32	3.217	32.1667		31
1.94		.32	3.233	32.3333		31
1.95		.33	3.250	32.5000		31
1.96		.33	3.267	32.6667		31
1.97		.33	3.283	32.8333		30
1.98		.33	3.300	33.0000		30

Estimated Min/Unit	Hr/1	Hr/10	Hr/100	Hr/1000	Hr/10,000	Pc/hr
1.99		.33	3.317	33.1667		30
2.00		.33	3.333	33.3333		30
2.04		.34	3.400	34.0000		29
2.06		.34	3.433	34.3333		29
2.08		.35	3.467	34.6667		29
2.10		.35	3.500	35.0000		29
2.12		.35	3.533	35.3333		28
2.14		.36	3.567	35.6667		28
2.16		.36	3.600	36.0000		28
2.18		.36	3.633	36.3333		28
2.20		.37	3.667	36.6667		27
2.22		.37	3.700	37.0000		27
2.24		.37	3.733	37.3333		27
2.26		.38	3.767	37.6667		27
2.28		.38	3.800	38.0000		26
2.30		.38	3.833	38.3333		26
2.32		.39	3.867	38.6667		26
2.34		.39	3.900	39.0000		26
2.36		.39	3.933	39.3333		25
2.38		.40	3.967	39.6667		25
2.40		.40	4.000	40.0000		25
2.42		.40	4.033	40.3333		25
2.44		.41	4.067	40.6667		25
2.46		.41	4.100	41.0000		24
2.48		.41	4.133	41.3333		24
2.50		.42	4.167	41.6667		24
2.52		.42	4.200	42.0000		24
2.54		.42	4.233	42.3333		24
2.56		.43	4.267	42.6667		23
2.58		.43	4.300	43.0000		23
2.60		.43	4.333	43.3333		23
2.62		.44	4.367	43.6667		23
2.64		.44	4.400	44.0000		23
2.66		.44	4.433	44.3333		23
2.68		.45	4.467	44.6667		22
2.70		.45	4.500	45.0000		22
2.72		.45	4.533	45.3333		22
2.74		.46	4.567	45.6667		22
2.76		.46	4.600	46.0000		22
2.78		.46	4.633	46.3333		22
2.80		.47	4.667	46.6667		21
2.82		.47	4.700	47.0000		21
2.84		.47	4.733	47.3333		21
2.86		.48	4.767	47.6667		21
2.88		.48	4.800	48.0000		21
2.90		.48	4.833	48.3333		21
2.92		.49	4.867	48.6667		21
2.94		.49	4.900	49.0000		20

Estimated Min/Unit	Hr/1	Hr/10	Hr/100	Hr/1000	Hr/10,000	Pc/hr
2.96		.49	4.933	49.3333		20
2.98		.50	4.967	49.6667		20
3.00	.1	.50	5.000	50.0000		20
3.05	.1	.51	5.083	50.8333		19.7
3.10	.1	.52	5.167	51.6667		19.4
3.15	.1	.53	5.250	52.5000		19.0
3.20	.1	.53	5.333	53.5555		18.8
3.25	.1	.54	5.417	54.1667		18.5
3.30	.1	.55	5.500	55.0000		18.2
3.35	.1	.56	5.583	55.8333		17.9
3.40	.1	.57	5.067	56.6667		17.6
3.45	.1	.58	5.750	57.5000		17.4
3.50	.1	.58	5.833	58.3333		17.1
3.55	.1	.59	5.917	59.1667		16.9
3.60	.1	.60	6.000	60.0000		16.7
3.65	.1	.61	6.083	60.8333		16.4
3.70	.1	.62	6.167	61.6667		162
3.75	.1	.63	6.250	62.5000		16.0
3.80	.1	.63	6.333	63.3333		15.8
3.85	.1	.64	6.417	64.1667		15.6
3.90	.1	.65	6.500	65.0000		15.4
3.95	.1	.66	6.583	65.8333		15.2
4.00	.1	.67	6.667	66.6667		15.0
4.05	.1	.68	6.750	67.5000		14.8
4.10	.1	.68	6.833	68.3333		14.6
4.15	.1	.69	6.917	69.1667		14.5
4.20	.1	.70	7.000	70.0000		14.3
4.25	.1	.71	7.083	70.8333		14.1
4.30	.1	.72	7.167	71.6667		14.0
4.35	.1	.73	7.250	72.5000		13.8
4.40	.1	.73	7.333	73.3333		13.6
4.45	.1	.74	7.417	74.1667		13.5
4.50	.1	.75	7.500	75.0000		13.3
4.55	.1	.76	7.583	75.8333		13.2
4.60	.1	.77	7.667	76.6667		13.0
4.65	.1	.78	7.750	77.5000		12.9
4.70	.1	.78	7.833	78.3333		12.8
4.75	.1	.79	7.917	79.1667		12.6
4.80	.1	.80	8.000	80.0000		12.5
4.85	.1	.81	8.083	80.8333		12.4
4.90	.1	.82	8.167	81.6667		12.2
4.95	.1	.83	8.250	82.5000		12.1
5.00	.1	.83	8.333	83.3333		12.0
5.10	.1	.85	8.500	85.0000		11.8
5.20	.1	.87	8.667	86.6667		11.5
5.30	.1	.88	8.833	88.3333		11.3
5.40	.1	.90	9.000	90.0000		11.1
5.50	.1	.92	9.167	91.6667		10.9

Estimated Min/Unit	Hr/1	Hr/10	Hr/100	Hr/1000	Hr/10,000	Pc/hr
5.60	.1	.93	9.333	93.3333		10.7
5.70	.1	.95	9.500	95.0000		10.5
5.80	.1	.97	9.667	96.6667		10.3
5.90	.1	.98	9.833	98.3333		10.2
6.00	.1	1.00	10.000	100.0000		10.0
6.20	.1	1.03	10.333			9.7
6.40	.1	1.07	10.667			9.4
6.60	.1	1.10	11.000			9.1
6.80	.1	1.13	11.333			8.8
7.00	.1	1.17	11.667			8.6
7.20	.1	1.20	12.000			8.3
7.40	.1	1.23	12.333			8.1
7.60	.1	1.27	12.667			7.9
7.80	.1	1.30	13.000			7.7
8.00	.1	1.33	13.333			7.5
8.20	.1	1.37	13.667			7.3
8.40	.1	1.40	14.000			7.1
8.60	.1	1.43	14.333			7.0
8.80	.1	1.47	14.667			6.8
9.00	.2	1.50	15.000			6.7
9.20	.2	1.53	15.333			6.5
9.40	.2	1.57	15.667			6.4
9.60	.2	1.60	16.000			6.3
9.80	.2	1.63	16.333			6.1
10.00	.2	1.67	16.667			6.0
11.00	.2	1.83	18.333			5.5
12.00	.2	2.00	20.000			5.0
13.00	.2	2.17	21.667			4.6
14.00	.2	2.33	23.333			4.3
15.00	.3	2.50	25.000			4.0
16.00	.3	2.67	26.667			3.8
17.00	.3	2.83	28.333			3.5
18.00	.3	3.00	30.000			3.3
19.00	.3	3.17	31.667			3.2
20.00	.3	3.33	33.333			3.0
25.00	.4	4.17	41.667			2.4
30.00	.5	5.00	50.000			2.0
35.00	.6	5.83	58.333			1.7
40.00	.7	6.67	66.667			1.5
45.00	.8	7.50	75.000			1.3
50.00	.8	8.33	83.333			1.2
60.00	1.0	10.00	100.000			1.0
70.00	1.2	11.67				0.86
80.00	1.3	13.33				0.75
90.00	1.5	15.00				0.67
100.00	1.7	16.67				0.60

Index

Abrasive belt machine deburring, 368–372
Abrasive deburring processes, 383–388
Abrasive and grinding machinery:
   element estimating, 287–322
      (*See also* entries for specific machines)
   PHC, 25, 31, 36, 42
Abrasive saw cutoff machines, 60–62
Abrasives:
   deburring, 383
   honing, 296
Assembly:
   element estimating, 439–442, 484–495
   PHC, 27, 32, 33, 38, 43, 44
Assembly machines, 442–448
Automatic lamination machines, 397–398
Automatic screw machines:
   multispindle, 182–185
   single spindle, 173–181
Automatic tool changer elements, 217–219
Axial-lead electrical component insertion machines, 466–469
Axial traverse grinding, 287–292, 299, 302–304

Bandsaw, power, cutoff machines, 55–60
Base quantity, 7
Bed milling machines, 194–198
Bench assembly:
   element estimating, 439–448, 484–490
   PHC, 27, 32, 33, 38, 43–44
Bench inspection, 449–451
Bench, mask and unmask, 413–417
Bender machines, hand-operated, 115–117
Bending machines, tube, 121–123
Bill of material (BOM), 18
Blade life, sawing, 56, 59
Blanking machines, 123–125
Blast-cleaning machines, 428–432
Blow molding machines, 85–86
BLS (*see* Bureau of Labor Statistics)

Bobbins, winding machines, 512
BOM (*see* Bill of material)
Booth painting, 417–425
Bore (machining), 255–261
Boring machines:
   element estimating, 247–261
      boring and facing machines, 252–254
   PHC, 25, 30, 36, 41
Box-packaging bench, foam-injected, 527–529
Box volume, 16
Brake machines:
   foot, 112–114
   hand operated, 115–117
   power press, 110–112
Brazing, 354–357
Break edges (machining), 255–261
Brinell hardness (Bhn), 386
Broaching, broaching machines:
   element estimating, 283–286
   PHC, 25, 30, 36, 42
Buffing and polishing, 372–375
   (*See also* Deburring)
Bureau of Labor Statistics (BLS), Standard Metropolitan Statistical Areas, 5–6

Cable connector assembly, flat, 490–492
Cardboard cartons, packaging in, 522–525
Case packing conveyor, 526–527
Centerless grinding machines, 293–296
Chemical cleaning, metals, 425–428
Chemical machining:
   element estimating, 389–396
   PHC, 26, 32, 37, 43
Chucking lathes, numerically controlled, 168–172
Circuit board drilling machines, printed, 499–502
Circuit board fabrication, printed, 389–396
Circuit board stuffing, printed, 473–476

Cities, representative, and PHC, 23–44
Cleaning, 425–432
   blast, 428–432
   chemical, 425–428
   steam, 426, 428
Cluster drilling machines, 237–240
Coil winding machines:
   electrical, 512–515
   hand operated, 115–117
Cold-chamber die-casting machines, 87–90
Component insertion machines, electrical, 463–473
   light-directed, 469–473
Component lead forming machines, 498–499
Component sequencing machines, electrical, 461–463
Computer-aided torch cutting processes, 66–68
Connector assembly:
   D-subminiature, 492–495
   flat cable, 490–492
Contour band sawing machines, 55–60
Conversion processes (painting), 417–425
Conversion table, 549–556
   instructions for using, 18–19
Conveyor:
   assembly, packaging, 522–526
   case packing, 526–527
Conveyor painting, 417–425
Conveyorized developing machines, 399–401
Corrugated-cardboard packaging conveyor assembly, 522–525
Cost estimate, total product, 5
Cost Summary sheet, 13–15
Counterboring, countersinking, 261–267
Cranes, 56, 58
Cutoff (machining), 255–261
Cutoff machining saw, 55–60

Cutting and sawing:
  element estimating, 55–68
  PHC, 23, 28, 34, 39
Cutting tools, replacement of, 275–281
Cylindrical grinding machines, 287–292

D-subminiature connector assembly, 492–495
Deburring:
  element estimating, 365–388
  (*See also* specific entries, machines and processes)
  PHC, 26, 31–32, 37, 42–43
Developing machines, conveyorized, 399–401
Dicing machines, wafer, 509–512
Die-casting machines, 86–90
Dies:
  element estimating, 531–548
  PHC, 28, 33, 39, 44
Dip brazing, 354–357
Dip painting, 417–425
Direct labor and material, 4–5
Disk grinding machines, 312–315
Drill press machine deburring, 365–368
Drilling, 261–267
Drilling machines:
  element estimating, 221–245
    cluster (multi-spindle), 237–240
    horizontal, 247–252
    radial, 241–245
    sensitive drill press machines, 223–227
    setup and layout, 221–223
    turret, 231–237
    upright, 227–231
  PHC, 24–25, 30, 35–36, 41
Drilling tool life and replacement, 275–281
Drop-forging machines, 137–146
Dwell, traverse grinding, 288–289, 291

EDM (*see* Electrical discharge machining)
Electrical component insertion machines, 463–469
Electrical component sequencing machines, 461–463
Electrical discharge machining (EDM), 405–407
  traveling wire, 408–411
Electrical soldering machines, 476–481
Electrical welding, 505–509
Electrolytes, 432
Electronic fabrication:
  element estimating, 461–515
  (*See also* entries for specific machines and processes)
  PHC, 27–28, 32–33, 38–39, 44
Electroplating metals, 432–437
Electrostatic spraying process, painting, 417–418
Element estimates, estimating data:
  adjustment, 12–13
  detailing, 11
  operations sheet, 9–11, 13
  productivity factors, 11
  setup and organization, 11–12
  tables, use of, 15
  time in handling, 12, 15–17
  (*See also* entries for specific machines and processes)

Elemental tables, rules for using, 15
End milling, 268–275
Engine lathes, 147–151
Engraving, engrave milling, 268–275
Epoxy ink, 134
Estimating data (*see* Element estimating data)
Estimator, use of (*see* Use)
Etching (*see* Chemical machining)
Explosive forging machines, 144–146
Extrusion molding machines, 82–84

Face milling, 268–275
Facing and boring machines, 252–254
FCAW (*see* Flux-cored arc welding)
Feedback devices, robot machines, 445
Ferrous products, material costs, 7, 47–52
Finishing:
  element estimating, 413–437
  pedestal machines, 372–375
  PHC, 26–27, 31, 32, 37, 38, 43
  (*See also* Cleaning; Deburring; Electroplating; Masking; Painting; Polishing)
Fixtures:
  element estimating, 531–548
  PHC, 28, 33, 39, 44
Flame cutting (*see* Oxygen cutting)
Flat cable connector assembly, 490–492
Flat ribbon cable, 490
Flux-cored arc welding (FCAW), 337–346
Foam-injected box-packaging bench, 527–529
FOB (*see* Free-on-board)
Foot brake machines, 112–114
Forging machines:
  crew, 137–138
  element estimating, 137–146
  PHC, 24, 29, 35, 40
Form (machining), 255–261
Form milling, 268–275
Form rolling:
  element estimating, 333–335
  PHC, 25, 31, 36, 42
Forming machines, 502–505
  component lead, 498–499
Free abrasive grinding machines, 309–312
Free-on-board (FOB) services and centers, 7–8
Furnace brazing, 354–357
Furnaces:
  element estimating, 361–363
  PHC, 26, 31, 37, 42

Gages, measuring, 449–451
Gas metal-arc welding (GMAW), 337, 346–350
Gas tungsten-arc welding (GTAW), 346–350
Gear cutting:
  element estimating, 323–332
  PHC, 25, 31, 36, 42
Gear shaper machines, 323–327
Glass-cloth layup table, 81–82
GMAW (*see* Gas metal-arc welding)
Grinding and abrasive machinery:
  element estimating, 287–322
  (*See also* entries for specific machines)
  PHC, 25, 31, 36, 42
GTAW (*see* Gas tungsten-arc welding)

Hacksaw machines, 55–60
Hammer-forging machines, 137–146
Hand deburring, 379–381
Hand milling machines, 209–212
Handheld portable-tool deburring, 375–378
Hand-operated machines, 115–117
Harness assembly, 484–490
Heat treatment, heat treat furnaces, 361–363
Hobbing, hobbing machines, 327–332
Honing machines, 296–298
Horizontal milling, drilling and boring machines, 247–252
Hot-chamber die-casting machines, 87–90
Hotworking machines, 137–146

In-line plating processes, 401–405
Insertion machines, electrical components, 463–466
  axial-lead, 466–469
  light-directed, 469–473
Inspection:
  element estimating, 449–459
  PHC, 27, 32, 38, 44
Inspection table and machines, 452–459
Instructions, general, 3–5
Internal grinding machines, 306–309
  numerically controlled, 319–322
  vertical, 316–319
Iron and steel products, material costs, 7, 47–52
Ironworker machines, 123–125
Isostatic molding machines, 91–92

Jigs:
  element estimating, 531–548
  PHC, 28, 33, 39, 44
Joining and welding:
  element estimating, 337–360
  electrical, 505–509
  (*See also* specific entries, machines and processes)
  PHC, 26, 31, 37, 42
Jump shear machines, 112–114

Kick press machines, 112–114
Knee and column milling machines, 189–193

Labor-estimating data, instructions, 3, 4, 8–11, 17–19
  computer software, 4
Lamination machines, automatic, 397–398
Laser marking machines, 131–134
Lathes:
  engine, 147–151
  numerically controlled chucking, 168–172
  numerically controlled turning, 161–168
  turret, 151–157
  vertical, 158–160
  (*See also* Screw machines; Turning machines)
Layup table, glass-cloth, 81–82
Light-directed component-selection insertion machines, 469–473
Loose-abrasive deburring, 383–385
Loss determination, 7
Lot estimate, 9

Machine inspection, 449–451
Machining, 255–281

Machining centers:
  element estimating, 213–219
  PHC, 24, 30, 35, 41
Machining and tool replacement, PHC, 25, 30, 36, 41–42
Marking machines:
  element estimating, 127–136
  PHC, 24, 29, 35, 40
Masking; mask and unmask bench, 413–417
Material costs, 3, 4, 6–8, 17
  base quantity, 6
  bill of, 17–18
  ferrous products, 7
    tables, 47–52
  FOB service, 7–8
  loss determination, 7
  nonferrous products, 7
    tables, 53–54
  rules for using tables, 7
  shipping cost, 7
  small requirement, 7
  unit cost, 7
Measuring instruments, 449–451
Media flow deburring, abrasive, 386–388
Metal cleaning, chemical, 425–428
Metal-cutting machines, various, 255–261
Metal inert gas (MIG) welding, 346–350
Metal polishing, 372
  (See also Deburring)
Metals, electroplating, 432–437
Metals, oxide coating processes, 432–437
Metropolitan Statistical Areas (BLS), 5–6
Microdrilling, 261–267
Micro-spot welding, resistance (RMSW), 505–507
MIG (see Metal inert gas)
Milling, 269–275
Milling machines:
  element estimating, 187–212, 268–275
    hand, 209–212
    horizontal, 247–252
    setup, 187–188
    special types and models, 205–209
    (See also specific entries)
  PHC, 24, 25, 30, 35, 36, 41, 42
Milling tool life and replacement, 275–281
Molding bench, plastics, 69–71
Molding machines:
  element estimating, 69–92
  PHC, 23, 28–29, 34, 39–40
  (See also entries for specific machines)
Multiform machines, hand-operated, 115–117
Multispindle automatic screw machines, 182–185
Multispindle drilling machines, 237–240

Nibbling machines, 118–119
Nonferrous products, material costs, 7, 53–54
Non-traditional machining:
  element estimating, 389–411
    chemical, 389–396
    electrical, 396–411
  PHC, 26, 32, 37–38, 43
Notching machines, 123–125
Numerically controlled chucking lathes, 168–172

Numerically controlled internal grinding machines, 319–322
Numerically controlled turning lathes, 161–168

Operations sheet, 9–11, 13
Organization of information, 12
Overhead, 5, 6, 11, 15
Oxide coating of metals, 432–437
Oxygen cutting process, 63–65

Packaging:
  element estimating, 517–529
  PHC, 28, 33, 39, 44
Packaging bench and machines, 517–529
Packaging materials, 517, 522–523
Pad printing machines, 134–136
Painting and conversion, 417–425
Parison, parison production, 85–86
PC boards (see Printed circuit boards)
Pedestal-machine deburring and finishing, 372–375
PHC (see Productive hour costs)
Pieces per hour estimation, 19
Piercing machines, 123–125
Planers (see Surface grinding machines)
Plastic material deburring, 381–383
Plastic preform molding machines, 71–72
Plastic welding machines, ultrasonic, 357–360
Plastics, plastic compounds, 69, 79
  (See also Molding)
Plastics molding bench, 69–71
Plating processes, in-line, 401–405
Plunge grinding, 287–292, 299, 303–304
  internal, 306–309, 318, 320
Polishing, polishing machines, 372–375
Portable-tool deburring, handheld, 375–378
Power bandsaw cutoff machines, 55–60
Power press brake machines, 110–112
Power shear machines, 93–97
Preform molding machines, plastic, 71–72
Press brake machines:
  element estimating, 110–112
  PHC, 23, 29, 34, 40
Presswork:
  element estimating, 93–125
  PHC, 23–24, 29, 34–35, 40
  (See also entries under specific machines)
Printed circuit (PC) board drilling machines, 499–502
Printed circuit board fabrication, 389–396
Printed circuit board stuffing, 473–476
Printed circuit lamination machines, 397–398
Printed circuit ultraviolet printing machines, 398–399
Printing machines, 134–136, 398–399
Process inspection, 449–451
Production systems, modern, 4–5
Productive hour costs (PHC), 3, 4, 5–6, 15
  and job description, 6
  tables—machine, process, bench, and city, 23–44
Productivity factor, 16–17
Punch press machines, 98–107
  first operations, 98–101
  hand-operated, 115–117

Punch press machines:
  secondary operations, 101–104
  turret machines, 105–107
Punching machines, 107–109, 502–505
  ironworking, 123, 125

Radial drilling machines, 241–245
Ram-type vertical-spindle milling machines, 198–202
Rapid travel tool changer elements, 217–219
Realization factors, 17
Reaming, 261–267
Replacement of tools, 275–281
Resistance micro-spot welding (RMSW), 505–507
Resistance spot welding (RSW), 337, 350–353
Ribbon cable, flat, 490
Riveting machines, 442–445
RMSW (see Resistance micro-spot welding)
Robot machines, 445–448
  cycle time, 4
Roller machines, hand-operated, 115–117
Rotary table grinding, 298–305
Router milling machines, 202–205
RSW (see Resistance spot welding)

Safety stock (machining), 255
SAW (see Submerged arc welding)
Saw milling, 268–275
Sawing and cutting:
  element estimating, 55–68
  PHC, 23, 28, 34, 39
Screen printing bench and machines, 129–131
Screw machines, automatic:
  multispindle, 182–185
  single spindle, 173–181
Sensitive drill press machines, 223–227
Sequencing machines, electrical components, 461–463
Setup and operation, 11–12
Shape-cutting machines, 55–68
Shearing machines, 502–505
  hand-operated, 115–117
  ironworker, 123–125
  jump shear, 112–114
  power, 93–97
Shielded metal-arc welding (SMAW), 337–346
Shipping costs, 7–8
Shop estimate (pc per hr), 19
Side milling, 268–275
Single-spindle automatic screw machines, 173–181
Single station punching machines, 107–109
Single-wire termination machines, 495–497
Slot milling, 268–275
SMAW (see Shielded metal-arc welding)
Solder coating machines, vertical, 481–484
Soldering, 354–357
  electrical (wave), 476–481
Solvents, metal cleaning, 425
Spark erosion (see Electrical discharge machining)
Sparkout, 287–289, 300, 306, 316
Special milling machines, 205–208
Spot ties, wire harness, 485

Spotwelding, 337, 350–353
  electrical, 505–509
Spray painting, 417–425
Start drill (machining), 255–261
Statistical areas, 5–6
Steam cleaning, 426, 428
Steel and iron products, material costs, 7, 47–52
Stick electrode arc welding (*see* SMAW)
Straddle milling, 268–275
Straightener machines, hand-operated, 115–117
Strapping (packaging), 522–523, 525
Submerged arc welding (SAW), 337–346
Surface grinding machines, 298–306

Tables, elemental, rules for using, 15
Tapping, 261–267
Tarry, traverse grinding, 288–289, 291
Termination machines, single-wire, 495–497
Thermoforming machines, 79–81
Thermoplastic injection molding machines, 72–76
Thermosetting plastic molding machines, 76–79
Thread cutting:
  element estimation, 333–335
  PHC, 25, 31, 36, 42
Threading (machining), 255–261
TIG (*see* Tungsten inert gas)
Time drivers, 12, 15–17
Tool-building data, 531–548
Tool changer elements, 217–219

Tool life, 275–281
Tools:
  element estimating, 531–548
  PHC, 28, 33, 39, 44
Torch brazing, 354–357
Torch cutting, 66–68
Toroids, winding machines, 512
Touch labor, 4–5
Transfer printing machines, 134–136
Traveling wire electrical discharge machining, 408–411
Traverse grinding, 287–292, 299, 302–304
  internal, 306–307, 308, 318
Tube bending machines, 120–123
Tungsten inert gas (TIG) welding, 346
Turn (machining), 255–261
Turning lathes, numerically controlled, 161–168
Turning machines:
  element estimating, 147–185
  PHC, 24, 29–30, 35, 40–41
  (*See also* entries for specific machines)
Turning tools, life of, 275–278
Turret drilling machines, 231–237
Turret lathes, 151–157
  vertical, 158–160
Turret punch press machines, 105–107
Turret welding machines, 507–509

Ultrasonic plastic welding machines, 357–360
Ultraviolet printing machines, 398–399
Unit estimate time conversion, 15–17

Upright drilling machines, 227–231
U.S. Department of Labor, Bureau of Labor Statistics, 5–6
Use of *AM Cost Estimator,* instructions for, 3–19
  cost summary, 13–15
  element estimating, 11–19
    (*See also* Element estimating data)
  labor estimating data, 3, 4, 8–11, 17–19
  material costs, 3, 4, 6–8, 17
    (*See also* Material costs)
  overhead, 5, 6, 11, 15
  productive hour costs (PHC), 3, 4, 5–6, 15
  tables, 23–44
  time drivers, 12, 15–17

Vertical bandsawing machines, 55–60
Vertical internal grinding machines, 316–319
Vertical solder coating machines, 481–484
Vertical-spindle ram-type milling machines, 198–202
Vertical turret lathes, 158–160

Wafer dicing machines, 509–512
Wave soldering machines, 476–481
Welding and joining:
  element estimating, 337–360
  electrical, 505–509
  PHC, 26, 31, 37, 42
  (*See also* specific entries)
Wheel wear, change time, grinding, 299–300
Winders, winding machines, 115–117, 512–515
Wire harness bench assembly, 484–490